Methods in Enzymology

Volume XXXV
LIPIDS
Part B

METHODS IN ENZYMOLOGY

EDITORS-IN-CHIEF

Sidney P. Colowick   Nathan O. Kaplan

Methods in Enzymology

Volume XXXV

Lipids

Part B

EDITED BY

John M. Lowenstein

GRADUATE DEPARTMENT OF BIOCHEMISTRY
BRANDEIS UNIVERSITY
WALTHAM, MASSACHUSETTS

1975

ACADEMIC PRESS   New York   San Francisco   London

A Subsidiary of Harcourt Brace Jovanovich, Publishers

ACADEMIC PRESS, INC.
111 Fifth Avenue, New York, New York 10003

United Kingdom Edition published by
ACADEMIC PRESS, INC. (LONDON) LTD.
24/28 Oval Road, London NW1

Library of Congress Cataloging in Publication Data
Main entry under title:

Lipids.

    (Methods in enzymology, v. 14,   )
    Includes bibliographical references.
    1.   Lipids.   I.   Lowenstein, John M., Date   ed.
[DNLM: 1.   Lipids.   I.   Lowenstein, John M., ed.
II.   Series.   W1ME9615K v. 14]
QP601.C733 no. 14, etc.   574.1'925'08s   [574.1'9247]
ISBN 0–12–181935–3 (v. 35)   77-26907

Table of Contents

Section I. Fatty Acid Synthesis

Section V. Miscellaneous Enzymes

Section VI. General Analytical Methods

Section VII. Synthesis or Preparation of Substrates

Section VIII. Preparation of Single Cells

Section IX. In Vivo and Perfusion Techniques

Contributors to Volume XXXV

Article numbers are in parentheses following the names of contributors.
Affiliations listed are current.

S. ABRAHAM (8), *Bruce Lyon Memorial Research Laboratory, Children's Hospital Medical Center of Northern California, Oakland, California*

D. O. ALLEN (57), *Department of Pharmacology, Indiana University School of Medicine, Indianapolis, Indiana*

MICHAEL J. ARSLANIAN (7a), *Department of Biochemistry, Baylor College of Medicine, Texas Medical Center, Houston, Texas*

J. ASHMORE (57), *Department of Pharmacology, Indiana University School of Medicine, Indianapolis, Indiana*

CLINTON E. BALLOU (11), *Department of Biochemistry, University of California at Berkeley, Berkeley, California*

YECHEZKEL BARENHOLZ (29, 30, 49), *Laboratory of Neurochemistry, Department of Biochemistry, The Hebrew University–Hadassah Medical School, Jerusalem, Israel*

EUGENE M. BARNES, JR. (13), *Department of Biochemistry, Baylor College of Medicine, Texas Medical Center, Houston, Texas*

J. BAR-TANA (15), *Department of Biochemistry, The Hebrew University–Hadassah Medical School, Jerusalem, Israel*

ROBERT L. BLAKELEY (26), *Department of Biochemistry, University of Queensland, St. Lucia, Queensland, Australia*

KONRAD BLOCH (10), *James Bryant Conant Laboratories, Harvard University, Cambridge, Massachusetts*

M. BOUTRY (56), *Laboratoire de Chimie Medicale, Université Libre de Bruxelles, Brussels, Belgium*

RALPH A. BRADSHAW (16), *Department of Biological Chemistry, Division of Biology and Biomedical Sciences, Washington University, St. Louis, Missouri*

ROSCOE O. BRADY (51), *Developmental and Metabolic Neurology Branch, National Institute of Neurological Diseases and Stroke, National Institutes of Health, Bethesda, Maryland*

H. BROCKERHOFF (40), *New York State Institute for Research in Mental Retardation, Staten Island, New York*

H. BRUNENGRABER (34, 56), *Laboratoire de Chimie Medicale, Université Libre de Bruxelles, Brussels, Belgium*

ERIC M. CAREY (9), *Department of Biochemistry, University of Sheffield, Sheffield, England*

THOMAS P. CARTER (52), *Eunice Kennedy Shriver Center for Mental Retardation at the Walter E. Fernald State School, Waltham, Massachusetts*

KENNETH D. CLINKENBEARD (20, 21), *Department of Physiological Chemistry, The Johns Hopkins University School of Medicine, Baltimore Maryland*

P. F. CRAIN (45), *Institute for Lipid Research and Department of Biochemistry, Baylor College of Medicine, Texas Medical Center, Houston, Texas*

Y. DAIKUHARA (56), *Department of Biochemistry, Dental School, Osaka University, Joan-Cho, Kita-Ku, Osaka, Japan*

DOMINIC M. DESIDERIO, JR. (45), *Institute for Lipid Research and Department of Biochemistry, Baylor College of Medicine, Texas Medical Center, Houston, Texas*

RAYMOND DILS (9), *Department of Biochemistry, The Medical School, University of Nottingham, Nottingham, England*

R. K. DINELLO (14), *Department of Biological Chemistry, Harvard Medical School, Boston, Massachusetts*

NICHOLAS P. B. DUDMAN (24), *Department of Biochemistry, University of*

Queensland, St. Lucia, Queensland, Australia

JOHN ELOVSON (12), *Department of Biology, University of California at San Diego, La Jolla, California*

M. L. ERNST-FONBERG (14), *Department of Biology, Yale University, New Haven, Connecticut*

JOHN N. FAIN (53), *Division of Biological and Medical Sciences, Brown University, Providence, Rhode Island*

R. RAY FALL (3), *Department of Chemistry, University of Colorado, Boulder, Colorado*

SHIMON GATT, *Laboratory of Neurochemistry, Department of Biochemistry, The Hebrew University–Hadassah Medical School, Jerusalem, Israel*

HILDA GJIKA (35), *Graduate Department of Biochemistry, Brandeis University, Waltham, Massachusetts*

GARY R. GRAY (11), *Department of Chemistry, University of Minnesota, Minneapolis, Minnesota*

RAS B. GUCHHAIT (4, 5), *Departments of Epidemiology and Biochemistry, The Johns Hopkins University School of Hygiene, Baltimore, Maryland*

ROSE MARY GUTIERREZ-CERNOSEK (35), *Graduate Department of Biochemistry, Brandeis University, Waltham, Massachusetts*

ROBERT W. GUYNN (37, 39), *Program in Psychiatry, The University of Texas Medical School at Houston, Texas Medical Center, Houston, Texas*

SVEN HAMMARSTRÖM (41), *Department of Chemistry, Karolinska Institutet, Stockholm, Sweden*

MICHAEL HELLER (27), *Department of Biochemistry, The Hebrew University–Hadassah Medical School, Jerusalem, Israel*

ROBERT L. HILL (18), *Department of Biochemistry, Duke University Medical Center, Durham, North Carolina*

RONALD A. HITES (44), *Department of Chemical Engineering, Massachusetts Institute of Technology, Cambridge, Massachusetts*

P. W. HOLLOWAY (31), *Department of Biochemistry, University of Virginia School of Medicine, Charlottesville, Virginia*

W. R. INGEBRETSEN, JR. (55), *Department of Medicine, University of California at San Diego School of Medicine, La Jolla, California*

HIDEO INOUE (1), *Department of Biochemistry, Dental School, Osaka University, Joan-Cho, Kita-Ku, Osaka, Japan*

L. W. JOHNSON (32), *Department of Obstetrics and Gynecology, Washington University School of Medicine, St. Louis, Missouri*

JULIAN N. KANFER (52), *Eunice Kennedy Shriver Center for Mental Retardation at the Walter E. Fernald State School, Waltham, Massachusetts*

A. S. KATOCS, JR. (57), *Department of Pharmacology, Indiana University School of Medicine, Indianapolis, Indiana*

JOHN C. KHOO (23), *Division of Metabolic Disease, Department of Medicine, University of California at San Diego School of Medicine, La Jolla, California*

EDWIN H. KOLODNY (51), *Eunice Kennedy Shriver Center for Mental Retardation at the Walter E. Fernald State School, Waltham, Massachusetts*

L. KOPELOVICH (56), *Division of Biochemistry, Sloan-Kettering Institute for Cancer Research, New York, New York*

M. R. LAKSHMANAN (6), *Lipid Metabolism Laboratory, Veterans Administration Hospital, Madison, Wisconsin*

M. DANIEL LANE (4, 5, 19, 20. 21), *Department of Physiological Chemistry, The Johns Hopkins University School of Medicine, Baltimore, Maryland*

E. E. LARGIS (57), *Department of Pharmacology, Indiana University School of Medicine, Indianapolis, Indiana*

LAWRENCE LEVINE (35), *Graduate Department of Biochemistry, Brandeis University, Waltham, Massachusetts*

H. RICHARD LEVY (2), *Biological Research Laboratories, Department of Biology, Syracuse University, Syracuse, New York*

JOHN M. LOWENSTEIN (1, 34, 38, 56), *Graduate Department of Biochemistry, Brandeis University, Waltham, Massachusetts*

JAMES A. MCCLOSKEY (43, 45), *Department of Biopharmaceutical Sciences, University of Utah, Salt Lake City, Utah*

EDDIE MAES (27), *Laboratoire de Chimie Generale I, Brussels, Belgium*

B. MIDDLETON (17, 22), *Department of Biochemistry, The University of Nottingham, Nottingham, England*

A. L. MILLER (2), *Department of Neurosciences, University of California at San Diego School of Medicine, La Jolla, California*

LEWIS C. MOKRASCH (42), *Department of Biochemistry, Louisiana State University Medical Center, New Orleans, Louisiana*

NAVA MOZES (27), *Department of Biochemistry, The Hebrew University–Hadassah Medical School, Jerusalem, Israel*

RICHARD A. MUESING (7), *Lipid Research Clinic, George Washington University, Washington, D.C.*

CARL M. NEPOKROEFF (6), *Lipid Metabolism Laboratory, Veterans Administration Hospital, Madison, Wisconsin*

WILLIAM T. NORTON (54), *The Saul R. Korey Department of Neurology, Albert Einstein College of Medicine of Yeshiva University, Bronx, New York*

BARBARA E. NOYES (16), *Department of Biochemistry, Stanford University School of Medicine, Palo Alto, California*

ERNST NYSTRÖM (46), *Department of Endocrinology, Sahlgren's Hospital, Göteborg, Sweden*

SHIRLEY E. PODUSLO (54), *Department of Neurology, The Johns Hopkins University School of Medicine, Baltimore, Maryland*

S. E. POLAKIS (4, 5), *Department of Physiological Chemistry, The Johns Hopkins University School of Medicine, Baltimore, Maryland*

JOHN W. PORTER (6, 7), *Lipid Metabolism Laboratory, Veterans Administration Hospital, and Department of Physiological Chemistry, University of Wisconsin, Madison, Wisconsin*

DAVID J. PRESCOTT (12), *Department of Biology, Bryn Mawr College, Bryn Mawr, Pennsylvania*

W. DOUGLAS REED (19), *Department of Pediatrics, University of Maryland School of Medicine, Baltimore, Maryland*

G. ROSE (15), *Department of Biochemistry, The Hebrew University–Hadassah Medical School, Jerusalem, Israel*

ARTHUR F. ROSENTHAL (48), *Department of Laboratories, Long Island Jewish–Hillside Medical Center, New Hyde Park, New York*

MARVIN SCHULMAN (28, 36), *Merck Sharp and Dohme Research Laboratory, Rahway, New Jersey*

KEITH SCOTT (25), *Department of Biochemistry, University of Queensland, St. Lucia, Queensland, Australia*

B. SHAPIRO (15), *Department of Biochemistry, The Hebrew University–Hadassah Medical School, Jerusalem, Israel*

JAN SJÖVALL (46), *Department of Chemistry, Karolinska Institutet, Stockholm, Sweden*

VLADIMIR P. SKIPSKI (47), *Donald S. Walker Laboratory, Sloan-Kettering Institute for Cancer Research, Rye, New York*

STUART SMITH (8), *Bruce Lyon Memorial Research Laboratory, Children's Hospital Medical Center of Northern California, Oakland, California*

DANIEL STEINBERG (23), *Division of Metabolic Disease, Department of Medicine, University of California at San*

Diego School of Medicine, La Jolla, California

HOWARD M. STEINMAN (18), *Department of Biochemistry, Duke University Medical Center, Durham, North Carolina*

WILHELM STOFFEL (50), *Institute of Physiological Chemistry, University of Cologne, Cologne, Germany*

TATSUO SUGIYAMA (20, 21), *Research Institute for Biochemical Regulation, School of Agriculture, Nagoya University, Furocho, Chikusa, Nagoya, Japan*

JOHN F. TALLMAN (51), *Developmental and Metabolic Neurology Branch, National Institute of Neurological Diseases and Stroke, National Institutes of Health, Bethesda, Maryland*

P. K. TUBBS (22), *Department of Biochemistry, University of Cambridge, Cambridge, England*

P. ROY VAGELOS (3, 12), *Department of Biological Chemistry, Washington University School of Medicine, St. Louis, Missouri*

RICHARD L. VEECH (33, 37, 39), *Laboratory of Alcohol Research, National Institutes of Alcoholism and Alcohol Abuse, Saint Elizabeths Hospital, Washington, D.C.*

D. VELOSO (33), *Laboratory of Neurochemistry, Division of Special Health Research, IRP, National Institute of Mental Health, Saint Elizabeths Hospital, Washington, D.C.*

M. WADKE (34, 38), *Graduate Department of Biochemistry, Brandeis University, Waltham, Massachusetts*

S. R. WAGLE (55), *Department of Pharmacology, Indiana University School of Medicine, Indianapolis, Indiana*

SALIH J. WAKIL (7a). *Department of Biochemistry, Baylor College of Medicine, Texas Medical Center, Houston, Texas*

HARLAND G. WOOD (28, 36), *Department of Biochemistry, Case Western Reserve University, Cleveland, Ohio*

BURT ZERNER (24, 25, 26), *Department of Biochemistry, University of Queensland, St. Lucia, Queensland, Australia*

D. B. ZILVERSMIT (32), *Graduate School of Nutrition, Cornell University, Ithaca, New York*

Preface

This volume covers recent developments in the enzymology of lipids. The enzymes included are in general those that have been purified to homogeneity or that have been purified highly.

The first section deals with enzymes of fatty acid synthesis, including the preparation of the separate components of acetyl-CoA carboxylase from *Escherichia coli*. The number of fatty acid synthases that have been purified highly has grown considerably, and most of the new preparations are included in this volume, among them fatty acid synthase from *Myco-bacterium phlei* and its polysaccharide activators. A section on β-hy-droxy-β-methylglutaryl-CoA enzymes is included, insofar as these enzymes are involved in acetoacetate synthesis. The section on general analytical methods is highly selective, since several volumes could have been filled on analytical methods alone. The topics for this section were chosen on the basis of their general usefulness to biochemists and to fill in some material not covered in Volume XIV. The section on the synthesis of substrates contains a large chapter on the chemical synthe-sis of phospholipids and analogues of phospholipids, which will, it is hoped, lead to a greater availability in pure form of many of these interesting compounds. Sections on the preparation of single cells and perfusion tech-niques are included as they have become of increasing importance to the biochemist studying the synthesis, interconversion, and degradation of lipids.

Suggestions for material to be included in future volumes are most welcome.

JOHN M. LOWENSTEIN

METHODS IN ENZYMOLOGY

EDITED BY

Sidney P. Colowick and Nathan O. Kaplan

VANDERBILT UNIVERSITY
SCHOOL OF MEDICINE
NASHVILLE, TENNESSEE

DEPARTMENT OF CHEMISTRY
UNIVERSITY OF CALIFORNIA
AT SAN DIEGO
LA JOLLA, CALIFORNIA

METHODS IN ENZYMOLOGY

EDITORS-IN-CHIEF

Sidney P. Colowick   Nathan O. Kaplan

Section I

Fatty Acid Synthesis

Vol. I [94]. β-Ketoreductase. F. Lynen and O. Wieland.

Vol. V [60]. Fatty Acid Synthesis from Malonyl-CoA. F. Lynen.

Vol. V [77]. Propionyl-CoA Carboxylase from Pig Heart. A. Tietz and S. Ochoa.

Vol. V [78]. Mitochondrial Propionyl Carboxylase. M. D. Lane and D. R. Halenz.

Vol. XIII [31]. Crystalline Propionyl-CoA Carboxylase from Pig Heart. Y. Kaziro.

Vol. XIII [32]. Methylmalonyl-CoA Racemase from Sheep Liver. R. Mazumder.

Vol. XIII [33]. Methylmalonyl-CoA Racemase from *Propionibacterium shermanii*. S. H. G. Allen, R. W. Kellermeyer, and H. G. Wood.

Vol. XIII [34]. Methylmalonyl-CoA Mutase from Sheep Liver. R. Mazumder and S. Ochoa.

Vol. XIII [35]. 2-Methylmalonyl-CoA Mutase from *Propionibacterium shermanii* (Methylmalonyl-CoA Isomerase). R. W. Kellermeyer and H. G. Wood.

Vol. XIII [36]. Oxaloacetate Transcarboxylase from *Propionibacterium*. H. G. Wood, B. Jacobson, B. I. Gerwin, and D. B. Northrop.

Vol. XIII [73]. Removal of Phenolic Compounds during the Isolation of Plant Enzymes. W. D. Loomis.

Vol. XIV [1]. Acetyl-CoA Carboxylase from Yeast. M. Matsuhashi.

Vol. XIV [2]. Acetyl-CoA Carboxylase from Chicken and Rat Liver. S. Numa.

Vol. XIV [3]. Yeast Fatty Acid Synthase. F. Lynen.

Vol. XIV [4]. Pigeon Liver Fatty Acid Synthase. R. Y. Hsu, P. H. W. Butterworth, and J. W. Porter.

Vol. XIV [5]. Mechanism of Saturated Fatty Acid Biosynthesis in *Escherichia coli*. P. R. Vagelos, A. W. Alberts and P. W. Majerus.

Vol. XIV [6]. Acyl Carrier Protein from *Escherichia coli*. P. W. Majerus, A. W. Alberts, and P. R. Vagelos.

Vol. XIV [7]. Acetyl-CoA Acyl Carrier Protein Transacylase. A. W. Alberts, P. W. Majerus, and P. R. Vagelos.

Vol. XIV [8]. Malonyl-CoA Acyl Carrier Protein Transacylase. A. W. Alberts, P. W. Majerus, and P. R. Vagelos.

Vol. XIV [9]. β-Ketoacyl Acyl Carrier Protein Synthase. A. W. Alberts, P. W. Majerus, and P. R. Vagelos.

Vol. XIV [10]. β-Ketoacyl Acyl Carrier Protein Reductase. P. R. Vagelos, A. W. Alberts, and P. W. Majerus.

Vol. XIV [11]. β-Hydroxybutyryl Acyl Carrier Protein Dehydrase from *Escherichia coli*. R. W. Majerus, A. W. Alberts, and P. R. Vagelos.

Vol. XIV [12]. Enoyl Acyl Carrier Protein Reductases from *Escherichia coli*. G. Weeks and S. J. Wakil.

Vol. XIV [13]. β-Hydroxydecanoyl Thioester Dehydrase from *Escherichia coli*. L. R. Kass.

Vol. XIV [14]. Acyl Carrier Protein Hydrolase. P. R. Vagelos and A. R. Larrabee.

Vol. XIV [15]. Isolation of Acyl Carrier Protein from Spinach Leaves. R. D. Simoni and P. K. Stumpf.

Vol. XVIIA [99]. α-Hydroxy-β-keto Acid Reductoisomerase (α-Acetohydroxy Acid Isomeroreductase) (*Neurospora crassa*). K. Kiritani and R. P. Wagner.

Vol. XVIIA [100]. α-Acetohydroxy Acid Isomeroreductase (*Salmonella typhimurium*). S. M. Arfin.

Vol. XVIIA [101]. α,β-Dihydroxy Acid Dehydratase (*Neurospora crassa* and Spinach). K. Kiritani and R. P. Wagner.

Vol. XVIIA [102]. α-Keto-β-hydroxy Acid Reductase (*Neurospora crassa*). R. P. Wagner and F. B. Armstrong.

[1] Acetyl Coenzyme A Carboxylase from Rat Liver

EC 6.4.1.2 Acetyl-CoA:carbon dioxide ligase (ADP)

By HIDEO INOUE and JOHN M. LOWENSTEIN

$$\text{Acetyl-CoA} + CO_2 + ATP \rightleftarrows \text{malonyl-CoA} + ADP + P_i$$

Assay Method

Principle. [^{14}C]Bicarbonate is converted into the carboxyl group of malonyl-CoA. As a result an acid-volatile compound is converted into an acid-stable compound. The reaction is stopped by addition of acid, and the reaction mixture is taken to dryness. Unreacted [^{14}C]bicarbonate escapes as $^{14}CO_2$, while the ^{14}C fixed into malonyl-CoA remains. The radioactivity of the residue is a measure of the activity of the enzyme. This assay gives linear results only during the initial portion of the reaction because the accumulation of malonyl-CoA causes the reaction to become inhibited. A difficulty is encountered when this assay is applied to crude extracts because such extracts usually contain the enzymes and substrates for other CO_2-fixing reactions. Even when all the substrates needed for such reactions are not present, exchange reactions can still cause an unwanted CO_2 fixation. This difficulty is readily overcome by subjecting the crude extracts to gel filtration on Sephadex G-25 prior to assay. This eliminates the contaminating, unwanted substrates and results in very low blanks.

Details of a number of other assays that have been employed to measure acetyl-CoA carboxylase activity have been discussed elsewhere.[1,2] We prefer the assay involving radioactive CO_2 fixation because of its simplicity, and because it does not require other enzymes. This simplifies studies of the regulatory properties of the enzyme.

In crude extracts acetyl-CoA carboxylase from rat must be activated by incubation with citrate or, better, with citrate and magnesium ions, before it can be assayed under conditions of maximum activity. This type of activation is inhibited by ATP.[3] Activation prior to assay appears to be unnecessary in the case of the highly purified enzyme, although the presence of citrate or a related activator is still required during the assay. It follows that if in the course of a purification the crude extracts are assayed without prior activation, then a spuriously low level of activity

[1] M. Matsuhashi, Vol. 14 [1].
[2] S. Numa, Vol. 14 [2].
[3] M. D. Greenspan and J. M. Lowenstein, *J. Biol. Chem.* **243**, 6273 (1968).

will be measured, and in subsequent steps, when activation prior to assay becomes unnecessary, a higher degree of purification will be calculated than was actually attained.

Reagents

Tris-HCl buffer, pH 7.5, 1.0 M
Dithiothreitol, 50 mM
Sodium citrate, 100 mM
Bovine serum albumin (fatty acid poor, or "fatty acid free"), 10 mg/ml
NaH^{14}CO$_3$ (0.25 μCi/μmole), 200 mM
ATP neutralized to pH 7.5 with NaOH, 100 mM
MgCl$_2$, 1.0 M
HCl, 4 N

Procedure for Gel Filtration and Activation of Crude Enzyme prior to Assay. For the purpose of assaying the crude enzyme the extract is first subjected to gel filtration to remove endogenous substrates. Between 1 and 5 ml of high-speed supernatant are placed on a column of Sephadex G-25 (gel bed 2.5 cm in diameter and 30 cm in height), which has been equilibrated with 20 mM Tris-HCl buffer, pH 7.5, and 1 mM dithiothreitol. Elution is carried out with the same buffer mixture, and fractions of 1 ml are collected. The fractions containing the highest protein concentrations are pooled. The resulting solution contains the crude enzyme.

Between 0.1 and 0.5 ml of crude enzyme is then activated in 20 mM sodium citrate, 20 mM MgCl$_2$, 1 mM dithiothreitol, 50 mM Tris-HCl buffer, pH 7.5, and 0.5 mg/ml bovine serum albumin (fatty acid poor or "free"), in a final volume of 1.0 ml, at 37°, for 30 minutes. The crude enzyme is now fully activated, or close to being fully activated, and is assayed for activity within a reasonably short period, say, within 20 minutes.

Activation prior to assay causes an increase in the assayable activity of the crude enzyme the magnitude of which depends on the nutritional history of the animal and the precise method of preparing the extract. The need for this type of activation is diminished by ammonium sulfate fractionation and abolished by DEAE-chromatography. In other words, activation prior to assay is unnecessary after the second step in the purification described below.

Procedure for Assaying Enzyme. The assay mixture contains 100 mM Tris-HCl buffer, pH 7.5, 1 mM dithiothreitol, 0.2 mM acetyl-CoA, 20 mM NaH^{14}CO$_3$ (0.25 μCi/μmole), 5 mM ATP, 20 mM citrate, 20 mM MgCl$_2$, 0.5 mg/ml of bovine serum albumin, and enzyme, 1–2 milliunits, in a final volume of 0.4 ml. The reaction, which is started by adding the enzyme, is run at 37° for 5 minutes. It is stopped by addition of

0.1 ml 4 N HCl. The mixture is taken to dryness in a gentle stream of air to expel unreacted $^{14}CO_2$. The residue is dissolved in 1 ml of water and counted. One unit of enzymatic activity is equal to 1 μmole of malonyl-CoA formed at 37° per minute.

Purification Procedure[4]

The purification procedure described below results in a final purification of the enzyme of about 520-fold. Table I shows a summary of the procedure. Unless otherwise indicated all steps were carried out in the cold room at 5°.

Step 1. Preparation of Crude Extract and High-Speed Supernatant. Frozen livers (500–600 g) of rats which had been starved for 2 days and had then been fed a diet high in glucose or fructose for 5 days were homogenized with 2 volumes by weight of Buffer A (0.05 M Tris-HCl buffer, pH 7.5, 20 mM sodium citrate, 0.5 mM EDTA, and 5 mM 2-mercaptoethanol) in a Waring blender at top speed for 10 seconds. The homogenate was centrifuged at 2000 g for 10 minutes. The packed residue was homogenized again with 200 ml of Buffer A using a glass homogenizer. The resulting homogenate was combined with the supernatant of the first centrifugation and centrifuged at 14,000 g for 45 minutes. The packed residue was washed with 300 ml of Buffer A. The supernatant and washings were combined and centrifuged at 105,000 g for 45 minutes.

Step 2. First Ammonium Sulfate Fractionation. Ammonium sulfate (175 g/liter) was added with stirring to the supernatant of Step 1. The pH was kept at 7.3 by cautious addition of 5 N KOH. The resulting precipitate was collected by centrifugation at 14,000 g for 60 minutes. It was homogenized with 200 ml of Buffer B (10 mM potassium phosphate buffer, 10 mM sodium citrate, pH 7.5, 0.5 mM EDTA, 5 mM mercaptoethanol) and dialyzed against 6 liters of Buffer B overnight. The dialyzed enzyme was then centrifuged at 105,000 g for 30 minutes, and the precipitate was homogenized with 50 ml of Buffer B, using a glass homogenizer with a Teflon piston. The resulting suspension was centrifuged at 105,000 g for 30 minutes. The precipitate was washed once more and the supernatants were combined.

Step 3. Second Ammonium Sulfate Fractionation. Ammonium sulfate (145 g/liter) was added with stirring to the clear supernatant from Step 2. The pH was kept at 7.3 as described before. The precipitate was harvested by centrifugation, dissolved in about 100 ml of Buffer B, and dialyzed against 4 liters of the same buffer overnight. The dialyzed en-

[4] H. Inoue and J. M. Lowenstein, *J. Biol. Chem.* **247**, 4825 (1972).

TABLE I
PURIFICATION OF ACETYL-CoA CARBOXYLASE FROM RAT LIVER[a]

Fraction	Volume (ml)	Total protein (mg)	Total activity (units)	Specific activity (units/mg protein)	Recovery (%)
1. Crude extract					
A	1450	47100	1350	0.0287	100
B	1595	56300	1495	0.0266	100
2. First ammonium sulfate fractionation					
A	367	9100	1230	0.135	91.1
B	325	10200	1350	0.132	90.3
3. Second ammonium sulfate fractionation					
A	140	3750	1160	0.309	85.9
B	174	5100	1270	0.249	84.9
4. DEAE-cellulose chromatography					
A	20	306	930	3.04	68.9
B	20	294	1110	3.78	74.2
5. Third ammonium sulfate fractionation					
A	5.0	188	920	4.89	68.1
B	11.5	169	1017	6.02	68.0
6. Calcium phosphate gel chromatography					
A	5.0	86.3	805	9.33	59.6
B[b]	—	—	—	—	—
7. Sepharose 2B gel filtration					
A	3.0	21.1	304	14.4	22.5
B	3.3	27.1	396	14.6	26.5

[a] The activity of the enzyme was assayed after prior activation up to and including Step 5, thereafter it was assayed without prior activation. Results of two preparations are shown, A was started with 535 g of liver, B with 600 g of liver. Activity units are micromoles per minute.

[b] This step was omitted in preparation B.

zyme was centrifuged at 105,000 g for 30 minutes and the residue was washed twice with 25 ml of Buffer B as described in Step 2.

Step 4. DEAE-Cellulose Column Chromatography. The combined supernatant and washings from Step 3 were applied to a DEAE-column (bed dimensions 3.6 cm in diameter and 70 cm in height) which had been equilibrated with 2 liters of 10 mM potassium phosphate buffer, pH 7.4. After the enzyme solution had soaked into the DEAE-cellulose, the column was washed with 50 ml of Buffer B. Elution of the material adsorbed on the column was conducted with a linear gradient from 1 liter of Buffer B (mixing vessel) to 1 liter of a solution containing 150 mM potassium phosphate and 150 mM sodium citrate, pH 7.0, 0.5 mM EDTA, and 5 mM mercaptoethanol (reservoir). The flow rate was 2 ml/minute. Active fractions containing more than 1 unit of enzymatic activity per ml were pooled and concentrated by precipitation with ammonium sulfate (175 g/liter). The precipitate was harvested by centrifugation at 14,000 g for 30 minutes, dissolved in Buffer C (10 mM potassium phosphate, 20 mM sodium citrate, pH 7.4, 1 mM EDTA, and 5 mM mercaptoethanol), and dialyzed against 2 liters of the same buffer.

Step 5. Third Ammonium Sulfate Fractionation. Saturated ammonium sulfate solution (300 ml/liter) was added dropwise to the pooled fractions from Step 4 (protein concentration about 15 mg/ml) with gentle stirring. The pH was kept at 7.3. The precipitated protein was harvested by centrifugation at 25,000 g for 30 minutes and dissolved in Buffer C. (If the specific activity of the enzyme is below five at this point, calcium phosphate gel chromatography can be used as an optional next step. If the specific activity is greater, this step is not necessary.)

Step 6. Calcium Phosphate Gel Column Chromatography. The solution of the third ammonium sulfate fraction obtained in Step 5 was dialyzed against 2 liters of 20 mM potassium phosphate buffer, pH 7.4, 1 mM EDTA, and 5 mM mercaptoethanol overnight. The precipitate, if any, was removed by centrifugation at 25,000 g for 20 minutes and washed with the same buffer. The combined supernatant and washings were diluted with an equal volume of 5 mM mercaptoethanol and applied to a calcium phosphate gel column (bed dimensions 2.7 cm in diameter and 23 cm in height). This was prepared from 25 g of cellulose powder and calcium phosphate gel (8.6 g dry weight) which had been equilibrated with 10 mM potassium phosphate buffer, pH 7.4, by the method of Massey.[5] Elution was conducted with a linear gradient from 400 ml of 10 mM potassium phosphate buffer, pH 7.4 (mixing vessel), to 400 ml of 500 mM potassium phosphate buffer, pH 7.4 (reservoir). Both solutions contained 5 mM mercaptoethanol. The flow rate was 0.4 ml/minute. Fractions containing more than 1 unit of enzymatic activity per ml were pooled and precipitated by addition of ammonium sulfate (175 g/liter). The precipitated protein was harvested by centrifugation at 25,000 g for 30 minutes and dissolved in Buffer C.

[5] V. Massey, *Biochim. Biophys. Acta* **37,** 310 (1960).

Step 7. Sepharose 2B Gel Filtration. The enzyme solution obtained from Step 5 or 6 was allowed to soak into a column packed with Sepharose 2B (bed dimensions 2.6 cm in diameter and 77 cm in height) which had been washed with Buffer C. The column was then eluted with Buffer C at a flow rate of 0.2 ml/minute. Fractions with a specific activity of over 14 were pooled and concentrated by addition of ammonium sulfate (175 g/liter). The protein precipitate was suspended in 2 ml of Buffer C and dialyzed against 100 ml of Buffer C (which contained 1 mM dithiothreitol instead of mercaptoethanol) at room temperature for 6 hours. The buffer was changed twice during this time. The enzyme can be stored frozen in small aliquots.

Step 8. Crystallization. The solution of enzyme obtained from Step 7 was diluted to 1 mg of protein per ml with Buffer C, and 40% saturated ammonium sulfate (pH 7.8) was added dropwise with mixing until the solution showed very slight turbidity. Usually this required about 0.7 ml of the ammonium sulfate solution per ml of enzyme solution. The mixture was then allowed to stand at 5° for 40 hours before the crystals were harvested by centrifugation.

Fractions with a specific activity of less than 14 can be pooled and concentrated by ammonium sulfate precipitation and then subjected to Sepharose gel filtration a second time.

Storage of Enzyme. The enzyme was stored at 5° under N_2 in 20 mM sodium citrate, 10 mM potassium phosphate buffer, pH 7.5, 1 mM EDTA, 1 mM dithiothreitol, at a protein concentration of greater than 2 mg/ml. Under these conditions no loss of activity occurred during 1 week. The same medium was used for prolonged storage of the enzyme, except that the mixture was frozen rapidly in a bath containing acetone and Dry Ice. Under these conditions about 20% of the activity was found to have been lost when the solution was thawed after 1 month. This loss could be restored by activating the enzyme as described above.[6] Each successive freezing and thawing led to further losses in activity; these losses of activity could not be restored by the activating procedure.

Properties

Activators and Inhibitors. The crude and the pure enzyme is activated by citrate and other tri- and dicarboxylic acids. As has already been mentioned, the crude but not the pure enzyme requires activation with citrate prior to assay. The activation of the crude enzyme prior to assay

---

[6] Procedure for activation of crude enzyme.

TABLE II
INHIBITION OF ACETYL-CoA CARBOXYLASE BY PALMITYL-CoA

[Palmityl-CoA] (μM)	[Bovine serum albumin] (mg/ml)	[Citrate] (mM)	Maximum activity (%)
0	1.0	20	100
25	1.0	20	53
0	10	20	88
25	10	20	85
0	1.0	0.5	48
25	1.0	0.5	12
0	10	0.5	44
25	10	0.5	27

can also be achieved with magnesium ions or, better, magnesium ions plus citrate; this type of activation is inhibited and can be prevented altogether by ATP. There is a complex relationship between the concentrations of citrate, magnesium ions, and ATP which the reader should be aware of, but which need be of no concern here since ATP is omitted from the activation.

The enzyme is inhibited by long-chain fatty acyl-CoA, for example, by palmityl- or oleyl-CoA. This inhibition can be prevented largely or altogether by adding bovine serum albumin. The enzyme is also inhibited by malonyl-CoA; this is the reason why the reaction rate versus time curve remains linear for only a relatively short span. This can be overcome by removal of the end product, for example, by coupling malonyl-CoA formation to the fatty acid synthase reaction.[1]

Other substances, such as free fatty acids, long-chain fatty acid esters of carnitine, and detergents also affect the activity of the enzyme.

Kinetic Properties. The pure enzyme is relatively unstable unless a protecting protein such as bovine serum albumin is added. When fully activated, the enzyme exhibits K_{app}[7] values as follows: acetyl-CoA, 30 μM; HCO_3^-, 6 mM; and ATP, 0.2 mM. The K_{app} for citrate is relatively meaningless unless the concentrations of Mg^{2+} ions and ATP are specified. At an ATP concentration of 5 mM, the K_{app} for citrate is 0.46 and 5.7 mM in the presence of 6 and 20 mM $MgCl_2$, respectively. The $K_{i_{app}}$ for palmityl-CoA is strongly dependent on the citrate concentration and on the amount of protein present in the reaction mixture. Addition to the reaction mixture of bovine serum albumin diminishes the inhibitory effect of long-chain fatty acyl-CoA. The results shown in Table II were obtained with the pure enzyme from rat liver and serve as an example.

[7] K_{app} is used by analogy with K_m.

The enzyme has a pH optimum of 7.65, and the activity remains within 90% of maximum over the pH range of 7.65 ± 0.30, at least when the enzyme is fully activated. This qualification is made because the pH behavior has not been investigated systematically under conditions where the enzyme is not fully activated.

For previous summaries of activation and inhibition behavior, as well as kinetic properties, the reader is referred to refs.[2,8,9,10]

Physical Properties. When the native enzyme is subjected to analytical ultracentrifugation it shows hypersharpening at protein concentrations of 0.9 and 3.2 mg/ml. The $s_{20,w}$ value is 45, and no other components are observed.

When the enzyme is subjected to ultracentrifugation in the presence of 6 M guanidine hydrochloride and 0.1 M 2-mercaptoethanol, a single peak is observed throughout the run. Sedimentation equilibrium runs in the same solution yielded a single peak which corresponded to a molecular weight of 111,100. A partial specific volume of 0.7360, calculated from the amino acid composition, was used in the calculations of the molecular weight.

When the enzyme is subjected to electrophoresis on 1% agarose–1% acrylamide, or 1% agarose–0.5% acrylamide, or 2% agarose, a single component is observed in each case.

When the pure enzyme is subjected to gel electrophoresis[11] in 25 mM phosphate buffer, pH 7.2, containing 0.1% sodium dodecyl sulfate, a sharp protein band is observed which has a molecular weight of about 215,000. Prolonged or repeated treatment of the enzyme with sodium dodecyl sulfate leads to the gradual dissociation of the enzyme into two bands which have molecular weights of about 118,000 and 125,000.

Biotin and Phosphate Content. The enzyme contains 4.23 nmoles of biotin per mg of protein (average of eight determinations performed on three separate preparations). This is equivalent to 0.91 mole of biotin per mole of dimer (MW 215,000).[4]

Earlier values for acetyl-CoA carboxylase from rat[12] and chicken[13,14] liver were 0.50 and 0.69[15] μg biotin per mg of protein. This corresponds

[8] M. D. Lane and J. Moss, *in* "Metabolic Pathways" (H. J. Vogel, ed.), Vol. 5, p. 23. Academic Press, New York, 1971.

[9] J. J. Volpe and P. R. Vagelos, *Ann. Rev. Biochem.* **42**, 21 (1973).

[10] J. Moss, M. Yamagashi, A. K. Kleinschmidt, and M. D. Lane, *Biochemistry* **11**, 3779 (1972).

[11] K. Weber and M. Osborn, *J. Biol. Chem.* **244**, 4406 (1969).

[12] S. Nakanishi and S. Numa, *Eur. J. Biochem.* **16**, 161 (1970).

[13] C. Gregolin, E. Ryder, A. K. Kleinschmidt, R. C. Warner, and M. D. Lane, *Proc. Nat. Acad. Sci.* **56**, 148 (1966).

[14] S. Numa, E. Ringelmann, and B. Riedel, *Biochem. Biophys. Res. Commun.* **24**, 750 (1966).

[15] Means of 0.63 and 0.74 μg/mg protein reported in refs. 13 and 14.

to 0.84 and 1.1 moles biotin per mole of enzyme (MW 410,000), respectively. Expressed in terms of the molecular weight of 215,000, obtained by us for the dimer, these results become 0.44 and 0.58 mole biotin per mole of dimer, respectively.

The enzyme contains 9.83 nmoles of phosphate per mg of protein (average of four determinations). This is equivalent to 2.1 moles phosphate per mole of dimer (MW 215,00).[4]

Acknowledgment

This work was supported by the National Science Foundation.

[2] Acetyl-CoA Carboxylase from Rat Mammary Gland

EC 6.4.1.2 Acetyl-CoA:carbon dioxide ligase (ADP)

By A. L. Miller and H. Richard Levy

$$\text{Acetyl-CoA} + \text{ATP} + \text{CO}_2 \rightleftharpoons \text{malonyl-CoA} + \text{ADP} + \text{P}_i$$

Assay Methods

Principle. The methods which can be used to assay acetyl-CoA carboxylase have been detailed in previous articles in this series.[1,2] Although the optical methods are most convenient they are unsuitable except for enzyme of high purity because of interference from turbidity and contaminating ATPase. The procedures employed in assaying the mammary enzyme are based on measuring the incorporation of $H^{14}CO_3^-$ into malonyl-CoA under carefully restricted conditions which give highly reproducible results. The enzyme is first activated by incubating with citrate for 15 minutes; substrates are then added and incubation is continued for 10 minutes; the reaction is halted with perchloric acid and residual $H^{14}CO_3^-$ is completely removed; finally, the [^{14}C]malonyl-CoA is determined in a liquid scintillation spectrometer. Two methods are used[3]: assay method A is employed throughout the purification procedure; assay method B provides conditions for maximum activity. The principal differences between these methods are in the concentrations of assay components, the use of Mn^{2+} in assay A and Mg^{2+} in assay B, and the inclusion of an ATP-generating system in assay A.

[1] M. Matsuhashi, Vol. 14 [1].
[2] S. Numa, Vol. 14 [2].
[3] A. L. Miller and H. R. Levy, *J. Biol. Chem.* **244**, 2334 (1969).

Assay Method A

Reagents

Preincubation medium:

Imidazole-HCl, 0.5 M, pH 7.0

Potassium citrate, 0.5 M

$MnCl_2 \cdot 4H_2O$, 0.1 M

Creatine phosphokinase, 45 units/ml, in 1% bovine serum albumin

2-Mercaptoethanol, 0.029 M

Incubation medium:

Creatine phosphate, 0.1 M

ADP, 0.01 M

Acetyl-CoA, 5.0 mM

$NaH^{14}CO_3$, 0.5 M, 0.1 mCi/ml

Other solutions:

Bovine serum albumin, 1%

$HClO_4$, 4 N

KCl, 2 M

The preincubation medium is prepared by mixing 1.32 ml of imidazole-HCl, 0.44 ml of potassium citrate, 0.44 ml of $MnCl_2$, 1.10 ml of creatine phosphokinase, 0.21 ml of 2-mercaptoethanol, and 0.89 ml of water. The incubation medium consists of 1.54 ml of creatine phosphate, 0.66 ml of ADP, 0.66 ml of acetyl-CoA, 0.50 ml of $NaH^{14}CO_3$ and 1.04 ml of water. These quantities are sufficient for 20 tubes. To small tubes (75 × 10 mm) are added: 0.2 ml of preincubation medium, 0.06 ml of 1% bovine serum albumin, and water to bring to 0.28 ml. The tubes are placed into a reciprocating water bath shaker at 37°. The preincubation is initiated by adding 20 μl of enzyme and allowed to proceed for 15 minutes. The assay is then initiated by adding 0.2 ml of incubation medium. After 10 minutes the reaction is terminated by adding 0.05 ml of 4 N $HClO_4$. The tubes are placed in an ice bath and excess ClO_4^- is precipitated by adding 0.1 ml of 2 M KCl. The precipitated protein and $KClO_4$ are allowed to settle out (or centrifuged for 10 minutes at 2000 g) and 0.1 ml samples of the clear supernatant solution are spotted on Whatman No. 1 filter paper disks of 2.2 cm diameter. The disks are completely dried with a stream of warm air, which removes any unreacted $H^{14}CO_3^-$ as $^{14}CO_2$. The disks are placed into scintillation vials and 15 ml of scintillation fluid {4 g of 2,5-diphenyloxazole and 0.1 g of 1,4-bis[2-(4 methyl-5-phenyloxazolyl)]benzene/liter of toluene} are added. The samples are then counted in a scintillation spectrometer. Quenching by the paper disks is negligible.

Assay Method B

Reagents

Preincubation medium:
  Imidazole-HCl, 0.5 M, pH 7.5
  Potassium citrate, 0.5 M
  $MgCl_2 \cdot 6H_2O$, 0.4 M
  2-Mercaptoethanol, 0.029 M
Incubation medium:
  Acetyl-CoA, 0.01 M
  ATP, 0.1 M, neutralized
  $NaH^{14}CO_3$, 0.5 M, 0.1 mCi/ml
Other solutions:
  As for assay A

The preincubation medium is made by mixing 0.66 ml of imidazole-HCl, 0.33 ml of potassium citrate, 0.33 ml of $MgCl_2$, 0.06 ml of 2-mercaptoethanol, and 0.82 ml of H_2O. The incubation medium contains 0.70 ml of acetyl-CoA, 0.80 ml of ATP, 0.50 ml of $NaH^{14}CO_3$, and 0.15 ml of water. These quantities are sufficient for 20 tubes. To small tubes (75 × 10 mm) are added: 0.2 ml of preincubation medium, 0.03 ml of 1% bovine serum albumin, and water to bring to 0.39 ml. The preincubation is initiated with 10 μl of enzyme and the assay with 0.1 ml of incubation medium; all other procedures are exactly the same as for assay A.

Precautions. Both assays give highly reproducible results provided proper precautions are used. The selection of the appropriate amount of enzyme is a principal factor. During the enzyme purification it is essential that the volume and dilution of enzyme are selected at each stage of purification to yield linearity in the assay procedure with respect to both incubation time and enzyme concentration.[3] Complete removal of unreacted $H^{14}CO_3^-$ from the paper disks is also important and is readily achieved under the conditions described, but appropriate controls are routinely included to establish this.

Units. One unit of acetyl-CoA carboxylase is defined as that quantity of protein which catalyzes the incorporation of 1 μmole of $H^{14}CO_3^-$ into malonyl-CoA per minute at 37°. Specific activity is expressed as units per milligram of protein; protein concentration is measured from the absorbance at 260 and 280 nm.[4,5]

[4] O. Warburg and W. Christian, *Biochem. Z.* **310**, 384 (1941).
[5] H. M. Kalckar, *J. Biol. Chem.* **167**, 461 (1967).

Purification Procedure

Abdominal and inguinal mammary glands from Sprague-Dawley–derived lactating rats, 14–20 days *postpartum*, each of which nursed 7 healthy pups, is the starting material. Occasionally an area of mammary tissue leading to one of the teats has an appearance which grossly resembles that of regressing mammary tissue following weaning. Such tissue, which is readily recognizable, contains little enzyme and is discarded. For the enzyme preparation illustrated in Table I, 48 rats were used.

The success of the procedure to be described depends on the use of buffers at pH 6.5 containing glycerol and the appropriate concentration of Mg^{2+}. All operations are carried out at 0–4° except where indicated otherwise. All three buffers used during the purification procedures contain 20% glycerol, 7 mM 2-mercaptoethanol, and are adjusted to pH 6.5 with HCl. In addition, buffer 1 contains 50 mM imidazole and 1 mM EDTA; buffer 2 contains 50 mM imidazole, 0.1 mM EDTA, and 20 mM potassium citrate; and buffer 3 contains 5.0 mM imidazole, 0.1 mM EDTA, and 50 mM $MgSO_4·7H_2O$.

Step 1. Preparation of Extract. The abdominal and inguinal mammary glands (250 ml) are excised and rinsed with buffer 1, finely minced with scissors, mixed with 2 volumes of buffer 1, and homogenized for

TABLE I

PURIFICATION OF ACETYL-CoA CARBOXYLASE FROM
LACTATING RAT MAMMARY GLANDS

Fraction	Total protein (mg)	Total activity (units)[a]	Specific activity (units/mg)[a]	Recovery	Purification
1. Crude extract	15,900	208	0.013	100	1.0
2. $(NH_4)_2SO_4$ precipitation and dialysis	7,180	182	0.025	87	1.9
3. DE-52, pH 6.5	2,400	154	0.064	74	4.9
4. $(NH_4)_2SO_4$ fractionation	248	74	0.298	35	23
5. Sepharose 2B	58	51	0.872	24	67
6. Postsucrose gradient[b]	(0.11)	(0.197)	1.95[c]	(0.095)	150

[a] Using assay A.

[b] Only 1.13 units of the enzyme preparation from Step 6 were used. Thus numbers in parentheses do not apply to whole preparation.

[c] The specific activity under optimum conditions (assay B) was 5.48 μmoles/minute/mg of protein.

1 minute in a Waring blender at full speed. The homogenate is centrifuged for 10 minutes at 2000 g, the fat layer and the precipitate discarded, and the supernatant solution filtered through gauze and cotton. This solution is centrifuged for 30 minutes at 23,000 g, the precipitate and the small amount of fat discarded, and the supernatant solution (385 ml) filtered as above.

Step 2. Ammonium Sulfate Precipitation. The enzyme is precipitated by the addition of ammonium sulfate to 50% of saturation. The suspension is allowed to stand overnight and the precipitate then sedimented by centrifuging for 30 minutes at 23,000 g. The precipitate is redissolved in buffer 1 to a final volume of 260 ml. The solution is dialyzed against 20 liters of buffer 3 for 18 hours with three changes of dialysis medium. The enzyme solution is centrifuged for 30 minutes at 23,000 g to remove any precipitate which forms during dialysis.

Step 3. Negative Absorption with DE-52. The supernatant solution (263 ml) is placed onto a column of DE-52 (40 $\times$ 500 mm) which has been equilibrated with buffer 3. The column is prepared and run at room temperature. Buffer 3 is added to wash through the enzyme, which does not adhere to the ion exchange resin. The enzyme (350 ml) is then concentrated to 11–12 mg of protein/ml in a Diaflo ultrafiltration device, model 400, containing a UM-10 membrane. This is essential in order to prevent undue losses as a result of dilution in the next step.

Step 4. Ammonium Sulfate Fractionation. To the concentrated enzyme solution (210 ml) an equal volume of neutralized saturated ammonium sulfate is added and the suspension allowed to stand overnight. The precipitate is sedimented at 23,000 g for 30 minutes and the supernatant solution discarded. To the precipitate is added approximately 25 ml of buffer 1 containing 40% ammonium sulfate. The precipitate is suspended in this solution, stirred for 10 minutes, and the suspension then centrifuged for 10 minutes at 23,000 g. The supernatant solution is saved for assays. This procedure is repeated 4 times, each time with 15–20 ml of buffer 1 containing 30, 20, 10, and 0% ammonium sulfate, respectively. The bulk of the enzymatic activity is always found in the samples eluted with the buffers containing 10 and 0% ammonium sulfate. These are combined (47 ml) and the enzyme precipitated by adding enzyme grade ammonium sulfate to 30% saturation.

Step 5. Gel Filtration on Sepharose 2B. After the enzyme suspension has stood for 1 hour it is sedimented for 30 minutes at 23,000 g. The supernatant solution is discarded. The precipitate is redissolved in buffer 2 to give a protein concentration of 20–30 mg/ml. There are always two visibly different components in the redissolving precipitate. One portion dissolves within a few hours whereas the other usually takes 3 days to dissolve completely. All the enzymatic activity is associated with this latter fraction. The enzyme solution (5.2 ml) is applied to a Sepharose 2B column. The column (25 × 450 mm) is prepared and run at room temperature under a hydrostatic head of 50 cm of buffer. The flow rate is approximately 40 ml/hour.

Fractions of 2.0 ml are collected and assayed immediately. The fractions of highest specific activity are combined (73 ml) and the enzyme is precipitated with enzyme grade ammonium sulfate at 50% of saturation. It is essential to precipitate the enzyme quickly in order to avoid extensive loss of activity which normally occurs at low protein concentrations. After the suspension has stood overnight it is centrifuged for 30 minutes at 23,000 g, and the precipitate is redissolved in the minimum volume of buffer 2 (2 ml).

Step 6. Sucrose Gradient Centrifugation. Linear gradients from 5 to 20% sucrose in buffer 2 are prepared at room temperature, and 0.1 to 0.2 ml of enzyme solution layered on top. The total volume is 5.0 ml in each tube. The tubes are centrifuged for 4 hours at 105,000 g and 20° in a Spinco model L2-65B ultracentrifuge with the SW-39 rotor. The tubes are pierced and 20-drop samples collected. The contents of the tubes containing enzyme of high specific activity are pooled (1.1 ml) and stored at −20°. Since the enzyme is much less stable following sucrose gradient centrifugation, it is routinely purified through Step 5 and stored at −20° in small aliquots which are thawed and placed onto sucrose gradients as needed. Under these circumstances very little activity is lost over a period of 6 months.

Properties

Effect of pH. When the pH is varied, maintaining all other assay conditions constant, maximum activity is found at pH 7.5.

Kinetic Constants. The apparent K_m for acetyl-CoA is 5.0×10^{-5} M, determined using conditions of assay B. The kinetics when ATP is varied give a biphasic reciprocal plot, suggesting negative cooperativity and yielding two apparent K_m values of 5.88×10^{-5} M and 2.27×10^{-4} M. The apparent K_m for bicarbonate is 1.0×10^{-2} M. Michaelis constants are not easily determined for citrate or Mg^{2+} since they give sigmoid saturation curves; high concentrations of Mg^{2+} inhibit the enzyme.

Specificity. Propionyl-CoA can replace acetyl-CoA. The apparent K_m is 4.5×10^{-5} M, i.e., very similar to that for acetyl-CoA, but pronounced substrate inhibition occurs. The maximum rate of the reaction is only about 20% of that for acetyl-CoA. The only nucleotide triphosphate which can replace ATP is deoxy-ATP. Mn^{2+} can be used instead of Mg^{2+}.

The optimum Mn^{2+} concentration is 8.0×10^{-3} M, and high concentrations of Mn^{2+} inhibit the enzyme. In contrast to Mg^{2+}, there is no evidence for sigmoidicity at low concentration of Mn^{2+}.

Activation. Although mammary acetyl-CoA carboxylase does not show an absolute requirement for tricarboxylic acids, the activity is enhanced about 6-fold by preincubating the enzyme with 0.03 M citrate. Phospholipids stimulate the enzyme, but no detailed studies have been carried out on this effect. Bovine serum albumin increases enzymatic activity when included in the assay mixture; 0.6 mg/ml gives an 80% stimulation.

Inhibition. Avidin inhibits the enzyme, and biotin prevents this inhibition. Malonyl-CoA is inhibitory. The enzyme is strongly inhibited by rat milk. This inhibition results from the presence of high concentrations of nonesterified fatty acids in the milk.[6] Inhibition is observed at low concentrations (0.2–20 μM) of several fatty acids under appropriate conditions, laurate and myristate being the most effective of the fatty acids tested. Palmitoyl-coenzyme A is a more effective inhibitor. The effects of fatty acids and palmitoyl-CoA can be prevented, but not reversed, by bovine serum albumin.

[6] A. L. Miller, M. E. Geroch, and H. R. Levy, *Biochem. J.* **118**, 645 (1970).

[3] Biotin Carboxyl Carrier Protein from *Escherichia coli*

By R. RAY FALL and P. ROY VAGELOS

The acetyl-CoA carboxylase of *Escherichia coli* has been resolved into three functionally distinct protein components, including biotin carboxyl carrier protein (BCCP), biotin carboxylase, and transcarboxylase.[1,2] The BCCP component plays a central role as the carrier for an activated carboxyl group as indicated in reactions (1)–(3):

$$ATP + HCO_3^- + BCCP \overset{\text{biotin carboxylase. } Mn^{2+}}{\rightleftharpoons} CO_2^- - BCCP + ADP + P_i \quad (1)$$

$$CO_2^- - BCCP + CH_3CO - SCoA \overset{\text{transcarboxylase}}{\rightleftharpoons} {}^- O_2CCH_2CO - SCoA + BCCP \quad (2)$$

$$\text{Sum: } ATP + HCO_3^- + CH_3CO - SCoA \rightleftharpoons {}^- O_2CCH_2CO - SCoA + ADP + P_i \quad (3)$$

[1] A. W. Alberts and P. R. Vagelos, *Proc. Nat. Acad. Sci. U.S.* **59**, 561 (1968).

[2] A. W. Alberts, A. M. Nervi, and P. R. Vagelos, *Proc. Nat. Acad. Sci. U.S.* **63**, 1319 (1969).

The biotin prosthetic group of BCCP is carboxylated in a Mn^{2+}- and ATP-dependent reaction catalyzed by the biotin carboxylase component to form CO_2^-–BCCP.[2,3] The third protein component, the transcarboxylase, catalyzes the transfer of the carboxyl group from CO_2^-–BCCP to acetyl-CoA, forming malonyl-CoA[4,5]; this reaction regenerates BCCP, which is then available to accept another carboxyl group.

Several species of BCCP, ranging in molecular weight from 45,000 to 8,900 and varying in activity in the biotin carboxylase and transcarboxylase reactions, have been isolated from *E. coli* extracts.[6–10] The largest of these species has been characterized as the apparent native form of BCCP in *E. coli*,[7,8] while the smaller species represent proteolytic fragments of BCCP.[10] This report will describe the isolation and properties of native BCCP.

Assay Method

Principle. BCCP is isolated from *E. coli* cells which have been grown in the presence of [^{3}H]biotin.[10] The [^{3}H]biotin is incorporated as the biotin prosthetic group of BCCP, and since BCCP accounts for greater than 90% of the protein-bound biotin in *E. coli*,[7] the purification procedure can be performed by following protein-bound radioactivity as a measure of BCCP.

Purified preparations of BCCP are assayed for CO_2 acceptor ability in the biotin carboxylase reaction [reaction (1)] or for CO_2 donor ability in the transcarboxylase reaction [reaction (2)] by recombining the three protein components of *E. coli* acetyl-CoA carboxylase and measuring malonyl-CoA formation spectrophotometrically by coupling the carboxylase reaction with purified fatty acid synthase from yeast and following the decrease in the absorption of NADPH.[6,7] The effect of BCCP concen-

[3] P. Dimroth, R. B. Guchait, E. Stoll, and M. D. Lane, *Proc. Nat. Acad. Sci. U.S.* **67**, 1353 (1970).

[4] A. W. Alberts, S. G. Gordon, and P. R. Vagelos, *Proc. Nat. Acad. Sci. U.S.* **68**, 1259 (1971).

[5] R. B. Guchait, J. Moss, W. Sokolski, and M. D. Lane, *Proc. Nat. Acad. Sci. U.S.* **68**, 653 (1971).

[6] A. M. Nervi, A. W. Alberts, and P. R. Vagelos, *Arch. Biochem. Biophys.* **143**, 401 (1971).

[7] R. R. Fall, A. M. Nervi, A. W. Alberts, and P. R. Vagelos, *Proc. Nat. Acad. Sci. U.S.* **68**, 1512 (1971).

[8] R. R. Fall and P. R. Vagelos, *J. Biol. Chem.* **247**, 8005 (1972).

[9] A. W. Alberts and P. R. Vagelos, *Enzymes* **6**, 37 (1972).

[10] R. R. Fall and P. R. Vagelos, *J. Biol. Chem.* **248**, 2078 (1973).

tration on biotin carboxylase activity [reaction (1)] is tested under conditions of limiting biotin carboxylase and excess transcarboxylase in the reaction mixtures. To test the effect of CO_2^-–BCCP concentration on the transcarboxylase reaction [reaction (2)], reaction mixtures are prepared with an excess of biotin carboxylase and limiting transcarboxylase.

Reagents

Imidazole-Cl, 1.0 M, pH 6.7
ATP, 0.01 M
$MnCl_2$, 0.02 M
$KHCO_3$, 0.5 M
Acetyl-CoA,[11] 6 mM
NADPH, 10 mg/ml
Fatty acid synthase (yeast),[12] 7 units/ml
Biotin carboxylase (*E. coli*)[6,13]
Transcarboxylase (*E. coli*)[4,14]

Procedure. For detection and quantitation of [³H]BCCP during the purification procedure, samples are dissolved in 0.5 ml of water, 10 ml of a scintillation fluid (capable of solubilizing protein samples)[15] is added, and counting is performed in a scintillation counter. For assays of the CO_2 acceptor–donor activity of purified BCCP preparations, reactions are carried out at 25° in microcuvettes with a path length of 1 cm. Reaction mixtures contain 5 μl of imidazole-Cl, pH 6.7, 5 μl ATP, 3 μl $MnCl_2$, 3 μl $KHCO_3$, 5 μl acetyl-CoA, 2 μl NADPH, 1 μl fatty acid synthase, variable amounts of biotin carboxylase, transcarboxylase, and BCCP, and water to a final volume of 90 μl. Reactions measuring BCCP as a CO_2 acceptor [reaction (1)] are carried out with 2 milliunits of biotin carboxylase and 40 milliunits of transcarboxylase; reactions measuring CO_2^-–BCCP as a CO_2 donor are carried out with 2 milliunits of transcarboxylase and 40 milliunits of biotin carboxylase.[16] Reactions are started by the addition of BCCP and the decrease in absorbancy at 340 nm is followed with a recording spectrophotometer.

---

[11] E. J. Simon and D. Shemin, *J. Amer. Chem. Soc.* **75**, 3520 (1953).
[12] See F. Lynen, Vol. 14 [3].
[13] See R. B. Guchhait, S. E. Polakis, and M. D. Lane [4].
[14] See R. B. Guchhait, S. E. Polakis, and M. D. Lane [5].
[15] 3a40 scintillation fluid, Research Products International, Elk Grove Village, Illinois.
[16] Units of biotin carboxylase and transcarboxylase are defined as the amount of each which catalyzes the production of 1 μmole of malonyl-CoA per minute at 25° in the presence of an excess of the other.

Purification of BCCP

Several lots of *E. coli* B and K-12 cells were obtained commercially,[17] and BCCP was purified from each through step 3 of the purification procedure (see Table I). Poor yields of BCCP were obtained with full log and unwashed cells. The best yields were obtained with washed *E. coli* K-12 cells grown in enriched medium and harvested at 3/4 log and these cells were used for the procedure described below.[8] All steps were carried out at 0–4°.

TABLE I
PURIFICATION OF [³H]BCCP FROM *Escherichia coli*[a]

Step	Total protein (mg)	Total dpm $\times 10^{-6}$	dpm/(mg)	Yield (%)
1. Crude extract	316,200	152.8	480	(100)
2. (NH₄)₂SO₄ I, II	145,320	150.7	1,040	98.6
3. Alumina Cγ eluate	3,245	89.3	27,500	58.4
4. (NH₄)₂SO₄ III	2,734	85.6	31,300	56.0
5. Protamine sulfate	2,218	81.2	36,600	53.2
6. (NH₄)₂SO₄ extraction	806	61.2	75,900	40.0
7. Sephadex G-200	166	28.0	168,700	18.3
8. Preparative electrophoresis I	40.4	18.1	448,000	11.8
9. Preparative electrophoresis II	9.8	11.9	1,214,000	7.8

[a] Represents the purification of BCCP from 5 pounds of *E. coli*.

Cell Extract and Ammonium Sulfate Fractionation I. Frozen *E. coli* K-12 cells (1 pound at a time) are thawed in 2.5 volumes of 0.02 M potassium phosphate, pH 7.0, containing 1 mM EDTA, 1 mM *o*-phenanthroline and 10 mM 2-mercaptoethanol, and then disrupted by a single passage through a Manton-Gaulin submicron disperser[18] (prechilled with ice) at 9000 psi. The homogenate is centrifuged for 60 minutes at 27,000 g. The supernatant fraction (crude extract) is carefully decanted and immediately brought to 45% saturation with ammonium sulfate by the addition of 258 g/liter of solid ammonium sulfate. The precipitate is collected by centrifugation at 27,000 g for 30 minutes. The precipitate is stored at −20° until a total of 5 pounds of *E. coli* cells have been brought through the first two steps. BCCP is stable under these conditions for several months.

[17] Grain Processing Corp., Muscatine, Iowa.
[18] See Vol. 22 [37].

Ammonium Sulfate Fractionation II. The ammonium sulfate precipitates from above are dissolved in 0.05 M imidazole-Cl, pH 6.7, and diluted with the same buffer to a protein concentration of 20 mg/ml. To this solution is added a similar ammonium sulfate fraction [approximately 1 g of protein, 153×10^6 disintegrations/minute (dpm) from [³H]biotin labeled cells of *E. coli* PA 502[10] which have been treated as above, except that the cells are ruptured in a French pressure cell. The combined fractions are centrifuged at 27,000 g for 30 minutes to remove insoluble material and the supernatant fraction is stirred with solid ammonium sulfate (258 g/liter) to bring the solution to 45% saturation. The precipitate is collected as above.

Alumina Cγ Elution. The ammonium sulfate precipitate from above is dissolved in a minimal volume of 0.05 M imidazole-Cl, pH 6.7, to give a protein concentration of 80–100 mg/ml. To this fraction a slurry of alumina Cγ gel is added with stirring to give a concentration of 1 mg of gel per 25 mg of protein; after stirring for 10 minutes the gel is collected by centrifugation at 10,000 g for 10 minutes. The supernatant solution is treated a second time with a similar amount of gel. The combined alumina Cγ gels are washed by resuspension and stirring for 30 minutes with 0.1 M Tris-Cl, pH 7.7, 0.5 M ammonium sulfate (1 ml per 5 mg of gel), followed by centrifugation at 10,000 g for 10 minutes. This washing is repeated 4 times. BCCP is eluted from the alumina Cγ gel with three washes with 0.4 M potassium phosphate, pH 7.7 (1 ml per 5 mg of gel); these washes are pooled and brought to 50% saturation with ammonium sulfate (291 g/liter), and the precipitate is collected by centrifugation.

Ammonium Sulfate Fractionation III. The precipitate from above is suspended in 0.02 M imidazole-Cl, pH 6.7, at a protein concentration of 10 mg/ml, and insoluble material is removed by centrifugation at 48,000 g for 30 minutes. The supernatant fraction is brought to 20% saturation with ammonium sulfate by the slow addition of solid ammonium sulfate (0.106 g/ml). Stirring is continued for 30 minutes. The supernatant fraction is collected after centrifugation at 27,000 g for 15 minutes and brought to 45% saturation with ammonium sulfate (0.143 g/ml). The resulting precipitate is collected by centrifugation. (With some preparations up to 30% of the BCCP precipitates at 20% saturation with ammonium sulfate; the majority of this BCCP can be recovered by extraction with 0.02 M imidazole-Cl, pH 6.7, and refractionation as above.)

Protamine Sulfate Fractionation. The fraction precipitating between 20 and 45% saturation with ammonium sulfate is dissolved in 0.02 M imidazole-Cl, pH 6.7, at a protein concentration of 10 mg/ml, and a suspension of protamine sulfate[19] (50 mg/ml, pH 6.5) is added dropwise with stirring to give a final concentration of 2 mg of protamine sulfate per 10 mg of protein. The pH is adjusted to 6.5 with cold 0.1 N HCl, and after stirring 10 minutes the turbid suspension is centrifuged at 27,000 g for 15 minutes. The clear supernatant solution is collected and slowly brought to pH 5.5 with the slow dropwise addition of cold 0.1 N HCl; the resulting turbid solution is centrifuged as above. The clear supernatant fraction is rapidly brought to pH 7–7.5 with cold 2 M Tris base, and an aliquot is removed for protein determination. Then solid ammonium sulfate is added to 50% saturation, and the precipitate is collected by centrifugation.

Ammonium Sulfate Fractionation IV. The precipitate from above is repeatedly resuspended and extracted with decreasing concentrations of ammonium sulfate (including 50, 45, 40, 35, 30, 25, 20, and 15% saturated solutions in 0.02 M potassium phosphate, pH 7.0). Extractions at each ammonium sulfate concentration are repeated twice; in each case the precipitate is resuspended and stirred for 10 minutes and then centrifuged at 15,000 g for 10 minutes. Each supernatant fraction is saved, as well as a final insoluble residue. BCCP is routinely distributed approximately as follows: 50% in the 30 and 25% ammonium sulfate fractions, 30% in the insoluble residue and 20% in the 40, 35, and 20% ammonium sulfate fractions. The insoluble residue is solubilized with a minimal volume of solution containing 6 M urea, 0.02 M potassium phosphate, pH 7.0, and 0.01 M 2-mercaptoethanol, and then diluted 10-fold with 0.02 M potassium phosphate, pH 7.0, and added to the pooled 40, 35, and 20% ammonium sulfate fractions. This pool is precipitated by the addition of solid ammonium sulfate to 50% saturation, and the precipitate is collected by centrifugation. The extraction procedure is then repeated as above, and the 30 and 25% ammonium sulfate fractions are pooled with the same fractions from the first extraction, and the pool is precipitated with ammonium sulfate as above.

Sephadex G-200 Chromatography. The precipitate from above is dissolved in 20 ml of 0.02 M Tris-Cl, pH 8.0, and one-half is applied to a 2.5×80 cm column of Sephadex G-200 which is equilibrated with the same buffer. The flow rate is 10–12 ml/hour, and 3–4 ml fractions are collected. The fractions eluting between 320 and 410 ml are pooled and concentrated by pressure filtration[20] to a small volume and stored at $-196°$. The second half of the precipitate from above is similarly treated.

---

[19] Protamine sulfate (grade I), Sigma Chemical Co., St. Louis, Missouri.
[20] Amicon Corp., Lexington, Massachusetts.

Preparative Polyacrylamide Gel Electrophoresis I. One-third of the Sephadex G-200 pool from above (4 ml) is thawed and adjusted to 0.01 M 2-mercaptoethanol, and then solid urea is added to a concentration of 6 M. A drop of 0.1% bromphenol blue tracking dye is added and the mixture is layered onto the top of a 2.5 × 8 cm preparative polyacrylamide gel,[21] pH 8.4,[22] prepared with a stacking gel containing 6 M urea. Electrophoresis is initiated at 5 mA with an elution buffer flow rate of 2 ml/minute until the tracking dye has entered the separating gel (approximately 2 hours); then the current is increased to 10 mA and the elution buffer flow rate reduced to 1 ml/minute; 4 ml fractions are collected. Two peaks of [³H]BCCP are detected: a minor, rapidly eluting peak, and a major slower eluting peak. Analysis of aliquots of the two peaks on a calibrated Sephadex G-100 column reveals approximate molecular weights of 24,000 for the faster migrating peak and 46,000 for the slower peak. SDS polyacrylamide gel electrophoresis of both fractions reveals a single radioactive band of approximate molecular weight 22,000. These results suggest that the faster migrating peak represents a monomeric form of [³H]BCCP, while the slower migrating peak represents the native dimer. The latter is pooled and concentrated by pressure filtration.

Preparative Polyacrylamide Gel Electrophoresis II. The pool of the concentrated slower migrating peak of [³H]BCCP from above is adjusted to 10% glycerol, and one-fourth (approximately 0.7 ml, 10 mg of protein) is layered onto the top of a 2.5 × 8 cm preparative polyacrylamide gel,[21] pH 8.4,[22] and electrophoresis is carried out as described above. Fractions (24–38) containing [³H]BCCP are pooled and concentrated by pressure filtration and stored at −196°. This procedure is repeated on the remaining three-fourths of the pool above. The protein is stable for several months when stored this way.

Comments on the Purification Procedure. The results of the purification procedure are summarized in Table I. The total purification achieved is approximately 2500-fold with a yield of 8%. During the course of studies on BCCP[s] several alternate purification procedures were tested in order to simplify the purification scheme and increase the yield of BCCP. For a variety of reasons none of the procedures tested was satisfactory (including chromatography on DEAE-, CM-, and P-cellulose, DEAE- and QAE-Sephadex, and hydroxylapatite, adsorption and elution from calcium phosphate gel, ethanol and acid precipitations, and isoelectric focusing). Ion-exchange chromatography under a variety of conditions was unsuitable as a result of irreversible adsorption of BCCP to

[21] A PD-2/320 "Prep-Disc" column was used. Canalco, Rockville, Maryland.
[22] D. H. Coy and T-C. Wuu, *Anal. Biochem.* **44**, 174 (1971).

anion exchangers, poor adsorption to cation exchangers, and the need for extensive dialysis which under some conditions led to apparent dissociation or aggregation of BCCP. Acid precipitation of BCCP below pH 5.5 gave excellent purification, but this procedure led to the appearance of three bands of BCCP on polyacrylamide gels containing 8 M urea, and therefore this procedure was avoided. BCCP precipitates at its isoelectric point, ruling out preparative isoelectric focusing as a purification procedure.

The purification procedure used here, although tedious, is reproducible and yields a preparation of BCCP which is indistinguishable chromatographically and electrophoretically from BCCP found in freshly prepared extracts.

Properties of BCCP

Purity. The BCCP obtained from the above procedure sediments as a single symmetrical peak in the analytical ultracentrifuge, and exhibits a single protein band, which is coincidental with applied [³H]biotin BCCP radioactivity, upon electrophoresis in polyacrylamide gels in the presence or absence of 8 M urea or 1% SDS.[8]

Molecular Properties. BCCP contains approximately 207 amino acid residues and one biotin residue per polypeptide chain for a residue molecular weight of 22,200.[8] Subunit molecular weights of 22,500 and 21,800 have been determined by SDS polyacrylamide gel elctrophoresis and gel filtration in 6 M guanidine hydrochloride, respectively.[8] The native form of BCCP appears to be a dimer (MW 45,000) based on gel filtration on Sephadex G-100 and determinations of the sedimentation coefficient ($s_{20,w} = 5.7$–5.9).[8] Isoelectric focusing in polyacrylamide gels reveals an isoelectric point of 4.5.[8] BCCP serves as a carboxyl acceptor in the biotin carboxylase reaction and as a carboxyl donor in the transcarboxylase reaction with K_m values of 2×10^{-7} M and 4×10^{-7} M, respectively.[7,8]

Susceptibility to Proteolysis. Native BCCP is very susceptible to proteolysis by subtilisin.[10] Limited proteolysis of crude or purified preparations of native BCCP with subtilisin Carlsberg produces a mixture of smaller BCCP species analogous to those previously isolated,[6,9] and after more extensive proteolysis, all of the intermediate species are converted to a small, stable form of BCCP, termed BCCP$_{(SC)}$.[10] BCCP$_{(SC)}$ has been isolated and crystallized, and represents a biotin-containing 80 amino acid residue peptide region of native BCCP. BCCP$_{(SC)}$ and previously isolated small forms of BCCP are active as CO_2 acceptors and donors in the biotin carboxylase and transcarboxylase reactions but with K_m values 50–100-fold higher than those for native BCCP.[6,9,10] The previ-

ously isolated small forms of BCCP were probably the result of a sub-
tilisin-type proteolysis of native BCCP during large-scale purification
procedures, and suitable precautions are necessary to avoid degradation
of native BCCP during purification.[8,10] BCCP isolated as described here
is indistinguishable, chromatographically and electrophoretically, from
BCCP in freshly prepared crude extracts.[8]

[4] Biotin Carboxylase Component of Acetyl-CoA Carboxylase from *Escherichia coli*

By Ras B. Guchhait, S. E. Polakis, and M. Daniel Lane

$$\text{ATP}^{4-} + \text{HCO}_3^- + \begin{matrix}\text{biotin}\\\text{or}\\\text{CCP}\\\text{(carboxyl}\\\text{carrier}\\\text{protein)}\end{matrix} \underset{}{\overset{\text{Mg}^{2+}}{\rightleftharpoons}} \text{ADP}^{3-} + \text{P}_i^{2-} + \begin{matrix}\text{Biotin-CO}_2^-\\\text{or}\\\text{CCP}-\text{CO}_2^-\end{matrix} + \text{H}^+$$

The acetyl-CoA carboxylase system from *E. coli* can be resolved into
three protein components: biotin carboxylase (BC) which catalyzes the
first half-reaction [reaction (1)], carboxyltransferase (CT) which catalyzes the second half-reaction [reaction (2)], and carboxyl carrier protein
(CCP-biotin) which contains the covalently bound biotinyl prosthetic
group.

$$\text{ATP} + \text{HCO}_3^- + \text{CCP-biotin} \underset{\text{BC}}{\overset{\text{Me}^{2+}}{\rightleftharpoons}} \text{CCP-biotin-CO}_2^- + \text{ADP} + \text{P}_i \quad (1)$$

$$\text{CCP-biotin-CO}_2^- + \text{acetyl-CoA} \underset{\text{CT}}{\rightleftharpoons} \text{malonyl-CoA} + \text{CCP-biotin} \quad (2)$$

$$\text{Net: ATP} + \text{HCO}_3^- + \text{acetyl-CoA} \underset{\substack{\text{BC, CT}\\\text{CCP-biotin}}}{\overset{\text{Me}^{2+}}{\rightleftharpoons}} \text{malonyl-CoA}$$

$$+ \text{ADP} + \text{P}_i \quad (3)$$

This paper describes the preparation, assay, and properties of the biotin
carboxylase component, while that for the carboxyltransferase component
is described in another chapter.[1] While the natural carboxyl acceptor for
biotin carboxylase [carboxyl carrier protein:CO₂ ligase (ADP)] is carboxyl carrier protein, the enzyme also catalyzes a model carboxylation

[1] This volume [5].

TABLE I

PURIFICATION OF BIOTIN CARBOXYLASE FROM *E. coli* B

Step	Total activity[a] (units)	Protein[b] (mg)	Specific activity (units/mg)	Yield (%)
1. Cell-free extract[c] (20,000 *g* supernatant solution)	334	111,000	0.0030	100
2. 25–42% saturated (NH₄)₂SO₄ fractionation	175	26,900	0.0065	52
3. Calcium phosphate gel fractionation	196	5,630	0.035	59
4. DEAE-cellulose chromatography	137	38	3.6	41
5. Cellulose phosphate	73	12.0	$6.1(3.8)^d$	22
6. Crystallization				
(a) first	53	9.6	$5.5(3.4)^d$	16
(b) second	40	5.3	$5.9(3.7)^d$	12

[a] $H^{14}CO_3^-$ fixation biotin carboxylase assay.

[b] Protein was determined by the biuret method[4] (steps 1–4) and by the spectro-photometric method of Warburg and Christian[5] (steps 5 and 6).

[c] From 1 kg of packed *E. coli* B cells.

[d] Specific activities in parentheses are calculated on the basis of refractometrically determined protein.

reaction with D-biotin and some of its analogues to yield their corresponding 1'-*N*-carboxy-D-biotin derivatives.[2,3]

Assay Methods

Principle. The enzymatic activity is conveniently determined[3] by following the rate of formation of 1'-*N*-[¹⁴C]carboxybiotin from $H^{14}CO_3^-$ and D-biotin. Excess [¹⁴C]bicarbonate is removed by gassing with unlabeled CO_2, carboxybiotin being stable to this treatment. Alternatively, after the first three steps of the purification procedure (Table I) biotin carboxylase activity can be assayed spectrophotometrically[3] by following

[2] P. Dimroth, R. B. Guchhait, E. Stoll, and M. D. Lane, *Proc. Nat. Acad. Sci. U.S.* **67**, 1353 (1970).

[3] R. B. Guchhait, S. E. Polakis, P. Dimroth, E. Stoll, J. Moss, and M. D. Lane, *J. Biol. Chem.* (1974) (in press).

[4] A. G. Gornall, C. J. Bardawill, and M. M. David, *J. Biol. Chem.* **177**, 751 (1949).

[5] O. Warburg and W. Christian, *Biochem. Z.* **310**, 384 (1942).

the rate of oxidation of NADH in the presence of phosphoenolpyruvate, NADH, pyruvate kinase, and lactate dehydrogenase.

Reagents

Triethanolamine hydrochloride buffer, pH 8.0, at 30° KH^{14}CO$_3$, 0.1 M; in order to obviate high blanks resulting from [^{14}C] contaminant(s) in commercial NaH^{14}CO$_3$ samples that cannot be discharged by gassing with CO$_2$ at the end of the assay, H^{14}CO$_3^-$ is purified in the following manner. Ba^{14}CO$_3$ (0.088 mmole; approximately 60 mCi/mmole) is suspended in 2 ml of water and allowed to stand for 2 hours at room temperature. After centrifugation, the Ba^{14}CO$_3$ precipitate is washed again in the same way and then transferred to the main compartment of a Warburg flask; 0.5 ml of 0.18 N NaOH (0.09 mmole) and 1 ml of 7% perchloric acid are placed in the center-well and sidearm, respectively. After evacuation of the flask using a water pump, the acid is tipped into the main compartment and ^{14}CO$_2$ transfer to the center-well allowed to occur at room temperature overnight. The center-well contents are then diluted with 0.1 M KHCO$_3$ to a specific activity of about 3×10^5 cpm/μmole of HCO$_3^-$. The specific activity must be determined accurately.

KHCO$_3$, 0.1 M

MgCl$_2$, 0.1 M

Bovine serum albumin (BSA), 6 mg/ml

Reduced glutathione (GSH), 0.1 M

Absolute ethanol

Phosphoenolpyruvate (PEP), 0.1 M

NADH, 0.01 M

Biotin (potassium salt pH 7.0), 0.25 M

ATP, 0.1 M

Pyruvate kinase, 10 mg/ml, ∼200 units/mg

Lactate dehydrogenase, 5 mg/ml, ∼550 units/mg

n-Octanol

NaOH, 0.1 M

Liquid scintillator (2668 ml toluene, 1332 ml Triton X-100, 22 g of 2,5-diphenyloxazole (PPO) and 0.4 g of 1,4-bis[2-(5-phenyloxazolyl)]-benzene.

Procedure

[^{14}C]*Bicarbonate Fixation Assay.* The complete reaction mixture contains 100 mM triethanolamine (Cl$^-$) buffer, pH 8.0; 1 mM ATP; 8 mM

$MgCl_2$; 8 mM $KHCO_3$ (300 cpm/nmole); 50 mM potassium D-biotin; 3 mM glutathione; 0.3 mg bovine serum albumin; 0.05 ml (10 vol%) ethanol; and 0.1–1.0 milliunit of biotin carboxylase in a total volume of 0.5 ml. The reaction is initiated by the addition of enzyme. After incubation for 10 minutes at 30°, carboxylation is terminated by rapid transferal of a 0.4-ml aliquot to 1 ml of ice-cold water containing 2 drops of n-octanol and by bubbling of CO_2 through the solution for 30 minutes at 0–2° to remove the excess $H^{14}CO_3^-$. No significant losses of [^{14}C]carboxybiotin occur during gassing with CO_2 ($t_{1/2}$ = 192 minutes for decarboxylation under these conditions). After gassing, 0.1 ml of 0.1 N NaOH is added and the entire volume is transferred to a scintillation counting vial, liquid scintillator added, and [^{14}C] activity determined.

Spectrophotometric Assay. A 1-ml reaction mixture is used containing the same components and concentrations as above except that $H^{14}CO_3^-$ is replaced by unlabeled bicarbonate and 0.5 mM phosphoenolpyruvate, 0.2 mM NADH, 5.4 units of crystalline lactate dehydrogenase, and 3 units of crystalline pyruvate kinase are added. The reaction is initiated with biotin carboxylase and the D-biotin–dependent oxidation of NADH is followed at 340 nm and 30°. The two assay methods are in good agreement; both follow zero-order kinetics and are proportional to enzyme concentration within the limits described.

Definition of Unit and Specific Activity. One unit of biotin carboxylase is defined as the amount of enzyme which catalyzes the formation 1 μmole of free carboxybiotin per minute under the conditions described above. During the purification procedure protein is determined by the methods indicated in Table I. The protein concentration of homogeneous biotin carboxylase is determined spectrophotometrically. An absorbance of 1.00 at 280 nm (1 cm light path) is equivalent to 1.6 μg/ml of enzyme. Specific activity is expressed as units per milligram of protein.

Purification Procedure

The purification procedure described is based on that of Guchhait *et al.*[3] The first two steps of the purification procedure for the biotin carboxylase component are identical to those for the purification of carboxyltransferase described in another chapter.[1] All manipulations are carried out at 2–4° and all solutions employed in the purification contain 1 mM EDTA and 5 mM 2-mercaptoethanol. The results of a typical biotin carboxylase purification are summarized in Table I.

One kilogram of *E. coli* B cells, $\frac{3}{4}$ log phase, grown on enriched medium, are suspended in 2 liters of 0.1 M potassium phosphate buffer, pH 7.0. The suspension is passed twice through a precooled Manton-Gaulin submicron disperser at 9000 psi. The cell breaker is then washed 2–3 times with 500 ml of the same buffer, the washes combined with the original extract, the pooled extract centrifuged at 20,000 g for 30 minutes, and the supernatant solution retained.

Ammonium Sulfate Fractionation (25–42% Saturation). The supernatant solution is brought to 25% saturation by addition of solid ammonium sulfate (144 g/liter), the pH being maintained at 7.0 by addition of dilute ammonium hydroxide. After 15 minutes, the mixture is centrifuged, the supernatant recovered and brought to 45% saturation with solid ammonium sulfate (125 g/liter). The precipitate is recovered by centrifugation and extracted with 4 liters of a 42% saturated ammonium sulfate solution containing 50 mM potassium phosphate buffer, pH 7.0. After decanting the supernatant, the pellet is either stored at $-90°$ in a tightly stoppered bottle or fractionated immediately with calcium phosphate gel for the purification of biotin carboxylase (below) or carboxyltransferase.

Calcium Phosphate Gel Fractionation. The precipitated protein from the preceding step (approximately 27 g of protein) is dissolved in 200 ml of 5 mM K$_2$HPO$_4$ containing 20% glycerol and dialyzed overnight against 10 liter of the same solution. The protein concentration of the dialyzed enzyme is determined spectrophotometrically[3] and the solution diluted to give a protein concentration of 20 mg/ml. Sufficient calcium phosphate gel (30 mg dry weight/ml) is added to absorb 30% of the protein; this usually requires bringing the gel–protein ratio to 1:1. After stirring for 30 minutes the mixture is centrifuged and the supernatant solution, which contained less than 10% of the original biotin carboxylase activity, is discarded; the gel is washed once with 2 liters of 5 mM potassium phosphate buffer, pH 7.0, and this wash is also discarded. Biotin carboxylase activity is eluted by resuspending the gel precipitate in 2 liters of 0.12 M potassium phosphate buffer, pH 7.0, stirring for 30 minutes, and centrifuging; this process is repeated two additional times, the eluates combined and brought to 50% saturation with solid ammonium sulfate (313 g/liter). While all of the biotin carboxylase activity is recovered in this eluate, carboxyltransferase activity remains adsorbed to the gel pellet. Approximately 6 g of protein are recovered in the 0.12 M buffer eluate which can be stored frozen at $-90°$ as a pellet after centrifugation or subjected immediately to fractionation by DEAE-cellulose chromatography.

DEAE-Cellulose Chromatography. The precipitated enzyme from the preceding step is dissolved in 75 ml of 10 mM potassium phosphate buffer, pH 7.0, containing 20% glycerol. After clarification of the dialyzed solution by centrifugation at 20,000 g for 20 minutes, the clear supernatant containing approximately 6 g of protein is applied to a 3-liter DEAE-cellulose column (9×50 cm) previously equilibrated with dialysis buffer. Elution is accomplished with 8 liters of the same 10 mM potassium phosphate buffer and the eluate collected fractionally and assayed for biotin carboxylase activity and protein. Biotin carboxylase activity appears with the first protein eluted from the column and precedes the major "breakthrough" protein peak; the bulk of the protein applied to the column remains adsorbed. This step achieves a purification of at least 100-fold, giving rise to approximately 30–40 mg of protein in the biotin carboxylase peak; the active fractions are pooled and immediately subjected to chromatography on cellulose phosphate. The DEAE-cellulose step has been successfully carried out with enzyme arising from the processing of 2 kg of cells using a larger column (5 liters, 9×80 cm).

Cellulose Phosphate Chromatography. The pooled fractions from DEAE-cellulose chromatography (about 40 mg of protein) are applied directly to a 50-ml cellulose phosphate column (2×30 cm) previously equilibrated with 10 mM potassium phosphate buffer, pH 7.0, containing 20% glycerol. Elution is carried out with a 1-liter linear potassium phosphate gradient (10–500 mM, pH 7.0, containing glycerol). The effluent is collected fractionally and is monitored for biotin carboxylase activity and protein; carboxylase activity is eluted at a phosphate concentration of approximately 0.1 M. The fractions containing maximal activity are combined, poured into dialysis bags, and dialyzed against a solution containing 50 mM potassium phosphate buffer, pH 7.0, 20% glycerol, and sufficient ammonium sulfate to bring the solution to 60% saturation at equilibrium. After 1–2 days the flocculated protein is recovered by centrifugation and stored as a pellet at $-90°$ without loss of activity for several months or is subjected immediately to the crystallization procedure.

Crystallization. The precipitated enzyme from the cellulose phosphate chromatographic procedure is dissolved in a minimal volume of 10 mM potassium phosphate buffer, pH 7, containing 1 mM EDTA and 2 mM dithiothreitol to produce a protein concentration of 4–6 mg/ml. The solution is then dialyzed against the same buffer and crystallization usually begins within 24 hours. Crystals appear as prisms ranging from 0.01 to 0.025 mm.[6] Crystallization can be accelerated by the use of seed crystals or by the presence of 10% ethanol in the medium; crystals prepared in the presence of ethanol are larger and appear as plates which crack spontaneously.

---

[6] P. Dimroth, R. B. Guchhait, and M. D. Lane, *Hoppe-Seyler's Z. Physiol. Chem.* **352**, 351 (1971).

Properties

Homogeneity and Molecular Characteristics.[2] Biotin carboxylase prepared as described above is homogeneous in the analytical ultracentrifuge and by disc gel electrophoresis. The sedimentation coefficient $(s_{20,w})$ of the enzyme is 5.7 S and its Stokes radius 35 Å. On the basis of these values the molecular weight of the enzyme is 95,000 assuming a partial specific volume of 0.75. The enzyme is composed of two subunits of identical weight as indicated by dodecyl sulfate polyacrylamide gel electrophoresis. The subunit molecular weight corresponds to 51,000. The relationship between absorbancy $(A_{280\,nm}^{1\,cm})$ at 280 nm and refractometrically determined protein concentration (c) in mg/ml is given by the equation: $c = 1.6 \times A_{280\,nm}^{1\,cm}$.

Reactions Catalyzed, Kinetics and Specificity. In addition to the carboxylation of carboxyl carrier protein and free D-biotin,[2,7] the enzyme catalyzes: (1) carboxyl carrier protein and bicarbonate-dependent ATP-[^{14}C]ADP and ATP-^{32}P exchanges[7] and (2) D-biotin (or CCP)–dependent phosphoryl transfer from carbamyl phosphate to ADP to form ATP.[8]

The enzymatic carboxylation of free D-biotin is markedly activated by certain organic solvents, 15 volumes% of ethanol being most effective; the activation is more pronounced at low temperatures.[2]

The metal requirement of the enzyme can be satisfied by Mg^{2+}, Mn^{2+}, or Co^{2+}. The V_{max} at the optimal metal concentration is independent of the nature of the metal.[3] The enzyme exhibits absolute specificity for the naturally occurring D isomer of biotin; 1-biotin is completely inactive. Various analogues and derivatives of the D form are active as substrates in the carboxylation reaction; however, the best acceptor is D-biotin.[2]

The K_m values at pH 8.0 are 170 mM, 0.11 mM, and 2.9 mM for D-biotin, ATP · Mg, and HCO_3^-, respectively.[2] The active species of "CO_2" in the reaction has been shown to be HCO_3^-.[7]

Finally, GTP is a poorer substrate than ATP, while ADP and, to a lesser extent, GDP are competitive inhibitors with respect to ATP.[7]

Acknowledgments

This work was supported by research grants from The National Institutes of Health, USPHS (AM-14574 and AM-14575), and The American Heart Association, Inc. The authors are indebted to Mr. Eberhard Zwergel for his superb technical assistance.

[7] S. E. Polakis, R. B. Guchhait, E. E. Zwergel, T. G. Cooper and M. D. Lane, *J. Biol. Chem.* (1974) (in press).

[8] S. E. Polakis, R. B. Guchhait, and M. D. Lane, *J. Biol. Chem.* **247**, 1335 (1972).

[5] Carboxyltransferase Component of Acetyl-CoA Carboxylase from *Escherichia Coli*

By Ras B. Guchhait, S. E. Polakis, and M. Daniel Lane

CCP-biotin-CO_2^-
(carboxyl carrier
protein)
or $+$ acetyl-CoA $\rightleftharpoons$ malonyl-CoA $+$
1'-N-Carboxy-D-
biotin methyl ester

CCP-biotin
or
D-biotin
methyl ester

This paper describes the preparation, assay, and properties of carboxyltransferase, a component of the *E. coli* acetyl-CoA carboxylase system (see preceding chapter, this volume). Carboxyltransferase catalyzes the second step in the carboxylation of acetyl-CoA, the first step being carried out by biotin carboxylase whose preparation is described in the preceding section.[1] Although the natural carboxyl donor for the transferase is the carboxylated form of the carboxyl carrier protein, the enzyme also catalyzes the reversible transcarboxylation reaction shown above in which D-biotin methyl ester replaces carboxyl carrier protein. The latter reaction serves as a convenient assay method for the enzyme.

Assay Methods

Principle. Carboxyltransferase catalyzes the reversible transfer of the free carboxyl group of malonyl-CoA to D-biotin (and certain of its derivatives) to form acetyl-CoA and the corresponding 1'-N-carboxy-D-biotin derivative. Enzymatic activity is determined by measuring the loss of acid-stable radioactivity in the D-biotin–dependent "decarboxylation" of [^{14}C]malonyl CoA, 1'-N-carboxy-D-biotin being acid-labile.[2] The assay can also be employed spectrophotometrically using partially purified preparations. This assay involves coupling the acetyl-CoA generated in the carboxyltransferase reaction to the combined citrate synthase–malate dehydrogenase reactions; thus, NAD^+ reduction can be followed at 340 nm.

---

[1] This volume [4].
[2] R. B. Guchhait, J. Moss, W. Sokolski, and M. D. Lane, *Proc. Nat. Acad. Sci. U.S.* **68**, 653 (1971).

Radioactive Assay

Reagents

Tris(Cl⁻) buffer, 1.0 M, pH 8.0 at 25°
Malonyl-CoA (2-^{14}C- or 3-^{14}C-), 0.5 mM, 4–6 × 10³ cpm/nmole
D-Biotin methyl ester, 0.05 M, in 40% ethanol
Bovine serum albumin, 6 mg/ml
HCl, 6 N
Liquid scintillator 5.5 g of 2,5-diphenyloxazole (PPO), 0.1 g of 1,4-bis-2-(4-methyl-5-phenyloxazoyl)-benzene (dimethyl POPOP), 667 ml of toluene and 333 ml of triton/liter.

Procedure. The complete reaction mixture contains the following components: 50 mM Tris (Cl⁻) buffer, pH 8.0; 100 μM [2-^{14}C]- or [3-^{14}C]malonyl-CoA (4–6 × 10³ cpm/nmole); 10 mM D-biotin methyl ester; 0.3 mg of bovine serum albumin, and up to 1 milliunit of enzyme in a total volume of 0.5 ml. After 5, 10, 15, and 20 minutes of incubation at 30°, 0.1 ml aliquots are transferred into scintillation vials containing 0.1 ml of 6 N HCl. The acidified mixture is taken to dryness at 95° in a forced-draft oven after which 0.3 ml of water and 3 ml of liquid scintillator are added and residual acid-stable radioactivity is determined. The reaction products, [2-^{14}C]acetyl-CoA, $^{14}CO_2$ or the [^{14}C]carboxyl group of N-carboxybiotin methyl ester generated from [2-^{14}C]malonyl-CoA or [3-^{14}C]malonyl-CoA are volatile under this treatment, whereas [^{14}C]malonyl-CoA is not. The extent of carboxyl transfer or decarboxylation is determined from the difference between acid-stable radioactivity at zero time and after reaction. Carboxyltransferase activity is equal to acid-stable radioactivity lost during incubation with D-biotin methyl ester minus that lost in the absence of the D-biotin derivative.

The assay follows zero-order kinetics under the conditions described and activity is proportional to enzyme concentration.

Spectrophotometric Assay

Reagents

Tris(Cl⁻) buffer, 1.0 M, pH 8.0 at 25°
Malonyl-CoA, 2.5 mM
L-Malate, 0.5 M
Bovine serum albumin, 6 mg/ml
NAD, 10 mM
D-Biotin methyl ester, 0.05 M, in 40% ethanol
Malate dehydrogenase (Boehringer), 10 mg/ml
Citrate synthase (Boehringer), 10 mg/ml

Procedure

The reaction mixture contains 100 mM Tris(Cl⁻) buffer, pH 8.0; 0.1 mM malonyl-CoA; 10 mM L-malate; 0.5 mM NAD⁺; 0.6 mg bovine serum albumin; 125 μg malate dehydrogenase; 50 μg citrate synthase; 1–10 milliunits of enzyme and 10 mM D-biotin methyl ester. The assay is conducted at 30° and is initiated by the addition of D-biotin methyl ester. NADH formation is followed at 340 nm for a few minutes.

Definition of Unit and Specific Activity. One unit of carboxyltransferase is defined as that amount of enzyme which catalyzes the formation of 1 μmole of free carboxybiotin methyl ester per minute under the conditions of the radioactive assay. Specific activity is expressed as units per milligram of protein which is determined by the methods of biuret[3] and Folin.[4]

Isolation and Purification of Carboxytransferase from *E. Coli* B

The preparation of the cell-free extract of *E. coli* B cells (¾ log phase) and its fractionation with ammonium sulfate (25–42% saturated ammonium sulfate fraction) is carried out as described in the preceding chapter[1] except that 2 kg of packed cells are used. All subsequent steps in the purification are carried out at 2–4° using solutions containing 1 mM EDTA and 5 mM 2-mercaptoethanol. The results of a typical purification of the carboxyltransferase are summarized in Table I.

Calcium Phosphate Gel Fractionation. The precipitate obtained by ammonium sulfate fractionation (25–42% saturated cut) obtained as described in the preceding chapter is dissolved in 250 ml of 2 mM K$_2$HPO$_4$ and dialyzed against 25 liters of the same buffer overnight. The dialyzed enzyme is diluted with the same buffer to a protein concentration of 20 mg/ml and sufficient calcium phosphate gel (30 mg dry weight/ml) is added to adsorb 65% of the protein in the extract; this usually requires a gel–protein ratio of approximately 2.5:1.0. After stirring the suspension for 30 minutes, the mixture is centrifuged and the supernatant solution which contains about 5% of the original carboxyltransferase activity is discarded. The gel pellet is extracted 3 times with 2 liters of 0.12 M potassium phosphate buffer, pH 7.0. These eluates, which contained virtually no carboxyltransferase activity, are discarded. The enzyme is then eluted from the gel by extracting the pellet 3 times with 3 liters of 0.5 M potas-

[3] A. G. Gornall, C. J. Bardawill, and M. M. David, *J. Biol. Chem.* **177**, 751 (1949).
[4] O. H. Lowry, N. J. Rosebrough, A. L. Farr, and R. J. Randall, *J. Biol. Chem.* **139**, 265 (1951).

TABLE I
PURIFICATION OF THE CARBOXYLTRANSFERASE FROM *E. coli* B CELLS

| | | Carboxyltransferase | |
Step	Protein[a]	Total activity (units)	Specific activity (units/ mg protein)
1. Cell-free extract[b]	210,000	—[c]	—[c]
2. 25–42% saturated ammonium sulfate fractionation	43,000	8,500	0.2
3. Calcium phosphate gel fractionation	5,176	6,910	1.33
4. First DEAE-cellulose (standard) chromatography	360	4,564	12.7
5. Phosphocellulose chromatography			
Peak I	20	900	45
Peak II	30	260	8.67
6. Second DEAE-cellulose (type 20) chromatography of Peak I	6	576	96

[a] Protein in steps 1–4 was determined by the biuret method[3] and in steps 5 and 6 by the method of Folin.[4]

[b] From 2 kg of packed *E. coli* B cells.

[c] Carboxyltransferase activity cannot be assayed accurately in the cell-free extract because of the presence of an inhibitor; this inhibitor is removed by ammonium sulfate fractionation (step 2).

sium phosphate buffer, pH 7.0, each time. The pooled 0.5 M eluate which contains 80–90% of the original carboxyltransferase activity is brought to 60% saturation with solid ammonium sulfate (390 g/liter) and allowed to stir for 20 minutes. After centrifugation at 20,000 g for 30 minutes, the precipitate is dissolved in 200 ml of 25 mM potassium phosphate buffer, pH 7.0, and then dialyzed against 30 liters of the same buffer overnight.

First DEAE-Cellulose Chromatography. The dialyzed enzyme (about 5 g of protein) is applied to a 3.8-liter DEAE-cellulose column (9 × 60 cm) previously equilibrated with 25 mM dialysis buffer. The column is washed with 4 liters of 25 mM potassium phosphate buffer, pH 7.0, after which the enzyme is eluted with an 8-liter linear phosphate gradient (25–500 mM, pH 7.0); carboxyltransferase is eluted at a phosphate concentration of approximately 0.2 M. The most active fractions are pooled and the enzyme precipitated by bringing the ammonium sulfate concentration to 60% saturation with solid ammonium sulfate. After collecting the precipitated protein by centrifugation the pellet is dissolved in 40 ml of 25 mM phosphate buffer, pH 7.0, and then dialyzed overnight against 6 liters of the same buffer.

Phosphocellulose Chromatography. The dialyzed enzyme (about 360 mg of protein) is applied to a 410-ml phosphocellulose column (2.5 × 90 cm) previously equilibrated with 25 mM buffer, pH 7.0. After passing 200 ml of the same buffer through the column, the enzyme is eluted with a 3-liter linear phosphate gradient (25–300 mM potassium phosphate, pH 7.0). The enzyme appears in the column eluate when the phosphate concentration reaches approximately 0.15 M. Two distinct peaks of carboxyltransferase activity are observed. The major activity peak (Peak I, at 1650 ml eluate volume) which comprised approximately 80% of the total activity is retained and further purified; although Peak II has not been further purified, its kinetic properties appear to be indistinguishable from those of Peak I. The pooled fractions from Peak I are placed in dialysis bags and the enzyme is precipitated by dialysis against a sufficient volume of 65% saturated ammonium sulfate containing 50 mM phosphate buffer, pH 7.0, such that at equilibrium the final percent saturation reaches 60%. The precipitate is recovered by centrifugation, dissolved in 2 ml of 50 mM potassium phosphate buffer, pH 7.0, containing 1 mM EDTA and 2 mM of dithiothreitol, and then dialyzed overnight against 1 liter of the same buffer.

Second DEAE-Cellulose (Type 20) Chromatography. The dialyzed enzyme (about 20 mg of protein) from the preceding step is applied to a 120-ml DEAE-cellulose (type 20) column (1.5 × 90 cm) and the carboxyltransferase eluted with a 500-ml linear potassium phosphate, pH 7.0, gradient (50–250 mM); the flow rate is maintained at 5 ml/hour. Carboxyltransferase activity appears in the eluate when the phosphate concentration reaches about 150 mM and is exactly coincident with the protein peak. The fractions having maximal enzymatic activity are pooled and precipitated by dialysis against 65% saturated ammonium sulfate as described in the preceding step. After centrifugation the enzyme pellet can either be stored at −90° in a tightly stoppered tube for several months without loss of activity or dissolved in 50 mM potassium phosphate buffer, pH 7.0, containing 1 mM EDTA and 2 mM dithiothreitol and dialyzed against the same buffer overnight.

Properties. Carboxyltransferase, prepared as described above, is homogeneous in the analytical ultracentrifuge and by polyacrylamide gel electrophoresis. The molecular weight of the enzyme determined by the sedimentation equilibrium method[5] is 131,000 and by gel filtration on

[5] R. B. Guchhait, S. E. Polakis, P. Dimroth, E. Stoll, J. Moss, and M. D. Lane, *J. Biol. Chem.* (1974) (in press).

Sephadex G-200 is 145,000. SDS-gel electrophoresis gives rise to two stained protein bands of 30,000 and 35,000 daltons which have approximately equal staining intensities. Thus, the carboxyltransferase appears to be composed of two polypeptide chains of differing size and to have an A_2B_2 subunit structure.

Maximal activity in the pH range 4.5–8.0 is achieved with D-biotin methyl ester and D-biotinol acetate; D-biotinol and biocytin are approximately 50% as active. D-Biotin, D-homobiotin, and D-norbiotin are less active exhibiting greatest activity at pH 4.5 and being far less active at pH 8.0. 1-Biotin and D-2'-thiobiotin are inactive.

Acknowledgments

This work was supported by research grants from The National Institutes of Health, USPHS (AM-14574 and AM-14575), and The American Heart Association, Inc. The authors are indebted to Mr. Eberhard Zwergel for his superb technical assistance.

[6] Fatty Acid Synthase from Rat Liver

By Carl M. Nepokroeff, M. R. Lakshmanan, and John W. Porter

The fatty acid synthase complex exists in mammalian as well as avian liver in the soluble portion of the cell.[1] Negligible activity for the *de novo* synthesis of fatty acids is associated with the microsomal and mitochondrial fractions. The purification of the soluble fatty acid synthase complex of rat liver to homogeneity has been achieved,[2] and this complex has been found to have a molecular weight of approximately 540,000.

Assay Method

The fatty acid synthase complex may be assayed by either radiochemical or spectrophotometric methods. In the radioisotopic method, incorporation of 1-^{14}C-labeled acetyl-CoA into fatty acids is measured in the presence of malonyl-CoA and NADPH. This method is reliable for either crude or purified enzyme preparations. (For the methods of prepa-

---

[1] J. W. Porter, S. Kumar, and R. E. Dugan, *Progr. Biochem. Pharmacol.* **6**, 1 (1971).

[2] D. N. Burton, A. G. Haavik, and J. W. Porter, *Arch. Biochem. Biophys.* **126**, 141 (1968).

ration of 1-^{14}C-labeled acetyl-CoA, unlabeled acetyl-CoA, malonyl-CoA, and details of the radioassay, see Hsu et al.[2a]) The spectrophotometric method measures the malonyl-CoA– and acetyl-CoA–dependent rate of oxidation of NADPH at 340 nm and is best suited for purified enzyme preparations. However, it can be used with crude preparations providing appropriate corrections are made.

Spectrophotometric Assay. The reaction mixture (total volume 1.0 ml; final pH 7.0) contains 500 μmoles potassium phosphate buffer, pH 7.0; 33 nmoles of acetyl-CoA; 100 nmoles of malonyl-CoA; 100 nmoles of NADPH; 1 μmole of EDTA; 1 μmole of β-mercaptoethanol; and enzyme protein (5–10 μg of DEAE-cellulose purified fatty acid synthase or 50–100 μg of the 105,000 g supernatant protein fraction). The reaction is initiated by the addition of enzyme to the mixture of substrates previously equilibrated at 30° for 5 minutes. It is important to preincubate the enzyme in order to obtain maximum enzymatic activity.[3] The oxidation of NADPH is followed at 340 nm, while the cuvette chamber is maintained at 30°. Full-scale deflection of the recorder tracing is set at 0.2 absorbance unit. The initial slope of the recorder tracing is used to calculate the rate of fatty acid synthesis. A correction is made for the rate of NADPH oxidation in the absence of malonyl-CoA.

It should be noted that butyryl-CoA has been reported to be a better primer than acetyl-CoA in the mammalian fatty acid synthase system.[4] The fatty acid synthase assay system has also been modified and successfully employed for the determination of malonyl-CoA in tissue extracts.[5]

Units. A unit of enzymatic activity is defined as the amount of enzyme protein required to synthesize 1 nmole of palmitic acid (equivalent to the oxidation of 14 nmoles of NADPH) per minute under the conditions of the assay. The specific activity is defined as the number of activity units per milligram of protein. Protein is determined by the biuret method of Gornall et al.[6]

Purification of the Enzyme System

All solutions used in the preparation of the enzyme system are prepared with deionized water and all buffers contain potassium phosphate,

[2a] R. Y. Hsu, P. H. W. Butterworth, and J. W. Porter, Vol. XIV [4], 33.

[3] Preincubation of either purified enzyme or crude liver homogenate fractions in 500 mM potassium phosphate buffer, pH 7.0, and 5 mM dithiothreitol at 37° for at least 15 minutes is essential for maximum enzymatic activity.

[4] C. H. Lin and S. Kumar, *J. Biol. Chem.* **247,** 604 (1972).

[5] R. W. Guynn, D. Veloso, and R. L. Veech, *J. Biol. Chem.* **247,** 7325 (1972).

[6] A. G. Gornall, C. J. Bardawill, and M. M. David, *J. Biol. Chem.* **117,** 751 (1949).

pH 7.0, 1 mM EDTA and either 1 mM dithiothreitol or 1 mM β-mercaptoethanol, unless otherwise stated.

Preparation of Liver Supernatant Solution. Since the concentration of fatty acid synthase in liver is related to the nutritional state of the rat[7] it is important to begin the preparation with rats which have been fasted and refed. Hence, male rats weighing about 150 g each are starved for 48 hours, then fed a fat-free diet[8] for 48 hours. The rats are killed by decapitation and the livers are quickly removed and placed on ice. The chilled livers are homogenized in 1.5 volumes of a phosphate-bicarbonate buffer (70 mM KHCO$_3$, 85 mM K$_2$HPO$_4$, 9 mM KH$_2$PO$_4$, and 1 mM dithiothreitol), pH 8.0, in a Potter-Elvehjem type of homogenizer with a motor-driven Teflon pestle. Large batches of liver are more conveniently homogenized in a Waring blender for 15 seconds at full speed. The homogenate is centrifuged at 20,000 g for 10 minutes; the supernatant solution (postmitochondrial fraction) is retained, and then it is recentrifuged at 105,000 g for 60 minutes. The supernatant solution (105,000 g fraction) contains the fatty acid synthase. This solution can be stored at $-15°$ for at least 2 months without appreciable loss of enzymatic activity.

The purification procedure for the fatty acid synthase that follows is a modification of the method originally described by Burton *et al.*[2] This procedure is carried out at room temperature unless otherwise specified.

First Ammonium Sulfate Fractionation. The frozen rat liver 105,000 g supernatant solution (40 ml) is thawed at room temperature. Saturated ammonium sulfate solution, pH 7.0, containing 3 mM EDTA and 1 mM β-mercaptoethanol is added to a saturation of 20%. After stirring for 15 minutes the mixture is centrifuged and the precipitate discarded. The supernatant solution is brought to 35% saturation with ammonium sulfate. The precipitated protein is collected by centrifugation and retained for further fractionation.

Calcium Phosphate Gel Adsorption. The protein fraction from the previous step (approximately 500 mg protein) is dissolved and then diluted to approximately 150 ml with 5 mM potassium phosphate buffer. The protein fraction is solubilized by gentle stirring with a glass rod in a small volume of buffer. Foaming is avoided insofar as possible since the enzyme is susceptible to surface denaturation.[2] Calcium phosphate gel equal to half the weight of the protein is added with stirring. The

[7] D. N. Burton, J. M. Collins, A. L. Kennan, and J. W. Porter, *J. Biol. Chem.* **244**, 4510 (1969).

[8] Fat-free diet is obtained from Nutritional Biochemical Corporation, Cleveland, Ohio.

suspension is centrifuged immediately for 2–3 minutes at 5000 g and the supernatant solution is retained. Since the fatty acid synthase is not stable in a low ionic strength buffer,[2] it is imperative that this procedure be performed rapidly and that the next purification step, DEAE-cellulose chromatography, follow immediately.

DEAE-Cellulose Chromatography. The supernatant solution from the previous step is adsorbed to a column of DEAE-cellulose (10.3 × 3.5 cm) which has previously been washed with 50 mM potassium phosphate buffer. Most of the protein is removed from the column by washing with the above buffer. Washing with the buffer is continued until the light absorption at 280 nm decreases to approximately 0.05 unit. The enzyme is then eluted from the column with 160 mM potassium phosphate buffer. This chromatographic procedure is performed at room temperature with a flow rate of approximately 5–10 ml/minute. All steps of the chromatographic procedure are performed rapidly since the enzyme is not stable in low ionic strength buffer. Elution of the protein is monitored spectrophotometrically at 280 nm and 5 ml eluate fractions are collected. Those fractions exhibiting absorbancies greater than 0.50 unit are combined and retained.

Second Ammonium Sulfate Fractionation. The material eluted from DEAE-cellulose is brought to 33% saturation with ammonium sulfate and the precipitated protein is collected by centrifugation. This protein fraction is dissolved in a minimum volume (1–2 ml) of cold 0.5 M potassium phosphate buffer, pH 7.0, and dialyzed against the same buffer overnight at 4°. The enzyme at this stage is referred to as DEAE-cellulose purified fatty acid synthase. A yield of 30–40 mg of the fatty acid synthase can be expected from 40 ml (approximately 2000 mg of total soluble rat liver protein) of the crude 105,000 g supernatant fraction. Since the rat liver enzyme is sensitive to surface denaturation,[2] as are the rat mammary gland and pigeon liver enzymes, this preparation should be handled with care.

The results of a typical purification of this enzyme are shown in Table I. The purification of DEAE-cellulose purified fatty acid synthase from the 105,000 g supernatant fraction requires approximately 5 hours.

Sucrose Density Gradient Centrifugation. The results of ultracentrifugation studies by Collins *et al.*[9] on DEAE-cellulose purified fatty acid synthase isolated from livers of fasted, diabetic, or normal rats showed the presence of a contaminating protein with a sedimentation coefficient of 7 S. The relative concentration of this component is negligible, however, in enzyme isolated from animals in the refed state and is therefore de-

[9] J. M. Collins, M. C. Craig, C. M. Nepokroeff, A. L. Kennan, and J. W. Porter, *Arch, Biochem. Biophys.* **143**, 343 (1971).

TABLE I
PURIFICATION OF RAT LIVER FATTY ACID SYNTHASE

Fraction	Total protein (mg)	Enzymatic activity (units[a])	Specific activity (units[a]/mg protein)
105,000 g supernatant	2000	10,000	5
$(NH_4)_2SO_4$, 20–33%	500	6,000	12
DEAE-cellulose	40	2,640	66
Sucrose density gradient	10[b]	750	75

[a] Units are reported in nanomoles of palmitate formed per minute at 30°.

[b] Recovery of protein after gradient centrifugation is actually not as low as indicated. In order to insure that the enzyme fraction was completely devoid of minor components, only the leading half of the protein profile was recovered after each centrifugation. Total recovery of enzyme from the gradients is usually 90–95%.

tected only by immunologic methods. This 7 S component can be removed by two successive sucrose density gradient centrifugations in a Spinco SW 25.1 rotor at 4° and 58,000 g for a period of 46 hours. The linear gradients contain 5–20% (w/v) of sucrose and 500 mM potassium phosphate buffer. After centrifugation the bottom of the tube is punctured and the solution is collected dropwise in 1 ml fractions. Protein in the gradient is monitored spectrophotometrically at 280 nm. The fatty acid synthase sediments as a single peak of protein. The faster sedimenting half of the protein profile from the gradient is concentrated by ammonium sulfate precipitation and the protein fraction is dialyzed against 500 mM potassium phosphate buffer. This fraction is again subjected to a second identical gradient centrifugation. The protein profile from the second centrifugation is fractionated as before and the faster sedimenting half of the profile is retained. This fraction is dialyzed overnight against 500 mM potassium phosphate buffer and then stored in the same buffer. This concentration of buffer is necessary to preserve the structural integrity of the rat liver multienzyme complex. The purified enzyme fraction is referred to as the sucrose density gradient (SDG) purified fatty acid synthase. This enzyme has been shown to be homogeneous by immunologic criteria.[10]

Storage of Enzyme. The purified rat liver fatty acid synthase can be stored in the 500 mM potassium phosphate buffer containing 1 mM dithiothreitol at 0–4° for 4–5 days with negligible loss of activity. The

[10] M. C. Craig, C. M. Nepokroeff, M. R. Lakshmanan, and J. W. Porter, *Arch. Biochem. Biophys.* **152**, 619 (1972).

TABLE II

PARTIAL AMINO ACID COMPOSITIONS OF RAT AND
PIGEON LIVER FATTY ACID SYNTHASES[a]

Amino acid	Rat (moles/mole of enzyme)	Pigeon (moles/mole of enzyme)
Aspartic acid	280	254
Threonine	189	137
Serine	248	193
Proline	194	137
Glutamic acid	355	344
Glycine	274	240
Alanine	306	250
Valine	256	240
Isoleucine	141	165
Leucine	456	340
Tyrosine	76	76
Phenylalanine	118	99

[a] Reproduced from Burton et al.[2]

synthase can also be stored frozen ($-20°$) in the same buffer. Under this condition the enzyme complex is stable for at least 1 month. Whether the synthase is stored at 4° or at $-20°$, it is imperative that the preparation be preincubated prior to assay.[3]

Characterization of the Fatty Acid Synthase

Physicochemical Characterization.[2] The DEAE-cellulose purified rat liver fatty acid synthase sediments essentially as a single component in the Spinco model E analytical ultracentrifuge when dissolved in 0.5 M potassium phosphate buffer. The $s_{20,w}$ value is 12.3 S. A trace of a slower sedimenting component represents either half-molecular weight subunits of the fatty acid synthase or the 7 S minor component. The DEAE-cellulose purified enzyme complex has also been shown to be homogeneous by SDG centrifugation. The molecular weight of the rat liver fatty acid synthase has been determined to be 5.4 × 10⁵ g/mole by sedimentation-diffusion analysis in 0.5 M potassium phosphate buffer.[2]

The amino acid composition of rat and pigeon liver fatty acid synthases are very similar (Table II), but these enzymes are immunologically dissimilar since they do not cross-react.[11] The rat liver enzyme has a high sulfhydryl content (90–92 groups/mole) and a 4'-phosphopante-

[11] M. C. Craig, C. M. Nepokroeff, and J. W. Porter, unpublished results.

theine content of 1.05 moles/mole of enzyme. Negligible amounts of flavin (0.006–0.024 mole of flavin per mole of enzyme) are found.

Kinetics. The pH optimum for the purified rat liver fatty acid synthase complex is 7.0. The K_m value for acetyl-CoA is 4.4 μM.[2] The apparent K_m value for malonyl-CoA is 10 μM.[11a]

Immunochemical Characterization[10]

(a) *Preparation of antiserum.* The SDG-purified fatty acid synthase is used as the antigen in order to obtain antiserum that is specific for only the fatty acid synthase. The enzyme purified by two successive SDG centrifugations is mixed with an equal volume of incomplete Freund's adjuvant and injected into the foot pads of rabbits. Each rabbit receives 2 mg of fatty acid synthase protein and a second injection is given 10 days later. The rabbits are bled at 2-week intervals after the second injection, sera are pooled, and the γ-globulin fraction is isolated as described by Levy and Sober.[12] The serum from rabbits which are injected only with incomplete Freund's adjuvant is prepared in the same way and used as control serum. All sera are stored at −15°.

(b) *Immunodiffusion and precipitation reactions.* Ouchterlony micro-double diffusion is carried out on microscope slides in 0.5% agarose containing 1.0% (w/v) sodium chloride. The agarose gels are developed for 24 hours at 24°. Immunoprecipitation reactions are performed for 1 hour at 37° in 10 mM potassium phosphate buffer, pH 7.0, containing 1 mM dithiothreitol. The samples are then incubated at 4° overnight. Precipitates are collected by centrifugation and then washed twice with cold 0.9% (w/v) sodium chloride containing 10 mM potassium phosphate buffer, pH 7.0. Quantitative precipitin analyses and equivalence point determinations are performed as outlined by Kabat and Mayer.[13]

Our recent studies have shown that rat liver fatty acid synthase is immunologically homogeneous only after the DEAE-cellulose purified enzyme is subjected to two successive SDG centrifugations.[10] Ouchterlony double diffusion patterns show that only one precipitin band is obtained with antiserum prepared against fatty acid synthase purified by SDG centrifugation. However, antiserum prepared against DEAE-cellulose purified enzyme gives a second minor band. The minor precipitin band

---

[11a] S. S. Katiyar, Dept. of Physiological Chemistry, University of Wisconsin, Madison (unpublished).

[12] H. B. Levy and H. A. Sober, *Proc. Soc. Exp. Biol. Med.* **103,** 250 (1960). The ammonium sulfate method of preparation of the γ-globulin fraction was also successfully employed. See M. Goldman, "Fluorescent Antibody Methods." Academic Press, New York, 1968.

[13] E. A. Kabat and M. M. Mayer, *in* "Experimental Immunochemistry," 2nd ed., p. 22. Thomas, Springfield, Illinois, 1971.

is the result of the reaction of antibody with the 7 S component present in the DEAE-cellulose purified fatty acid synthase.[9] The immunodiffusion studies also show that the same species of fatty acid synthase is present in the livers of rats fed a normal diet, fat-free diet, or fasted.

The rat liver fatty acid synthase reacts with its antiserum to produce a typical quantitative precipitin reaction. At the equivalence point, 33.5 mg of antiserum protein were required to precipitate 1 mg of purified fatty acid synthase. The equivalence point is essentially the same for enzyme obtained from rats fed a normal diet or refed a fat-free diet.

Antiserum prepared against fatty acid synthase has been successfully employed in immunochemical studies to evaluate the relative importance of synthetic and degradative rates on changes in content of rat liver fatty acid synthase under different nutritional conditions.[10] The results of these immunochemical studies are discussed in the next section.

Control of the Rate of Synthesis of the Fatty Acid Synthase

Rat liver fatty acid synthase is markedly affected by the nutritional and hormonal states of the animal. Thus, the level of the enzyme is greatly reduced on fasting and increased to supranormal levels on refeeding a fat-free diet to previously fasted animals. This increase has been shown to result from an adaptive increase in the rate of synthesis of the enzyme.[7]

The rates of synthesis and degradation of rat liver fatty acid synthase have been determined as a function of nutritional state in order to determine the relative importance of each in controlling the quantity of this enzyme in the liver.[10] The amount of fatty acid synthase has been determined in these studies by enzyme isolation and by the immunochemical technique. The results of these studies have established the major factor controlling the level of this enzyme to be the rate of its synthesis. However, changes in the rate of degradation may be important in the early stages of fasting.

Recently, it has been shown that insulin is essential for the dietary induction of rat liver fatty acid synthase to supranormal levels.[14] Diabetic rats synthesize negligible amounts of this enzyme whereas insulin administration brings about a dramatic increase in its synthesis. Glucagon and cyclic AMP block the dietary induction of the fatty acid synthase. Hence, it is important to consider both the hormonal and dietary status of animals in studies which concern the isolation and the properties of rat liver fatty acid synthase.

[14] M. R. Lakshmanan, C. M. Nepokroeff, and J. W. Porter, *Proc. Nat. Acad. Sci. U.S.* **69**, 3516 (1972).

[7] Fatty Acid Synthase from Pigeon Liver

By Richard A. Muesing and John W. Porter

$$CH_3\text{—}CO\text{—}SCoA + 7\ HOOC\text{—}CH_2\text{—}CO\text{—}SCoA$$
$$+ 14\ NADPH + 14\ H^+ \rightarrow CH_3(CH_2)_{14}CO_2H$$
$$+ 8\ CoASH + 14\ NADP^+ + 7\ CO_2 + 6\ H_2O$$

The fatty acid synthase enzyme systems of pigeon,[1] chicken,[2] and rat liver,[3] and of rat lactating mammary gland[4] have been purified to homogeneity. These systems exist in the soluble portion of the cell as multienzyme complexes with molecular weights of approximately 5×10^5 g/mole. The constituent enzymes catalyzing the component reactions of fatty acid synthesis are tightly bound in these complexes, whereas the fatty acid synthesizing enzymes of *Escherichia coli* are isolated in the dissociated state.[5]

Assay of Overall Activity

Principle. Avian and mammalian fatty acid synthase complexes catalyze the conversion of acetyl- and malonyl-CoA to long-chain saturated fatty acids in the presence of NADPH. The products and the stoichiometry for the overall reaction are given above.[6] In the absence of NADPH triacetic acid lactone is the product.[7]

An integrated mechanism proposed for the synthesis of palmitic acid by the pigeon liver enzyme system is given in Fig. 1.[8,9] In this mechanism fatty acid synthesis is initiated by the transfer of the acetyl group of acetyl-CoA to a hydroxyl "loading site" (B_1). The acyl group is then successively transferred to the 4'-phosphopantetheine (A_2) and the cysteine (B_2) sites. A malonyl moiety from malonyl-CoA is simultaneously transferred to 4'-phosphopantetheine via the hydroxyl site. Con-

[1] R. Y. Hsu, G. Wasson, and J. W. Porter, *J. Biol. Chem.* **240**, 3736 (1965).

[2] R. Y. Hsu and S. L. Yun, *Biochemistry* **9**, 239 (1970).

[3] D. N. Burton, A. G. Haavik, and J. W. Porter, *Arch. Biochem. Biophys.* **126**, 141 (1968).

[4] S. Smith and S. Abraham, *J. Biol. Chem.* **245**, 3209 (1970).

[5] P. R. Vagelos, P. W. Majerus, A. W. Alberts, A. R. Larabee, and G. P. Ailhaud, *Fed. Proc., Fed. Amer. Soc. Exp. Biol.* **25**, 1485 (1966).

[6] R. Bressler and S. J. Wakil, *J. Biol. Chem.* **236**, 1643 (1961).

[7] J. E. Nixon, G. R. Putz, and J. W. Porter, *J. Biol. Chem.* **243**, 5471 (1968).

[8] J. W. Porter, S. Kumar, and R. E. Dugan, *Progr. Biochem. Pharmacol.* **6**, 1–101 (1971).

[9] S. Kumar, G. T. Phillips, and J. W. Porter, *Int. J. Biochem.* **3**, 15 (1972).

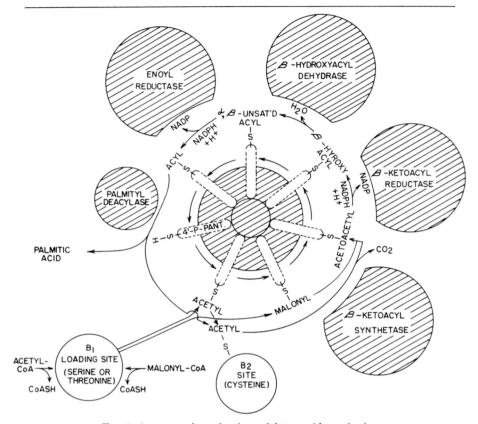

Fig. 1. A proposed mechanism of fatty acid synthesis.

densation between the acetyl moiety at the cysteine site and the malonyl group results in the formation of acetoacetyl-4'-phosphopantetheine enzyme with the simultaneous release of CO_2. The condensation is followed by reduction of the β-ketoacyl derivative, dehydration of the newly formed β-hydroxyacyl group, and then reduction of the α,β-unsaturated enoyl derivative. The saturated acyl group arising from these successive reactions is then transferred to the cysteine site. The 4'-phosphopantetheine site is thus free to bind another malonyl moiety. The cycle is repeated with successive additions of two carbon units from malonyl-CoA to finally yield palmityl-4'-phosphopantetheine enzyme. This acyl group is then released by deacylation.

     The overall activity for fatty acid synthesis may be measured either by determining the incorporation of the [1-^{14}C]acetyl moiety of labeled acetyl-CoA into long-chain fatty acids or by following the rate of

NADPH reduction spectrophotometrically at 340 nm. The former method is reliable for the assay of crude and purified preparations, while the spectrophotometric method is most useful with purified enzyme. Details of the radioactive assay have been previously reported.[1] The spectrophotometric assay is carried out as follows.[10]

Reagents

Potassium phosphate buffer, pH 6.8, 1 M
Dithiothreitol, prepared daily, 0.1 M
EDTA, disodium salt, pH 6.8, 0.1 M
Malonyl-CoA, 3.3 mM
Acetyl-CoA, 1.67 mM
NADPH, pH 8.4, in 5 mM Tris, 35 mM glycine, prepared daily and
  kept at 0°

Spectrophotometric Assay. The reaction mixture contains 100 μM malonyl-CoA, 33 μM acetyl-CoA, 100 μM NADPH, 0.2 M potassium phosphate, pH 6.8, 1 mM EDTA, 1 mM dithiothreitol, and fatty acid synthase complex, 5–10 μg of purified protein, in a total reaction volume of 1.0 ml. The reaction is started by the addition of enzyme to the mixture of substrates previously equilibrated at the desired assay temperature for 10 minutes. Determinations are routinely performed at 25 or 30°. The recorder scale is expanded to give a full-scale deflection for an optical density increment of 0.2 unit. The oxidation of NADPH is followed at 340 nm and the initial slope of the recorder tracing is used to calculate the rate of fatty acid synthesis. The slope of the decrease in NADPH absorption in the absence of malonyl-CoA is subtracted from that obtained in the presence of malonyl-CoA.

Specific activity is expressed as nanomoles NADPH oxidized per minute per milligram of protein. Division by 14 converts the value to nanomoles of palmitate formed per minute per milligram of protein. The concentration of protein in crude preparations is determined by the method of Gornall *et al.*[11] Concentrations of purified enzyme protein are determined from the absorbance at 280 nm using 3.87×10^5 M^{-1} cm^{-1} as the molar absorption index.[12]

Preparation of Enzyme for Assay. Although the pigeon liver enzyme is routinely stored at −20° and full activity is obtained after warming

[10] S. Kumar, J. A. Dorsey, R. A. Muesing, and J. W. Porter, *J. Biol. Chem.* **245**, 4732 (1970).
[11] A. G. Gornall, C. J. Bardawill, and M. M. David, *J. Biol. Chem.* **117**, 751 (1949).
[12] P. C. Yang, P. H. W. Butterworth, R. M. Bock, and J. W. Porter, *J. Biol. Chem.* **242**, 3501 (1967).

the enzyme solution to 25°, it rapidly inactivates when kept at 0°.[13] Therefore it is imperative that aliquots of the frozen enzyme be thawed and then kept at room temperature for at least 30 minutes prior to assay for activity. Thus, unlike procedures routinely employed with many enzymes, temperatures near 0° cannot be used for routine daily use of the enzyme if optimal activities are required.

Preparation of Substrates for the Component Reactions of Fatty Acid Synthesis[10]

Preparation of CoA Esters. [1-^{14}C]Acetyl-CoA, caproyl-CoA, and [2-^{14}C]- or [1,3-^{14}C]malonyl-CoA are prepared according to the procedures of Simon and Shemin[14] and Trams and Brady,[15] respectively, and each ester is purified by the paper chromatographic methods of Brodie and Porter.[16] [1-^{14}C]Palmityl-CoA is synthesized enzymatically by the procedure of Kornberg and Pricer.[17]

Synthesis of S-Acyl-N- acetylcysteamine Derivatives. N-Acetylcysteamine is prepared from acetyl chloride and cysteamine according to the procedure of Martin *et al.*[18] The S-acetoacetyl derivative of N-acetylcysteamine is then synthesized by the method of Lynen and Wieland,[19] m.p. 60°.

DL-S-β-Hydroxybutyryl-N-acetylcysteamine is synthesized by the general mixed anhydride method described by Kass *et al.*,[20] except that the ratio of tetrahydrofuran to water is 1:2. Evaporation of the ether extract obtained from the reaction mixture yields a thick oily residue. The ether extract is further purified on a silicic acid column, 26 × 1.5 cm, by elution with 400 ml of a linear gradient of 0–5% methanol in methylene chloride, and then with 500 ml of 5% methanol in methylene chloride. The eluted fractions are assayed for ester by the method of Lipmann and Tuttle.[21] Three distinct peaks are obtained. The third peak has been characterized as the β-hydroxybutyryl derivative.

S-Crotonyl-N-acetylcysteamine is synthesized from crotonyl chloride and the lead salt of N-acetylcysteamine in anhydrous benzene by the

[13] S. Kumar, R. A. Muesing, and J. W. Porter, *J. Biol. Chem.* **247**, 4749 (1972).
[14] E. J. Simon and D. Shemin, *J. Amer. Chem. Soc.* **75**, 2520 (1953).
[15] E. G. Trams and R. O. Brady, *J. Amer. Chem. Soc.* **82**, 2972 (1960).
[16] J. D. Brodie and J. W. Porter, *Biochem. Biophys. Res. Commun.* **3**, 173 (1960).
[17] A. Kornberg and W. E. Pricer, Jr., *J. Biol. Chem.* **204**, 329 (1953).
[18] R. B. Martin, S. Lowey, E. L. Elson, and J. T. Edsall, *J. Amer. Chem. Soc.* **81**, 5089 (1959).
[19] F. Lynen and O. Wieland, Vol. 1 [94].
[20] L. R. Kass, D. J. H. Brock, and K. Bloch, *J. Biol. Chem.* **242**, 4418 (1967).
[21] F. Lipmann and L. C. Tuttle, *Biochim. Biophys. Acta* **4**, 301 (1950).

method of Kass *et al.*[20] The compound, m.p. 61–62°, is then crystallized from benzene-petroleum ether.

D-Pantetheine is prepared by reduction of D-pantethine with sodium amalgam. The concentration of the product is determined by the method of Ellman.[22]

Assays for Component Reaction Activities of the Fatty Acid Synthase Complex[10]

Acetyl- and Malonyl-CoA-Pantetheine Transacylase Activities. The rate of transfer of acetyl and malonyl moieties from CoA esters to D-pantetheine is measured by the following assay procedure. The reaction mixture contains 0.1 mM [1-^{14}C]acetyl-CoA, 7570 disintegrations/minute/nmole, or 0.1 mM [2-^{14}C]malonyl-CoA, 3000 disintegrations/minute/nmole; 3.4 mM D-pantetheine, 0.2 M potassium phosphate, pH 6.8, 1 mM EDTA, 1 mM dithiothreitol, and fatty acid synthase in a total volume of 0.1 ml. The reaction is started by the addition of 0.05–0.1 μg of purified fatty acid synthase, and then stopped after 1 minute by the addition of 0.015 ml of 12% HClO$_4$.

The reaction mixture is chromatographed on Whatman No. 3 MM paper in a descending system of isobutyric acid–water–25% ammonia–0.1 M EDTA, pH 4.5 (62:37.7:2.3:1). The CoA and pantetheine esters are quantitatively separated. Values of R_f for malonyl-CoA, malonyl-pantetheine, acetyl-CoA, and acetyl-pantetheine are approximately 0.46, 0.74, 0.66, and 0.9, respectively. The paper chromatograms are cut into 1 cm strips, placed in counting vials overnight with 3.0 ml of 50% CH$_3$OH. Extraction of CoA and pantetheine esters is quantitative under these conditions. Fifteen milliliters of scintillator solution is then added and the sample is assayed for radioactivity in a liquid scintillation spectrometer. The scintillator solution is comprised of 6% naphthalene, 10% methanol (v/v), 2% ethylene glycol (v/v), 0.4% 2,5-diphenyloxazole, and 0.02% 2,5-bis[2-(5-*tert*-butylbenzoxazoyl)] thiophene in dioxane.

Condensation-CO$_2$ Exchange Activity. The CO$_2$ exchange reaction, originally demonstrated by Vagelos and Alberts[23] with an enzyme system from *Clostridium kluyveri*, is modified as follows. Assays are carried out at 30° in 0.18 M potassium phosphate buffer, pH 6.8, with 1 mM EDTA, 2 mM dithiothreitol, 1.0 mM CoA, 0.2 mM caproyl-CoA, 0.3 mM malonyl-CoA, 32.5 mM NaH^{14}CO$_3$ (specific activity 1 μCi/μmole), and 250 μg of purified enzyme protein in a final volume of 0.2 ml. The reaction

[22] G. L. Ellman, *Arch. Biochem. Biophys.* **74**, 443 (1958).
[23] P. R. Vagelos and A. W. Alberts, *J. Biol. Chem.* **235**, 2786 (1960).

is initiated by the addition of enzyme and stopped after 10–15 minutes by the addition of 0.05 ml of 5% $HClO_4$.

Reaction mixtures are quantitatively transferred to 1×4 cm strips of Whatman No. 3 MM paper. The papers are dried with hot air to remove radioactive CO_2. The strip is cut into small segments which are placed in a counting vial and then assayed for radioactivity.

Spectrophotometric Assays for Reductase and Dehydrase Activities. All spectrophotometric assays are carried out at 30° in 0.18 M potassium phosphate, pH 6.8, containing 1 mM EDTA and 2 mM dithiothreitol. The full scale on the recorder chart paper is calibrated to give an optical density change of 0.2 unit. Initial rates are obtained from tracings of change in light absorbance with time. The rate of change of absorbance in the absence of enzyme is then subtracted from the rate obtained in the presence of enzyme. Initial rates are expressed as nanomoles of product formed per minute per milligram of protein.

(a) *S*-Acetoacetyl-*N*-acetylcysteamine reductase activity. This assay measures the decrease in NADPH concentration at 340 nm in the presence of *S*-acetoacetyl-*N*-acetylcysteamine.

$$
\begin{array}{c}
\text{O} \qquad\quad \text{O} \qquad\qquad\qquad\qquad \text{O} \\
\| \qquad\qquad \| \qquad\qquad\qquad\qquad\quad \| \\
CH_3-C-CH_2-C-S-CH_2-CH_2-NH-C-CH_3 + \text{NADPH} + \text{H}^+ \\
\downarrow\uparrow \;\;\text{β-ketoacyl reductase} \\
\text{OH} \qquad\quad \text{O} \qquad\qquad\qquad\qquad \text{O} \\
| \qquad\qquad\; \| \qquad\qquad\qquad\qquad\quad \| \\
CH_3-CH-CH_2-C-S-CH_2-CH_2-NH-C-CH_3 + \text{NADP}^+ \quad (1)
\end{array}
$$

Besides the reagents given above, the reaction mixture contains 100 μM NADPH, 14.8 mM substrate, and 5 μg of purified enzyme protein, in a total volume of 1.0 ml. The reaction is started by the addition of enzyme, and the activity is obtained from the corrected initial rate of decrease of absorbance at 340 nm.

(b) DL-*S*-β-Hydroxybutyryl-*N*-acetylcysteamine dehydrase activity. The kinetics of the dehydration reaction [Eq. (2)] catalyzed by the fatty acid synthase is determined by measurement of the increase in absorbance at 263 nm.

$$
\begin{array}{c}
\text{OH} \qquad\quad \text{O} \qquad\qquad\qquad\qquad \text{O} \qquad\quad \text{β-hydroxyacyl dehydrase} \\
| \qquad\qquad\; \| \qquad\qquad\qquad\qquad\quad \| \\
CH_3-CH-CH_2-C-S-CH_2-CH_2-NH-C-CH_3 \rightleftharpoons \\
\text{O} \qquad\qquad\qquad\qquad \text{O} \\
\| \qquad\qquad\qquad\qquad\quad \| \\
CH_3-CH=CH-C-S-CH_2-CH_2-NH-C-CH_3 + H_2O \quad (2)
\end{array}
$$

In addition to the reagents given above, the assay mixture contains 9.0 mM DL-S-β-hydroxybutyryl-N-acetylcysteamine and approximately 600 μg of purified enzyme protein in a reaction volume of 1.0 ml. When substrate concentrations are greater than 5 mM, high background light absorbance at 263 nm ($\epsilon = 2.9 \times 10^2$ M^{-1} cm^{-1}), due to substrate, leads to inaccurate determinations of initial rates; therefore, under these conditions measurements are made at 270 nm. Extinction coefficients at 270 nm for substrate and product (S-crotonyl-N-acetylcysteamine) were found to be 1.5×10^2 M^{-1} cm^{-1} and 5.7×10^3 M^{-1} cm^{-1}, respectively. Initial rates are then calculated from the optical density change at 263 or 270 nm/unit of time.

(c) S-Crotonyl-N-acetylcysteamine reductase activity. The reduction of S-crotonyl-N-acetylcysteamine [Eq. (3)] and the concomitant oxidation of NADPH is measured spectrophometrically at 340 nm.

$$CH_3-CH{=}CH-\overset{\overset{\displaystyle O}{\|}}{C}-S-CH_2-CH_2-NH-\overset{\overset{\displaystyle O}{\|}}{C}-CH_3 + NADPH + H^+$$

$$\xrightarrow[\text{reductase}]{\text{enoyl}}$$

$$CH_3-CH_2-CH_2-\overset{\overset{\displaystyle O}{\|}}{C}-S-CH_2-CH_2-NH-\overset{\overset{\displaystyle O}{\|}}{C}-CH_3 + NADP^+ \quad (3)$$

In addition to the reagents given above, the reaction mixture contains 100 μM NADPH, 13.5 mM substrate, and 1 mg purified enzyme protein, in a total reaction volume of 1.0 ml. The initial rates of reductase activity are calculated from the recorder tracings of optical density plotted against time.

Palmityl-CoA Deacylase Activity. The palmityl-CoA deacylase activity of fatty acid synthase is measured at 30° in 0.2 M potassium phosphate, pH 6.8, 1 mM EDTA, 2 mM dithiothreitol, 1-4 mM[1-^{14}C]palmityl-CoA (specific activity, 1.62×10^4 disintegrations/minute/nmole), and 10 μg of purified protein, in a total reaction volume of 1.0 ml. The deacylation reaction is stopped after 5 minutes by the addition of 0.03 ml of 60% HClO$_4$. After the addition of 1.0 ml of ethanol to each reaction mixture, palmitic acid is extracted with petroleum ether (b.p., 30–60°). The extract is transferred to a counting vial, the solvent is evaporated under nitrogen, and the palmitic acid is assayed for radioactivity.

Sites of Covalent Binding of Acetyl and Malonyl Groups to the Fatty Acid Synthase. The method of Jacob et al.[24] is used with slight modifications to determine the binding of acetyl and malonyl groups to the en-

[24] E. J. Jacob, P. H. W. Butterworth, and J. W. Porter, *Arch. Biochem. Biophys.* **124**, 392 (1968).

zyme. Fatty acid synthase, 10 mg and 22 nmoles, in 0.2 M potassium phosphate, pH 6.8, containing 1 mM EDTA and 2 mM dithiothreitol is incubated with 220 nmoles of [1-^{14}C]acetyl-CoA, 8×10^3 disintegrations/minute/nmole, at 0–4°. After 20 seconds the reaction is stopped with 0.04 ml of 60% perchloric acid. In the reaction with [2-^{14}C]malonyl-CoA, 4×10^3 disintegrations/minute/nmole, 440 nmoles of substrate are used with a reaction time of 2 minutes. The precipitated protein is washed with 0.2 N acetic acid to remove unbound substrate.

The precipitated protein containing covalently bound acetyl or malonyl groups is suspended in 2.0 ml of 0.01 N HCl; then 1.0 ml of pepsin in 0.01 N HCl is added (0.25 mg of pepsin/mg of protein precipitate). Digestion is continued at 30° for 18 hours. Reaction mixtures are freeze-dried and the resulting residues are extracted with 0.2 ml of water. The peptides in the aqueous extract are separated on Whatman No. 3 MM paper by high voltage electrophoresis at pH 3.7 (acetic acid–pyridine–H$_2$O, 10:1:289). Separation is achieved at 2500 V and 2.7 mA/cm in 1.5 hours. Radioactive peptides are located by scanning the paper with a strip scanner. The paper is then cut into 1 cm strips and assayed for radioactivity in a liquid scintillation spectrometer. Controls of [1-^{14}C]acetyl-CoA and [2-^{14}C]malonyl-CoA are also subjected to high voltage electrophoresis.

Purification of the Enzyme System

It is important to begin the preparation of enzyme with well-fed birds since the level of fatty acid synthase in the liver is related to the nutritional state.[25] Routinely, 6 or 12 dozen pigeons are killed. All solutions used in the extraction and subsequent purification of the enzyme are made up in deionized water. The saturated ammonium sulfate solution, adjusted to pH 7.0 with KOH, contains 3 mM EDTA and 1 mM dithiothreitol (freshly added). All buffers are potassium phosphate, pH 7.0, containing 1 mM EDTA and 1 mM dithiothreitol, except as stated otherwise.

Preparation of Supernatant Solution. Preparation of the pigeon liver supernatant solution is accomplished by the procedure of Wakil *et al.*,[26] and all manipulations in this procedure are carried out at 0°. Immediately following the decapitation of the pigeons, livers are removed, cleaned, and placed on ice. The chilled livers are then homogenized in 1.5 volumes

[25] P. H. W. Butterworth, R. B. Guchhait, H. Baum, E. B. Olson, S. A. Margolis, and J. W. Porter, *Arch. Biochem. Biophys.* **116**, 453 (1966).
[26] S. J. Wakil, J. W. Porter, and D. M. Gibson, *Biochim. Biophys. Acta* **24**, 453 (1957).

of a phosphate-bicarbonate buffer (70 mM KHCO$_3$, 85 mM K$_2$HPO$_4$, and 9 mM KH$_2$PO$_4$, pH 8.0) containing 1 mM EDTA and 1 mM dithiothreitol in a Waring blender for 30 seconds at full speed. The homogenate is filtered through a single layer of cheesecloth and centrifuged at 23,000 g for 30 minutes. The sediment is discarded and the supernatant solution is recentrifuged at 100,000 g for 45 minutes. The resulting supernatant solution is stored under nitrogen in sealed cellulose nitrate tubes (40 ml) at −20°. The protein concentration of this solution is about 50 mg/ml. In this state, the fatty acid synthase is stable for at least 1–2 months.

Ammonium Sulfate Fractionation. A 40-ml sample of supernatant solution is thawed and warmed to ambient temperature. Purification is carried out at room temperature (22–24°). Centrifugations may be done at 4°. Saturated ammonium sulfate is added dropwise to the solution, which is stirred gently with a magnetic stirrer under a stream of nitrogen, until a final concentration of 25% saturation is reached. Stirring is continued for an additional 15 minutes and then the suspension is centrifuged at 40,000 g for 10 minutes. The pellet is discarded. The supernatant solution is brought to 40% of saturation with saturated ammonium sulfate, stirred for 15 minutes, and then centrifuged as before. The supernatant solution is discarded. The pellet is dissolved in 20 ml of 5 mM potassium phosphate buffer, pH 6.8, and the protein concentration determined. It is necessary to proceed rapidly, after solubilization of the enzyme in the 5 mM buffer, through the calcium phosphate gel and DEAE-cellulose chromatography purification steps. The enzyme is unstable at low ionic strength and it gradually dissociates to form inactive half-molecular weight subunits. It is also especially susceptible to oxidative inactivation at low ionic concentrations when 2-mercaptoethanol is used.[13]

Calcium Phosphate Gel Adsorption. The above enzyme preparation and an amount of calcium phosphate gel equal to half the weight of the protein are simultaneously added to 150 ml of 5 mM phosphate buffer. This solution is stirred with a magnetic stirrer. The suspension is centrifuged immediately for 3 minutes at 4000 g. Since the fatty acid synthase is not stable in a medium of low ionic strength the next step in the purification procedure, DEAE-cellulose chromatography, must follow immediately.

DEAE-Cellulose Chromatography. A column of DEAE-cellulose (10.3 × 3.5 cm) is prepared from material that has been prewashed and equilibrated with approximately 400 ml of 40 mM phosphate buffer; the column flow rate should be at least 5 ml/minute. Dithiothreitol may be omitted from the first portion of the wash. The gel-treated enzyme solution is applied to the column, and the latter is washed with 40 mM phosphate buffer until the absorbance of the eluate, at 280 nm, is less than

0.05. Further elution is carried out with 0.25 M phosphate buffer. A single protein peak emerges from the column after about 60 ml of eluate volume. An additional 30–40 ml of eluate is collected. This eluate contains the fatty acid synthase at a concentration of about 3 mg of protein/ml. This preparation is stable.

Second Ammonium Sulfate Fractionation. Saturated ammonium sulfate is added to the DEAE-cellulose eluate to 26% of saturation with stirring under an atmosphere of nitrogen. The suspension is centrifuged as before, and the precipitate is discarded. The supernatant solution is brought to 32% saturation with saturated ammonium sulfate, stirred, and centrifuged. The precipitate contains the active enzyme. This enzyme is dissolved in about 1 ml of 0.2 M phosphate buffer, pH 6.8, containing 1 mM dithiothreitol. At this point in the procedure the preparation usually has less than 5% of contaminating proteins and the preparation may be used for most studies after dialysis against 0.2 M phosphte buffer. However, the purity is highly dependent on the nutritional state of the birds. Even a short period of fasting gives rise to the presence of one or more contaminating proteins of molecular weight less than that of the fatty acid synthase. These proteins, if present, may be separated from the fatty acid synthase by sucrose density gradient centrifugation (see Chapter 6 on rat liver fatty acid synthase). The smaller contaminating proteins (less than 100,000 molecular weight) may also be separated by Sephadex G-100 gel filtration.

Sephadex G-100 Gel Filtration. A column of Sephadex G-100, 22.0 × 1.0 cm (bed volume 18.0 ml), is prepared and equilibrated with 0.20 M phosphate buffer, pH 6.8, containing 1 mM dithiothreitol. The flow rate is maintained at about 15 ml/hour. Enzyme, in a volume of approximately 1.0 ml, is applied to the column and the enzyme is eluted with the equilibrating buffer. Fractions of 0.5 ml are collected. A single protein peak of high specific activity for fatty acid synthesis is obtained.

A typical purification of the fatty acid synthase is summarized in Table I.[26a] The complete purification, starting with pigeon liver supernatant solution, can be accomplished in less than 6 hours. This procedure is highly reproducible. The specific activity of the purified enzyme is 1120–1400 nmoles of NADPH oxidized per minute per milligram of protein when assayed at 30°. This is equivalent to 80–100 nmoles of palmitate formed/min/mg of protein. The purified enzyme complex is homogeneous in the ultracentrifuge, in moving boundary electrophoresis, on gradient chromatography on a DEAE-cellulose column, and by starch gel electrophoresis.[1,11,12]

Storage of the Enzyme. Dithiothreitol, 10 mM, is added to the prepa-

[26a] R. Y. Hsu, P. H. W. Butterworth, and J. W. Porter, Vol. 14 [4].

TABLE I
PURIFICATION OF PIGEON LIVER FATTY ACID SYNTHASE[a]

Fraction	Volume (ml)	Total activity[b] (units)	Total protein (mg)	Specific activity[b] (units/mg protein)
Pigeon liver supernatant solution	38.0	10,246	1,807.0	5.6
$(NH_4)_2SO_4$, 25–40%	14.0	9,639	607.0	15.9
DEAE-cellulose	45.0	5,022	83.7	60.0
$(NH_4)_2SO_4$, 26–32%	1.5	4,660	67.0	69.6
Sephadex G-100	2.5	3,856	49.2	78.4

[a] This purification was taken from Hsu et al.[26a]
[b] Units are reported as nanomoles palmitic acid formed per minute.

ration recovered after dialysis or after chromatography on Sephadex G-100 and the preparation is then divided into aliquots which may be stored frozen at −20° for at least a month with no appreciable loss in activity. Before use, the enzyme solution is thawed and then kept at room temperature for at least 30 minutes. The latter temperature equilibration is essential since recent studies have demonstrated the rapid inactivation of the enzyme at 0°. These studies have also established the requirements for reactivation of the enzyme.[27]

Properties of the Enzyme Complex

Inactivation and Dissociation. Purified fatty acid synthase may be dissociated to half-molecular weight subunits which are inactive for fatty acid synthesis. Inactivation and partial or complete dissociation of the complex may be accomplished by one of the following procedures:

(a) Aging of the enzyme at 4° in 0.2 M potassium phosphate buffer, pH 7.0, containing 1 mM EDTA and 1 mM 2-mercaptoethanol. Ninety percent of the enzymatic activity is lost in 8 days, and 40% of the complex is dissociated.[28] Seventy-two percent of the activity is recovered by incubation with 10 mM dithiothreitol at 38° for 2 hours. Similar inactivation, dissociation, and reactivation results are obtained by reaction of the enzyme with carboxymethyl disulfide and subsequent reduction of the carboxymethylated enzyme.

(b) Treatment of the complex with maleate[28] or the detergent action

[27] R. A. Muesing, F. A. Lornitzo, and J. W. Porter, unpublished results.
[28] P. H. W. Butterworth, P. C. Yang, R. M. Bock, and J. W. Porter, *J. Biol. Chem.* **242**, 3508 (1967).

TABLE II

INACTIVATION OF PIGEON LIVER FATTY ACID SYNTHASE

Buffer	Ionic strength	Temperature (°C)	pH	$t_{1/2}$ (minute)
Tris, 5 mM; glycine, 35 mM	0.008	25	8.35	210
Tris, 0.44 mM; glycine, 35 mM	0.009	0	8.35	20
Tris, 5.0 mM; glycine, 35 mM	0.009	25	8.35	240
Tris, 12.6 mM; glycine, 35 mM	0.009	37	8.35	45
Histidine-HCl, 22 mM	0.023	25	6.0	54
Histidine-HCl, 22 mM	0.023	25	7.0	800
2-Amino-2-methyl 1,3-propanediol-HCl, 22 mM	0.023	25	9.5	28

of palmityl-CoA[29] results in complete and irreversible inactivation and dissociation of the enzyme.

(c) Treatment of the enzyme with low ionic strength buffers at mildly alkaline pH results in inactivation and concomitant dissociation of the complex.[13,30] Enzyme is either dialyzed against 5 mM Tris, 35 mM glycine, 1 mM EDTA and reducing agent, or desalted by passage of a 200-μl aliquot of enzyme solution through a 1×23 cm Sephadex G-25 column employing the same buffer. At 25°, 50% of the activity is lost in 3½ hours, with a corresponding conversion of complex to subunits. Identical rates of inactivation and dissociation to half-molecular weight subunits are obtained with 1 mM 2-mercaptoethanol and with 5 mM dithiothreitol.[13] After complete dissociation, high ionic strength and dithiothreitol are necessary for reactivation when dissociation is performed in the presence of 2-mercaptoethanol. When dithiothreitol is used in the dissociation, only high ionic strength is necessary for reactivation at ambient temperature.[27]

The rate of inactivation and dissociation of the complex is highly dependent on ionic concentration, pH, and temperature. Some variation in the rate is also observed with different ions.[13] Table II summarizes differences in rates of inactivation under various conditions.

Inactivation of the enzyme at pH 8.35 is greatly decreased by the addition of KCl, 0.2 M, or by low concentrations of either NADP or NADPH, 20 μM.[31]

(d) Treatment of pigeon liver fatty acid synthase at 0° rapidly inactivates and dissociates the complex. Inactivation readily occurs in 0.2

[29] J. A. Dorsey and J. W. Porter, *J. Biol. Chem.* **243**, 3512 (1968).
[30] S. Kumar, J. K. Dorsey, and J. W. Porter, *Biochem. Biophys. Res. Commun.* **40**, 825 (1970).
[31] S. Kumar and J. W. Porter, *J. Biol. Chem.* **246**, 7780 (1971).

M potassium phosphate buffer, pH 7.0, 1 mM EDTA, and 10 mM dithiothreitol; 50% of the activity is lost in approximately 2 hours and less than 5% remains after 24 hours.[13,27] There is a corresponding conversion of complex to subunits. The rate of inactivation increases with a decrease in potassium phosphate concentration. Reactivation is accomplished merely by warming to ambient temperature.

Reactivation and Reassociation. Reassociation studies have been made with enzyme routinely inactivated and dissociated by overnight incubation in 0.44 mM Tris, 35 mM glycine, 1 mM EDTA, and 10 mM dithiothreitol, pH 8.35 at 0°.[27] (The pH of Tris buffer varies considerably with temperature; hence, when dissociation is carried out at 25° the pH is 7.6.) Reactivation is effected by an increase in ionic strength or the presence of low concentrations of NADP or NADPH. The extent of reactivation (85% of control) is nearly the same with 0.2 M KCl at pH 8.35, 0.2 M potassium phosphate at pH 7.0, and 50 μM NADP or NADPH at pH 8.35; neither NAD nor NADH is effective for reactivation. A temperature of at least 12° is necessary for maximum reactivation, and no appreciable reactivation occurs at 0°.

With chicken[32] and pigeon[27] liver fatty acid synthases, reactivation follows first-order kinetics, thereby suggesting an inactive complex as an intermediate. When 0.2 M KCl is used for reactivation of the pigeon liver enzyme, maximum rates are obtained between pH 7.5 and 8.4. When reactivation is performed with 0.2 M KCl at 25° and pH 8.35, the time to regain half of the activity is about 2 minutes.

Since pigeon liver fatty acid synthase subunits are reassociated and reactivated not only at high ionic strength but also by NADPH, and since the reactivation at 25° is extremely rapid, special precautions must be taken when assaying dissociated or partially dissociated preparations. Therefore, assays for overall activity are performed at 2° in a cell compartment that is cooled with circulating ice water. All manipulations are performed with pipettes chilled on ice. Cells containing the reaction mixture without enzyme are first placed in an ice bath for approximately 10 minutes, then allowed to equilibrate in the cell compartment for 5 minutes. Enzyme is added and the solution is mixed either by gentle passage of air through it or with a small mixing rod, without removing the cell from the cell compartment. Approximately 100 μg of purified enzyme is used in the assay.

Physicochemical Constants. Table III lists the physiocochemical constants of the purified enzyme complex.[1,12,33]

[32] S. L. Yun and R. Y. Hsu, *J. Biol. Chem.* **247**, 2689 (1972).
[33] P. C. Yang, R. M. Bock, R. Y. Hsu, and J. W. Porter, *Biochim. Biophys. Acta* **110**, 608 (1965).

TABLE III
PHYSICOCHEMICAL CONSTANTS OF PIGEON LIVER FATTY ACID SYNTHASE[a]

Property	Value
$s^{\circ}_{20,w} \times 10^{13}$	14.7 sec
$s^{\circ}_{20,b} \times 10^{13}$	12.14 sec
$D^{\circ}_{20,b} \times 10^{7}$	2.50 cm$^2 \times$ sec^{-1}
$\bar{v}$	0.744 cm$^3 \times g^{-1}$
Molecular weight $\times 10^{-5}$	$4.5 \pm 0.23\ g \times$ mole^{-1}
E_{280} nm $\times 10^{-5}$	3.87 M^{-1} cm^{-1}
Electrophoretic mobility at pH 7.0 $\times 10^5$	3.12 cm^2 sec^{-1} V^{-1}

[a] These values were previously reported by Hsu et al.[26a]

TABLE IV
KINETIC VALUES OF PIGEON LIVER FATTY ACID SYNTHASE

Reaction	Substrate	K_m (app) (μM)	V_{max} (nmoles/minute/ mg protein)
β-Keto reduction	S-Acetoacetyl-N-acetylcysteamine +	30,000	3,500
	NADPH	20	—
Dehydration	DL-S-β-Hydroxybutyryl-N-acetylcysteamine	31,000	35
Deacylation	Palmityl-CoA	7	100
Transacylation	D-Pantetheine +	500	—
	acetyl-CoA or malonyl-CoA	5–20	—

Kinetics. Table IV lists kinetic constants for the component reactions with the purified enzyme complex.[10]

Specificity. This enzyme is specific for malonyl-CoA. The requirement for acetyl-CoA is less specific. Propionyl-CoA or butyryl-CoA can be incorporated into fatty acids in the presence of malonyl-CoA at lower rates.[6] Fatty acid synthesis also occurs in the presence of only malonyl-CoA and NADPH, but at a much slower rate. The rate of fatty acid synthesis using NADH as the reductant is only 10–15% of that using NADPH.[1]

Components. The purified enzyme does not contain flavin or other groups that absorb visible light. It does contain 64–66 sulfhydryl groups per mole.[34] One of the sulfhydryl groups is derived from the sole 4'-phos-

[34] P. H. W. Butterworth, H. Baum, and J. W. Porter, *Arch. Biochem. Biophys.* **118**, 716 (1967).

phopantetheine moiety of the synthase complex.[35,36] The acetyl moiety of acetyl-CoA is bound covalently to the 4′-phosphopantetheine site, to a hydroxyl site, and to a cysteine sulfhydryl site. The malonyl moiety is covalently bound only at the 4′-phosphopantetheine and hydroxyl sites.[37–40]

Stability. Pigeon liver fatty acid synthase is relatively stable in 0.2 *M* potassium phosphate buffer at neutral pH and moderate temperatures (12–25°) when adequately protected by dithiothreitol. The enzyme is also stable in the frozen state, and it can be stored for at least a month at —20° in the above buffered solution with little loss in activity. The pigeon liver enzyme is, however, highly sensitive to inactivation at low ionic concentrations, especially at mildly alkaline pH and in the presence of 2-mercaptoethanol. The enzyme is also susceptible to surface denaturation and thus foaming must be avoided and unscratched glassware must be used in routine manipulations. In addition, pigeon liver fatty acid synthase, along with rat liver and mammary gland synthase[41] is cold-labile.

[35] I. P. Williamson, J. K. Goldman, and S. J. Wakil, *Fed. Proc., Fed. Amer. Soc. Exp. Biol.* **25**, 340 (1966).

[36] C. J. Chesterton, P. H. W. Butterworth, A. S. Abramovitz, E. J. Jacob, and J. W. Porter, *Arch. Biochem. Biophys.* **124**, 386 (1968).

[37] C. J. Chesterton, P. H. W. Butterworth, and J. W. Porter, *Arch. Biochem. Biophys.* **126**, 864 (1968).

[38] G. T. Phillips, J. E. Nixon, A. S. Abramovitz, and J. W. Porter, *Arch. Biochem. Biophys.* **138**, 357 (1970).

[39] J. E. Nixon, G. T. Phillips, A. S. Abramovitz, and J. W. Porter, *Arch. Biochem. Biophys.* **138**, 372 (1970).

[40] G. T. Phillips, J. E. Nixon, J. A. Dorsey, P. H. W. Butterworth, C. J. Chesterton, and J. W. Porter, *Arch. Biochem. Biophys.* **138**, 380 (1970).

[41] S. Smith and S. Abraham, *J. Biol. Chem.* **246**, 6428 (1971).

[7a] Fatty Acid Synthase from Chicken Liver

By MICHAEL J. ARSLANIAN and SALIH J. WAKIL

The fatty acid synthase of animal tissues is a tight complex of many enzymes that catalyzes the synthesis of palmitate from acetyl-CoA, malonyl-CoA, and NADPH according to the following overall reaction:

$$CH_3-\overset{O}{\overset{\|}{C}}-S-CoA + 7 \ HOOC-CH_2-\overset{O}{\overset{\|}{C}}-S-CoA$$
$$+ \ 14 \ NADPH + 14 \ H^+ \rightarrow CH_3-(CH_2)_{14}-COOH$$
$$+ \ 8 \ CoA-S-H + 14 \ NADPH^+ + 7 \ CO_2 + 6 \ H_2O$$

Highly purified complexes have been isolated from various sources.[1-8] This communication describes a procedure routinely used in our laboratory for the isolation of fatty acid synthase (FAS) from chicken livers. The FAS preparations thus prepared oxidize 1.2 to 1.6 μmoles of NADPH/mg of protein/min, a specific activity higher than many preparations reported for this enzyme complex.

Assay Methods

Spectrophotometric as well as radioactive assays may be used to measure fatty acid synthase activity based on oxidation of NADPH, or incorporation of radioactive acetyl- or malonyl-CoA, respectively. Assay of the enzyme complex in crude homogenates, especially those derived from livers of starved animals or from tissues containing low enzyme activity or high NADPH oxidase levels, were usually performed by the radioactive procedure because of its high sensitivity and direct measure of palmitate synthesis.

Reagents

Potassium phosphate buffer	1 M, pH 6.5
Acetyl-CoA	3 mM
Malonyl-CoA	3 mM
Dithiothreitol (DTT)	0.2 M
NADPH	2.8 mM (2.5 mg/ml)
Enzyme	1–50 μg protein

Spectrophotometric Assay. Measurements are carried out in 1.0 cm light path cuvettes at 340 nm using a recording spectrophotometer at 25°. The enzyme sample (1–50 μg protein) is added to 0.4 or 0.5 ml of a solution containing the following components at the indicated concentrations: 0.075 M potassium phosphate buffer, pH 6.5; 5 mM DTT, 25 μM acetyl CoA; and 150 μM NADPH. After 3–5 minutes of preincubating the reaction mixture, malonyl-CoA (final concentration 250 μM) is added and the initial velocity recorded. The rate of NADPH oxidation

[1] D. B. Martin, M. G. Horning, and P. R. Vagelos, *J. Biol. Chem.* **236**, 663 (1961).
[2] R. Bressler and S. J. Wakil, *J. Biol. Chem.* **236**, 1643 (1961).
[3] R. Y. Hsu, G. Wasson, and J. W. Porter, *J. Biol. Chem.* **240**, 3736 (1965).
[4] D. N. Burton, A. G. Haavik, and J. W. Porter, *Arch. Biochem. Biophys.* **126**, 141 (1968).
[5] S. Smith and S. Abraham, *J. Biol. Chem.* **245**, 3209 (1970).
[6] R. Y. Hsu and S. L. Yun, *Biochemistry* **9**, 239 (1970).
[7] E. M. Carey and R. Dils, *Biochem. Biophys. Acta* **210**, 371 (1970).
[8] H. Dutler M. J. Coon, A. Kull, H. Vogel, G. Waldvogel, and V. Prelog, *Eur. J. Biochem.* **22**, 203 (1971).

prior to malonyl-CoA addition serves as a blank value which is subtracted from the overall rate. It is advisable to check linearity of the activity with respect to protein concentration, especially when measuring rates approaching the limits of sensitivity of the spectrophotometer.

Radioactive Assays. The reaction mixtures used in these assays are the same as those utilized in the spectrophotometric assay, except that [1-^{14}C]acetyl-CoA or [1,3-^{14}C]malonyl-CoA is used (0.05 μCi per assay), depending upon the amount of enzyme to be measured. When labeled malonyl-CoA is employed, the sensitivity of the assay increases sevenfold, as compared to assays utilizing labeled acetyl-CoA. Assays are carried out for 10 minutes at 37°. The labeled substrate is added to start the reaction after five minutes preincubation of the enzyme with the reaction mixture. Then 0.1 ml of 0.5 N NaOH is added to terminate the reaction, and samples are saponified for 15 minutes in boiling water. The reaction mixtures are acidified with 0.1 ml of 1 N HCl, and the fatty acids are extracted three times with 2 ml pentane. The pentane extracts are combined, washed with water (acidified with acetic acid), and evaporated under nitrogen. The residue is redissolved in 0.5 ml of pentane and aliquots of each sample are counted in a liquid scintillation spectrometer.

Definition of Units. One unit of enzyme activity is defined as the amount of enzyme required to oxidize 1 nmole of NADPH per minute at 25°, under the conditions described for the spectrophotometric assay. Specific activity is defined as units of enzyme per milligram of protein.

Protein concentration is related to the absorbance at 280 nm. For the purified chicken FAS a 0.1% solution gave $A_{279\ nm}$ = 0.965 (light path 1 cm).[6]

Values obtained from radioactive assays can be converted to defined units by multiplying by 14 in the case of acetyl-CoA, and by 2 in the case of malonyl-CoA.

Purification of the Chicken Fatty Acid Synthase

Animals. Young adult chickens (White Leghorn) weighing about 5 lb are acclimatized for about a week to insure adequate nutritional intake. Then a dozen animals are starved for 3 days, after which they are fed a 1:1 mixture of sucrose-bread crumbs (low fat diet). Sucrose solution (1 M) is substituted for drinking water. The chickens are maintainted on this diet for 48 hours, then sacrificed by decapitation and the livers are excised and packed in ice. Typically, livers of starved–refed animals look gray and weigh 30–40 g.

Precautions. Prior to utilization, all glassware, utensils, and polycarbonate or polypropylene tubes are treated with 0.1% EDTA overnight and then rinsed with deionized water. Columns and fraction collector tubes are washed with chromic sulfuric acid, rinsed thoroughly with deionized water, and then treated with boiling water or steam. Unless otherwise specified all operations are carried out at 4°.

Preparation of High-Speed Supernatant Solution (HSS). The livers are washed with ice-cold 0.14 M NaCl, blotted with filter paper, weighed, and placed in a precooled Waring blender. For each g of liver, 1.5 ml of 0.05 M potassium phosphate buffer, 1 mM DTT, 1 mM EDTA, pH 7.4 (buffer 1) is added to the blender, and the livers are homogenized for 45 seconds at full speed. The resulting mixture is centrifuged at 100,000 g for 1 hour, in the model L 2-65B Beckman Ultracentrifuge. The supernatant solution (HSS) is filtered through glass wool to get rid of floating fat particles. Total volume of the HSS obtained is 600–700 ml.

Chromatography on Dowex-1-phosphate. The clear solution of HSS is applied to a Dowex-1-phosphate column (30 $\times$ 5 cm) which is equilibrated with buffer 1. The column is then washed with 150 ml of the same buffer under a pressure of 15 psi.

DEAE-Cellulose Column Chromatography. The Dowex-treated HSS solution is applied to a DEAE-cellulose column (90 $\times$ 5.5 cm) which is equilibrated with buffer 1. The protein concentration is adjusted to 10–20 mg/ml by diluting the HSS with 2 volumes of buffer 1. The dilution is done automatically while the sample is being applied to the column. Batch dilution, in our experience, causes inactivation of the enzyme and is avoided. The sample is applied at a slow rate (flow rate 750 ml/hour) and the column is washed with 4 liters of buffer 1 at a flow rate of 1 liter/hour until the optical density of the eluate at 280 nm drops to 0.150. Then, a linear gradient of 2 liters of buffer 1 (mixing vessel) and 2 liters of 0.15 M potassium phosphate, 1 mM DTT, 1 mM EDTA, pH 7.4 (reservoir) is applied, and fractions of 20–22 ml are collected. The contents of tubes with enzyme specific activities of 500 nmoles mg^{-1} min^{-1} or above are pooled in a precooled graduated cylinder, and the protein is carefully precipitated with ammonium sulfate.

Ammonium Sulfate Precipitation. A saturated solution of ammonium sulfate at 0° containing 1 mM potassium phosphate (pH 7.8), 1 mM DTT, and 1 mM EDTA is used to precipitate the enzyme. Direct addition of solid salt causes changes in pH, and leads to denaturation of the protein with concomitant loss of enzyme activity. The white, cloudy suspension of protein is immediately centrifuged in a Sorvall RC-2B using a GS3 rotor at 10,000 g for 30 minutes. The supernatant solution is decanted first and is then aspirated completely. The precipitate is dissolved in a minimum volume of buffer 1 (e.g., 1000–1200 mg precipitate requires the addition of about 16 ml of buffer to give a final protein concentration of 30–40 mg/ml). The solution is clarified by centrifugation at 48,000 g for 10 minutes.

Sepharose 6B Gel Filtration. The clear enzyme solution (30–50 ml) is layered slowly on a Sepharose 6B column (85 $\times$ 7.2 cm), packed in and equilibrated with 0.25 M potassium phosphate, pH 7.5, and 1 mM DTT, and 1 mM EDTA. The enzyme is eluted overnight with the same buffer. About 200 fractions of 20–22 ml each are collected. Fractions containing enzyme of specific activities of 1.0 μmole mg^{-1} min^{-1} or above are pooled and precipitated at 50% ammonium sulfate saturation. Extreme care is taken during pooling and subsequent addition of the ammonium sulfate to avoid surface denaturation of the protein which becomes visible, with the formation of long threads of insoluble inactive enzyme protein.

Yield. The above procedure yields routinely about 0.5 g of enzyme with an overall yield of 50% of total enzyme activity. The specific activity of a pooled precipitated sample usually varies between 1.2–1.6 μmoles of NADPH oxidized mg^{-1} min^{-1}. A typical preparation is summarized in Table I.

Storage and Stability of Chicken FAS. The precipitate obtained in the final step is usually dissolved in 20% glycerol solution that has the following components at the indicated concentration: 0.8 M potassium phosphate buffer, pH 7.4, 10 mM DTT, and 1 mM EDTA. The solution is then stored in liquid nitrogen. Under these conditions fatty acid synthase loses activity at a very slow rate (half-life about 6 months), provided freezing and thawing is avoided and the enzyme taken out from storage is dialyzed briefly in a phosphate, EDTA, DTT buffer, like buffer 1.

Properties

Purity. The enzyme preparation is apparently homogenous in the ultracentrifuge in the presence of a high concentration of phosphate buffer, and has a molecular weight of about half a million. Dissociation of the enzyme into two nonidentical 250,000 molecular weight species occurs upon dialysis for a few days against a buffer containing the following components at the indicated concentrations: 5 mM Tris base, 33 mM glycine, pH 8.3, 1 mM EDTA, and 1 mM DTT. Reassociation of the two subunits takes place upon dialysis against high concentration of phosphate buffers.[9]

Product. Under our assay conditions, 95% of the product formed is

[9] S. L. Yun and R. Y. Hsu, *J. Biol. Chem.* **247**, 2689 (1972).

TABLE I
PREPARATION OF FATTY ACID SYNTHASE FROM CHICKEN LIVER

Purification step	Volume (ml)	Protein (mg ml^{-1})	Total protein (mg)	Activity (nmoles ml^{-1} min^{-1})	Total activity (nmoles min^{-1})	Specific activity (nmoles mg^{-1} min^{-1})	Yield (%)
HSS	680	46.0	31,280	2,282	1,551,488	49.6	100
Dowex-1-phosphate	850	28.7	24,395	1,429	1,214,871	49.8	78
DEAE-cellulose	1,170	1.2	1,404	942	1,102,140	785.0	71
Ammonium sulfate	34	30.3	1,030	32,270	1,097,163	1,065.0	71
Sepharose 6B	177	3.1	549	4,616	817,014	1,489.0	53

palmitate and 5% stearate. Hsu and Yun[6] report 60% palmitate and 19% stearate (the remaining 21% is not accounted for).

Physiochemical Properties. The enzyme has a pH optimum of 6.7.[9] FAS has an absolute requirement for malonyl-CoA. However, propionyl- and butyryl-CoA can substitute for acetyl-CoA as an acceptor, albeit at a lower efficiency. Longer-chain acyl-CoA derivatives do not act as primers for the condensation of malonyl-CoA. As in the case of the pigeon enzyme, there are three acetyl- and two malonyl-CoA binding sites on the FAS complex.[10] The chicken enzyme is NADPH-dependent and NADH is only 3–5% as efficient as NADPH. Four moles of the nucleotide bind per mole of enzyme.[11]

In the presence of 6 M guanidine-HCl, 70 $\pm$ 5 sulfhydryl groups can be titrated per mole of FAS.[9] Sulfhydryl groups in the reduced state are essential for activity and integrity of the enzyme complex, since oxidation or blockage by alkylating agents leads to complete loss of enzymatic activity.

Immunologic Identity. Antisera from both the rabbit and the goat against chicken liver FAS cross-react with pigeon liver but fail to react with the rat or hamster enzymes.[9]

Acknowledgment

This article is based upon work supported by National Institutes of Health, Grant GM-19091-03.

[10] V. C. Joshi, C. A. Plate, and S. J. Wakil, *J. Biol. Chem.* **245**, 2857 (1970).
[11] R. Y. Hsu and B. J. Wagner, *Biochemistry* **9**, 245 (1970).

[8] Fatty Acid Synthase from Lactating Rat Mammary Gland[1]

By Stuart Smith and S. Abraham

$$CH_3\text{—}CO\text{—}SCoA + n\ COOH\text{—}CH_2\text{—}CO\text{—}SCoA + 2n\ NADPH$$
$$+ 2n\ H^+ \rightarrow CH_3\text{—}CH_2(CH_2\text{—}CH_2)_{n-1}\text{—}CH_2\text{—}COOH$$
$$+ (n+1)\ CoASH + 2n\ NADP^+ + n\ CO_2 + (n-1)\ H_2O$$

Assay Method

Principle. The most convenient method is based on measuring the decrease in extinction at 340 nm which accompanies the oxidation of

[1] Supported by grants from the National Science Foundation (GB-32087) and the National Cancer Institute (CA-11736).

NADPH during the reaction. This method is applicable at all stages of purification of the enzyme since in the lactating rat mammary gland there is very little oxidation of NADPH by contaminating enzymes. Alternatively the enzyme can be assayed by measuring the incorporation of [1,3-$^{14}C_2$]malonyl-CoA into fatty acids.

Preparation of Substrates. Acetyl-CoA is prepared from acetic anhydride,[2] and malonyl-CoA from S-malonyl-N-caprylcysteamine.[3] The material (usually 50 μmole batches) is desalted by passage through a column (31 $\times$ 2 cm) of Sephadex G-15 at 0–2° and purified by chromatography on a column of DEAE-cellulose (23 $\times$ 2.2 cm) at 0–2°. The DEAE-cellulose is previously equilibrated with 0.01 M LiCl in 3 mM HCl, and material is eluted from the column with a gradient of LiCl (0–0.2 M) in 3 mM HCl. The $E_{265\ nm}$ is recorded automatically and the fractions containing ultraviolet-absorbing material are assayed for acyl-CoA by measuring the release of sulfhydryl groups[4] following mild alkaline hydrolysis (0.1 N NaOH, room temperature for 10 minutes). Fractions containing acyl-CoA are pooled and lyophilized. The residue is then dissolved in a minimum volume of water and desalted by gel filtration as before. The final concentration of acetyl-CoA or malonyl-CoA is calculated from the extinction at 260 and 232 nm,[5] and from the sulfhydryl release on hydrolysis. Purity is usually about 95%. Samples can be stored frozen at —20° in water (pH 5–6) for several months without decomposition.

Reagents. Convenient concentrations are

  Potassium phosphate buffer, pH 6.6, 1.0 M
  NADPH, 3.75 mM
  Acetyl-CoA, 1.25 mM
  Malonyl-CoA, 0.54 mM

Procedure. Potassium phosphate (50 μmoles), acetyl-CoA (25 nmoles), and NADPH (75 nmoles) are normally dispensed together as a "cocktail" in 0.1 ml portions per assay. The cocktail is added to a semimicro absorption cuvette having a 1-cm light path; water and enzyme solution is added to a volume of 0.45 ml and the cuvettes placed in a recording spectrophotometer equipped with dual thermoplates for efficient and uniform heat transfer. Temperature is maintained at 30°. Five minutes later, the reaction is started by the addition of 50 λ malonyl-CoA solution (27 nmoles) and the rate of decrease in extinction at 340

[2] E. J. Simon and D. Shemin, *J. Amer. Chem. Soc.* **75**, 2520 (1953).
[3] F. Lynen, Vol. 5, p. 443.
[4] G. L. Ellman, *Arch. Biochem. Biophys.* **882**, 70 (1959).
[5] E. R. Stadtman, Vol. 3, p. 931.

nm is recorded. A decrease in extinction at 340 nm of 1.0 corresponds to the oxidation of 80.5 nmoles NADPH. The rate of decrease in extinction in the absence of malonyl-CoA is usually very low.

Units. A unit of enzyme is defined as that amount catalyzing the malonyl-CoA–dependent oxidation of 1 μmole NADPH per minute at 30°.

Purification of the Enzyme

Animals. Lactating rats that are suckling 6–12 pups are routinely used. For maximum recovery of fatty acid synthase it is preferable to use rats between 10 and 18 days postpartum. The level of the mammary gland fatty acid synthase is not elevated by feeding a fat-free diet,[6] and animals are given a chow diet and water *ad libitum*. The rats are killed by cervical fracture and the entire axillary and inguinal glands excised and placed in 0.25 M sucrose at 0–2°. Care is taken to remove as much of the surrounding connective tissue as possible in order to facilitate homogenization. Between 10 and 20 g of tissue are normally obtained from a single rat; up to 40 g of tissue can conveniently be handled in each purification, although as little as 2 g can be used successfully.

Preparation of High-Speed Supernatant. The mammary glands are cut into strips about 0.5 cm wide and washed with 0.25 M sucrose at 0–2° until no milk can be seen in the washings. The tissue is then minced to a pulp with scissors and homogenized in 0.25 M sucrose (3 volumes/g) in a Potter-Elvehjem homogenizer with a loose fitting Teflon pestle. The homogenate is centrifuged at 100,000 g for 1 hour at 0–2°. The supernatant fraction is decanted through a loose plug of glass wool to remove floating fat. Normally the purification procedure is continued during the same day, but the high-speed supernatant fraction can be stored at —80° for several days without significant loss in activity.

Ammonium Sulfate Fractionation. This procedure is carried out at 0–2°. The high-speed supernatant is stirred gently with a magnet and solid ammonium sulfate added over a period of about 10 minutes, to a concentration of 25% saturation. The suspension is stirred for 10–15 minutes longer and then centrifuged at 12,000 g for 10 minutes. The precipitate is discarded and ammonium sulfate is added to the supernatant, as before, until a concentration of 40% saturation is reached. After stirring for 10–15 minutes, the precipitate is collected by centrifugation and the supernatant discarded. The precipitate is dissolved in 4 mM potassium phosphate buffer, pH 7.0, containing 1 mM EDTA and 1 mM DTT[7]

---

[6] S. Smith, H. T. Gagné, D. R. Pitelka, and S. Abraham, *Biochem. J.* **115**, 807 (1969).

[7] Here DTT stands for dithiothreitol.

(1.5–2.0 times the volume of the original high-speed supernatant fraction). The next steps of the purification are carried out as quickly as possible since in this low ionic strength buffer the enzyme is unstable, having a half-life of about 5 hours.

Calcium Phosphate Gel Adsorption. This procedure is carried out at 0–2°. New batches of calcium phosphate gel are aged for at least 3 months and then their adsorptive properties are determined.[8] Maximum purification is usually achieved with 0.4–0.5 mg of gel per milligram of protein, depending on the batch of gel. The protein solution is gently mixed with calcium phosphate gel and stirred for 2 minutes. The gel is removed by centrifugation at 1500 g for 5 minutes. Most of the fatty acid synthase activity remains in the supernatant, which is then brought to room temperature (usually 20°).

DEAE-Cellulose Chromatography. This procedure is carried out at 20°. The DEAE-cellulose (12 g) is washed successively with 0.5 N NaOH, water, 0.5 N H$_3$PO$_4$, water, and 0.025 M potassium phosphate buffer, pH 7.0, containing 1 mM EDTA, 1 mM DTT. A slurry is poured into a glass column, 2.2 cm in diameter, to a bed height of about 23 cm. The supernatant from the calcium phosphate gel stage is applied to the column. Normally not more than 350 mg of protein is chromatographed on the column. The proteins are eluted with a phosphate gradient generated from two vessels, one containing 0.025 M potassium phosphate buffer pH 7.0 (225 ml), the other containing 0.025 M potassium phosphate buffer, pH 7.0 (450 ml); both buffers contain 1 mM EDTA and 1 mM DTT. A flow rate of about 150 ml/hour is usually maintained, and fractions of between 5 and 10 ml are collected. A typical DEAE-cellulose elution profile is shown in Fig. 1. It is fairly common to observe two protein peaks, each of which contains fatty acid synthase activity. The exact nature of the two species has not been elucidated.[9]

Peaks I and II are commonly eluted at phosphate concentrations of 50–70 and 80–100 mM, respectively. The fractions with enzymatic activity from both peaks are pooled.

Ammonium Sulfate Precipitation. This procedure is carried out at 0–2°. Solid ammonium sulfate is added to the pooled fractions from the DEAE-cellulose column to a concentration of 33% saturation. The suspension is magnetically stirred for 10–15 minutes and the protein precipitate collected by centrifugation as before. Potassium phosphate buffer pH 7.0, 0.25 M containing 1 mM EDTA, 1 mM DTT (1–2 ml) is added to the precipitate at 20° and the tube gently swirled to dissolve the protein.

[8] D. Keilin and E. F. Hartree, *Proc. Roy. Soc., Ser. B* **124**, 397 (1938).
[9] See section entitled "Reproducibility of the Procedure."

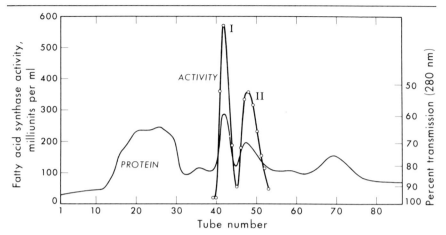

FIG. 1. Elution profile for DEAE-cellulose chromatography step.

Sephadex G-200 Chromatography. This procedure is carried out at 20°. A column of Sephadex G-200 (36 × 2.5 cm), equilibrated with 0.25 M potassium phosphate buffer, pH 7.0, containing 1 mM EDTA, 1 mM DTT, is prepared and the enzyme solution applied to it. Since the enzyme is relatively stable in high ionic strength buffers at 20°, the Sephadex column can be conveniently developed overnight with 0.25 M potassium phosphate buffer pH 7.0 containing 1 mM EDTA, 1 mM DTT. A flow rate of 5–8 ml/hour is maintained and fractions of 2–3 ml are collected. Fatty acid synthase activity is eluted with a relative retention volume (elution volume/void volume) of 1.15. The active fractions from the leading edge and the main section of the peak are pooled; fractions from the trailing edge of the peak sometimes contain fatty acid synthase of lower specific activity and are therefore routinely discarded.

Reproducibility of the Procedure. The only variable feature of the purification is the DEAE-cellulose profile. The distribution of fatty acid synthase activity in the peaks I and II (Fig. 1) can vary considerably from one experiment to another. If the protein in each peak is separately precipitated with ammonium sulfate at 0–2° to 33% saturation and redissolved at 20° in 0.25 M potassium phosphate, pH 7.0, containing 1 mM EDTA, 1 mM DTT, the two samples usually have the same specific activity. In addition, both sediment as 13 S species on sucrose density gradients. These two components eluted from the DEAE-cellulose column may result from exposure of the enzyme to low ionic strength (4 mM phosphate) and low temperatures (0–2°) during the early stages of purification. Such conditions are known to induce dissociation of the native enzyme into its 9 S subunits, a process which can readily be reversed by exposing the subunits to high ionic strength buffers at room temperature.[10] It is conceivable that these two components eluted from the DEAE-cellulose column may be the native enzyme and the dissociated enzyme, but that on exposure to high ionic strength buffer at room temperature the subunits rapidly reassociate.

The specific activity of the purified enzyme usually lies in the range of 0.7–1.0 unit/mg of protein. A flow sheet from a typical purification is given in Table I. The same procedure can be used to obtain homogeneous preparations of fatty acid synthase from lactating mouse mammary gland.

Homogeneity. The purified enzyme is homogeneous according to the following criteria: it sediments as a single component in the analytical ultracentrifuge[11]; it migrates as a single component during electrophoresis on cellulose acetate[12]; and the antibodies prepared against the purified enzyme when subjected to immunodiffusion or immunoelectrophoresis with the high-speed supernatant fraction from lactating rat mammary gland give a single precipitin line.[13]

Stability. In common with other fatty acid synthase multienzyme complexes, the lactating rat mammary gland enzyme is readily susceptible to surface denaturation: extra care should be exercised during dissolution of ammonium sulfate precipitates and transfer of the enzyme from one vessel to another.

The enzyme undergoes dissociation into its two component subunits on exposure to low temperatures (0–2°) and/or low ionic strength solutions (below 0.25 M). The dissociation process is a first-order reaction. At 0° in 0.25 M potassium phosphate buffer, pH 7.0, the rate constant is 0.56 day^{-1}; the rate constant is decreased to 0.15/day when deuterium oxide is substituted for H_2O in the buffer.[14] In the absence of DTT, enzyme activity is lost rapidly, but the inactivation is readily reversed by incubating the enzyme with 10 mM DTT at 30° for 30 minutes.[10] The most effective stabilizing agent for the enzyme is glycerol. Samples of enzyme stored in 0.25 M potassium phosphate, pH 7.0, 1 mM EDTA, 10 mM DTT containing 20% glycerol at —20° for 3 months, when preincubated at 30° for 45 minutes and then assayed, retain over 80% of the original activity; samples stored in the absence of glycerol lose most of their activity.

[10] S. Smith and S. Abraham, *J. Biol. Chem.* **246**, 6428 (1971).
[11] S. Smith and S. Abraham, *J. Biol. Chem.* **245**, 3209 (1970).
[12] S. Smith, unpublished information.
[13] S. Smith and S. Abraham, *J. Biol. Chem.* **246**, 2537 (1971).
[14] S. Smith, *Biochim. Biophys. Acta* **251**, 477 (1971).

TABLE I

PURIFICATION OF FATTY ACID SYNTHASE FROM LACTATING RAT MAMMARY GLAND

Fraction	Volume (ml)	Activity (milli-units/ml)	Total activity (milliunits)	Protein (mg/ml)	Total protein (mg)	Specific activity (milli-units/mg)	Yield (%)
High-speed supernatant	50	1,151	57,550	12.2	610	94.3	100
25–40% ammonium sulfate	100	512	51,200	2.97	297	172	88.9
Calcium phosphate gel	105	464	48,720	1.91	200	243	84.6
DEAE-cellulose Peak I	20	331	6,620	0.64	12.8	517	33
Peak II	47	264	12,400	0.42	19.7	629	—
Sephadex G-200	14.5	955	13,850	1.21	17.5	791	24

Properties

Physicochemical Properties. Some characteristics of the purified enzyme are listed in Table II. From the hydrodynamic data, it can be predicted that the native enzyme is highly asymmetric; assuming a degree of hydration in the range 10–30% the axial ratio for an assumed ellipsoidal model would be approximately 10.[11]

Kinetic Properties. These are presented in Table III. Kinetic data reported in the literature for fatty acid synthases from various sources are based on assays carried out at 30° or 37°. Comparisons must take into account the temperature coefficient of the enzymes. The Q_{10} for the lactating rat mammary gland enzyme decreases with increasing temperature; Q_{10} 0–10° = 7.5, 10–20° = 3.5, 20–30° = 2.0, and 30–40° = 1.6.

Specificity. The enzyme shows high specificity for NADPH as hydrogen donor. The K_m for NADH (1.5 mM) is much higher than for NADPH and the V_{max} is considerably lower. The enzyme exhibits rather less specificity toward the starter substrate; acetyl-CoA can be replaced by

TABLE II

PHYSICOCHEMICAL PROPERTIES OF LACTATING RAT MAMMARY GLAND
FATTY ACID SYNTHASE[a]

Property	Value
MW	478,000 g
$s_{20,w}$	12.9×10^{-13} sec
$D_{20,w}$	2.56×10^{-7} cm^1 sec^{-1}
α	8.46×10^{-7} cm
f/f_0	1.63
$\bar{v}$	0.740

[a] Data taken from Smith and Abraham.[11]

TABLE III

KINETIC PROPERTIES OF LACTATING RAT MAMMARY GLAND FATTY ACID SYNTHASE[a]

Property	Value
Turnover number	3.4×10^7 μmoles acetyl-CoA utilized per min per mole enzyme
pH optimum	6.5–6.8
Activation energy (0–20°)	28 kcal
K_m acetyl-CoA	22 μM
K_m malonyl-CoA	13 μM
K_m NADPH	34 μM

[a] Data taken from Smith and Abraham.[11]

butyryl-CoA, for which the enzyme has a lower K_m (3 μM). Longer acyl-CoA's up to 10 carbon atoms in chain length can substitute for acetyl-CoA, but less effectively both in terms of K_m and V_{max}.[13] Thus the K_m's for hexanoyl-CoA, octanoyl-CoA, and decanoyl-CoA are 50, 80, and 80 μM, and the V_{max}'s, 33%, 28% and 17%, respectively, of that observed with acetyl-CoA and butyryl-CoA.

Inhibitors. The enzyme is inhibited by arsenite (50% inhibition at 3.5 mM), iodoacetamide (50% inhibition at 1.0 mM), and *p*-chloromer-curibenzoate (50% inhibition at 14 μM).

Binding Sites. Two moles of malonate and three moles of acetate are bound per mole of enzyme.[12] The nature of the binding sites has not been investigated in the case of the lactating rat mammary gland enzyme but extensive studies have been reported with enzymes from other sources.[15,16]

Components of the Enzyme. The enzyme contains one 4'-phosphopan-tetheine equivalent and 56 cysteine equivalents per mole.[11] Maintenance of these sulfur-containing residues in the reduced state is essential for full enzymatic activity. The fully active, native enzyme contains no disulfide bridges. The amino acids with hydrophobic side chains are important in the maintenance of the quaternary structure of the enzyme. Weakening of the hydrophobic interactions results in dissociation of the enzyme into its two 9 S subunits, which are presumably nonidentical, one of them containing the 4'-phosphopantetheine residue.[10,14]

Products of the Reaction. The products of the reaction can conveniently be determined using labeled malonyl- or acetyl-CoA. Volatile fatty acids are analyzed by silicic acid chromatography,[17] longer chain fatty acids by gas–liquid chromatography.[18] Under cofactor conditions favoring the maximum rate of fatty acid synthesis, palmitate is the main product; only trace amounts of butyrate are produced.

The enzyme is subject to substrate control of chain length. When the ratio of malonyl-CoA:acetyl-CoA concentration is lowered, the overall rate of fatty acid synthesis is decreased but the proportion of medium chain length fatty acids synthesized is increased.[11]

Immunochemical Cross-Reactivity of Fatty Acid Synthases. No difference in cross-reactivity, determined by immunodiffusion and immunoprecipitin tests, is demonstrable between the mammary gland and liver

[15] C. J. Chesterton, P. H. W. Butterworth, and J. W. Porter, *Arch. Biochem. Biophys.* **126**, 864 (1968).
[16] E. M. Carey, H. J. M. Hansen, and R. Dils, *Biochim. Biophys. Acta* **260**, 527 (1972).
[17] S. Kumar, V. N. Singh, and R. Keren-Paz, *Biochim. Biophys. Acta* **98**, 221 (1965).
[18] S. Abraham, K. J. Matthes, and I. L. Chaikoff, *Biochim. Biophys. Acta* **49**, 268 (1961).

enzymes from either the rat or the mouse. Fatty acid synthases from tissues of the rat give reactions of partial identity with synthases from tissues of the mouse. Fatty acid synthases from pigeon liver and rabbit mammary gland do not react with rabbit antisera prepared against the rat enzymes.[19]

[19] S. Smith, *Arch. Biochem. Biophys.* **156**, 751 (1973).

[9] Fatty Acid Synthase from Rabbit Mammary Gland

By RAYMOND DILS and ERIC M. CAREY

The fatty acid synthase complex isolated from lactating rabbit mammary gland catalyzes the synthesis of saturated even-numbered fatty acids from acetyl-CoA, malonyl-CoA, and NADPH.

Most of the information reported is taken from our previous papers.[1,2]

Preparation of Substrates

Acetyl-CoA is prepared by modifying the method of Ochoa.[2a] Ten milligrams of CoA are dissolved in 1.0 ml of 0.1 M KHCO$_3$. Twenty microliters of acetic anhydride (redistilled under anhydrous conditions) are added, the mixture reacted at 4° for 5 minutes and the pH brought to 2 with HCl. Malonyl-CoA can be prepared via monothiophenyl-malonate.[3] Neither CoA ester contains free thiol groups as measured by the method of Jocelyn.[4] [1-^{14}C]Acetyl-CoA and [2-^{14}C]malonyl-CoA can be purchased from New England Nuclear Chemicals.

If unreacted CoA is present in these substrates, they can be purified by adapting the method of Ryder *et al.*[5] An aqueous solution of the thioester (1–3 μmoles) is applied to a column (20 × 1 cm) of DEAE-50 cellulose at 4° previously equilibrated with 50 mM LiCl in 5 mM HCl, and eluted using a linear gradient of LiCl (50–150 mM in 5 mM HCl). $A_{260\ nm}$ or radioactivity is measured on each 10 ml fraction, and the peak of acetyl-CoA or malonyl-CoA (eluted at about 80 mM LiCl and 100–110 mM LiCl, respectively) collected. CoA elutes at about 50 mM LiCl. The

[1] E. M. Carey and R. Dils, *Biochim. Biophys. Acta* **210**, 371 (1970).
[2] E. M. Carey and R. Dils, *Biochim. Biophys. Acta* **210**, 388 (1970).
[2a] S. Ochoa, Vol. 1, [114].
[3] E. G. Trams and R. O. Brady, *J. Amer. Chem. Soc.* **82**, 2972 (1960).
[4] P. C. Jocelyn, *Biochem. J.* **85**, 480 (1962).
[5] E. Ryder, C. Gregolin, H. Chang, and M. D. Lane, *Proc. Nat. Acad. Sci. U.S.* **57**, 1455 (1967).

LiCl can be removed from these fractions by passing them through a column (20 × 1 cm) of Sephadex G-10. Alternatively, the malonyl-CoA fraction can be lyophilized and dissolved in the minimum volume of methanol–acetone (1:4, v/v) at room temperature. It precipitates out when stored overnight at −20°. The precipitate is collected by centrifugation and made up to 1 μmole/ml in water. Acetyl-CoA and malonyl-CoA are stable at pH 5.0 for at least 1 month at −20°.

Assay Methods

Spectrophotometric Assay. The activity of purified fatty acid synthase and of the enzyme during purification can be measured via the oxidation of NADPH. Optimum assay conditions at 37° are 200 mM potassium phosphate buffer, pH 6.6, 1 mM dithiolthreitol, 1 mM EDTA, 0.24 mM NADPH, 30 μM acetyl-CoA, 40–50 μM malonyl-CoA, and enzyme protein to produce an absorbance change of 0.05–0.15 unit/minute in a final volume of 1.0 ml. After measuring NADPH oxidation without added acetyl-CoA and malonyl-CoA, the reaction is started by adding these substrates. The reaction rate should be linear with time for at least 3 minutes. One unit of enzymatic activity oxidizes 1 μmole of NADPH per minute.

Radioactive Assay. Fatty acid synthase can be assayed by measuring the incorporation of [1-^{14}C]acetyl-CoA (5 μCi/μmole) or [2-^{14}C]malonyl-CoA (1 μCi/μmole) into synthesized fatty acids. The incorporation of malonyl-CoA is a more sensitive assay since 7 times as much malonyl-CoA is incorporated as is acetyl-CoA. The optimum conditions are those described for the spectrophotometric assay except that the incubation time is increased to 15 minutes. 0.48 mM NADPH and 8 μg of purified enzyme are used to ensure linearity of reaction rate during the incubation. The reaction is stopped with alcoholic KOH. After saponification, acidification, and extracting twice with 3 volumes of light petroleum (b.p. 40–60°), the combined extracts are thoroughly dried under N_2. Scintillator solution is added and the radioactivity determined. This procedure does not assay the volatile fatty acids ($C_{4:0}$ and $C_{6:0}$) synthesized since they are lost during the evaporation step.

There is a good correlation between the activity of the purified enzyme measured by the spectrophotometric method and by the radioactive assay using [2-^{14}C]malonyl-CoA. However, with [1-^{14}C]acetyl-CoA there is some enzymatic decarboxylation of nonradioactive malonyl-CoA to acetyl-CoA (see later section entitled "Further Characteristics"). This results in a lower apparent activity of the enzyme measured by the rate of [1-^{14}C]acetyl-CoA incorporation compared with the activity measured by NADPH oxidation.

Assay by Gas–Liquid Chromatography of the Individual ^{14}C-*Labeled Fatty Acids Synthesized.* The individual fatty acids synthesized (including volatile fatty acids) are assayed by the following method.[6] The reaction is stopped with alcoholic KOH and the mixture saponified. After cooling to 4°, 1.5 ml of diethyl ether–pentane (3:7, v/v) is added and the mixture acidified with HCl. Fatty acids are extracted 3 times with 1.5 ml of the diethyl ether–pentane mixture. The pooled extracts are made up to 5.0 ml with pentane and portions transferred to vials fitted with silicone-rubber seals. Carrier fatty acids ($C_{8:0}$ to $C_{16:0}$) are added to give suitable mass peaks during gas–liquid chromatography. A mixture of equal weights of butyric, hexanoic, and nonanoic acid (48 μg in all) is added and the volume slowly decreased to 0.2 ml under N_2. Boron trifluoride–methanol reagent (0.2 ml) is added, and the vial stoppered and heated at 65° for 10 minutes. Quantitative methylation occurs even if two phases remain during heating. The vials are cooled to 4° and 0.3 ml of water added. The upper layer is transferred to a small tube that has been drawn out at the bottom. The lower phase is extracted with 0.2 ml and then 0.1 ml of pentane and the combined extracts slowly evaporated to 10 μl under N_2 and taken up into a microsyringe. The tube is washed with 20–30 μl of pentane, the washings evaporated to 10 μl and taken up into a second syringe. The extract and washings are injected from the two syringes onto a 1.52-m (5 feet) column of 10% (w/w) diethylene–glycol adipate plus 3% (w/w) phosphoric acid on 100–200 mesh Diatomite C. The injection port is maintained at 130° and the oven temperature increased linearly from 40 to 190° at 8°/minute. The individual fatty acids can be trapped and their radioactivity determined by scintillation counting. Alternatively, the radioactivity in the effluent can be monitored continuously by flow counting. A suitable apparatus which is commercially available is the sensitive radiogas detector system interfaced with gas–liquid chromatograph described by Strong *et al.*[7]

Losses of butyric acid (about 50–60%) and hexanoic acid (about 20%) relative to that nonanoic acid can be calculated from their mass peaks on the gas chromatograph. The overall recovery of longer chain fatty acids is about 80%.

Purification of the Enzyme

Preparation of the Particle-Free Supernatant Solution. All operations are at 0–4°. Mammary tissue (about 100 g wet weight) is obtained from

[6] C. R. Strong, I. Forsyth, and R. Dils, *Biochem. J.* **128**, 509 (1972).

[7] C. R. Strong, R. Dils, and T. Galliard, *in* "Column" (P. Ridgeon, ed.), No. 13, p. 2. Pye Unicam Publ., Cambridge, England, 1971.

a lactating rabbit 4–20 days *postpartum*. Fatty acid synthase with maximum specific activity is obtained from animals at 4 days *postpartum*. From 4 to 20 days *postpartum* there is a gradual decline in the specific activity of the enzyme isolated. The tissue should be *very thoroughly* chopped into a "paste." Otherwise, the connective tissue will make homogenization difficult. The chopped tissue is freed from excess milk by washing twice with homogenization medium. Portions of the chopped tissue are homogenized with 3–6 volumes/g wet weight of 0.25 M sucrose containing 1 mM dithiolthreitol and 1 mM EDTA (final pH of the medium about 6.8). The homogenizer is loose fitting (about 0.02 inch difference in diameter) and consists of a Teflon plunger on a stainless steel rod and a thick-walled glass or stainless steel tube. After 12–15 up and down strokes at 600 rpm, the homogenate is strained through nylon gauze. Particulate material is removed from the homogenate by centrifugation at 1.4×10^5 g_{av} for 45 minutes. The floating fat layer is carefully removed. The particle-free supernatant contains the fatty acid synthase activity of the homogenate and can be stored at $-20°$ under N_2 for up to 1 month without loss of fatty acid synthase activity.

The following purification steps[1] are based on those described by Hsu *et al.*[7a] for the pigeon liver enzyme, except that all solutions are kept 1 mM with respect to dithiolthreitol and EDTA and at pH 7.0. All operations are 4°.

First Ammonium Sulfate Fractionation. The supernatant solution (containing about 1.5–2.0 g of protein and $3–4 \times 10^4$ activity units/100 ml) is thawed slowly to 4°. Solid ammonium sulfate is added at the rate of about 1 g/minute/100 ml to bring the saturation to 25%. The sediment is removed by centrifugation at 3.5×10^5 g/min. Protein precipitating between 25 and 40% saturation with ammonium sulfate is redissolved in about 10 ml of homogenization medium and dialyzed for 8–10 hours against the same medium. Occasionally, a slight increase in specific activity is observed after dialysis. Since the enzyme is unstable in low molarity phosphate buffer the following steps are carried out rapidly.

Calcium Phosphate Gel Treatment. The buffer concentration of the dialyzed fraction is reduced to 25 mM, calcium phosphate gel[8] added (dry weight of gel to protein, 1:1 w/w), and the suspension stirred for 1 minute. A higher gel to protein ratio leads to decreased purification and lower enzyme recovery. The suspension is centrifuged for 7 minutes at 1.7×10^3 g_{av}, the gel washed twice with 50 mM potassium phosphate buffer, and the three supernatants combined. The degree of purification and the recovery of synthase activity during this treatment are variable.

[7a] R. Y. Hsu *et al.*, Vol. 14 [4].

[8] D. Keilin and E. F. Hartree, *Proc. Roy. Soc., Ser. B* **124**, 397 (1938).

TABLE I
PURIFICATION OF RABBIT MAMMARY GLAND FATTY ACID SYNTHASE[a]

Fraction	Protein (mg)	Total activity (units $\times 10^{-3}$)	Specific activity	Enzyme recovery (%)
Supernatant	1750	97	56 ± 20	100
First $(NH_4)_2SO_4$ fraction	350	77	220 ± 30	79
Gel eluate	140	44	315 ± 50	45
DEAE-52 pooled fractions	80	42	505 ± 35	43
Second $(NH_4)_2SO_4$ fraction	20	16	880 ± 30	16

[a] The results are mean values obtained during 15 purifications; the specific activities are quoted ± S.D. for the 15 preparations.

DEAE-52 Cellulose Chromatography. The combined supernatants are applied to a column (12 × 2.5 cm) of DEAE-52 cellulose which has been previously equilibrated for at least 24 hours with 50 mM potassium phosphate buffer. The enzyme is eluted with a linear gradient of 50–250 mM of this buffer (500 ml) at a flow rate of 1.5–2.0 ml/minute. Seven to ten milliliter fractions are collected. The peak of synthase activity corresponding to the major protein peak is eluted at 110 mM. Fractions containing high synthase activity are carefully pooled. Care should be taken to minimize disturbing the solutions; otherwise, rapid precipitation of long strands of denatured protein results.

Second Ammonium Sulfate Fractionation. Protein in these pooled fractions which precipitates between 26 and 32% saturation with ammonium-sulfate is dissolved in 2–5 ml of 0.25 M potassium phosphate buffer (pH 7.0) containing 1 mM dithiolthreitol, 1 mM EDTA, and 20% glycerol. A small (4 ml) Potter-Elvehjem glass homogenizer is used and gently rotated by hand to avoid denaturing the enzyme. Any denatured protein is removed by centrifugation. The purified enzyme can be stored at −20° under N_2.

Purification takes 7–9 hours to complete (excluding dialysis time). The purified enzyme has a specific activity at 37° of 880 ± 30 mμmoles of NADPH oxidized per minute per milligram of protein (mean of 15 preparations ± S.E.M.). The maximum purification achieved from the supernatant solution is 22-fold. The composite results of 15 purifications are given in Table I.

Criteria of Purity

Ultracentrifugation. Freshly prepared fatty acid synthase sediments as a single symmetrical peak on analytical centrifugation at 42,040 rpm in 0.25 M potassium phosphate buffer (pH 7.0) containing 1 mM dithiol-threitol and 1 mM EDTA at 4°.

Sephadex Chromatography. When the freshly prepared enzyme is eluted from a column (22 × 1 cm) of Sephadex G-200 at 4° with 0.25 M potassium phosphate buffer (pH 7.0) containing 1 mM dithiolthreitol and 1 mM EDTA, a single protein peak is obtained. The peak fractions have a constant specific activity within the accuracy of measurement of enzymatic activity (±2%) and protein concentration (±3%). The peak has a slight retention volume of 0.5 ± 0.2 ml (void volume 6.1 ± 0.2 ml).

DEAE-52 Cellulose Chromatography. The freshly prepared dialyzed enzyme is applied to a column (25 × 1 cm) of DEAE-52 cellulose at 4° which has been previously equilibrated with 50 mM potassium phosphate buffer (pH 7.0) containing 1 mM dithiolthreitol and 1 mM EDTA. Using a linear gradient of this buffer (50–150 mM phosphate, total volume 500 ml), the enzyme is eluted as a single protein peak with constant specific activity at a phosphate concentration of 110 mM.

Polyacrylamide Gel Electrophoresis. Disks (60 × 4 mm) of acryl-amide are prepared.[9] The purified enzyme (100–200 μg) in saturated sucrose solution is applied to the cathode end of the gel rod. Electrophoresis is at room temperature in 30 mM Tris-glycine buffer (pH 8.4) for 60–80 minutes with a current maintained at 2.5 mA per tube. On 7% gels the migration of the enzyme is severely restricted. Decreasing the acrylamide concentration stepwise to 4% increased the mobility. In all cases a major protein band is obtained, though two very narrow bands are observed at the counterion front with mobilites unaffected by the acrylamide concentration. These may result from dissociation of the enzyme into fragments at the low phosphate concentration used.

Stability of the Enzyme

The purified enzyme loses only 5% of its activity when stored at −20° for 1 month in 0.25 M potassium phosphate buffer (pH 7.0) containing 1 mM dithiolthreitol, 1 mM EDTA, and 20% glycerol. The enzyme is less stable in the absence of glycerol and in media of low ionic strength. Freezing and thawing inactivates the enzyme.

Properties

Physicochemical Constants.[1] Values for $s^\circ_{20,w}$ of 16.5 ± 0.2 S (mean of 3 preparations ± S.E.M.) and for $D^\circ_{20,w}$ of 1.6 × 10^{-7} sec^2 cm^{-1} are ob-

---

[9] J. L. Hendrick and A. J. Smith, *Arch. Biochem. Biophys.* **126**, 155 (1968).

TABLE II
PARTIAL AMINO ACID COMPOSITION OF RABBIT MAMMARY
GLAND FATTY ACID SYNTHASE[a]

Amino acid	Moles amino acid/mole enzyme
Asp	428
Thr	241
Ser	323
Glu	646
Pro	387
Gly	446
Ala	564
Val	410
Ile	173
Leu	641
Tyr	91
Phe	170
His	127
Lys	232
Arg	309

[a] Data from Carey and Dils.[1]

tained by extrapolating (using the least squares fit) the results obtained for the sedimentation coefficient and diffusion constant at protein concentrations between 3 and 11 mg/ml. The molecular weight is calculated to be 9.1×10^5, assuming a partial specific volume of 0.725 ml/g.[9a] The molecular weight when calculated from the relative mobility of the enzyme during polyacrylamide gel electrophoresis is 1.0×10^6.

The ratio $(f/f_{minimum})$ of the frictional coefficient (f) of the hydrated molecule to the minimum frictional coefficient $(f_{minimum})$ of the anhydrous molecule is 2.06. This gives a hydrated radius for the enzyme of 120 Å and indicates that it is either elliptical in shape or extensively hydrated or both.

Amino Acid Composition. The partial amino acid composition of the purified enzyme is shown in Table II.

Kinetics. The pH optimum of the purified enzyme is 6.6. The K_m for acetyl-CoA and malonyl-CoA are 0.9×10^{-5} M and 2.9×10^{-5} M, respectively. At concentrations higher than the optimum (30 μM) acetyl-CoA is strongly inhibitory. Even at 100 μM, malonyl-CoA only slightly ($<10\%$) inhibits the enzyme in the presence of 200 mM phosphate buffer.

[9a] H. K. Schachman, Vol. 4 [2].

Specificity. Oxidation of NADH (0–400 μM) is very low compared with that of NADPH. The addition of optimum concentrations of NADPH (240 μM) to the enzyme which had been preincubated for 3 minutes at 37° with 120 or 240 μM NADH did not give the optimum rate of oxidation of NADPH. The inhibition is dependent on NADH concentration.

Further Characteristics. The purified enzyme contains 58–59 thiol groups per mole. Iodoacetamide inhibits the enzyme by 47 and 62% at 3 and 5 μM, respectively. During purification, malonyl-CoA decarboxylase activity copurifies with fatty acid synthase. The specific activity of malonyl-CoA decarboxylase associated with the purified synthase is, however, very low (7 mμmoles malonyl-CoA decarboxylated per minute per milligram of protein). Examination of the purified fatty acid synthase by electron microscopy, using the negative staining technique with phosphotungstic acid, reveals spherical particles with a radius of about 50 Å within the dark areas of negative stain. This anhydrous radius agrees well with the value of 60 Å determined from the anhydrous molecular weight.

Chain Lengths of the Fatty Acids Synthesized. With optimum concentrations of substrates the main products are unesterified palmitic acid ($C_{16:0}$), and butyric acid ($C_{4:0}$) that is mainly in the form of butyryl-CoA. Decreasing the malonyl-CoA concentration decreases the proportion of long-chain ($C_{14:0}$ and $C_{16:0}$) and increases the proportion of short-chain ($C_{4:0}$ and $C_{6:0}$) acids formed (Table III). Medium-chain fatty acids

TABLE III

Effect of Malonyl-CoA on the Chain Length of Fatty Acid Synthesized[a]

Malonyl-CoA (μM)	Percent distribution of radioactivity in fatty acids								Total mμomoles [1-^{14}C] acetyl-CoA incorporated[b]
	4:0	6:0	8:0	10:0	12:0	14:0	16:0	18:0	
0	82	12	3	3	—	—	—	—	0.25
2	28	7	6	7	7	18	27	—	0.40
5	36	4	2	3	2	15	38	—	1.7
10	48	2	—	1	1	10	38	—	2.3
25	23	2	—	1	1	9	63	1	3.2
50	19	1	1	—	1	2	75	1	3.4

[a] Data from Carey and Dils.[2]

[b] With optimum incubation conditions for radioactivity assay using [1-^{14}C]acetyl-CoA (20 μCi/μmole).

TABLE IV

Fatty Acid Synthase Complexes Isolated from Mammary Gland of Lactating Rabbit, Guinea Pig, Rat, and Cow

Properties	Rabbit[a]	Guinea pig[b]	Rat[c]	Cow[d]
Specific activity (mμmoles NADPH oxidized/minute/mg protein) in particle-free supernatant (mean ± S.E.M.)	52 ± 20	67 ± 4	51	72
Specific activity (mμmoles NADPH oxidized/minute/mg protein) of purified enzyme (mean ± S.E.M.)	880 ± 30	634 ± 56	760	875 ± 82
$s^\circ_{20,w}$	16.5 ± 0.2	12.3	12.9	15.4 ± 0.3
$D^\circ_{20,w}$	1.6	2.8	2.6	
Molecular weight	9.1×10^5	4.0×10^5	4.8×10^5	4.5×10^5
Thiol groups per molecule	58–59	54–59	56	
pH optimum	6.6	6.6	6.5–6.8	6.6
K_m for acetyl-CoA	$0.9 \times 10^{-5}\ M$	$2.4 \times 10^{-5}\ M$	$2.2 \times 10^{-5}\ M$	
K_m for malonyl-CoA	$2.9 \times 10^{-5}\ M$	$2.7 \times 10^{-5}\ M$	$1.3 \times 10^{-5}\ M$	
Preferred pyridine nucleotide	NADPH	NADPH	NADPH	NADPH ($K_m = 1.4 \times 10^{-5}\ M$)
Major fatty acids synthesized with optimum assay conditions	$C_{16:0}$ and $C_{4:0}$	$C_{16:0}$ and $C_{4:0}$	$C_{16:0}$ ($C_{4:0}$ not investigated)	$C_{16:0}$ and $C_{4:0}$

[a] From Carey and Dils.[1]
[b] From Strong and Dils.[13]
[c] From Smith and Abraham.[14]
[d] From Knudsen.[15]

($C_{8:0}$ and $C_{10:0}$) that are characteristic of rabbit milk are only formed in amounts approximately equimolar to the fatty acid synthase protein. This is in striking contrast to the predominant synthesis of $C_{8:0}$ and $C_{10:0}$ acids by intact epithelial tissue from lactating rabbit mammary gland. A factor or factors is present in the particle-free supernatant fraction of homogenates of this tissue which can increase the proportion of $C_{8:0}$ and $C_{10:0}$ acids synthesized by the purified synthetase from 1 to 24%.[10] It appears that this factor, which can alter the specificity of the synthase for chain termination, is lost during purification of the synthase from the particle-free supernatant fraction.

Butyrate is formed both from acetyl-CoA and from acetyl-CoA plus malonyl-CoA. This indicates two pathways of butyrate synthesis operate with the purified enzyme. In the absence of malonyl-CoA and NADPH, butyrate and triacetic lactone are formed showing that self-condensation of acetyl units is catalyzed under these conditions.

At the same high rates of overall synthesis, a lower proportion of longer chain fatty acids are synthesized by the purified enzyme from enzymatically carboxylated [$1\text{-}^{14}C$]acetyl-CoA than from added [^{14}C]malonyl-CoA.[2,11] One explanation suggested[11] is that acetyl-CoA carboxylase may act directly on acetate groups covalently bound to fatty acid synthase. Acetate can be covalently bound to the synthase by enzymatic transacylation from acetyl-CoA. Per mole of synthase 2 moles of acetate bind to thiol groups and up to 1 mole binds to nonthiol groups. Acetate can also be covalently bound to the synthase by chemical acetylation with acetic anhydride in the absence of CoA. Sixty moles of acetate are bound per mole of synthase, with 20–25 moles being bound to thiol groups. However, these acetyl–fatty acid synthase complexes hydrolyze so rapidly even at 0° that they cannot be used to demonstrate any direct enzymatic carboxylation of acetyl-synthase to malonyl-synthase.[12]

Comparison of Fatty Acid Synthase Complexes Isolated from Mammary Glands. Table IV[1,13–15] summarizes some of the properties of the enzyme purified from lactating mammary gland of rabbit, guinea pig, rat, and cow. The most striking feature is their similarity. The enzyme complex from rabbit mammary gland has approximately twice the molecular weight of the other complexes and may therefore be a dimer.

[10] C. R. Strong, E. M. Carey, and R. Dils, *Biochem. J.* **132**, 121 (1973).
[11] H. J. M. Hansen, E. M. Carey, and R. Dils, *Biochim. Biophys. Acta* **210**, 400 (1970).
[12] E. M. Carey, H. J. M. Hansen, and R. Dils, *Biochim. Biophys. Acta* **260**, 527 (1972).
[13] C. R. Strong and R. Dils, *Int. J. Biochem.* **3**, 369 (1972).
[14] S. Smith and S. Abraham, *J. Biol. Chem.* **245**, 3209 (1970).
[15] J. Knudsen, *Biochim. Biophys. Acta* **280**, 408 (1972).

[10] Fatty Acid Synthases from *Mycobacterium phlei*

By KONRAD BLOCH

The synthesis of long-chain fatty acids is catalyzed in biologic systems by enzyme systems of two types. Those designated as Type I synthases[1] are multienzyme complexes in which all the component catalytic activities and acyl carrier protein (ACP) form tight aggregates of high molecular weight. They occur in yeast, all animal tissues, and in a few bacterial species. By contrast, fatty acid synthases of Type II consist of individual, separable enzymes and require for activity the external addition of ACP. Type II systems are found in most bacteria and in plants.

Type I Fatty Acid Synthase of *Mycobacterium phlei*

The *de novo* synthesis of long-chain fatty acids in *Mycobacterium phlei* is catalyzed by a multienzyme complex according to Eq. (1):

$$CH_3COSCoA + n \; HOOCCH_2COSCoA + x \; TPNH + y \; DPNH$$
$$+ (x + y) \; H^+ \rightarrow CH_3(CH_2CH_2)_nCOSCoA + n \; CoASH + n \; CO_2$$
$$+ x \; TPN^+ + y \; DPN^+ + n \; H_2O \quad (1)$$

The synthase products are the CoA derivatives of fatty acids ranging in chain length from C_{14} to C_{26}. In the above equation n may, therefore, vary from 6 to 12. Under normal assay conditions (see below) the principal products are stearoyl-CoA and tetracosanoyl-CoA, i.e., the chain length pattern is bimodal. Conditions affecting the relative abundance of the shorter chain (C_{14}–C_{18}) and longer chain (C_{20}–C_{26}) acids will be described below.

Under conditions and at substrate concentrations that afford optimal rates of fatty acid synthesis in most systems[1a] the *M. phlei* fatty acid synthase exhibits very low activity. The system has an unusually high K_m for acetyl-CoA and is not saturated by concentrations of this substrate less than 2 mM. Addition of FMN and of DPNH as well as TPNH markedly increase the synthetic rate. A further pronounced stimulation of the *M. phlei* synthase is caused by certain polysaccharides also isolated from extracts of *M. phlei*. The rate-enhancing polysaccharides are of two types. One, designated as MMP, MW 2100, contains 10 consecutive

[1] D. Brindley, S. Matsumura, and K. Bloch, *Nature* (*London*) **224**, 666 (1969).
[1a] See, e.g., F. Lynen, Vol. 14 [18].

3-*O*-methylmannose residues and 2 residues of mannose.[2] The second type of mycobacterial polysaccharide, MGLP, MW 3700, is composed of 7 glucose, 1 3-*O*-methylglucose, and 10 consecutive 6-*O*-methylglucose residues, 1 glyceric acid moiety, 6 short-chain acyl groups, and either 0, 1, 2, or 3 succinate residues (MGLP-I, MGLP-II, MGLP-III, MGLP-IV).[3]

Materials for Enzyme Assay

Synthase is assayed by measuring the incorporation of [2-^{14}C]malonyl-CoA into pentane-soluble acidic products. Acetyl-CoA is prepared by the method of Simon and Shemin.[4] All other acyl-CoA derivatives are commercial products (P. L. Biochemicals) as is [2-^{14}C]malonyl-CoA (New England Nuclear Corp.). The polysaccharides MMP and MGLP are isolated from *M. phlei* ATCC-356 grown to stationary phase (48 hours) as described[5] up to and including Bio-Gel P-10 chromatography. The resulting mixture of polysaccharides is chromatographed on DEAE-Sephadex according to Keller and Ballou.[6] MMP is eluted with distilled water and the various species of MGLP (MGLP-0, -I, -II, and -III) with a 0–0.15 M NH$_4$HCO$_3$ gradient. The separated polysaccharides are lyophilized and further purified by passage through a Bio-Gel P-6 column with distilled water as an eluting solvent. For characterization, the polysaccharide fractions are hydrolyzed with acid and identified by gas chromatography of the trimethylsilyl derivatives of the methylglycosides.[5]

Assay for *de Novo* Fatty Acid Synthesis

The reaction mixture contains, in a final volume of 0.5 ml, 0.1 M potassium phosphate buffer, pH 7.0; 5 mM dithiothreitol (DTT); 30 μM DPNH; 30 μM TPNH; 1 μM FMN; 20 μM [2-^{14}C]malonyl-CoA (2 mCi/millimole); 300 μM acetyl-CoA; 200 μg of mixed polysaccharides (PS, fraction after Bio-Gel P-10 step[5]), and 1 μg of purified enzyme. After 15 minutes of incubation at 37°, reactions are stopped by addition of 0.15 ml of 50% KOH, the tubes are heated at 100° for 30 minutes,

[2] G. R. Gray and C. E. Ballou, *J. Biol. Chem.* **246**, 6855 (1971).
[3] M. H. Saier and C. E. Ballou, *J. Biol. Chem.* **243**, 4332 (1968).
[4] E. J. Simon and D. Shemin, *J. Amer. Chem. Soc.* **75**, 2520 (1953).
[5] M. Ilton, A. W. Jevans, E. D. McCarthy, D. E. Vance, H. B. White, III, and K. Bloch, *Proc. Nat. Acad. Sci. U.S.* **68**, 87 (1971).
[6] J. M. Keller and C. E. Ballou, *J. Biol. Chem.* **243**, 2905 (1968).

HCl is added to pH 1 and the fatty acids extracted into petroleum ether (b.p. 30–40°). One drop of glacial acetic acid is added and the solvent and acetic acid removed at 55°. The radioactivities of the samples are measured by scintillation counting.

One unit of enzymatic activity is defined as the amount of enzyme required to incorporate 1 nmole of malonate into fatty acids per minute.

Purification of the Synthase[7]

All steps are carried out at 0–4°. Protein is determined spectrophotometrically at 260 and 280 nm.[8] Buffers contain 1 mM DTT and 1 mM EDTA and are adjusted to pH 7.0. Eighty grams of *M. phlei* cells (see above) are suspended in 300 ml of 0.1 M potassium phosphate buffer. The cells are broken by passage through a French pressure cell operated at 8000 psi and the disrupted cells centrifuged at 17,000 g for 20 minutes. The resulting supernatant is centrifuged at 105,000 g for 90 minutes and then brought to 35% saturation with ammonium sulfate. After 15 minutes of stirring, the precipitate is removed by centrifugation at 37,000 g for 20 minutes and discarded. The supernatant is slowly brought to 55% ammonium sulfate saturation, stirred for 30 minutes, and centrifuged at 37,000 g. The precipitate is dissolved in 0.1 M potassium phosphate buffer and 32.8 ml of a 1% protamine sulfate solution (0.12 mg/mg of protein) are added dropwise with stirring. After 30 minutes, the solution is centrifuged at 37,000 g and the precipitate discarded. The supernatant is applied to a DEAE-cellulose column (3 × 20 cm) that had previously been equilibrated with 0.1 M potassium phosphate buffer. After washing of the column with 1 liter of 0.25 M phosphate buffer, the enzyme is eluted by a linear gradient of 650 ml of 0.25 M and 0.70 M potassium phosphate buffer. Fractions containing active enzyme are concentrated to 50 ml on a Diaflo apparatus with an XM 50 membrane. This concentrated solution is slowly brought to 60% saturation with ammonium sulfate and centrifuged at 37,000 g for 20 minutes. The precipitate is dissolved in 2 ml of 0.5 M potassium phosphate buffer and applied to a Bio-Gel A-5m column (70 × 2.8 cm) that had been equilibrated with the same buffer. Enzymatic activity elutes with the first protein peak. Peak fractions are combined and concentrated to 2.5 ml on a Diaflo apparatus with an XM-50 membrane. The results of the purification are summarized in Table I.

[7] D. E. Vance, O. Mitsuhashi, and K. Bloch, *J. Biol. Chem.* **248**, 2303 (1973).
[8] O. Warburg and W. Christian, *Biochem. Z.* **310**, 384 (1941).

TABLE I
PURIFICATION OF FATTY ACID SYNTHASE

	Total activity (units)	Total protein (mg)	Specific activity (units/mg)	Yield (%)	Purification (fold)
Crude supernatant	3829	3466	1.10	(100)	(1)
Ammonium sulfate (35–55%)	2739	1259	2.18	71.5	2.0
Protamine sulfate	3063	1231	2.49	80.0	2.3
DEAE-cellulose	1161	43.5	26.7	30.3	24.3
Ammonium sulfate (0–60%)	1109	30.5	36.4	29.0	33.1
Bio-Gel A-5m	656	1.75	375	17.1	341

Properties

In the absence of polysaccharide (MMP or MGLP), acetyl-CoA does not saturate the synthases until its concentration is about 2 mM. In the presence of 200 μg per 0.5 ml of crude polysaccharide (PS, Bio-Gel P-10 fraction) the K_m value for acetyl-CoA is lowered to 90 μM. For malonyl-CoA the K_m values in the absence and presence of polysaccharide are 50 and 9.6 μM, respectively. While TPNH as the sole source of pyridine nucleotide supports fatty acid synthesis ($K_m = 600$ μM) DPNH does not ($K_m = 3$ mM). Optimal rates of malonyl-CoA incorporation occur with 30 μM TPNH + 30 μM DPNH.[9] TPNH appears to be the specific electron donor for the β-ketoacyl reductase step and DPNH is specifically utilized for α, β-enoyl reduction.[10] FMN (1 μM) raises the activity of the synthase up to threefold. The magnitude of this effect depends on the concentration of acetyl-CoA.

The 341-fold purified synthetase (Table I) is homogeneous as judged by the sedimentation pattern in the analytical ultracentrifuge. In 0.5 M phosphate buffer, pH 7.0, the S value is 23.61 corresponding to a molecular weight of 1.39 $\times$ 10^6. The specific activity of this preparation is 375 (nanomoles of malonyl-CoA incorporated per milligram of protein per minute) when assayed under the conditions described above.[7]

Enzyme isolated from *M. phlei* cells grown on medium containing [³H]β-alanine yields on alkaline hydrolysis a radioactive fraction characterized as 4-phosphopantetheine.[1] It is, therefore, assumed that synthase contains ACP integrated into the multienzyme complex.

[9] H. B. White, III, O. Mitsuhashi, and K. Bloch, *J. Biol. Chem.* **246**, 4751 (1971).
[10] P. Flick, unpublished.

TABLE II
ACTIVITY OF ACYL-CoA PRIMERS IN FATTY ACID SYNTHESIS CATALYZED
BY *M. phlei* SYNTHASE I[a]

Acyl-CoA	Fatty acid synthesis (units/mg)[b]	
	No polysaccharide	+MMP (100 μg)
Acetyl-CoA	13.7	120
Propionyl-CoA	1.7	36.5
Butyryl-CoA	0.5	11.5
Hexanoyl-CoA	0.15	4.5

[a] Data from Vance *et al.*[7]

[b] Fatty acid synthesis was assayed with enzyme of specific activity of 120 with 600 μM acyl-CoA and 100 μM malonyl-CoA and other additions as described under section entitled "Assay for *de Novo* Fatty Acid Synthesis."

Enzyme dissolved in 0.5 M phosphate buffer when dialyzed at 4° against 0.005 M potassium phosphate buffer is inactivated with a half-time of about 50 minutes. No activity remains after 10 hours of dialysis. When enzyme so inactivated is dialyzed against 0.5 M potassium phosphate buffer, activity returns to about 35% of the original within 2 hours but does not increase further on continued dialysis against 0.5 M potassium phosphate buffer. The inactivated enzyme has an S value of 7.3 and the reactivated synthase has S = 23.1.[7]

Primer Specificity

Enzymatic activity is highest with acetyl-CoA and decreases with increasing chain length of the acyl-CoA primer as shown in Table II. Polysaccharide increases synthase activity not only with acetyl-CoA but also, and by an even greater factor, with the longer chain acyl-CoA derivatives.

Enzyme Inhibitors and Activators

Palmityl-CoA in a concentration of 20 μM inhibits the synthase completely.[10] Bovine serum albumin (0.1 mg/ml) when simultaneously incubated with enzyme and palmityl-CoA restores the enzyme to full activity i.e., it relieves palmityl-CoA inhibition. The polysaccharides MMP or MGLP (0.1–0.2 mg/ml) also prevent enzyme inhibition by palmityl-CoA. The latter effect appears to result from complex formation between polysaccharide and long-chain acyl-CoA derivatives.[11] It has been suggested that the increase in activity of the *M. phlei* synthase caused by

[11] Y. Machida and K. Bloch, *Proc. Nat. Acad. Sci. U.S.* **248**, 2317 (1973).

MMP or MGLP in the absence of added palmityl-CoA may result from complexation of long-chain acyl-CoA derivatives generated in the course of synthesis.[12] According to this hypothesis, palmityl-CoA inhibits the synthase by negative feedback, the polysaccharides providing relief from this inhibition by sequestering palmityl-CoA.

The various partial reactions catalyzed by the synthase when studied with the *N*-acetylcysteamine derivatives as substrates are not stimulated by either MMP or MGLP.[10] Coenzyme A enhances the *de novo* fatty acid synthesis 4-fold at low (50 μM) but not at higher (300 μM) acetyl-CoA concentrations.

End Products of the Synthase

Under all assay conditions for *de novo* synthesis the enzyme produces a mixture of fatty acyl-CoA derivatives ranging in length from C_{14} to C_{26}.[1] Stearate and tetracosanoate are always the two principal products. The relative abundance of the shorter chain (C_{14}–C_{18}, S) and the longer chain acids (C_{20}–C_{26}, L) depends on assay conditions. In the absence of polysaccharide and at high acetyl-CoA concentrations (500 μM) the ratio of S:L is about 1:3. The presence of MMP or MGLP (0.2 mg/ml) changes the S:L ratio to about 1:1. Inclusion of bovine serum albumin (0.2–1 mg/ml) affords an S:L ratio of 3:1.[12a]

M. phlei Synthase I as an Elongating System

The *M. phlei* multienzyme complex catalyzes the further elongation of acyl-CoA derivatives with carbon chains up to C_{20} in addition to *de novo* synthesis. Throughout purification (Table I) the relative activities of the enzyme with acetyl-CoA and, e.g., palmityl-CoA as primers remain constant. With acyl-CoA derivatives up to C_{16}, the end product pattern is bimodal affording stearate and tetracosanoate as the principal fatty acids. Elongation activity has the same cofactor requirements, including polysaccharides, as the *de novo* synthesis from acetyl-CoA. However, under these assay conditions (see above), the elongation of decanoyl-CoA and its higher homologues is inefficient and enzymatic activity decreases progressively as the carbon chain of the primer is lengthened. Addition of a partially purified protein from *M. phlei* extracts markedly increases elongation activity.[13] This "elongation factor," at 6 μg of protein per mil-

---

[12] H. Knoche, T. W. Esders, K. Koths, and K. Bloch, *J. Biol. Chem.* **248**, 2317 (1973).
[12a] P. K. Flick and K. Bloch, *J. Biol. Chem.* **249**, 1031 (1974).
[13] D. E. Vance, T. W. Esders, and K. Bloch, *J. Biol. Chem.* **248**, 2310 (1973).

liliter, stimulates the utilization of various acyl-CoA derivatives assayed at 100 μM concentration as follows: C_{10}, 2.8-fold; C_{12}, 4-fold; C_{14}, 5.5-fold; C_{16}, 6.7-fold; and C_{18}, 24-fold.[13]

The elongating factor is inseparable from long-chain acyl-CoA thioesterase activity and appears to function by releasing free coenzyme A from substrates. In an assay system for elongating palmityl-CoA, 50–100 μM CoA enhances enzymatic activity to the same extent as 6 μg of partially purified elongating factor (palmityl-CoA thioesterase).[13]

Synthase II from *M. phlei* [14]

Extracts of *M. phlei* contain a fatty acid synthase (malonyl-CoA incorporation) which requires ACP and is, therefore, distinct from the Type I enzyme (multienzyme complex) described above. The ACP-dependent activity (enzyme II) is retained by Sephadex G-150 in contrast to the Type I synthetase which emerges in the column void volume. Enzyme system II has an estimated average molecular weight of $<250,000$ suggesting that it is nonaggregated synthase similar to that of *Escherichia coli*.[14a] Enzyme II functions only as an elongating system and not as a *de novo* synthase. For chain initiation, enzyme II can use palmityl-CoA and stearyl-CoA but not acetyl-CoA or octanoyl-CoA. *Mycobacterium phlei* synthase II has no requirements for FMN, CoA, or polysaccharide (MGLP or MMP). Both $ACP_{E.\ coli}$ and an ACP isolated from *M. phlei* extracts[15] support System II activity. $ACP_{M.\ phlei}$ has a molecular weight of 10,450 and a similar amino acid composition as $ACP_{E.\ coli}$[15a] except that it contains four instead of one proline residue per mole.

[14] S. Matsumura, D. Brindley, and K. Bloch, *Biochem. Biophys. Res. Commun.* **38**, 369 (1970).
[14a] See P. R. Vagelos, A. W. Alberts, and P. W. Majerus, Vol. 14 [5].
[15] S. Matsumura, *Biochem. Biophys. Res. Commun.* **38**, 238 (1970).
[15a] P. W. Majerus, A. W. Alberts, and P. R. Vagelos, Vol. 14 [6].

[11] Methylated Polysaccharide Activators of Fatty Acid Synthase from *Mycobacterium phlei*

By GARY R. GRAY and CLINTON E. BALLOU

Mycobacterium species contain in the cytosol two highly methylated polysaccharides,[1,2] both of which have been shown to activate the fatty

[1] C. E. Ballou, *Accounts Chem. Res.* **1**, 366 (1968).
[2] G. R. Gray and C. E. Ballou, *J. Biol. Chem.* **246**, 6835 (1971).

acid synthase complex of *Mycobacterium phlei* in a manner leading to a reduction in the K_m for acetyl-CoA.[3] The most active substance is composed of 10 3-*O*-methyl-D-mannose units in $\alpha(1 \rightarrow 4)$ linkage, forming a chain to which is attached at position 6 of one of the 3-*O*-methylmannose units a side chain of two D-mannose units in $\alpha(1 \rightarrow 2)$ linkage. It is called MMP for methylmannose polysaccharide. The other substance is a lipo-polysaccharide containing 7 moles of D-glucose, 10 moles of 6-*O*-methyl-D-glucose and 1 mole of 3-*O*-methyl-D-glucose. The polymer is linked to D-glyceric acid and is acylated by acetyl, propionyl, isobutyryl, octanoyl, and succinyl groups. Four forms of the lipopolysaccharide can be isolated depending on the net negative charge which reflects different contents of succinic acid. These are called MGLP-I, II, III, and IV for methylglu-cose lipopolysaccharide.

Principle. Acetone-dried *M. phlei* cells are refluxed with 70% ethanol, and the ethanol extract is evaporated to dryness. The extract is partitioned between chloroform and water, and the water-soluble material is passed through a Sephadex G-50 column which separates MGLP and MMP from small and large contaminants. Methylglucose lipopolysaccharide is then fractionated into the four species by gradient elution from DEAE-Sepha-dex. The MMP, which is eluted in the water wash, is purified by treat-ment with β-amylase, which removes contaminating $\alpha(1 \rightarrow 4)$-glucan, followed by an additional gel filtration step. While the polysaccharides can be detected by the carbohydrate assay, it is most convenient to add an extract of cells labeled with L-[methyl-³H]methionine and then follow the radioactivity of the methyl groups of the MGLP and MMP during the purification.

Purification Procedure

Cultivation of Bacteria. Mycobacterium phlei (ATCC 356) is grown for 2 days at 37° in a medium containing 2% glycerol, 0.5% Casamino acid, 0.1% fumaric acid, 0.1% $K_2HPO_4 \cdot 3H_2O$, 0.03% $MgSO_4 \cdot 7H_2O$, and 0.002% $FeSO_4 \cdot 7H_2O$ adjusted to pH 7.0 with KOH. The cells are har-vested by centrifugation and stored at $-10°$. One liter of medium yields about 15 g of wet cell paste.

The bacterium is grown in the presence of L-[methyl-³H]methionine in a medium containing 2% glycerol, 0.5% glutamic acid, 0.1% fumaric acid, 0.65% $K_2HPO_4 \cdot 3H_2O$, 0.1% NH_4Cl, 0.005% $MgSO_4 \cdot 7H_2O$, 0.005% $FeSO_4 \cdot 7H_2O$, 0.00006% $CuSO_4$, 0.00003% $MnCl_2 \cdot 4H_2O$, and

[3] M. Ilton, A. W. Jevans, E. D. McCarthy, D. Vance, H. B. White, III, and K. Bloch, *Proc. Nat. Acad. Sci. U.S.* **68**, 87 (1971).

0.00008% ZnSO₄, adjusted to pH 7.0 with KOH. A 10% inoculum from a 48-hour culture is used to inoculate 1 liter of fresh medium containing 50 μCi of L-[methyl-³H]methionine, the cells are grown for 24 hours at 37°, and are harvested by centrifugation.

Isolation of MMP and MGLP. Frozen *M. phlei* cells (250 g) are extracted twice at room temperature with 5 volumes of acetone and air dried, and the residue is extracted twice with 15 volumes (based on the volume of wet cells) of refluxing 70% ethanol. After evaporation of solvents, the ethanol extract is partitioned between the two layers of chloroform–methanol–water (8:4:3, 750 ml). The upper aqueous layer is removed and centrifuged at 10,000 *g* to remove the remaining chloroform layer, evaporated to remove methanol and traces of chloroform, and then lyophilized.

The residue from the lyophilized water layer (22 g) is combined with the same fraction from cells grown in the presence of L-[methyl-³H] methionine and dialyzed 6 hours against water to remove trehalose and small oligosaccharides. The contents of the dialysis bag are lyophilized (2.5 g), then applied in 0.5 g portions to a Sephadex G-50 column (4 × 180 cm) in 0.1 *M* acetic acid (Fig. 1).[4] The major radioactive fractions (peak B) are combined and lyophilized, then applied to a DEAE-Sephadex A-25 column (2 × 10 cm) in the bicarbonate form (Fig. 2). The column is eluted at 4° with water (ten 8-ml fractions) to remove MMP and other neutral material, then with a linear gradient of 0.0 to 0.15 *M* NH₄HCO₃ (1 liter) to separate the four forms of MGLP. These fractions are combined and evaporated under vacuum several times from water to remove NH₄HCO₃. Alternatively, NH₄HCO₃ can be removed by desalting on a small Sephadex G-10 column. Fractions containing MGLP are lyophilized to yield a fluffy white powder. Small amounts of arabinose and mannose may be observed on hydrolysis of these fractions, indicative of a contaminating polysaccharide. Pure MGLP can be obtained by extracting peak B from the Sephadex G-50 column with chloroform–methanol (4:1, 100 ml) prior to the ion exchange step. Methylglucose lipopolysaccharide is soluble in this solvent, but some loss of the more highly succinylated forms occurs because of their decreased solubility.

The neutral fraction contains MMP and varying amounts of a glucan which is removed by digestion with β-amylase in 0.1 *M* sodium acetate buffer at pH 4.0. After 48 hours, the digest is heated and centrifuged to remove protein, reduced in volume, then fractionated on a Sephadex

---

[4] On one occasion, this elution was done with water and led to resolution of MGLP from MMP, apparently because the completely ionized MGLP was more effectively excluded from the gel.

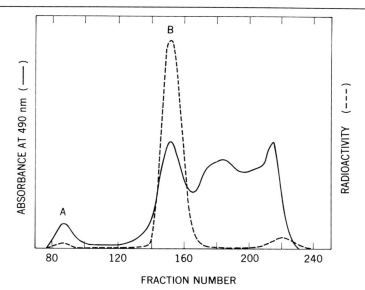

Fig. 1. Elution from a Sephadex G-50 column of the partially purified 70% ethanol extract of *M. phlei* cells which contains MMP and MGLP. The latter two are obtained in peak B.

G-25 column (2.5 × 95 cm) in water (Fig. 3). Peak A (125 mg) is MMP and peak B is maltose, the expected product from β-amylase digestion of an $\alpha(1 \to 4)$-glucan.

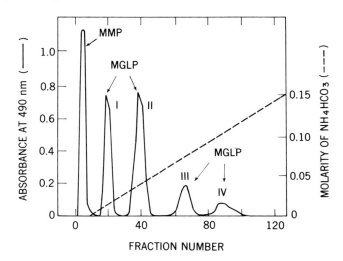

Fig. 2. Fractionation on DEAE-Sephadex of the different forms of MGLP.

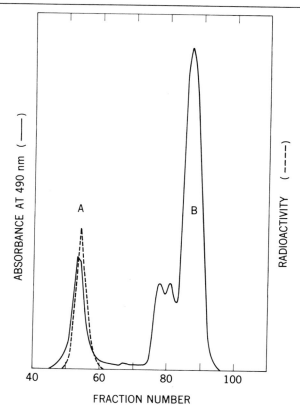

Fig. 3. Recovery of purified MMP by gel filtration on Sephadex G-25. The MMP is in peak A.

Properties

Methylmannose polysaccharide is a water-soluble, neutral, nonreducing polysaccharide with a molecular weight of about 2084.[2] The presence of the methyl groups causes the substance to migrate rapidly on thin-layer and paper chromatography (like a di- to tetrasaccharide), and to travel on gel filtration with the properties of an amylooligosaccharide of about 25 glucose units.[5] In a good preparation, the ratio of 3-*O*-methyl-mannose to mannose should be about 5, and the acid hydrolysate should be free of glucose. Some heterogeneity is usually found in the degree of methylation, and it is evident by the separation of purified MMP into 3 peaks by gel filtration on Bio-Gel P-4. Such material becomes labeled

[5] For example, see E. A. Grellert and C. E. Ballou, *J. Biol. Chem.* **247**, 3236 (1972).

upon incubation with L-[methyl-^{14}C]S-adenosylmethionine and a particulate enzyme from *M. phlei,* suggesting that incomplete molecules are present.[6]

Methylglucose lipopolysaccharide-I, -II, -III, and -IV are water-soluble, acidic, acylated, nonreducing polysaccharides with molecular weights around 3500. The dry substance will dissolve in chloroform–methanol. Like MMP, the chromatographic and gel filtration properties are dramatically affected by the high degree of methylation and acylation. The polysaccharide component does not appear to possess any microheterogeneity, but heterogeneity does exist in the content of acyl groups.[7] A succinylated arabinomannan is a common contaminant of the MGLP fractions obtained by elution from DEAE-Sephadex. The acyl composition may be checked by paper chromatography of the acyl hydroxamates,[8] and an acid hydrolysate of the polysaccharide should be free of mannose and have a 6-*O*-methylglucose to glucose ratio of about 1.4.

It was reported by Ilton *et al.*[3] that MMP is eluted from Bio-Gel P-6 ahead of MGLP, which implies that it is the larger of the two. However, we have compared a sample of MMP supplied by Dr. Bloch with our own preparations, and we find that they are essentially identical and that MMP is eluted from Bio-Gel P-6 later than MGLP. The results of Ilton *et al.* could be rationalized if they were dealing with a separation based on ion exchange rather than on simple gel filtration. The observation of Ilton *et al.*[3] that two forms of MGLP were obtained by gel filtration again suggests that their separation was affected by ion exchange.

[6] E. A. Grellert and C. E. Ballou, unpublished.
[7] W. L. Smith and C. E. Ballou, *J. Biol. Chem.* **248,** 7118 (1973).
[8] J. M. Keller and C. E. Ballou, *J. Biol. Chem.* **243,** 2905 (1968).

[12] Acyl Carrier Protein Synthetase[1]

By David J. Prescott, John Elovson, and P. Roy Vagelos

$$\text{Apo-ACP} + \text{CoASH} \rightleftharpoons \text{ACP} + 3',5'\text{-adenosine diphosphate}$$

Assay Method

Principle. The conversion of apo-ACP to ACP is measured in a two-stage assay. In the first-stage synthesis of ACP is allowed to proceed, catalyzed by ACP synthetase. This stage is terminated, and the amount

[1] J. Elovson and P. R. Vagelos, *J. Biol. Chem.* **243,** 3603 (1968).

of ACP synthesized is determined in the second stage by means of the malonyl-CoA–CO_2 exchange reaction.

Reagents for first stage		Amount per assay (μl)
Tris-HCl buffer 0.1 M pH 8.0		10.0
CoASH	0.01 M	2.4
$MgCl_2$	0.08 M	1.0
$(NH_4)_2SO_4$	0.04 M	1.0
Dithiothreitol	0.02 M	1.0
H_2O		22.6
Apo-ACP	0.5 mg/ml (57 pmole/μl)	1.0

Apo-ACP may be prepared enzymatically[1] from ACP[2] or chemically using 60% HF to cleave the phosphodiester bond linking the hydroxyl of serine-36 to 4'-phosphopantetheine.[3]

Enzyme: 0.0002 to 0.003 units dissolved in 1 μl.

The first-stage incubation is carried out at 33° for 10 minutes.

Reagents for second stage		Amount per assay (μl)
Imidazole-HCl	2.0 M pH 6.2	5.5
EDTA	0.5 M pH 7.0	4.0
Malonyl pantetheine	0.01 M	8.5

Malonyl pantetheine may be synthesized chemically by thioester exchange between malonyl thiophenol and pantetheine.[4]

Alternatively, malonyl CoA, which is available commercially (Sigma Chemical Co.) may be substituted at the same concentration.

Caproyl pantetheine	0.02 M	2.0

Caproyl pantetheine may be synthesized chemically from caproic anhydride and pantetheine.[5]

KH$^{14}CO_3$ 0.2 mCi/mmole 0.25 M	10
Fraction A 15 mg of protein/ml	20

Fraction A is prepared by a modification of the method previously published[1] in order to further reduce blank values. The batching step from DEAE-cellulose is replaced with a linear gradient from

[2] P. W. Majerus, A. W. Alberts, and P. R. Vagelos, *Biochem. Prep.* **12**, 56 (1968).
[3] D. J. Prescott, J. Elovson, and P. R. Vagelos, *J. Biol. Chem.* **244**, 4517 (1969).
[4] E. G. Trams and R. O. Brady, *J. Amer. Chem. Soc.* **82**, 2972 (1960).
[5] E. J. Simon and D. Shemin, *J. Amer. Chem. Soc.* **75**, 2520 (1953).

0.075 M KCl to 0.35 M KCl in the same buffer.[1] A malonyl-CoA–CO_2 exchange assay, utilizing excess ACP, is employed to assay the column fractions.[2] The peak of CO_2 exchange activity is combined, precipitated by addition of ammonium sulfate to 80% of saturation and stored as a slurry in liquid N_2. Blank assays performed with fraction A prepared in this manner show virtual background radio-activity indicating absence of endogenous ACP. The fraction A is desalted over a small G-25 Sephadex column immediately before use in the two-stage malonyl-CoA–CO_2 assay for ACP synthetase.

HCl   4 N 10

Purification

All operations are carried out at 4°.

Step 1. Preparation of Crude Extract. Escherichia coli strain B (880 g, early log phase, Grain Processing Corp., Muscatine, Iowa) are thawed and suspended in 2 liters of 0.01 M Tris-HCl, pH 7.4 containing 0.01 M 2-mercaptoethanol and 0.0001 M EDTA. The suspension is ruptured in a Manton-Gaulin submicron dispersor at 9000 psi. To the viscous stirred suspension is added 800 ml of a 10% streptomycin sulfate solution and the suspension is centrifuged for 30 minutes at 27,500 g. The supernate is saved (3400 ml, 55 g protein) (see Table I).

Step 2. Ammonium Sulfate Precipitation. The streptomycin sulfate supernate is brought to 75% saturation by the careful addition of solid ammoninum sulfate. The mixture is centrifuged and the pellet is slurried

TABLE I
PURIFICATION OF ACP SYNTHETASE OF *Escherichia coli*

Fraction	Total protein (mg)	Total activity (units)	Specific activity (units per mg)	Recovery of activity (%)	Purifica-tion (-fold)
Crude extract	55,000	6,100	0.11	100	—
Heat step	8,800	3,500	0.40	57	3.5
Calcium phosphate gel	1,100	1,390	1.21	23	11
Ammonium sulfate, 0–60%	499	1,170	2.4	19	21
Sephadex G-200	56	920	16.4	15	148
DEAE-Sephadex pooled peak	7.2	570	7.9	9.3	680

in a small amount of buffer (0.01 M Tris-HCl, pH 7.4, 0.01 M 2-mercaptoethanol, and 0.0001 M EDTA) and dialyzed overnight against 25 liters of the same buffer.

Step 3. Heating. The clear, dark brown solution (635 ml, 61 mg protein/ml) is divided into four stainless steel centrifuge bottles. The solution is brought to 54° in a water bath which is preheated sufficiently to compensate for the decrease caused by immersion of the bottles. The temperature is maintained for 12 minutes after which the mixture is chilled in ice and centrifuged at 27,000 g for 15 minutes.

Step 4. Calcium Phosphate Gel Elution. The viscous amber supernate (510 ml, 17.2 mg protein/ml) is adjusted to pH 7.8 with 10 N NaOH. Glycerol (140 g), MgCl$_2$ (1 M, 12 ml), 2-mercaptoethanol (1.7 ml), and reduced CoASH (Li salt 0.05 M, 2.5 ml) are then added and thoroughly mixed. Calcium phosphate gel suspension (400 ml, 37 mg of gel per milliliter in water) is added and stirred for a few minutes and centrifuged. The gel is washed with ammonium sulfate (60 ml, 0.05 M in a buffer containing 10% glycerol, v/v, 0.05 M Tris-HCl, pH 7.8, 0.01 M 2-mercaptoethanol, 0.01 M MgCl$_2$, and 0.0001 M reduced CoASH). The enzyme is eluted with four successive washes of 0.25 M ammonium sulfate (70 ml in the same buffer mixture). The combined eluates (285 ml, 1.1 g of protein) are taken to 60% saturation of ammonium sulfate. The precipitate is recovered by centrifugation and suspended in a minimal volume of buffer (0.02 M Tris-HCl, pH 7.3, 0.01 M MgCl$_2$, 0.08 M ammonium sulfate, 0.0003 M dithiothreitol, 0.001 M EDTA, and 0.0001 M reduced CoASH). Inactive protein is removed by centrifugation and the supernate (8 ml, 490 mg of protein) is saved.

Step 5. Sephadex G-200 Gel Filtration. The column (2.5 × 100 cm) is first equilibrated with the buffer described in the final portion of Step 4, except that CoASH is omitted. Immediately before applying the sample from Step 4, the column is equilibrated with 300 ml of the buffer containing 0.0001 M CoASH in reverse flow. The column is then eluted with the buffer without CoASH. Using this procedure, the enzyme is protected by the presence of CoASH during the gel filtration step. Fractions of 3 ml are collected and the concentration of protein, absorbance at 260 nm and enzymatic activity are determined. The enzyme elutes in the last part of the protein peak. Active fractions are pooled and concentrated by ultrafiltration of 1 mg protein/ml (if no further purification is desired). The enzyme at this stage is stable for at least one year when stored in liquid N_2 and in the presence of CoA. The purity at this stage is usually adequate for most purposes.

Step 6. DEAE-Sephadex A-50. The final step is a negative purification on DEAE-Sephadex in which the enzyme remains unabsorbed. Active fractions from the G-200 step are pooled and diluted to 240 ml with 0.0001 M CoASH containing 0.005 M MgCl$_2$. This solution is passed over a column of DEAE-Sephadex A-50 (2 $\times$ 6 cm) which had been equilibrated previously with a solution containing 0.025 M ammonium sulfate, 0.01 M Tris-HCl, pH 7.3, 0.005 M MgCl$_2$, and 0.0001 M CoASH. After washing the column with another 50 ml portion of the same buffer, the combined eluate and wash, containing the unabsorbed enzyme, are concentrated by pressure filtration to a volume of 1 ml and stored in liquid N$_2$.

Procedure

The reaction is carried out in glass test tubes (5 $\times$ 50 mm). A stock solution consisting of the first six reagents for the first stage may be prepared and kept in an ice bath before use. This stock solution (38 μl/tube) is then added to each test tube. Apo-ACP is added next and, finally, the enzyme. It is convenient to use microliter syringes equipped with repeating dispensers (Hamilton, Model PB600-1) to add the apo-ACP and enzyme. Incubation of the complete reaction mixture is carried out at 33° for 10 minutes. A stock solution containing all reagents for the second stage is prepared fresh and kept on ice until used. The second stage stock solution is added (50 μl/tube) with agitation to insure mixing. Excess EDTA immediately terminates the first stage since the ACP synthetase demonstrates a divalent metal requirement. Further incubation is carried out at 33° for 15 minutes. Appropriate blanks are carried through the entire procedure in order to correct for the presence of endogenous ACP either in the apo-ACP preparation, the ACP synthetase preparation, or the fraction A. The second stage of the assay is terminated by the addition of 4 N HCl. An aliquot (50 μl) of the acidified reaction mixture is transferred to scintillation vials and dried in an oven. Water (0.5 ml) is added first to each vial to redissolve the radioactive malonyl pantetheine (or CoA) and then 10 ml of scintillation cocktail suitable for counting aqueous systems are added. Acid stable radioactivity is determined, and from standards containing known amounts of ACP the conversion of apo-ACP to ACP may be calculated.

The malonyl-CoA—CO$_2$ exchange reaction is linear up to 40 ρmoles of ACP per assay. The synthetase assay (first stage) is linear for at least 30 minutes and may be extended to this time when low enzymatic activity or high levels of ACP, or both, are encountered in crude preparations. If the amount of holo-ACP formed in the synthetase assay is expected to fall outside the linear range of the malonyl-CoA—CO$_2$ exchange reaction, an appropriate aliquot of the first stage reaction may be diluted to 0.04 ml with 0.05 M Tris-HCl (pH 7.4) containing 0.005 M dithiothrei-tol and then used for the second stage.

Units. A unit of enzyme is defined as the amount required to synthesize 1 nmole of ACP per minute in the standard two-stage assay.

Properties

Enzymatic Purity. ACP synthetase prepared in the manner described above is free of ACP, apo-ACP, and ACP hydrolase activity. Low phosphatase activity is detectable. Several major bands are apparent in standard 7% polyacrylamide gel electrophoresis. Gel exclusion chromatography and sucrose density gradient sedimentation indicate a molecular weight of approximately 50,000 daltons.

Stability. The enzyme is stable when CoASH is present at half-saturating concentrations, and it is necessary to include this substrate in all steps of purification beyond the 54° treatment. When stored as indicated, the enzyme maintains full activity for at least one year.

pH Optimum. The peak of enzymatic activity is observed between pH 7.5 and 9.0.

Reversibility. Formation of radioactive CoASH from ACP labeled with ^{14}C in the 4'-phosphopantetheine prosthetic group has been observed.[1] 3',5'-ADP is necessary or the reverse reaction to occur.

Activation and Inhibitors. ACP synthetase requires either Mn^{2+} or Mn^{2+} for activity, but the response to these two cations is different. Mn^{2+} activates up to approximately 10^{-4} M. Further increase in concentration leads to inhibition up to 5×10^{-4} M followed by slow but extensive recovery of activity. The activity observed at 0.1 M is twice that found at 1.0×10^{-4} M.[1]

In the presence of increasing amounts of Mg^{2+}, the apparent activity rises in the usual hyperbolic manner with a K_m of 3×10^{-3} M. Further increasing the concentration beyond the expected saturating value leads to a slower but continuous rise in enzymatic activity which finally levels off at about 0.1 M. These effects are not a result of ionic strength. Ca^{2+}, Fe^{2+}, Co^{2+}, and Zn^{2+} are ineffective, inhibitory, or both, at intermediate and high concentrations.[1]

Substrate Specificity

4'-Phosphopantetheine Donor. The K_m for CoASH was found to be 1.5×10^{-4} M. Oxidized CoA is inactive at 1×10^{-3} M concentration, as is reduced dephospho-CoA at 2×10^{-3} M levels.

Apo-peptide Acceptor. A large number of natural and synthetic peptides, some of which contain amino acid analogues, have been tested for activity, and the results will be summarized briefly.[3,6-9] The amino terminal hexapeptide of apo-ACP is important for activity with ACP synthetase. Analogues of apo-ACP(1 → 74) have been prepared in which amino terminal residues have been progressively deleted. Thus, a progressive loss of activity is observed in which the ultimate peptide of the series (7 → 74) is essentially inactive.[9] The arginine at position 6 in apo-ACP-(1 → 74) is not necessary for activity with ACP synthetase, since [norleucine-6,44]-apo-ACP-(1 → 74) exhibits prosthetic group acceptor activity similar to apo-ACP-(1 → 74). However, [norleucine-6,44]-holo-ACP-(1 → 74) is inactive in the malonyl-CoA–CO_2 exchange reaction. In all other cases studied, apo-peptides which accept the prosthetic group exhibit activity in the malonyl-CoA–CO_2 exchange reaction as the holo-peptide. Lysine residues at positions 8, 9, and 18 as well as methionine at position 44 may be replaced by norleucine without loss of activity. A large part of the carboxyl terminal residues do not seem to be essential since the peptide (1 → 61) retains significant activity[9] and, of course, peptide (1 → 74) functions equally as well as (1 → 77). Apo-peptide (1 → 74) exhibits a K_m value of 2.0×10^{-6} M and no substrate inhibition, while apo-ACP has a K_m of 5.5×10^{-7} M and exhibits strong substrate inhibition with ACP synthetase.[1,3]

In Vivo ACP Prosthetic Group Turnover

Both ACP synthetase and ACP hydrolase, the enzyme which specifically cleaves the phosphodiester linkage of the prosthetic group,[10,11] are active in *in vivo* turnover of the prosthetic group.[12] Studies have shown that CoASH is the most likely prosthetic group donor for ACP synthetase *in vivo*.[12] The rate of turnover of 4′-phosphopantetheine of ACP is 4 times faster than the rate of growth of the ACP pool.[12]

[6] D. J. Prescott and P. R. Vagelos, *Advan. Enzymol.* **36**, 269 (1972).

[7] W. S. Hancock, D. J. Prescott, G. R. Marshall, and P. R. Vagelos, *J. Biol. Chem.* **247**, 6224 (1972).

[8] W. S. Hancock, D. J. Prescott, W. L. Nulty, J. Weintraub, P. R. Vagelos, and G. R. Marshall, *J. Amer. Chem. Soc.* **93**, 1799 (1971).

[9] W. S. Hancock, G. R. Marshall, and P. R. Vagelos, *J. Biol. Chem.* **248**, 2424 (1973).

[10] P. R. Vagelos and A. R. Larrabee, *J. Biol. Chem.* **242**, 1776 (1967).

[11] See P. R. Vagelos and A. R. Larrabee, Vol. 14 [14].

[12] G. L. Powell, J. Elovson, and P. R. Vagelos, *J. Biol. Chem.* **244**, 5616 (1969).

[13] Long-Chain Fatty Acyl Thioesterases I and II from *Escherichia coli*

$$\text{Palmityl-}S\text{-CoA} + H_2O \rightarrow \text{Palmitate} + \text{CoA–SH}$$

By Eugene M. Barnes, Jr.

Two enzymes are present in extracts of *Escherichia coli* which catalyze the hydrolytic cleavage of fatty acyl thioesters. Palmityl thioesterase I is specific for long-chain (C_{12}–C_{18}) thioesters and is sensitive to inhibition by diisopropyl fluorophosphate (DFP).[1] Palmityl thioesterase II has a broader substrate specificity and is not blocked by DFP.[2,3]

Palmityl Thioesterase I

Assay Method

Principle. The assay is based on the liberation of the thiol group of CoA as measured by reduction of 5,5'-dithiobis(2-nitrobenzoic acid) (DTNB).[4] The reaction is measured by the increase in optical density at 412 nm.

Reagents

Potassium phosphate, 1.0 M, pH 7.4
Palmityl-CoA, 50 μM
Bovine serum albumin, 2.0 mg/ml
DTNB, 10 mM (in 0.10 M potassium phosphate, pH 7.0)

Procedure. The complete assay mixture contains 30 μmoles of potassium phosphate, pH 7.4, 7.0 nmoles of palmityl-CoA, 40 μg of serum albumin, 50 nmoles of DTNB, and 0.2–2.0 milliunits of palmityl thioesterase I in a final volume of 0.50 ml. All assay mixtures are incubated for 5 minutes at 25° before initiation of the reaction by addition of thioesterase; initial rates are measured at 25°. A Gilford recording spectrophotometer is convenient for these measurements. A molar extinction coefficient of 13,600 is used to quantitate the DTNB reduction.[4] Since DTNB also reacts slowly with the thiol groups of proteins, it is necessary to

[1] E. M. Barnes, Jr. and S. J. Wakil, *J. Biol. Chem.* **243**, 2955 (1968).
[2] E. M. Barnes, Jr., A. C. Swindell, and S. J. Wakil, *J. Biol. Chem.* **245**, 3122 (1970).
[3] W. M. Bonner and K. Bloch, *J. Biol. Chem.* **247**, 3123 (1972).
[4] G. L. Ellman, *Arch. Biochem. Biophys.* **82**, 70 (1959).

correct the initial rates of reduction by subtraction of controls in which palmityl-CoA is omitted. However, the level of palmityl thioesterase activity in crude extracts of *E. coli* is sufficiently high that these controls seldom exceed 10% of the hydrolytic rate resulting from palmityl-CoA cleavage.

Units. An enzyme unit is defined as the amount of activity necessary to catalyze the cleavage of 1 μmole of palmityl-CoA per minute under the above conditions. Specific activity is expressed as units per milligrams of protein.

Alternate Assay Procedures. Palmityl thioesterase activity may also be assayed by measuring release of pentane-soluble radioactivity from [^{14}C]palmityl-CoA[1] or by disappearance of thioester absorption at 232 nm.[3]

Purification Procedure

The following procedure, summarized in Table I, yields a 260-fold purification of palmityl thioesterase I. Resolution of thioesterase I from contaminating thioesterase II is achieved by DEAE-cellulose chromatography.[2] All operations are carried out at 4° unless otherwise noted.

Step 1. Preparation of Crude Extract. A frozen cell paste of *E. coli* E-26 (500 g, wet weight) is allowed to partially thaw and is then suspended in 1 liter of 50 mM potassium phosphate, pH 7.4. This suspension is sonically irradiated in 150 ml batches for 10 minutes with a Branson

TABLE I
PURIFICATION OF PALMITYL THIOESTERASE I

Step	Protein (mg)	Specific activity (μmole/min/mg)	Total activity (μmole/min)	Yield (%)
1. Crude extract	36,000	0.014	500	
2. DEAE-cellulose column	24,000	0.019	460	92
3. Freezing–thawing supernatant	17,000	0.023	400	80
4. DEAE-cellulose column	3,300	0.046	150	30
5. Hydroxylapatite column	330	0.290	96	19
6. Sephadex G-100 column	9	3.70	33	6.7

sonifier, maintaining the extract at temperatures below 4° in an ice–salt bath. The resulting suspension is centrifuged at 37,000 g for 30 minutes. The supernatant solution is decanted and dialyzed overnight against 8 liters of 10 mM potassium phosphate, pH 7.4, with one change of dialyzing buffer.

Step 2. First DEAE-Cellulose Chromatography. The dialyzed solution is pumped (25 ml/minute) onto a DEAE-cellulose column (6.5 × 42 cm) which has been equilibrated with 10 mM potassium phosphate, pH 7.4. The column is eluted with the same buffer and the column eluate is continuously monitored by measuring absorbance at 280 nm with a LKB Uvicord. The peak of absorbing material (about 2200 ml) is collected and solid ammonium sulfate is added with stirring to 90% saturation (60.3 g/100 ml eluate). The precipitate is collected by centrifugation, redissolved in 10 mM potassium phosphate, pH 7.4, and dialyzed overnight against the same buffer. At this point the protein concentration is about 20 mg/ml.

Step 3. Freezing–Thawing Supernatant. After the overnight dialysis, the protein solution is faintly turbid. The particulate material is coagulated by freezing the solution in an ethanol–Dry Ice bath, followed by thawing in a 10° bath. This freezing–thawing procedure is repeated twice more and the suspension is centrifuged at 78,000 g for 60 minutes and the pellet discarded. The activity in the supernatant fluid is precipitated with ammonium sulfate and dialyzed as in Step 2.

Step 4. Second DEAE-Cellulose Chromatography. The dialyzed solution is applied to a second DEAE-cellulose column (6.5 × 45 cm) and eluted as in Step 2. Fractions of 25 ml are collected and assayed for thioesterase activity. The fractions with high activity are pooled, and the activity precipitated with ammonium sulfate and dialyzed as in Step 2.

Step 5. Hydroxylapatite Chromatography. The dialyzed eluate is absorbed onto a hydroxylapatite column (4 × 45 cm) previously equilibrated with 10 mM potassium phosphate, pH 7.4. The column is eluted with a linear gradient made from 900 ml of 10 mM potassium phosphate, pH 7.4, and 900 ml of 150 mM potassium phosphate, pH 7.4. Fractions of 10 ml are collected and assayed for activity. The fractions containing high activity (about the last 400 ml of the gradient) are pooled, and the activity precipitated with ammonium sulfate and dialyzed as in Step 2.

Step 6. Sephadex G-100 Gel Filtration. The dialyzed eluate (10 ml) is layered onto a Sephadex G-100 column (4 × 75 cm, 340 ml void volume) and eluted with 10 mM potassium phosphate, pH 7.4. Fractions of 10 ml are collected and assayed for palmityl thioesterase activity. The fractions containing high activity (centered around an elution volume of 700 ml) are pooled and adjusted to 95% ammonium sulfate saturation.

The precipitate is collected by centrifugation at 78,000 g for 60 minutes. The precipitate is dissolved in 2 ml of 10 mM potassium phosphate, pH 7.4, and dialyzed for 3 hours against 4 liters of the same buffer. At this and all other stages of purification the enzyme is stored by freezing at $-20°$. The activity is stable under these conditions.

Properties

Purity. Preparations of palmityl thioesterase I with a specific activity of 3.7 units/mg are free of palmityl thioesterase II and give a single band in the ultracentrifuge. Disc gel electrophoresis (nondenaturing) on polyacrylamide[5] reveals one major component and two minor components.

Molecular Weight. The enzyme has an estimated $s^o_{20,w}$ of 2.35 $\times$ 10^{-13} second and a molecular weight of 22,000 was derived from the s value assuming a globular protein. The elution volume from Sephadex G-100 is consistent with a molecular weight of 20,000–23,000.

Requirements and Extent of Hydrolysis. At the final stage of purification, dilution of the thioesterase in the assay mixture causes a rapid loss of activity which can be prevented by the addition of bovine serum albumin or other nonenzymatic protein.[1] Incubation of palmityl-CoA with the protecting protein prior to addition of palmityl thioesterase I is necessary for the protective effect. Inactivation of the enzyme is probably the result of the surface-active properties of palmityl-CoA. This effect is not observed at stages of purification prior to Step 6. Thioesterase preparations from the earlier steps or at the final stage of purification in the presence of serum albumin will catalyze the complete conversion of palmityl-CoA to stoichiometric amounts of CoA and palmitate. Since no detectable palmityl-CoA remains after completion of enzymatic hydrolysis, the reaction is an essentially irreversible process. The DTNB usually employed in the continuous assay of the enzyme has no effect on the equilibrium position of the reaction.

Kinetic Parameters and pH Optimum. Palmityl thioesterase I has an estimated K_m of 3.9 μM and a V_{max} of 4.8 μmoles/minute per milligram of protein for palmityl-CoA substrate. These parameters are tentative because of the detergent effects of this substrate. The kinetic analysis is complicated by the fact that optimal enzymatic activity is observed at levels above the critical micellar concentration (3–4 μM) for palmityl-CoA. The optimal pH for thioesterase activity is 8.5. However,

[5] B. J. Davis, *Ann. N.Y. Acad. Sci.* **121,** 404 (1964).

assays are not routinely run at this pH owing to the slow, spontaneous hydrolysis of palmityl-CoA under these conditions.

Specificity. At saturating levels of substrate, maximal initial rates of hydrolysis are found for palmityl-CoA, with activity decreasing with either increasing or decreasing length of the saturated fatty acyl chain. At equal, but limiting concentrations, rates of cleavage of myristyl, palmityl, and stearyl thioesters are the same. Much lower activity is observed for lauryl-CoA, and no hydrolysis of saturated fatty acyl-CoA thioesters of C_{10} chain length or less is detectable. Long-chain unsaturated thioesters are also cleaved by palmityl thioesterase I. Palmitoleyl-CoA and *cis*-vaccenyl-CoA are hydrolyzed slightly more rapidly than palmityl-CoA at equivalent concentrations. Oleyl-CoA is hydrolyzed at 25% of the rate observed for the other unsaturated substrates. β-Hydroxyacyl-CoA thioesters and saturated fatty acyl glycerides are not substrates for the enzyme. Chemically acylated palmityl-acyl carrier protein (ACP) is cleaved by the thioesterase at rates about 10% of those found for the corresponding CoA derivative.

Inhibitors. Incubation of palmityl thioesterase I with 1 mM DFP (40 minutes, 25°) produces a 95% inhibition of activity. Phenylmethyl sulfonylfluoride is less effective, producing a 27% loss of activity under similar conditions. Potassium fluoride, EDTA, or N-ethylmaleimide give no significant inhibition.

Enzymatic Determination of Fatty Acyl Thioesters

Since the thioesterase-catalyzed hydrolysis of fatty acyl thioesters proceeds to quantitative completion, preparations of this enzyme are useful for assay of fatty acyl thioesters of CoA and ACP. The assay procedure described utilizing palmityl thioesterase I purified through Step 4 can be employed for determination of palmityl-CoA (or other thioesters) in amounts as low as 1 nmole. With the DTNB assay care must be taken to remove interfering free thiols; if this is not possible, the less sensitive 232 nm assay[3] can be used.

Nomenclature of Thioesterases in *E. coli*

After the initial description of the purification and properties of palmityl thioesterases I and II from *E. coli*,[1,2] Bonner and Bloch[3] reported the purification to homogeneity and properties of an enzyme they termed "fatty acyl thioesterase I." This appears to be the same enzyme that we originally termed "palmityl thioesterase II."[2]

Palmityl Thioesterase II

Assay Method

Reagents

Tris-chloride, 1.0 M, pH 8.0
Palmityl-CoA or DL-β-hydroxymyristyl-CoA, 100 μM
DTNB, 10 mM (in 0.10 M potassium phosphate, pH 7.0)

Procedure. The complete reaction mixture contains 50 μmoles of Tris-Cl, pH 8.0, 10 nmoles of palmityl-CoA or DL-β-hydroxymyristyl-CoA, 50 nmoles of DTNB, and 0.2–2 milliunits of palmityl thioesterase II in a final volume of 0.50 ml. The assay is carried out as discussed for thioesterase I. Since both thioesterases cleave palmityl-CoA, DL-β-hydroxymyristyl-CoA is utilized as substrate when both enzymes are present because this latter thioester is cleaved only by palmityl thioesterase II. A unit of palmityl thioesterase II is defined as the amount of activity necessary to catalyze the hydrolysis of 1 μmole of DL-β-hydroxymyristyl-CoA per minute under these conditions. Specific activity is defined as units per milligram of protein.

Procedure for Partial Purification

The procedure outlined in Table II yields a 20-fold purification of palmityl thioesterase II. Recently, the purification to homogeneity of an enzyme termed "fatty acyl thioesterase I" has been reported by Bonner and Bloch.[3] This appears to be the same enzyme described here as palmityl thioesterase II. The reader should refer to the scheme of Bonner and Bloch[3] if a more highly purified preparation is desired.

A crude extract of *E. coli* from 200 g (wet weight) of cells in 400 ml of buffer is prepared as described in Step 1 of the preceding scheme. The crude extract is centrifuged at 105,000 g for 1 hour and the supernatant fluid decanted. A 1% solution of protamine sulfate is added with stirring to the supernatant solution to yield a final protamine sulfate to protein ratio of 1:5 (w/w). The heavy precipitate is removed by centrifugation and the supernatant solution is adjusted to 0.10 M in potassium phosphate, pH 7.4. Solid ammonium sulfate is added to yield 50% saturation. The precipitate is collected by centrifugation (discarding supernatant solution) and dissolved in 10 mM potassium phosphate, pH 7.4. This solution is dialyzed overnight against 4 liters of the same buffer.

The dialyzed solution is applied to a DEAE-cellulose column (4 $\times$ 42 cm) which has been equilibrated with 10 mM potassium phosphate, pH 7.4. The column is eluted with a linear gradient made from 750 ml of

TABLE II
Partial Purification of Palmityl Thioesterase II

Step	Protein (mg)	Specific activity[a] (μmole/ min/mg)	Total activity[b] (μmole/min)	Yield (%)
1. Crude extract	12,000	0.0050	60	
2. High-speed supernatant	7,800	0.0062	48	80
3. Protamine sulfate supernatant	6,200	0.0073	45	75
4. Ammonium sulfate fractionation	2,600	0.014	37	62
5. DEAE-cellulose column	1,200	0.016	20	33
6. Calcium phosphate gel eluate	440	0.026	12	19
7. Sephadex G-200 column[b]	(8.8)[b]	0.102	(0.90)[b]	

[a] Determined with DL-β-hydroxymyristyl-CoA substrate.
[b] A 125-mg portion of the protein from the previous step was used as starting material for this step.

10 mM potassium phosphate (pH 7.4) and 750 ml of 0.75 M sodium chloride in the same buffer. Fractions of 20 ml are collected and assayed for thioesterase II activity (hydroxymyristyl-CoA substrate). The fractions containing high activity (about the last third of the gradient) are pooled and the activity precipitated by addition of ammonium sulfate to 80% saturation. The precipitate is collected and dialyzed as before. Calcium phosphate gel is added (2 mg gel per milligram of protein) and the mixture is stirred for 10 minutes. The gel is collected by centrifugation and washed twice with 10 mM potassium phosphate, pH 7.4. The gel is eluted with 100 ml of 75 mM potassium phosphate, pH 7.4, and collected by centrifugation. This elution procedure is repeated and the combined eluates adjusted to 85% ammonium sulfate saturation. The precipitate is dissolved in 0.10 M potassium phosphate, pH 7.4, and dialyzed 1.5 hours against 4 liters of 10 mM potassium phosphate, pH 7.4.

A portion of the dialyzed solution (5.0 ml) equivalent to 125 mg of protein is made 50 mM in potassium phosphate, pH 7.4, and 1.0 M in guanidine hydrochloride and incubated at 25° for 20 minutes. This solution is cooled to 4° in an ice bath and applied to a Sephadex G-200 column (2.5 × 86 cm, 150 ml void volume) which has been equilibrated with 1.0 M guanidine hydrochloride in 50 mM potassium phosphate, pH

7.4. The column is eluted with this same solution and fractions of 5 ml are collected. Fractions 30–35 containing material excluded from the gel matrix have low activity and are discarded. Fractions 49–74 contain high activity eluted from the gel matrix. These are pooled, dialyzed for 1.5 hours against 8 liters of 50 mM potassium phosphate, pH 7.5, and adjusted to 90% ammonium sulfate saturation. The precipitate is collected by centrifugation, redissolved in 50 mM potassium phosphate, pH 7.4, and dialyzed for 1 hour against 4 liters of the same buffer. A small amount of precipitate is removed by centrifugation and the supernatant solution is stored at −20°. The activity is stable under these conditions.

Properties

Molecular Weight. A molecular weight of 90,000 is estimated for 20-fold purified palmityl thioesterase II by the elution volume from Sephadex G-100 and G-200. Bonner and Block[3] reported that their homogeneous preparation of the native enzyme exists as a tetramer (122,000 daltons), each subunit having a molecular weight of 30,000.

Substrate Specificity. Palmityl thioesterase II will catalyze the hydrolysis of saturated, unsaturated, and β-hydroxyacyl thioesters of CoA. Palmityl, palmitoleyl, and D(−)-β-hydroxymyristyl derivatives give the highest V_{max} values in each respective series of homologues. With palmityl-CoA as substrate, an estimated K_m of 3.9 μM and a V_{max} of 0.53 μmole/minute/mg is observed. As opposed to palmityl thioesterase I, palmityl thioesterase II will cleave short-chain (C_6–C_{10}) saturated acyl-CoA derivatives and β-hydroxyacyl thioesters. Chemically acylated thioesters of ACP are cleaved at 3–7% of rates observed for corresponding CoA derivatives. Thioesters of N-acetylcysteamine and O-esters are not substrates for palmityl thioesterase II.[2,3]

pH Optimum. Palmityl thioesterase II has a pH optimum of 8.5, identical to that of thioesterase I. However, the activity of palmityl thioesterase II is stimulated to a greater degree by increases in pH from 7.0 to 8.5 than is palmityl thioesterase I; the fold increases in activity are 1.5 and 5.6, respectively, for thioesterase I and II for this pH shift.[6]

Inhibitors. Palmityl thioesterase II is unaffected by DFP, phenylmethyl sulfonylfluoride, potassium fluoride, N-ethylmaleimide, or EDTA. Bonner and Bloch[3] reported that their preparations are inhibited by photoactivated methylene blue and iodoacetamide. The iodoacetamide inactivation appears to result from esterification of an essential carboxyl group.

[6] E. M. Barnes, Jr. (1969) Ph.D. Thesis, Duke University.

[14] Acyl Carrier Protein from *Euglena gracilis*

By R. K. DiNello and M. L. Ernst-Fonberg

Assay Method

Principle. Acyl carrier protein (ACP) which carries the carboxylic acid substrates of fatty acid synthesis as thiolesters during the sequence of fatty acid biosynthesis was first described by Vagelos and co-workers.[1] ACP is measured by the malonyl-pantetheine-$^{14}CO_2$ exchange reaction[2,3] which uses the first three enzymes in fatty acid biosynthesis to fix free $^{14}CO_2$ into an acid-stable radioactive product, malonyl-pantetheine.

$$\text{Acyl-pantotheine} + \text{ACP-SH} \underset{\text{acyl transacylase}}{\overset{}{\rightleftarrows}} \text{acyl-ACP} + \text{pantotheine-SH}$$

$$\text{Malonyl-pantotheine} + \text{ACP-SH} \underset{\text{malonyl transacylase}}{\overset{}{\rightleftarrows}} \text{malonyl-ACP} + \text{pantotheine-SH}$$

$$\text{Acyl-ACP} + \text{malonyl-ACP} \underset{\text{condensing enzyme}}{\overset{}{\rightleftarrows}} \beta\text{-ketoacyl-ACP} + {}^{14}CO_2 + \text{ACP-SH}$$

Reagents

Imidazole-HCl, 1.0 M, pH 6.2, 0.2 M in 2-mercaptoethanol
Euglena fatty acid synthase, type II
ACP solution (1–4 μM)
NaH$^{14}CO_3$, 250 mM (0.2–0.6 mCi/mmole)
Perchloric acid, 10%
Malonyl pantetheine,[2] 9.0 mM
Caproyl pantetheine,[2] 2.0 mM

Procedure. To test tubes secured in a 30° circulating water bath add 25 μl each of imidazole buffer, caproyl-pantetheine, and malonyl-pantetheine; *Euglena* fatty acid synthase, type II, 25 or 50 μl (an amount which fixes 500–1000 cpm without added ACP); ACP, 25–50 μl; water sufficient for a final volume of 250 μl; and NaH$^{14}CO_3$, 25 μl, to initiate reaction. The reaction is done in a fume hood, and the tubes are stoppered during the assay with serum caps containing two 18 gauge steel needles which are used at the termination of reaction to purge the test tubes with

[1] P. W. Majerus, A. W. Alberts, and P. R. Vagelos, *Proc. Nat. Acad. Sci. U.S.* **51**, 1231 (1964).
[2] P. W. Majerus, A. W. Alberts, and P. R. Vagelos, Vol. XIV [6].
[3] R. K. DiNello and M. L. Ernst-Fonberg, *J. Biol. Chem.* **248**, 1707 (1973).

air for 15 minutes to remove the $^{14}CO_2$ which is trapped in 40% KOH. Reaction is halted after 15 minutes by the injection of 25 μl of 10% perchlorate through the serum caps. The tubes are centrifuged in a clinical centrifuge for 10 minutes, and 25 μl aliquots of the supernatants are spotted on filter paper strips (Whatman No. 1, 1×2 cm). Following drying, the papers are placed in counting vials with 15 ml of toluene containing 15.1 g of 2,5-diphenyloxazole/gallon and measured in a liquid scintillation counter.

Euglena Fatty Acid Synthase, Type II. All procedures are carried out at 4°. *Euglena gracilis* strain Z³, 50 g, are suspended in 50 ml of 0.05 M imidazole-HCl, pH 7.0, 0.01 M in 2-mercaptoethanol in a 100-ml beaker. Cells are disrupted by ultrasound in 30 second bursts while the temperature is maintained at 2–5° with a methanol-ice bath. Cell breakage is monitored microscopically. The mixture is centrifuged in a Sorvall RC-2B centrifuge, 45,000 g for 70 minutes, and the supernatant is collected and adjusted to 35% $(NH_4)_2SO_4$ saturation by the addition of 20.84 g/100 ml over a 15-minute period with stirring. After stirring for an additional 30 minutes, the precipitate is removed by centrifugation for 20 minutes at 20,000 g. The supernatant is adjusted to 70% $(NH_4)_2SO_4$ saturation by the addition of 23.61 g/100 ml over 15 minutes. After 30 minutes of stirring, the precipitate containing the desired enzymes is collected as before while the supernatant is discarded. The precipitate is dissolved in 8 ml of the imidazole buffer described above and stored in 0.6 ml aliquots at −50° until use. At the time of use, the preparation is desalted on a small Sephadex G-25 column equilibrated in and eluted with 0.05 M imidazole-HCl, pH 7.0, 0.01 M in 2-mercaptoethanol.

Purification of Acyl Carries Protein

All procedures are carried out at 4° unless noted otherwise. A total of 2045 g of *Euglena gracilis* strain Z are disrupted in eight portions. About 250 g of frozen cells are thawed in 250 ml of 0.05 M imidazole-HCl, pH 7.1, 0.07 M in 2-mercaptoethanol. The suspension is saturated with N_2 and kept at 2–5° in a methanol-ice bath. The cells are broken by ultrasound in about 14 bursts of 1 minute. Disruption is monitored microscopically. The mixture is centrifuged in a Sorvall RC-2 centrifuge at 15,000 g for 80 minutes.

The combined supernatants are adjusted to 70% $(NH_4)_2SO_4$ saturation by the addition of 47.23 g/100 ml. Following 1 hour of stirring, the precipitate is removed by centrifugation at 22,000 g for 20 minutes. The ACP containing supernatant is brought to 95% $(NH_4)_2SO_4$ saturation by the slow addition of 17.9 g/100 ml over a period of 2 hours. After stirring the covered container for 12 hours, the precipitate is collected as described above and dissolved in 48 ml of 0.01 M potassium phosphate buffer, pH 6.2, containing 0.01 M 2-mercaptoethanol (buffer A). The solution is dialyzed against 1200 ml of the same buffer through three changes of 2 hours each.

The protein solution is centrifuged to remove any denatured protein and allowed to come to room temperature. It is then applied to a DEAE-cellulose column (2.5 × 29 cm) equilibrated with buffer A at room temperature (10 mg of protein/ml of column bed). After the protein solution has soaked in, the column is washed with 4 column volumes of buffer A, and is then eluted with 9 column volumes of a linear gradient from 0.1 to 0.4 M LiCl in the same buffer. Starting with the application of the gradient, 10 ml fractions are collected and assayed for ACP (fractions 73 through 83 under these conditions) and for absorbance at 280 and 416[4] nm. The appropriate fractions are pooled and assayed for protein. The solution is divided into 2 ml aliquots in test tubes which are thrust into a 90° water bath for 4 minutes and are then cooled immediately in an ice bath. The resulting suspension is centrifuged to remove any precipitate and is then assayed for protein.

The solution (104 ml) is dialyzed against three changes of 1200 ml each of 0.01 M potassium phosphate buffer, pH 7.5, 0.01 M in 2-mercaptoethanol (buffer B). A DEAE-cellulose column, 1.2 × 5.0 cm (1 ml column bed/mg protein), is equilibrated in the same buffer. Immediately prior to application the protein solution is brought to room temperature, and the column is run at room temperature. After the sample has soaked in, the column is washed with 50 ml of buffer B. It is then eluted with 50 ml of buffer B adjusted to contain 0.1 M LiCl. Finally, a 250-ml linear gradient from 0.1 to 0.15 M LiCl in buffer B is applied and 4 ml fractions are collected. Fractions are assayed as before, and the broad peak of ACP activity (fractions 28 through 51 under these conditions) is pooled. The pooled fractions are adjusted to pH 6.2 with 1 N HCl and dialyzed against three changes of buffer A, 1 liter each. The solution is concentrated on a DEAE-cellulose column (0.8 × 1.2 cm) equilibrated with buffer A, and eluted with the same buffer adjusted to 0.35 M LiCl. The protein containing fractions are pooled (4.5 ml under these conditions).

The final step in the purification is acrylamide gel electrophoresis. Gels of 17.5% are run using the pH 9.4 system described by Davis.[5] This

---

[4] *Euglena* ACP tends to copurify with cytochrome 552, a heme protein with a molecular weight of 13,500 and an isoelectric point near 5.5.

[5] B. J. Davis, *Ann. N.Y. Acad. Sci.* **121**, 404 (1964).

TABLE I
ACP PURIFICATION

Step	Total protein (mg)	Total activity (unit)[a]	Specific activity	Purification (-fold)
Crude extract	74,000	5,904[b]	0.08	0
70–95% $(NH_4)_2SO_2$ fraction	1,440	5,904	4.1	51
DEAE-cellulose, pH 6.2	59	6,962	118.00	1,475
90° for 4 minutes	48.5	4,800	99.00	1,238
DEAE-cellulose, pH 7.5	6.3	3,070	487.00	6,088
Acrylamide gel electrophoresis	2.8	2,340	835.00	10,438

[a] A unit of activity is defined as the incorporation of 1 nmole of $^{14}CO_2$ in 30 minutes.
[b] The activity in the crude extract was too dilute to measure. For the sake of estimating the extent of purification of the protein, the total activity found in the $(NH_4)_2SO_4$ precipitate was taken as representing that present in the crude extract.

procedure may be conducted in a preparative or analytical apparatus. When analytical gels in glass tubes are run, the rigid gel is extracted most easily from the tube by wrapping the tube in cloth and gently breaking it with a hammer. In all instances, it is important to remove the ACP from the pH 9.4 environment as soon as possible either by altering the pH of the eluted protein in a preparative gel or by immediately extracting the protein from an analytical gel. In the latter case, the appropriate gel bands are excised and homogenized in buffer A. The homogenate is centrifuged and washed several times with buffer A. All supernatants are combined and dialyzed against 10 volumes of buffer A for three changes. The protein solution is concentrated on a DEAE-cellulose column as described earlier and stored at −20° or below in buffer A (see Table I).

Properties of Acyl Carrier Protein

Purity. Acrylamide (17.5%) gel electrophoresis of the final material shows a single band of protein in a pH 9.4 system[5] and in a pH 6.5 system.[6] The preparation shows three bands after the second DEAE-cellulose column. The purified ACP must be stored in the presence of 0.01 M 2-mercaptoethanol, otherwise disulfide bond formation occurs with the formation of ACP dimer.

[6] D. Rodbard and A. Chrambach, *Anal. Biochem.* **40**, 95 (1971).

Properties. The molecular weight is estimated to be 10,400 from amino acid composition and from gel filtration. There is one mole of 4'-phosphopantetheine per mole of ACP. It contains 92 amino acids (trytophan was not measured) among which there are no histidine, cystine, or tyrosine residues. *Euglena* ACP is not stable to acid precipitation nor can it tolerate long periods at pH values above 8.

Section II

Fatty Acid Activation and Oxidation

[15] Long-Chain Fatty Acyl-CoA Synthetase from Rat Liver Microsomes

EC 6.2.1.3 Fatty Acyl-CoA Ligase (ATP)

By J. BAR-TANA, G. ROSE, and B. SHAPIRO

$$ATP + RCOOH + CoASH = RCOSCoA + PP_i + AMP$$

Assay Methods

The overall activation reaction may be assayed by trapping the acyl-CoA formed either as an hydroxamate[1] or as an acyl-carnitine,[2] by CoA sulfhydryl disappearance,[3] by PP_i release,[1] by AMP release,[4] or by following the appearance of the thio-ester bond at 232 nm.[5-7]

In the present assay the formation of [³H]palmitoyl-CoA from [9,10-³H]palmitate in the presence of purified palmitoyl-CoA synthase is described.[3]

Reagents

Tris-HCl buffer M/1, pH 7.4
$MgCl_2$, 0.5 M
Triton X-100, 5%
EDTA, 0.1 M
ATP potassium salt, 0.2 M, pH 7.0
[³H]palmitate potassium salt, 0.02 M (0.25 mCi/μmole)
CoA, 20 mM
Dithiothreitol, 0.1 M
Enzyme solution
Dole's medium[8] (isopropanol, *n*-heptane, N/1 H_2SO_4, 40:10:1)
n-Heptane

Procedure. The reagent mixture is prepared as follows: Tris-HCl buffer, 0.30 ml; Triton, 0.04 ml; EDTA, 0.04 ml; $MgCl_2$, 0.20 ml; ATP,

[1] A. Kornberg and W. E. Pricer, Jr., *J. Biol. Chem.* **204**, 329–343 (1953).
[2] M. Farstad, J. Bremer, and K. R. Norum, ·*Biochim. Biophys. Acta* **132**, 492–502 (1967).
[3] J. Bar-Tana, G. Rose, and B. Shapiro, *Biochem. J.* **122**, 353–362 (1971).
[4] J. Bar-Tana, G. Rose, and B. Shapiro, *Biochem. J.* **129**, 1101–1107 (1972).
[5] B. Sjöberg, *Z. Phys. Chem., Abt. B* **52**, 209 (1942).
[6] H. P. Koch, *J. Chem. Soc., London* p. 387 (1949).
[7] D. W. Yates and P. B. Garland, *in* "Methods in Enzymology" (J. M. Lowenstein, ed.), Vol. 14, pp. 622–625. Academic Press, New York, 1969.
[8] V. P. Dole, *J. Clin. Invest.* **35**, 150–159 (1956).

0.20 ml; [³H]K-palmitate 0.02 ml; CoA, 0.06 ml; dithiothreitol, 0.02 ml; H_2O, 0.12 ml.

The reaction is carried out by adding 0.1 ml of the reagent mixture to 1–5 μg of purified palmitoyl-CoA synthase (or 20–100 μg of microsomal preparation or partially purified fraction) in a final volume of 0.2 ml at 30° for 5–10 minutes. Incubation is terminated by the addition of 1 ml of Dole's medium, 0.4 ml H_2O, and 0.6 ml of heptane. After mixing thoroughly and subsequent centrifugation, the upper phase, containing the unreacted palmitate, is withdrawn by suction and the lower phase is washed 6 times successively with 0.6 ml of n-heptane, always discarding the washings (upper phases).

The lower phase is then assayed for radioactivity by scintillation counting in toluene–ethanol 2:1 scintillating fluid. In the control experiment enzyme is added after the addition of Dole's medium.

Activity is expressed in nanomoles of palmitoyl-CoA formed per milligram of protein per minute. Using the purified enzyme fraction the rate of the forward reaction is linear up to 15 minutes in the presence of 10 μg of enzyme protein. With the freeze-dried microsomal preparation the rate is linear up to 10 minutes in the presence of 20 μg of microsomal protein. Microsomal palmitoyl-CoA hydrolase is found to be strongly inhibited under the assay conditions used and therefore does not interfere with palmitoyl-CoA synthesis.

Preparation of Rat Liver Microsomes

Enzyme Purification.[3] Unless otherwise specified all operations are carried out at 0–4°. Male Wistar rats (150–200 g wt) fed *ad libitum* are decapitated and their livers removed and homogenized for 20 seconds in the cold in 4 volumes of 0.25 M sucrose in a Waring blender. After centrifugation of the homogenate for 20 minutes at 15,000 g, the supernatant is further centrifuged at 100,000 g for 60 minutes. The precipitate obtained is suspended in 2 ml of water per gram of liver and centrifuged again at 100,000 g for 60 minutes. The washed microsomal pellet is resuspended in 0.5 ml of water per gram of liver, freeze-dried, and stored at $-15°$ (protein content from 45–55% of the dry weight).

Lipid Extraction of Freeze-Dried Microsomal Preparation

The following procedure is carried out at $-15°$. One gram portion of freeze-dried microsomal preparation is extracted with 100 ml of dry butan-1-ol in a Waring blender. The residue obtained upon centrifugation at 15,000 g for 10 minutes is treated once more with 100 ml butan-1-ol followed by 100 ml of dry acetone and is finally washed twice with 100 ml of ether. The ether washed residue is dried well in a stream of N_2 and stored at $-15°$. The protein content of this "lipid depleted microsomal preparation" ranges from 65 to 70% of the dry weight.

Deoxycholate Solubilization

A 1-g portion of lipid depleted microsomal preparation is suspended in 20 ml ice-cold 0.1 M glycyl-glycine NaOH buffer, pH 8.0, containing 0.125 M sucrose and 2.5% w/v of sodium deoxycholate. After homogenization in a Potter-Elvehjem type of homogenizer the clear solution is left in ice for 10 minutes after which it is diluted with 6 volumes of ice-cold 1.5 mM dithiothreitol and centrifuged at 100,000 for 60 minutes. The clear supernatant is dialyzed overnight against 75–100 volumes of 50 mM tris-HCl buffer pH 7.9 containing 0.1 M sucrose, 1 mM EDTA, and 0.5 mM dithiothreitol (deoxycholate extract).

Ammonium Sulfate Fractionation

To 100 ml of the above extract 4.5 g of $(NH_4)_2SO_4$ is added slowly, the pH being kept at 8.2–8.3. After centrifugation at 15,000 g for 2 minutes, the supernatant is slowly brought to pH 7.4 with 1 N HCl and centrifuged at 15,000 g for a further 20 minutes. Another 25 g of $(NH_4)_2SO_4$ are added slowly to the clear supernatant, maintaining the pH at 7.4 and centrifuged as before. The residue obtained after centrifugation is suspended in 25 ml of 0.4 M KCl in 20% "glycerol buffer" composed of 20% w/v glycerol, 50 mM tris-HCl buffer, pH 7.9 (at 4°), 0.025% (w/v) of deoxycholate. 1 mM EDTA, and 1.0 mM dithiothreitol, and is dialyzed well against 100 volumes of 0.4 M KCl in 20% w/v glycerol buffer. The dialyzed clear solution is stored at $-15°$ (ammonium sulfate fraction).

DEAE Sephadex Fractionation

The ammonium sulfate fraction is diluted with 0.4 M KCl in 20% (v/v) glycerol buffer to give a protein concentration of 15 mg/ml, and mixed for 30 minutes with a suspension of DEAE-Sephadex A-50 previously equilibrated with the same buffer. The proportion of anion exchanger to protein is kept at 0.25 ml bed volume of resin per milligram of protein. The suspension is filtered on a Büchner funnel and the residue washed with the same buffer. The filtrate and washings are then diluted with 20% (v/v) glycerol buffer to a final concentration of 0.2 M KCl and applied on a column of DEAE-Sephadex A-25 previously equilibrated with 0.2 M KCl in 20% (v/v) glycerol buffer. The activity is eluted with 0.2 M KCl in 20 % (v/v) glycerol buffer ("DEAE-Sephadex fraction").

Hydroxyapatite Fractionation

The DEAE-Sephadex fraction is then applied onto a column of hydroxyapatite packed with a 1:1 mixture (w/w) of hydroxyapatite and cellulose and equilibrated with 0.2 M KCl in glycerol buffer; 20 mg of hydroxyapatite is used per milligram of protein. Enzymatic activity is eluted stepwise by washing successively with 0.05, 0.1, 0.2, 0.4, and 0.5 M K_2HPO_4 in 3.5% v/v glycerol buffer. The fractions with the highest specific activity obtained by the 0.2–0.5 M K_2HPO_4 elution steps are pooled and dialyzed against 3.5% v/v glycerol buffer and concentrated by precipitation with 40% w/v $(NH_4)_2SO_4$. The precipitate is dissolved in a minimal volume of 3.5% glycerol buffer.

Gel Filtration

The hydroxyapatite fraction is applied onto a column (150 $\times$ 2.5 cm) of Sepharose 6B previously equilibrated with 3.5% glycerol buffer. The enzymatic activity is eluted with 3.5% v/v glycerol buffer as the eluent. The fractions containing the highest specific activity are pooled and concentrated by precipitation with ammonium sulfate (38% w/v). The gel filtrated fraction is then reapplied onto the Sepharose 6B column and eluted with 20% v/v glycerol buffer minus the deoxycholate. The fractions with the highest specific activities are pooled, dialyzed extensively against 100 volumes of the same buffer, divided into small portions, and stored at −15° ("purified enzyme fraction").

The results of a typical purification procedure are summarized in Table I.[9,10]

Properties

Stability. The Sepharose 6B fraction (purified enzyme fraction) is stable at protein concentrations greater than 5 mg/ml at −15°. Activity

[9] O. H. Lowry, N. J. Rosebrough, A. L. Farr, and R. J. Randall, *J. Biol. Chem.* **193**, 265 (1951).

[10] O. Werburg and W. Christian, *Biochem. J.* **310**, 384 (1972).

TABLE I

PURIFICATION OF MICROSOMAL PALMITOYL-CoA SYNTHASE

	Total protein[a] (mg)	Specific activity (nmoles of palmitoyl-CoA formed/min/mg of protein)	Total activity (μmoles of palmitoyl-CoA formed/min)
Freeze-dried microsomal preparation	4400	17.7	78.5
Lipid-depleted microsomal preparation	4300	17.7	76.0
Deoxycholate extract	4200	18.0	76.0
Ammonium sulfate fraction	2250	35	78.5
DEAE-Sephadex fraction	650	107	69.5
Hydroxyapatite fraction	210	171	36
Sepharose 6B fraction	51	450	23

[a] Protein is estimated by Lowry et al.[9] and in the more purified fractions using the ratio of extinctions E_{280}/E_{260} based on a method by Warburg and Christian.[10]

decreases progressively on storage at 2–4°. Thus, speedy purification procedure results in maximal enzyme specific activity values.

Purity. The purified enzyme fraction is homogeneous, showing a single peak of constant specific activity upon gel filtration with Sepharose 6B. Electrofocusing in the 1–9 and 4–7 pH range similarly results in a single peak of constant specific activity with an isoelectric point of 5.5. The latter method may be used to remove traces of deoxycholate bound to the enzyme protein. Palmitoyl-CoA hydrolase or deacylase, ATPase, pyrophosphatase, and α-glycerophosphate palmitoyl-CoA transacylase activities are virtually absent in the purified enzyme fraction.

Structural Properties

The molecular weight of the active protein unit is 168,000 with extensive self-association into higher multiples of the above MW. In the presence of 6.4 M guanidine-HCl and 0.1 M mercaptoethanol the protein dissociates into subunits of 28,000 MW as well as multiples of this MW up to 168,000. Amino acid analysis reveals one half-cystine for every 28,000 MW subunit. Free fatty acid as well as phospholipid are bound to the purified enzyme fraction.[11]

pH Range. The pH activity of the purified enzyme fraction ranges from 6.4–10.6 with an optimum value of pH 8.2–8.4.

[11] J. Bar-Tana and G. Rose, *Biochem. J.* **131**, 443–449 (1973).

Substrate Requirements and Specificity

The forward reaction is strictly specific for ATP.[3] With regard to CoA, only phosphopantetheine is an active analogue to the extent of less than 10%.[3,4] Mg^{2+} and the presence of detergent (Triton X-100) are essential for activity. The optimum ATP/Mg^{2+} ratio is 1:2.5. Specificity for the fatty acid substrate covers the range of C_{12}–C_{18} saturated as well as unsaturated fatty acids, palmitate exhibiting maximal activity.[3] K_m values for the reactants of the forward and the reverse reactions are as follows: ATP, 4650 μM; palmitate, 42 μM; CoA, 50 μM; palmitoyl-CoA, 2600 μM; AMP, 630 μM (apparent); and PP, 890 μM (apparent).

[16] L-3-Hydroxyacyl Coenzyme A Dehydrogenase from Pig Heart Muscle

EC 1.1.1.35 L-3-Hydroxyacyl-CoA:NAD oxidoreductase

By RALPH A. BRADSHAW and BARBARA E. NOYES

$$L(+) \; R—\overset{\overset{\displaystyle OH}{|}}{C}H—CH_2—\overset{\overset{\displaystyle O}{||}}{C}—S—CoA + NAD^+ \rightleftharpoons$$

$$R—\overset{\overset{\displaystyle O}{||}}{C}—CH_2—\overset{\overset{\displaystyle O}{||}}{C}—S—CoA + NADH + H^+$$

R = alkyl group.

This mitochondrial enzyme participates in the β-oxidation of fatty acids. It can be distinguished from acetoacetyl-CoA reductase (D-3-hydroxyacyl-CoA:NADP oxidoreductase, EC 1.1.1. 36) by the stereochemical configuration of the hydroxyacyl substrate and the principal cofactor utilized.[1] Purified preparations of the enzyme from sheep[2] and beef liver[3] and pig heart[4-6] have been reported.

[1] S. J. Wakil, *in* "The Enzymes" (P. D. Boyer, H. Lardy, and K. Myrbäck, eds.), 2nd ed., Vol. 7, p. 97. Academic Press, New York, 1963.

[2] F. Lynen and O. Wieland, Vol. 1 [94].

[3] S. J. Wakil, D. E. Green, S. Mii, and H. R. Mahler, *J. Biol. Chem.* **207**, 631 (1954).

[4] J. R. Stern, *Biochim. Biophys. Acta* **26**, 448 (1957).

[5] M. Grassl, Ph.D. Thesis, Ludwig-Maximillian University, Munich, Germany (1957).

[6] B. E. Noyes and R. A. Bradshaw, *J. Biol. Chem.* **248**, 3052 (1973).

Assay Method

Principle. The activity of the enzyme is routinely measured by monitoring the decrease in absorption of NADH at 340 nm which accompanys the conversion of the S-acetoacetyl ester to the corresponding β-hydroxy compound. Either S-acetoacetylpantetheine or S-acetoacetyl-CoA can be used as substrate. However, because of expense and the more pronounced product inhibition encountered with the CoA ester,[6] S-acetoacetylpantetheine is the substrate choice.

Reagents

S-Acetoacetylpantetheine,[7] .025 M, pH 5
NADH, 5 μmoles/ml in 10^{-5} M NaOH
Sodium pyrophosphate buffer, 0.0125 M, pH 7.3

Procedure. Buffer (4.2 ml) and substrate (0.2 ml) for 10 assays are mixed and equilibrated in a constant temperature bath at 25°. An aliquot of 0.44 ml of this mixture is pipetted into a quartz cuvette with a 2-mm path length, and 0.05 ml of a freshly prepared solution of 5 mM NADH is added. The reaction is initiated with the addition of 0.01 ml of appropriately diluted enzyme solution. The decrease in absorption at 340 nm remains linear with time for at least 4 minutes. Best results are obtained utilizing a spectrophotometer and recorder equipped with a scale expander to give a full range scale of 0.5 optical density unit. The final concentration of S-acetoacetyl pantetheine is 1 mM and NADH, 0.5 mM. Assays utilizing S-acetoacetyl-CoA are carried out in a final volume of 1 ml with a final concentration of NADH of 0.25 mM.

Units. One unit of enzymatic activity is equivalent to 1 μmole of S-acetoacetylpantetheine reduced per minute, based on a molar extinction coefficient for NADH of 6.22 $\times$ 10^3 at 340 nm.[8] Specific activity is expressed as units of activity per milligram of protein.

Purification Procedure

The purification scheme described[6] can be carried out with 80 hearts (20 kg). However, the initial stages (steps 1–4) are best performed in batches of 12–15 hearts. All buffers contained 1 mM EDTA and 2 mM β-mercaptoethanol, and the entire preparation was carried out at

---

[7] S-Acetoacetylpantetheine can be prepared by the reaction of pantetheine and diketene as described in ref. 6. The substrate solution is stored at −20°, pH 5. S-Acetoacetyl-CoA and NADH are available from Sigma Chemical Co., St. Louis, Mo.

[8] B. L. Horecker and A. Kornberg, *J. Biol. Chem.* **175**, 385 (1948).

0–4°. Specific activities are based on protein determined by the micro-biuret reaction.[9]

Step 1. Homogenization. Fresh pig hearts are washed free of blood, trimmed of fat and connective tissue, and passed through a commercial meat grinder in lots of 5–6. The ground tissue is suspended in 2.25–2.5 liters of cold 0.01 M potassium phosphate buffer, pH 7.0, and homogen-ized in a 1-gallon Waring blender for approximately 2 minutes at low speed and 3–5 minutes at high speed. The temperature of the homogenate is kept below 20°. This material is acidified with 1 N acetate buffer pH 4.0 (usually 30–40 ml) to pH 5.0 to 5.5 and centrifuged for 20 minutes at 10,000 g. The supernatant is filtered through cheesecloth and adjusted to pH 7.0 with 1 N NaOH. The precipitate contains insignificant amounts of enzyme and is discarded without further treatment. In a representative experiment, the total volume of supernatant from 80 hearts was 35.9 liters and contained 528×10^3 units of enzymatic activity. The specific activity of this stage is 1.2 (Table I).

Step 2. Ammonium Sulfate Fractionation. The clear red supernatant is taken to 90% saturation by the slow addition (over 6 hours) of 603 g of solid $(NH_4)_2SO_4$ per liter of supernatant, and the mixture is stirred slowly overnight. The precipitate is collected by centrifugation at 10,000 g for 20 minutes, and the resulting supernatant, which contains less than 5% of the activity, is discarded. The precipitate from 12 to 15 hearts is dissolved in about 1 liter of 0.02 M potassium phosphate buffer, pH 7.0, and dialyzed against several changes (about 5) of the same buffer followed by 2 changes of 0.02 M potassium phosphate buffer, pH 6.0. The precipitate formed is removed by centrifugation prior to chromatography.

Step 3. First CM-Cellulose Chromatography. The dialyzed sample, containing approximately 60 g of protein in less than 1 liter, is pumped onto a column (5.3×60 cm)[10] containing fibrous CM-cellulose (medium mesh) equilibrated at 4° with 0.02 M potassium phosphate buffer, pH 6.0. The sample is applied and the column developed at a flow rate of 120 ml/hour maintained by a peristaltic pump. Following sample applica-tion, 0.02 M potassium phosphate buffer, pH 6.0 (about 350 ml), is pumped through the column until all of the sample, as judged by the red color, has entered the bed. The column is eluted stepwise with three linear gradients, the first two differing only in the volume of buffer used. The first gradient consists of 500 ml each of 0.02 M potassium phosphate,

---

[9] R. F. Itzhaki and D. M. Gill, *Anal. Biochem.* **9**, 401 (1964).
[10] The column used in this step had a bed volume of 900 ml, leaving a 300–400-ml buffer reservoir above the bed. This geometry facilitates the loading of the large sample volume.

pH 6.0, and 0.02 M potassium phosphate, pH 7.0, and the second gradient of 400 ml each of the same two buffers. The third gradient is formed from 400 ml each of 0.04 M potassium phosphate, pH 7.0, and 0.04 M potassium phosphate, pH 7.5. At the end of the third gradient, the column is washed with 0.04 M potassium phosphate, pH 7.5, for an additional 16–20 hours. The enzyme, which elutes between volumes of 3 and 4.5 liters, is usually only incompletely resolved from mitochondrial malate dehydrogenase and mitochondrial aspartate aminotransferase. However, if the enzymatic activity is pooled to omit the leading shoulder observed, most of these contaminants are eliminated.[11] Depending on how the fractions are pooled, the yield of enzyme after this step is about 50% with a specific activity of 26.

Step 4. Gel Filtration. Pooled enzyme from step 3 is concentrated to 50 mg/ml (∼35 ml) (using an Amicon concentrator with PM-30 membrane or equivalent procedure) and applied to a column (5 × 125 cm) of Sephadex G-100. The column is eluted at 30 mg/hour with 0.04 M potassium phosphate, pH 7.0. The enzyme is eluted between 300 and 500 ml and is generally freed of the red proteins, which emerge later on the gel filtration column. Enzyme, which possesses a specific activity at this stage of 110, is dialyzed against 0.02 M potassium phosphate buffer, pH 7.0.

Step 5. Second CM-Cellulose Chromatography. At this step, enzyme purified from 40 hearts (half the preparation) can be pooled. The dialyzed sample is applied to a column (2.5 × 45 cm) of CM-cellulose (Whatman CM-52) equilibrated in 0.02 M potassium phosphate, pH 7.0. Buffer is pumped through the column at a rate of 30 ml/hour for 5 hours after the sample is applied. Elution is accomplished with a linear gradient of 250 ml each of 0.02 M potassium phosphate, pH 7.0, and 0.05 M potassium phosphate, pH 7.5. As in step 3, a leading shoulder is often observed. The main peak, which is the last to be eluted, has a specific activity of about 173 with an overall recovery (main peak) of 23%.

Step 6. Third CM-Cellulose Chromatography. Enzyme from the entire preparation (80 hearts) can be combined at this stage for the final purification step. The pooled fractions from step 5 are dialyzed against 0.02 M potassium phosphate, pH 7.0, and chromatographed on CM-cellulose (Whatman CM-52) under the same conditions used in step 5. Enzyme obtained from this step has a specific activity of 176. Some variation

---

[11] The mitochondrial and cytoplasmic forms of malate dehydrogenase and aspartate aminotransferase can be isolated through procedures employing the same initial steps as outlined here for L-3-hydroxyacyl CoA dehydrogenase [B. E. Glatthaar, G. R. Barbarash, B. E. Noyes, L. J. Banaszak, and R. A. Bradshaw, *Anal. Biochem.* **57**, 432 (1974)].

TABLE I

PURIFICATION OF L-3-HYDROXYACYL-CoA DEHYDROGENASE
FROM 20 KG OF PIG HEART MUSCLE

Procedure	Volume (ml)	Total activity[a] ($\times 10^{-3}$ units)	Protein[b] (mg/ml)	Specific activity (unit/mg)	Yield (%)
1. Supernatant of acid precipitation	35,930	528	12.3	1.2	100
2. 90% $(NH_4)_2SO_4$					
(a) Precipitate	4,660	470	86	1.2	89
(b) Supernatant	44,345	25	1.7	0.3	5
3. First CM-cellulose chromatography	3,530	269	2.88	26	51
4. Gel filtration	622	195	2.85	110	37
5. Second CM-cellulose chromatography	115	119	5.98	173	23
6. Third CM-cellulose chromatography	100	86.4	4.91	176	16

[a] One unit = 1 μmole S-acetoacetylpantetheine reduced per minute.
[b] Determined using the microbiuret method.[9]

in the improvement in specific activity rendered by the third CM-cellulose column has been noted in different preparations. However, this step is recommended for experiments requiring completely homogeneous enzyme. In a typical experiment, 490 mg of L-3-hydroxyacyl-CoA dehydrogenase was recovered from 80 hearts (Table I). Long-term storage is achieved by bringing the pooled enzyme to 90% saturation with $(NH_4)_2SO_4$ and storing at 4°.

Crystallization of the enzyme, which is described below, is somewhat variable and has not been carried out on a routine basis because it results in insignificant increases in specific activity.

Step 7. Crystallization. Crystals can be obtained following either step 5 or 6. The enzyme is concentrated to 8 mg/ml as in step 4 and taken to 50% saturation by adding saturated $(NH_4)_2SO_4$ solution (0.1 M potassium phosphate, pH 6.8; 1 mM EDTA; and 2 mM β-mercaptoethanol) dropwise with stirring. Upon standing in the cold (4°) overnight, a small amount of amorphous precipitate forms which can be removed by centrifugation. An additional 0.5 ml of the saturated $(NH_4)_2SO_4$ solution is added to the protein solution (22 ml) and after 2 days at 4°, crystals are observed.[6] Crystals more suitable for X-ray analysis have been obtained by vapor-phase equilibration under similar conditions.[12]

[12] M. S. Weininger, B. E. Noyes, R. A. Bradshaw, and L. J. Banaszak, in preparation.

Properties of the Enzyme

Stability. Enzyme stored in 90% $(NH_4)_2SO_4$ at 4° retains activity for at least 6 months, although some decrease in specific activity is observed after long periods of storage. The maximum specific activity which has been observed for freshly prepared enzyme assayed with acetoacetyl-panetheine is 220 units/mg. This value may vary slightly with different batches of substrate.

Specificity. L-3-Hydroxyacyl-CoA dehydrogenase utilizes the *N*-acetylcysteamine, pantetheine, and CoA esters of acetoacetate, as well as various deoxy and dephospho derivatives of the last two, as substrates.[5,6] The K_m values calculated from double reciprocal plots are 10, 0.4, and 0.06 mM, respectively, for the *N*-acetylcysteamine, pantetheine, and CoA derivatives. From these observations, Grassl[5] concluded that the 2,4-dihydroxy-3-dimethyl butyric acid portion of the pantetheine moiety is the most important part of the substrate with respect to binding. In addition, the enzyme shows no specificity for chain length, acting on C_4 to C_{12} and higher length substrates at about equal rates.[1-4]

Molecular Weight and Subunit Structure. The molecular weight of the enzyme has been estimated to be 65,000 ± 4,000 by sedimentation equilibrium measurements.[13] Comparable although somewhat higher values were obtained from gel filtration techniques. In the presence of denaturing agents (guanidine-HCl or sodium dodecyl sulfate), a value of about 32,000 was obtained, indicating the presence of two polypeptide chains. End group analysis and tryptic peptide mapping established that the primary structure of the subunits is very similar or identical.[13]

Amino Acid Composition. Amino acid analysis of the enzyme as determined from acid hydrolysates indicated the presence of 618 amino acids.[6] By other techniques, only 2 residues each of half-cystine and tryptophan (or one per subunit) were found. Titration with *p*-chloromercuribenzoate showed that both half-cystinyl residues are present in the reduced form, thus eliminating the possibility of any disulfide bridges.[6] The low tryptophan content is consistent with an observed absorption maximum at 278 nm and an $E_{278\ nm}^{1\%} = 4.39$.[6] As can be judged from its behavior on CM-cellulose, the protein is basic in character. The isoelectric point measured by isoelectric focusing is 8.95.

Inhibition. The enzyme is completely inactivated by exposure to urea and guanidine · HCl. In the former case, it has been shown[13] that inactivation accompanies polypeptide chain unfolding between concentrations of 2 and 3 M. Up to 94% of the original activity of the enzyme could be recovered by dialyzing the protein at 4° against 0.1 M sodium pyro-

[13] B. E. Noyes and R. A. Bradshaw, *J. Biol. Chem.* **248**, 3060 (1973).

phosphate, pH 7.3, containing 0.01 M β-mercaptoethanol. Under these conditions, substrate and cofactor are without effect.

Modification of the single thiol group of each subunit with any reagent tested results in inactivation of the enzyme.[14] Specifically, iodoacetamide, iodoacetic acid, N-ethylmaleimide, p-chloromercuribenzoate, and 5,5'-dithiobis(2-nitrobenzoic acid) produce inactive enzyme. With the exception of N-ethylmaleimide, modification was accompanied by precipitation, suggesting that loss of activity results from denaturation of the protein rather than alteration of an active site residue. The single thiol peptide of L-3-hydroxyacyl-CoA dehydrogenase has been isolated from a tryptic digest of protein labeled with iodo-[14C]acetic acid and shown to have the sequence: His–Pro–Val–Ser–Cys–Lys. This sequence shows no obvious relationship to any cysteinyl-containing peptide isolated from any other dehydrogenase.[13]

Modification of the single tryptophan residue with N-bromosuccinimide results in inactivation and precipitation of the enzyme. However, modification with (2-hydroxy-5-nitrobenzyl) sulfonium bromide results in the incorporation of two HNB groups per mole tryptophan with no loss of enzymatic activity, indicating that this residue also has no catalytic role in the enzyme.[14]

Acknowledgments

The studies described were supported by a research grant from the National Institutes of Health (AM-13362). B.E.N. was supported by a U.S. Public Health Service traineeship (GM-1311) and R.A.B. by a Research Career Development Award (AM-23968).

[14] B. E. Noyes, Ph.D. Thesis, Washington University, St. Louis, Missouri (1973).

[17] 3-Ketoacyl-CoA Thiolases of Mammalian Tissues

By B. Middleton

3-Ketoacyl-CoA thiolase activity is widespread in animal tissues[1] reflecting the widespread involvement of the reaction RCH$_2$COCH$_2$CO—SCoA + CoA $\rightleftharpoons$ RCH$_2$CO—SCoA + CH$_3$CO—SCoA in the β-oxidation of fatty acids, in ketone body metabolism, and in steroid biosynthesis.

[1] G. Hartmann and F. Lynen in "The Enzymes" (P. D. Boyer, H. Lardy and K. Myrback, eds.), Vol. 5, p. 381. Academic Press, New York, 1961.

In accordance with this widespread involvement it has been shown[2] that more than one subcellular location exists for 3-ketoacyl-CoA thiolase activity in most mammalian tissues. Cytoplasmic thiolase activity is widely distributed among different tissues and is the result of an aceto-acetyl-CoA-specific thiolase. Mitochondrial thiolase activity results from two different types of enzyme whose relative amounts vary with the tissue. A 3-ketoacyl-CoA thiolase of general specificity for the ketoacyl-CoA substrate constitutes one type, the other being a specific acetoacetyl-CoA thiolase which differs from its cytoplasmic counterpart in being greatly stimulated by K+. This article describes a general method[2] for the separation, quantitation, and partial purification of these different enzymes and describes their properties.

Assay Method

Principle. In general, 3-ketoacyl-CoA thiolase activity is determined by following the stimulation of 3-ketoacyl-CoA breakdown (measured by the absorption of the enol form at 303 nm[3]) caused by the addition of CoA. Mg^{2+} is included to increase the apparent extinction of the 3-ketoacyl-CoA substrate[4] and thus increase the sensitivity of the assay. Mg^{2+} itself has no effect on 3-ketoacyl-CoA thiolase activity. As mentioned above all the ketoacyl-CoA thiolases commonly present in a given tissue can use acetoacetyl-CoA as a common substrate but they differ in other, easily measurable parameters. The parameters used here to give a rapid qualitative identification of the types of thiolase present in a tissue extract are (a) the ratio of activity with 3-ketohexanoyl-CoA to activity with acetoacetyl-CoA (in the presence of K+), and (b) the ratio of activity (with acetoacetyl-CoA) in the presence of K+ to the activity (with the same substrate) in the presence of Na+.

Reagents

Tris-HCl, pH 8.1, 100 mM containing 25 mM $MgCl_2$ and 50 mM KCl or NaCl (see below)

Acetoacetyl-CoA or longer chain 3-ketoacyl-CoA, 4 mM

Acetoacetyl-CoA is prepared by the reaction of CoA with diketen,[5] the longer chain 3-ketoacyl-CoA substrates are prepared by the method of Vagelos and Alberts.[6]

[2] B. Middleton, *Biochem. J.* **132**, 717 (1973).
[3] F. Lynen, *Fed. Proc., Fed. Amer. Soc. Exp. Biol.* **12**, 683 (1953).
[4] J. R. Stern, *J. Biol. Chem.* **221**, 33 (1956).
[5] T. Wieland and L. Rueff, *Angew. Chem.* **65**, 186 (1953).
[6] P. R. Vagelos and A. W. Alberts, *Anal. Biochem.* **1**, 8 (1960).

CoA, 10 mM, freshly prepared by dissolving the free acid form in deionized water.

Procedure.[2] (a) Standard assay. A 1-cm light path silica cuvette is set up containing the following reagents at 30°: 100 mM tris-HCl, pH 8.1, 25 mM MgCl$_2$, 50 mM KCl, 10 μM 3-ketoacyl-CoA, and enzyme in a total volume of 1.99 ml. The absorption of the enol form of the 3-ketoacyl-CoA is observed at 303 nm and any breakdown resulting from thiolesterase activity in the enzyme preparation is measured. This activity is generally less than 1% of the 3-ketoacyl-CoA thiolase activity in tissue extracts. On the addition of CoA (10 μl of 10 mM) the increase in 3-ketoacyl-CoA breakdown measures the thiolase activity of the preparation. The equilibrium position of the reaction under these conditions favors the thiolysis of the 3-ketoacyl-CoA ($K_{eq} = 7.8 \times 10^3$ in the direction of thiolysis[7]).

Under these standard assay conditions the apparent extinction coefficient of the 3-ketoacyl-CoA substrates are acetoacetyl-CoA, 16.9×10^3 liter, mole^{-1}; 3-ketohexanoyl-CoA, 15.6×10^3 liter mole^{-1}; 3-ketodecanoyl-CoA, 13.6×10^3 liter mole^{-1}.

(b) Measurement of activation by K$^+$. Ketoacyl-CoA thiolases that can use 3-ketohexanoyl-CoA and longer chain substrates are unaffected by the type of monovalent cation present. The effect of K$^+$ on thiolase activity is therefore measured using acetoacetyl-CoA as substrate. The thiolase activity is first measured in the presence of 50 mM KCl (as above) and then redetermined in an assay system in which the KCl is replaced by an equal concentration of NaCl. The activation (if any) is expressed as a ratio of the rate in the presence of K$^+$ to the rate in the presence of Na$^+$. When no activation occurs the ratio is unity. Na$^+$ is used as reference cation because it neither activates nor inhibits the K$^+$ sensitive thiolase of animal tissues. It is particularly important to insure that K$^+$ (or NH$_4^+$, which also activates) is completely absent from all solutions, substrates, or enzyme because the activation occurs at relatively low cation concentrations (see below).

Chromatographic Separation of 3-Ketoacyl-CoA Thiolases[2]

Principle. The method involves the preparation of a whole tissue extract containing all the mitochondrial and cytoplasmic thiolase activity followed by the quantitative adsorption of the cytoplasmic thiolase activity on DEAE-cellulose. This cytoplasmic thiolase can then be collected on elution. The mitochondrial thiolases are not adsorbed on to DEAE-

[7] B. Middleton, *Biochem. J.* (1974) (in press).

cellulose under the conditions used and, after passage through the DEAE-cellulose, they are adsorbed onto a cellulose phosphate column. Gradient elution separates the two mitochondrial thiolases and enables their quantitative recovery from the cellulose phosphate.

This chromatographic sequence is a general method, having given good results for the separation of thiolases from five different tissues obtained from three animal species. The chromatographic separation of mitochondrial and cytoplasmic thiolases is preferred to a subcellular fractionation because it gives no possibility of the cross contamination (as a result of mitochondrial damage) often associated with subcellular fractionation.

Preparation of Tissue Extracts. In general, the 3-ketoacyl-CoA thiolase activities of mammalian tissues are not affected by freeze-thawing and this is therefore carried out as a preliminary disruptive step. Tissue is then homogenized in 5 volumes of 100 mM sodium phosphate, 0.5 mM dithiothreitol, pH 7.2, in a Polytron overhead blender[8] run at full speed for 3×1 minute periods with cooling to below 10°. If the tissue has not been previously freeze-thawed then complete disruption is insured by the addition of Triton X-100 to 0.5% (w/v). This detergent has no effect upon any of the thiolases and does not interfere with the subsequent chromatographic procedures. The homogenate is then centrifuged at 100,000 g for 30 minutes and the clear supernatant retained. The supernatant is then gel-filtered (Sephadex G-25) at 18° into 100 mM tris-HCl, pH 8.2, containing 30% (v/v) glycerol. This extract is used for the separation of mitochondrial and cytoplasmic thiolases.

Quantitative Isolation of Cytoplasmic Thiolase. This enzyme is separated from the mitochondrial thiolases by passing the gel-filtered tissue extract down a column of DEAE-cellulose equilibrated at 18° with the same tris-HCl buffer used before. Not more than 15 mg of protein (at 25 mg/ml) is applied per milliliter of packed DEAE-cellulose. The column has a length to diameter ratio of 2:1. After loading the sample the column is washed with the same buffer and the unretarded material is collected; this contains all the mitochondrial thiolase activity. The cytoplasmic thiolase is then eluted from the column by 300 mM tris-HCl, pH 8.2, containing 0.5 mM dithiothreitol and 30% (v/v) glycerol. The overall yield of cytoplasmic thiolase is better than 90% but may drop below this value if the glycerol and dithiothreitol are omitted. In general, the cytoplasmic thiolase is purified 4- to 10-fold by this procedure depending on the source used. It contains no mitochondrial thiolase activities as measured by the lack of activation by K[+] and the absence of any activity with 3-ketohexanoyl-CoA as substrate.

[8] Kinematica G.m.b.H., Lucerne, Switzerland.

Quantitative Isolation of Individual Mitochondrial Thiolases. The unretarded material from the DEAE-cellulose column is first gel-filtered (Sephadex G-25) into 2 mM sodium phosphate, pH 6.6, containing 30% glycerol. This is then applied to a short column (length:diameter = 2:1) of cellulose phosphate equilibrated at 18° with the same buffer. The maximum loading used is 15 mg of protein/ml of packed volume of exchanger, the protein being applied at a concentration of up to 20 mg/ml. The column is washed with the same buffer before starting a linear gradient of 20–500 mM sodium phosphate, pH 6.6, containing 30% (v/v) glycerol. The gradient volume is 6 times that of the packed volume of the exchanger. The general 3-ketoacyl-CoA thiolase (identified by high relative activity with 3-ketohexanoyl-CoA and by its lack of activation by K$^+$ when acetoacetyl-CoA is the substrate) is eluted as a sharp peak emerging at 180–200 mM phosphate. The other mitochondrial enzyme, the K$^+$-activated acetoacetyl-CoA–specific thiolase emerges later at 300–400 mM phosphate. Thus, the mitochondrial thiolase elution profile from cellulose phosphate appears in general as two separate peaks when assayed with acetoacetyl-CoA in the presence of K$^+$. The total recovery of thiolase activity is better than 90%. The mitochondrial thiolases are considerably purified by these two sequential chromatographic procedures.

Thus, the general ketoacyl-CoA thiolase from ox liver is obtained with a specific activity of 7.9 μmoles of 3-ketohexanoyl-CoA removed per minute per milligram[9] and the acetoacetyl-CoA–specific mitochondrial thiolase from the same source is purified to a specific activity of 12.4 μmoles of acetoacetyl-CoA removed per minute per milligram.[9] These represent purifications of 17-fold and 53-fold, respectively, for the two mitochondrial enzymes when compared with their activity in liver extracts. The mitochondrial thiolases can be stored at −20° for several weeks without activity loss provided that the glycerol concentration is sufficiently high to prevent freezing.

Determination of the Tissue Levels of Individual 3-Ketoacyl-CoA Thiolases. The chromatographic methods described above give good yields of the individual thiolases and can be used for determining their tissue levels. Table I gives their tissue activities with acetoacetyl-CoA since this is the substrate common to all the thiolases. It can be seen that, in terms of the assay described here, the cytoplasmic acetoacetyl-CoA thiolase can contribute up to 81% of the total tissue thiolase activity in the brains of newborn rats and yet in other tissues (e.g., rat heart) the activity of this enzyme is <0.25% of the total tissue thiolase. The contributions of the two mitochondrial ketoacyl-CoA thiolases to the

[9] B. Middleton, *Biochem. Biophys. Res. Commun.* **46**, 508 (1972).

TABLE I

TISSUE CONTENTS OF INDIVIDUAL THIOLASES ISOLATED AND
SEPARATED BY COLUMN CHROMATOGRAPHY[a]

| Tissue | Cytoplasmic acetoacetyl-CoA thiolase[b] | Mitochondrial thiolases | |
		Acetoacetyl-CoA thiolase[b]	General ketoacyl-CoA thiolase[b]
Rat brain (0–1 day)	4.0	0.95	<0.1
Rat brain (adult)	1.50 ± 0.33(3)	1.93 ± 0.16(3)	<0.1
Rat muscle (hind limb)	0.10 ± 0.04(3)		
Rat stomach	0.63 ± 0.05(3)		
Rat heart	<0.05 (3)	18.6	2.7
Rat kidney	0.40 ± 0.12(4)	18.2	2.5
Rat ileum	1.33 ± 0.07(4)		
Rat liver	3.55 ± 0.43(10)	15.9 ± 2.8(3)	7.6 ± 1.7(3)
Rat adrenal	4.46 ± 0.77(4)		
Rat testis	0.66 ± 0.10(3)		
Pig heart	<0.05		
Ox heart	<0.05		
Ox liver	5.50 ± 0.57(6)	19.8 ± 2.5(3)	16.7 ± 2.1(3)
Ram sperm[c]		6.5	3.7

[a] From Middleton.[2]

[b] For purposes of comparison all the tissue activities are expressed as micromoles of acetoacetyl-CoA removed per minute per gram fresh weight of tissue and where applicable are means ± S.E.M. with the number of observations in parentheses.

[c] B. Middleton and T. Mann, unpublished results.

total thiolase activity shown by different tissues are also very variable and their relative activities are by no means related. In general, the highest tissue activities of all three thiolases are found in liver.

It must be noted that the concentrations of acetoacetyl-CoA and CoA used in the standard assay have been chosen to give good activities for all the three types of thiolase. For kinetic reasons (see the section below) it is not possible to fix the substrate concentrations in order to measure all three enzymes at their maximum velocities. Thus, the tissue activities quoted in Table I are not measures of maximum tissue thiolase capacities. They are, however, perfectly valid for the comparison of relative tissue thiolase activities.

Properties of the Individual 3-Ketoacyl-CoA Thiolases

These will be given under the headings of the particular type of thiolase.

(a) *Cytoplasmic Acetoacetyl-CoA Thiolase*

This enzyme has been isolated and partly purified from the sources shown in Table I and, in addition, the enzymes from chicken[10] and rat liver[7] have recently been purified to homogeneity.

Substrate Specificity. As summarized in Table II cytoplasmic thiolases are absolutely specific for acetoacetyl-CoA; longer chain ketoacyl-CoA compounds are not substrates. Specificity for the CoA moiety is not absolute; the enzyme from rat liver will use pantetheine and acetoacetyl-pantetheine as substrates.[7]

Cation Activation. None

Kinetic Properties. All the cytoplasmic thiolases listed in Table I, together with the enzyme from the chicken liver,[10] are strongly inhibited by CoA (substrate inhibition) at concentrations above 50 μM. For the rat liver enzyme,[7] in the direction of acetoacetyl-CoA breakdown the kinetic parameters (at infinite concentration of the invariant substrate) are K_m (acetoacetyl-CoA) = 33 μM, K_m (CoA) = 15 μM, and K_i (CoA, competitive with acetoacetyl-CoA) = 73 μM. In the direction of acetoacetyl-CoA synthesis the K_m (acetyl-CoA) = 115 μM.

Irreversible Inhibitors. Cytoplasmic thiolase is very rapidly inactivated by low concentrations of iodoacetamide but preincubation with acetoacetyl-CoA gives complete protection.[7]

Molecular Weight. The rat liver cytoplasmic thiolase has a molecular weight of 170,000 and consists of four subunits each of molecular weight 44,000.[7] These values are almost identical to those determined for the chicken liver enzyme.[10]

Isoelectric Point. Cytoplasmic thiolases from ox liver, rat liver, rat brain, and chicken liver have isoelectric points of 5.2,[9] 5.8,[7] 5.1,[11] and 7.1,[10] respectively.

(b) *Mitochondrial Acetoacetyl-CoA Thiolase*

This enzyme has been isolated and partially purified from the sources shown in Table I and the pig heart enzyme has been crystallized.[12]

Substrate Specificity. See Table II. 3-Ketoacyl-CoA homologues of acetoacetyl-CoA are not substrates. In the reverse direction of the reaction only acetyl-CoA is a substrate.[12] As with the cytoplasmic acetoacetyl-CoA thiolase the specificity for CoA moiety is not absolute; the mitochondrial acetoacetyl-CoA thiolase from pig heart uses pantetheine and acetoacetylpantetheine as substrates.[12]

[10] K. D. Clinkenbeard, T. Sugiyama, J. Moss, W. D. Reed, and M. D. Lane, *J. Biol. Chem.* **248**, 2275 (1973).

[11] B. Middleton, *Biochem. J.* **125**, 70P (1971).

[12] U. Gehring, C. Riepertinger, and F. Lynen, *Eur. J. Biochem.* **6**, 264 (1968).

TABLE II
SOME PROPERTIES OF MITOCHONDRIAL KETOACYL-CoA THIOLASES ISOLATED
FROM DIFFERENT TISSUES[a]

Source of enzyme	Acetoacetyl-CoA thiolase		General ketoacyl-CoA thiolase	
	Relative activity with 3-ketoacyl-CoA substrates $C_4:C_6$	Activation by K^+	Relative activity with 3-ketoacyl-CoA substrates $C_4:C_6:C_{10}$	Activation by K^+
Rat brain	1: <0.01	4.1	—	—
Rat heart	1: <0.01	4.3	1:4.2:4.1	1.0
Rat kidney	1: <0.01	4.0	1:4.0:3.2	1.0
Rat liver	1: <0.01	4.2	1:3.9:3.7	1.0
Pig heart	1: <0.01	4.8	1:3.4:3.4	1.0
Ox liver	1: <0.01	5.5	1:4.9:4.8	1.0
Ram sperm[b]	1: <0.01	5.2	1:3.3:3.0	1.0

[a] From Middleton.[2]
[b] B. Middleton and T. Mann, unpublished results.

Cation Activation. See Table II. K^+ activates in all cases and Na^+ and $Tris^+$ have no effect upon this activation. The K_a values for activation by K^+ are 1–2 mM.[2,9]

Kinetic Properties. All the mitochondrial acetoacetyl-CoA thiolases listed in Table I are strongly inhibited by acetoacetyl-CoA (substrate inhibition) at concentrations above 20 μM.[2] No substrate inhibition is found with CoA. Under the conditions of the standard assay the apparent K_m values (rat liver enzyme) are as follows: K_m (acetoacetyl-CoA), CoA at 50 μM, = 7 μM; K_m (CoA), acetoacetyl-CoA at 10 μM, = 21 μM.[2]

Irreversible Inhibitors. The pig heart enzyme is inhibited by thiol group reagents such as iodoacetamide and N-ethylmaleimide; preincubation with acetyl- or acetoacetyl-CoA gives protection.[12] Many specific inhibitors for this enzyme are known: bromoacetyl-CoA,[13] bromoacetyl-alka-3-ynoyl-CoA[14] compounds are all potent inhibitors.

Molecular Weight. The pig heart enzyme has a molecular weight of 169,000 and is composed of four subunits of molecular weight 44,000.[15]

Isoelectric Point. Mitochondrial acetoacetyl-CoA thiolases have pI values above 7: rat brain, 8.3[11]; rat liver, 8.4[16]; ox liver, 8.0[9]; and pig heart, 7.3.[12]

[13] J. F. A. Chase and P. K. Tubbs, *Biochem. J.* 100, 47P (1966).
[14] P. C. Holland, M. G. Clark, and D. P. Bloxham, *Biochemistry*, 12, 3309 (1973).
[15] U. Gehring and C. Riepertinger, *Eur. J. Biochem.* 6, 281 (1968).
[16] B. Middleton, *Biochem. J.* 125, 69P (1971).

(c) Mitochondrial General Ketoacyl-CoA Thiolase

This enzyme has been isolated and partially purified from the sources shown in Table I and the ox liver enzyme has been crystallized.[17]

Substrate Specificity. See Table II. 3-Ketoacyl-CoA homologues from acetoacetyl- to 3-ketopalmitoyl-CoA are substrates.[2,17] The highest activity is found with 3-ketohexanoyl-CoA.

Cation Activation. None.

Kinetic Properties. No high substrate inhibition is observed with 3-ketoacyl-CoA or CoA at concentrations up to 100 μM.[2] This contrasts with the properties of the acetoacetyl-CoA thiolases. Under the conditions of the standard assay the apparent K_m values (rat liver enzyme) are as follows: K_m (acetoacetyl-CoA), CoA at 50 μM, = 10 μM; K_m (CoA), acetoacetyl-CoA at 10 μM, = 18 μM.[2]

Isoelectric Point. The enzymes isolated from ox and rat liver both show multiple forms with different pI values: ox liver, 5.7 and 6.7[9]; rat liver, 6.95 and 8.2.[2]

[17] W. Seubert, I. Lamberts, R. Kramer, and B. Ohly, *Biochim. Biophys. Acta* **164**, 598 (1968).

[18] Bovine Liver Crotonase (Enoyl Coenzyme A Hydratase)

EC 4.2.1.17 L-3-Hydroxyacyl-CoA hydrolyase

By HOWARD M. STEINMAN and ROBERT L. HILL

$$R-CH{=}CH-\overset{\overset{\displaystyle O}{\|}}{C}-S-CoA + H_2O \leftrightarrows R-\overset{\overset{\displaystyle OH}{|}}{CH}-CH_2-\overset{\overset{\displaystyle O}{\|}}{C}-S-CoA$$

The purification of crotonase from ox liver was first described by Joseph R. Stern in the first volume of this series,[1] and the properties of the crystalline enzyme were reported in subsequent publications by Stern and co-workers.[2-4] The purification procedure reported in this chapter has been adopted from the Stern methodology, with very minor modifications.

[1] J. R. Stern, Vol. 1 [93], p. 559.

[2] J. R. Stern, A. del Campillo, and I. Raw, *J. Biol. Chem.* **218**, 971 (1956).

[3] J. R. Stern and A. del Campillo, *J. Biol. Chem.* **218**, 985 (1956).

[4] J. R. Stern, *in* "The Enzymes" (P. D. Boyer, H. Lardy, and K. Myrbäck, eds.), 2nd ed., Vol. 5, p. 511. Academic Press, New York, 1961.

Assay Method

Principle. In the β-oxidation pathway of fatty acids, crotonase catalyzes the hydration of *trans-α,β*-unsaturated acyl-CoA derivatives to the corresponding L(+)-β-hydroxyacyl-CoA thioesters.[5,6] Two spectrophotometric methods have been used to monitor this hydration continuously and quantitatively. (1) Addition of the elements of water across the double bond abolishes the characteristic ultraviolet absorption of the α,β-unsaturated system, which is conjugated with the carbonyl group of the thio ester.[7] This change provides the basis of the direct spectrophotometric method.[1,2] (2) Alternatively, the oxidation of the β-hydroxyacyl-CoA thio ester and the accompanying reduction of NAD may be followed in the presence of β-hydroxyacyl-CoA dehydrogenase.[6] The spectral changes associated with NAD reduction provide the quantitative basis of this coupled assay method.

Each of these two methods has its own merits. The direct method has been used exclusively in this laboratory and has the advantages of experimental simplicity, the absence of complexities that may complicate the kinetics of coupled assay systems, and the ready adaptability to the assay of nonphysiological substrates, which might not be oxidized by the β-hydroxyacyl-CoA dehydrogenase. The coupled assay method is attractive because of the certainty of quantitation, which is based upon NAD reduction; when using CoA esters containing unusual acyl groups, the extinction changes accompanying hydration may not be known. In addition, the coupled assay is potentially applicable as a means of detecting crotonase activity in tissues or in electrophoretic gels, through further coupling with a dye system, analogous to those used for the NAD-dependent dehydrogenases.[8]

Procedure.[1,2] A double beam recording spectrophotometer, with full-scale absorbance of 0.1–0.2 unit, is used for assay by the direct method. To the sample and to the reference quartz cuvette (1.0 cm path length, 1 ml working volume) are added 1.0 ml of assay buffer, 0.3 M Tris-HCl, pH 7.4, containing 0.005 M EDTA and ovalbumin (0.05 mg/ml).[9] A

---

[5] J. R. Stern and A. del Campillo, *J. Amer. Chem. Soc.* **75,** 2277 (1953).

[6] S. J. Wakil and H. R. Mahler, *J. Biol. Chem.* **207,** 125 (1954).

[7] F. Lynen and S. Ochoa, *Biochim. Biophys. Acta* **12,** 299 (1953).

[8] I. H. Fine and L. A. Costello, Vol. 6 [127], p. 958.

[9] The buffer concentration in the assay mixture 0.3 M Tris-HCl has been increased from that used previously,[1,2] approximately 0.05 M, to insure a constant pH value. Thus, solutions of thio ester substrates are most stable at acid pH[10]; since large amounts of substrates with a low K_m value must be used, the assay mixture must be strongly buffered to compensate for the acid added with the substrate.

[10] E. R. Stadtman, Vol. 3 [137], p. 931.

small volume of crotonyl-CoA solution (0.005–0.050 ml) is added to the sample cuvette to produce a final concentration of approximately 200 μM, which is about 10 times the K_m for this substrate. To the reference cuvette is added a small volume of a solution of AMP, ADP, or ATP in water (about 10 mg/ml), to optically balance the adenine absorbance of crotonyl-CoA. A base line trace, which should be perfectly flat, is recorded for 3–4 minutes, monitoring at 280 nm.[11] The reaction is initiated by addition of a small volume of enzyme solution (0.005–0.050 ml) to the sample cuvette. The decrease in absorbance at 280 nm is recorded for 5–6 minutes. The spectrophotometer trace is usually linear, and the initial slope, $-\Delta A_{280}$ per minute, is calculated with moderate precision ($\pm 10\%$), if the decrease in absorbance is less than about 0.02 per minute.

The amount of protein necessary for a significant absorbance change ranges from about 1 μg of the crude homogenate (purification step 1) to 1–2 ng of recrystallized enzyme. The purified enzyme is unstable at low protein concentrations, and when it is necessary to dilute crotonase solutions prior to assaying, 0.02 M dibasic potassium phosphate, 0.015 M EDTA, pH 7.4, containing ovalbumin (1 mg/ml) is used. The maximum in the difference spectrum of hydration of aliphatic α,β-unsaturated acyl-CoA thio esters occurs at 263 nm,[7] a wavelength which has been used for monitoring crotonyl-CoA hydration.[1,2] This wavelength is close to the absorption maximum of adenine nucleotide-containing substances. When substrates with a high K_m value are studied, the very large background absorbance at 263 nm frequently cannot be accommodated by ordinary laboratory spectrophotometers. However, assays may be performed monitoring absorbance changes at 280 nm, where nucleotide absorbance is less. This wavelength has been used for assay of all crotonase substrates, even crotonyl-CoA, for purposes of standardization.

One unit of activity is defined as that which produces a decrease of one absorbance unit per minute at 280 nm in a volume of 1.0 ml. Using 3600 M^{-1} cm^{-1} as the value for the $\Delta\epsilon$ (280 nm) accompanying hydration of an aliphatic α,β-unsaturated acyl-CoA,[7] one unit corresponds to 0.278 μmole/minute. Specific activity is defined as units of activity per microgram of protein. The protein concentration of crystalline crotonase is determined from the extinction at 280 nm ($E_{1\,cm}^{0.1\%} = 0.576$).[12] For all earlier fractions from the purification (steps 1–5), the concentration is

[11] The base line trace of crotonyl-CoA solution without enzyme vs. reference solution is advised as a check for trace amounts of crotonase in the cuvettes. A cuvette used for protein concentration measurements must be cleaned thoroughly before use in an activity assay since the amounts of protein used in the former are about 5 orders of magnitude greater than necessary for assay (of the crystalline enzyme).

[12] G. M. Hass and R. L. Hill, *J. Biol. Chem.* **244**, 6080 (1969).

estimated by the optical method of Warburg and Christian.[13] The specific activity is converted to a turnover number (moles hydrated per minute per mole of enzyme) by multiplication by 4.56×10^4, using 164,000 as the molecular weight of crotonase.[12] Thus, once or twice recrystallized crotonase has a specific activity of about 9 units/μg, with crotonyl-CoA, at pH 7.4, 25°, corresponding to a turnover number of 410,000. This number is in accord with the value of 730,000 reported earlier by Stern,[1,2] after correction for the currently accepted molecular weight of liver crotonase, and adjustment for determination of protein concentration by the Warburg–Christian method, which underestimates the amount of crotonase, because of its low extinction at 280 nm.

Preparation of Crotonyl-CoA[1,3]

Thio ester substrates of crotonase are most easily synthesized by reaction of the coenzyme A thiol with the symmetrical acid anhydride of the acyl group. This method, originated by Simon and Shemin,[14] is applicable to crotonyl-CoA synthesis. When the desired symmetrical anhydride is not readily available, the mixed anhydride with ethyl hydrogen carbonate can be prepared by simple laboratory procedures and employed under similar reaction conditions. This mixed anhydride method has found extensive use in preparation of thio ester substrates of fatty acid enzymes.[15-17] CoA thio esters have been purified by chromatography on DEAE-cellulose, at mildly acid pH, with lithium chloride gradients.[16-18] A general discussion of synthetic methods for CoA thio esters is found in an earlier volume of this series,[10] detailing the use of symmetrical anhydrides and thio ester exchange, and providing references to several other procedures.

Crotonyl-CoA is now commercially available (P–L Biochemicals), but it is quite easily prepared in the laboratory. Free coenzyme A, CoA–SH (60 mg, P–L Biochemicals), is dissolved to a concentration of 1% (w/v) in nitrogen-saturated 0.1 M potassium bicarbonate, cooled at 0°. Immediately thereafter, 3–4 aliquots of neat crotonic anhydride (0.015 ml, Eastman) are added, with efficient stirring, at 5 minute intervals. The reaction must be maintained between pH 6.5 and 7.5 by addition of ·2 M potassium bicarbonate. After the third or fourth reaction interval,

[13] O. Warburg and W. Christian, *Biochem. Z.* 310, 384 (1941–1942).
[14] E. J. Simon and D. Shemin, *J. Amer. Chem. Soc.* 75, 2520 (1953).
[15] P. Goldman and P. R. Vagelos, *J. Biol. Chem.* 236, 2620 (1961).
[16] R. M. Waterson and R. L. Hill, *J. Biol. Chem.* 247, 5258 (1972).
[17] H. M. Steinman and R. L. Hill, *J. Biol. Chem.* 248, 892 (1973).
[18] J. G. Moffatt and H. G. Khorana, *J. Amer. Chem. Soc.* 83, 633 (1961).

the free thiol content reaches a minimum, judged by visual examination of spot tests, performed by dilution of aliquots into buffered solutions of 5,5'-dithiobis(2-nitrobenzoic acid), DTNB.[19] The solution is then acidified with dilute hydrochloric acid to pH 3–4, and extracted 3 times with 1–2 volumes of diethyl ether. The residual ether is evaporated under a stream of nitrogen gas. The aqueous phase is then divided into portions corresponding to 10–20 mg of original CoA and freeze-dried. Such dried powders are stable on storage in the deep freeze for over a year. As required for assay, individual portions are dissolved in water to an apparent concentration of 5–10 mg of reactant CoA per milliliter; such solutions, kept at 0° during use, and frozen when not in use, are stable for at least 3–4 weeks.

The concentration of crotonyl-CoA and of other aliphatic α,β-unsaturated CoA thio esters may be determined by several means. (1) The most rapid and simple method is direct absorbance measurement. The ϵ_m for crotonyl-CoA at the 232-nm absorption peak characteristic of thioesters[20,21] has been determined to be $1.92 \times 10^4\ M^{-1}\ cm^{-1}$,[16] a value which may be used for other aliphatic α,β-unsaturated acyl-CoA derivatives. (2) The coupled enzyme assay may be used, under conditions where the oxidation catalyzed by β-hydroxyacyl-CoA dehydrogenase is practically quantitative; the spectral measure of formation of NADH is used for quantitation.[1,3] (3) There are two methods, not contingent upon α,β-unsaturation, which are used for determination of thio ester concentrations in general. Both depend upon the ability of hydroxylamine to quantitatively cleave the thio ester bond, forming CoA–SH and the hydroxamate of the acyl group.[10,22,23] Quantitation of the hydroxamate by the ferric chloride color reaction is unattractive because of the low extinction of the complex formed.[24] However, paper chromatography of hydroxamates[24] is an effective means of qualitative product identification when the products of crotonase action are to be demonstrated.[25] The free thiol group of CoA which is generated by hydroxylamine cleavage of thio esters may be quantitated by reaction with DTNB.[19] This sulfhydryl titration procedure is both simple and sensitive, and would seem to be suited for determining the concentration of CoA thio esters whose ultraviolet extinction values are not known.[17] It is advisable to analyze the

[19] G. L. Ellman, *Arch. Biochem. Biophys.* **82**, 70 (1959).
[20] B. Sjöberg, *Z. Phys. Chem., Abt. B* **52**, 209 (1942).
[21] E. R. Stadtman, *Abstr., 122nd Meet., Amer. Chem. Soc.* Abstr. 79, 32C (1952).
[22] E. R. Stadtman, Vol. 3 [39], p. 228.
[23] E. R. Stadtman and H. A. Barker, *J. Biol. Chem.* **184**, 769 (1950).
[24] W. N. Fishbein, J. Daley, and C. L. Streeter, *Anal. Biochem.* **28**, 13 (1969).
[25] R. M. Waterson, G. M. Hass, and R. L. Hill, *J. Biol. Chem.* **247**, 5252 (1972).

initial thio ester preparation with DTNB to determine the amount of free thiol present, which then is subtracted from the total sulfhydryl titrated after hydroxylamine treatment.

Purification Procedure[1,2]

A summary of the purification procedure is given in Table I.

Step 1. Crude Homogenate. Ox liver is obtained from a local abattoir shortly after the animal is slaughtered, and is transported on ice to the laboratory. The remaining operations of this step are performed in a 4° room. The liver is trimmed of any visible connective tissue, chopped, and wrapped in portions of convenient weight, usually 500 g, then stored in a freezer until use.

Frozen tissue (750 g) is partially thawed in water at room temperature, broken or chopped into small pieces, and then mixed in a Waring blender with 500 ml of cold 0.2 M potassium bicarbonate, 0.005 M 2-mercaptoethanol (unadjusted pH of 8.1–8.2). Two 5-minute periods of homogenization are separated by a 10-minute interval, and then an addi-

TABLE I
SUMMARY OF PURIFICATION PROCEDURE[a]

Step	Volume (ml)	Units $\times 10^{-6}$	Protein[b] (mg)	Specific activity (units/μg)	Yield of units (%)
1. Crude homogenate	1,800	5.69	243,000	0.0234 ± 0.0001	100
2. Acid heat supernate	10,600	3.19	74,100	0.043 ± 0.003	56
3. Acetone precipitate	301	1.81	20,900	0.0866 ± 0.0003	32
4. Ammonium sulfate fraction	153	1.36	6,010	0.226 ± 0.003	24
5. DE 32 eluate	64	1.26	1,510	0.836 ± 0.009	22
6. Crystallization	28.6	1.21	155	7.8 ± 0.7	21
7. First recrystallization	11.8	1.17	117	10.0 ± 0.7	21

[a] Data for 750 g ox liver (wet weight).

[b] Protein determination for steps 1–5, by Warburg–Christian method,[13] for steps 6 and 7 by $E_{1\,cm}^{0.1\%}$ (280) = 0.576.[12]

tional 1000 ml of buffer are added, followed by 5 minutes' homogenization at low speed. A Büchner funnel is lined with a layer of gauze, and the homogenate is filtered with aspirator vacuum. The filtrate is then centrifuged at 5200 rpm (7120 g) for 60 minutes at 2°. The cloudy red-brown supernate is decanted through a funnel, covered with a layer of gauze to trap any low density material floating on the surface. Additional bicarbonate-mercaptoethanol buffer, as necessary, is added to make a total volume of 1.8 liters (2.4 liters/kg of original liver). Step 1 has usually been performed in the late afternoon of day 1, and the homogenate stored overnight at 4°.

Step 2. Acid Heat Treatment. To 0.8 liter of homogenate are added 64 ml of 1 M dibasic potassium phosphate, pH 7.4, and 3.14 liters of cold water. With ice-cooling and efficient stirring, the suspension is titrated from an initial pH 7.5–7.6 to pH 5.5 with cold 1 M acetic acid. During slow addition of the acid (about 200 ml) over a 10-minute period, no significant precipitation should be observed. The cold, acidified solution is transferred to two 3-liter beakers, which are immersed in a hot water bath (57–58°) and mixed continuously by overhead stirrers. Over a period of about 20 minutes, the temperature increases to 49°, and heating is continued for an additional 3 minutes, during which the temperature rises to 51°. The beakers are quickly removed from the bath, and the contents, containing a copious flocculent, beige precipitate, are transferred to a 6-liter flask and cooled at 0° with stirring. Only when the acidified and heated homogenate has been cooled below 15° may it be neutralized. The cold solution is cooled in ice water, and cold 2 M potassium bicarbonate is added with efficient stirring over a 10-minute period to raise the pH from 5.5 to 7.0 (about 500 ml per 3.5 liters of acidified solution). The neutralized solution is centrifuged (5200 rpm, 7120 g, 20 minutes, 2°); the clear dark brown supernates are combined, stored at 4°, and the accumulated beige precipitates are discarded. It is convenient to overlap heating, cooling, and centrifugation operations of successive portions of the homogenate. A second 0.8-liter batch of the homogenate is treated identically to the first, and the remaining 0.2 liter is treated with proportionately smaller volumes. The total volume of the combined supernates is 10.6 liters.

Step 3. Acetone Precipitation. Low temperatures are critical for this step, and all operations are performed in a 4° room. With mixing by an overhead stirrer, 2.2 liter portions of the supernate from step 2 are cooled in a 4-liter beaker immersed in a Dry Ice–alcohol bath. The solution is allowed to cool to −1°, by which time a thin shell has begun to freeze on the beaker walls. With continued cooling and stirring, 1.8 liters (9/11 volumes) of cold (−15°) acetone is added to make a final concentration of 45% (v/v). The first 250 ml is added in 50 ml portions at 1 minute intervals. Spaced addition is necessary since the heat of mixing at this stage initially produces a temperature rise from −1° to about 0°. The temperature is allowed to drop to −3°, and the remaining 1.55 liters of acetone are added slowly and continuously, while the temperature drops further. Precipitation is first evident after addition of about 550 ml, corresponding to 20% (v/v) acetone. Each 4 liter portion of 45% acetone solution, thus obtained, is centrifuged (5200 rpm, 7120 g, 20 minutes, −6°); the pink-brown supernates are discarded, and the precipitates allowed to accumulate through repeated centrifugations of successive acetone-precipitated solutions, in the same centrifuge bottles. While one 4-liter batch is being centrifuged, another 2.2 liters may be treated with acetone. Any remaining volume, less than 2.2 liters, is treated in an identical fashion with a proportionately smaller volume of acetone. The total mass of precipitates is resuspended in about 200 ml of cold 0.02 M dibasic potassium phosphate, 0.003 M EDTA, pH 7.4, and dialyzed overnight, with one change of outer solution, vs. 12–16 liters of the phosphate–EDTA buffer.

Step 4. Ammonium Sulfate Fractionation. The dialyzed solution from step 3 is centrifuged (9,000 rpm, 13,200 g, 30 minutes, 3°) to remove a small amount of precipitate. The protein concentration of the dark brown supernate (301 ml, usually 60–70 mg/ml) is estimated by the Warburg–Christian method,[13] and cold pH 7.4 phosphate-EDTA (1.1 liters) is added to lower the concentration to 15 mg/ml. The ammonium sulfate precipitation is performed in a 4° room. With magnetic stirring, ammonium sulfate (Schwarz–Mann, enzyme grade pulverized with mortar and pestle) is added, in portions, over a 20-minute period (280 g/liter, 40% saturation). After stirring for an additional 10–15 minutes, a precipitate is removed from the suspension by centrifugation (5200 rpm, 7120 g, 25 minutes, 3°). To the clear brown supernate, by identical procedures, is added additional ammonium sulfate (180 g/liter of original solution), achieving 60% saturation. The clear tan supernate obtained by centrifugation is discarded. The precipitate is resuspended in 75 ml of 0.005 M dibasic potassium phosphate, 0.002 M EDTA, pH 7.0, to which 2-mercaptoethanol has been added to a concentration of 0.005 M, and dialyzed overnight at 4°, with 2–3 changes, vs. 2 liters of 0.0025 M dibasic potassium phosphate, 0.001 M EDTA, pH 7.0, containing no mercaptoethanol.

Step 5. Elution through DEAE-Cellulose Column. This step is performed at 0–4°. The dialyzed solution from step 4 is centrifuged (9,000 rpm, 13,200 g, 30 minutes, 3°) to remove a small amount of brown precipitate. The clear green-brown supernate (153 ml) is applied to a column of DEAE-cellulose (Whatman DE 32, 3.5 × 42 cm) equilibrated with

0.0025 M dibasic potassium phosphate, 0.001 M EDTA, pH 7.0. A bed volume of this size is capable of processing material from at least 1 kg of liver. The column is developed with the equilibrium buffer at a rapid flow rate (105–110 ml/hour), and fractions are collected at 6 minute intervals. Under these conditions, crotonase passes through the column unretarded, and a green contaminant is removed by adsorption to the upper one-fourth to one-third of the DEAE-cellulose. The column breakthrough peak, containing virtually all of the crotonase activity plus some contaminating proteins, is located by A_{280} measurements. The absorbance increases abruptly at the void volume of the column, remains elevated as the protein is eluted, and decreases with slight tailing to a plateau value of 0.4–0.6. To the pooled fractions constituting the breakthrough peak (262 ml of milky brown solution) are added 4.58 ml of 1 M dibasic potassium phosphate, pH 7.4, to increase the phosphate concentration to 0.02 M. The protein is precipitated by addition of ammonium sulfate (600 g/liter) by the procedures outlined in step 4, and separated from a clear colorless supernate, as a beige precipitate, by centrifugat'on (5200 rpm, 7120 g, 25 minutes, 3°). The precipitate is resuspended in 35 ml of 0.02 M dibasic potassium phosphate, 0.003 M EDTA, pH 7.4, and dialyzed vs. 2 liters of this buffer overnight with two changes.

Step 6. Crystallization. Since the volumes are small at this stage, crystallization and recrystallizations may be performed at the bench, with ice bucket cooling. The dialyzed solution is centrifuged (15,000 rpm, 27,000 g, 25 minutes, 3°) to remove a small amount of brown precipitate from a slightly milky, pink-brown supernate (64 ml). This solution is briefly cooled in Dry Ice–alcohol until it just begins to freeze, and is then swirled in ice water while cold absolute ethanol (0.11 volume, 7.0 ml) is added dropwise. If precipitation is not apparent 2–3 minutes after addition, the solution may be briefly cooled again in Dry Ice–alcohol. The characteristically crystalline sheen is usually readily apparent by swirling the solution and briefly removing it from the ice bath. The crystalline suspension is stored at 4°.

After 20–24 hours the protein is recrystallized. Centrifugation of the suspension (12,000 rpm, 17,300 g, 10 minutes, 3°) separates a gray precipitate from a milky red-brown supernate. The precipitate is resuspended in 25 ml of cold pH 7.4 phosphate–EDTA, and centrifuged (15,000 rpm, 27,000 g, 25 minutes, 3°). A clear, faintly brown supernate is decanted from a brown precipitate, which is resuspended in 3 ml of buffer, and centrifuged again to provide a second supernate that is combined with the first. By the same procedure as above, cold ethanol (0.11 volume,

3.1 ml) is added to the cooled protein solution (28.6 ml). Usually crystallization is evident before the entire volume of ethanol is added, often
by 0.07 volume. The protein may be recrystallized a second time, after
20–24 hours at 4°. As reported earlier by Stern,[1,2] the second recrystallization does not further increase the specific activity toward crotonyl-CoA. The solubility of crystalline crotonase, in pH 7.4 phosphate–EDTA at 0°, appears to be about 13 mg/ml.

Comments about the Purification Procedure

Comparison with the Stern Procedure. The purification scheme described here is essentially identical to that devised by Stern[1,2] except
that one additional step, the elution through a DEAE column, has been
introduced after ammonium sulfate fractionation. Without this step, it
was occasionally found that repeated recrystallizations (more than five)
were required to remove all of the colored contaminants from the crotonase. With this column step included, the purification scheme usually
affords crotonase of high specific activity more reliably after only one
or two recrystallizations. During the 17 years since the report of Stern's
procedure, a plethora of new protein purification techniques of high resolving power has become part and parcel of the biochemist's isolation
tools. It remains a moot question whether some of the more cumbersome
and tricky steps in the present scheme could be replaced by procedures
technically simpler or more efficacious with respect to recovery of
activity.

Stability of Crotonase Activity. Crotonyl-CoA hydration activity is
not extremely stable in the crude homogenate or in the supernate from
the acid heat treatment. Even the dialyzed acetone precipitate loses activity over a period of days on refrigerated storage. The dialyzed ammonium sulfate fraction seems moderately stable to storage as a solution
in the cold. The purified crystalline protein is most stable when stored
frozen in the crystallization medium [phosphate–EDTA, containing
10% (v/v) ethanol]; the crotonyl-CoA activity of crotonase has been
maintained in crystalline preparations kept frozen for over a year and
does not appear to be extremely sensitive to a small number of thawings
and refreezings. Moderately concentrated solutions of the enzyme (5–10
mg/ml) have been stored refrigerated for a period of weeks in triethanolamine–EDTA and tris–EDTA buffers, at near-neutral pH values,
without major activity losses.

Yield of Crystalline Protein. Portions of the same ox liver have been
found to yield variable amounts of crystalline crotonase when subjected

to apparently identical purification schemes. For unknown reasons, the yield ranges from 100 to 300 mg of crystalline enzyme per kilogram of liver. The original Stern procedure[1,2] produced 275 mg of crotonase per kilogram after the second recrystallization, when the reported yield is corrected for the currently accepted extinction value of crotonase.[12]

Scheduling of Purification Steps. With the amounts of liver ordinarily used in crotonase preparations (0.75–1 kg), the repeated centrifugations of steps 2 and 3 (acid heat treatment and acetone precipitation) require an entire day. Homogenization (step 1) has thus been performed in the afternoon of day 1; thus, steps 2 and 3 may be begun in the morning of day 2. Crotonase activity is not very stable in the crude homogenate, and nuances of the overnight refrigerated storage may be related to the observed variable yields of crystalline enzyme (*vide supra*). It may be as expedient, in terms of the amount of crystalline crotonase obtained, to commence with a smaller weight of ox liver and proceed through steps 1, 2, and 3 in one, albeit long, day, as was apparently done in the original procedure of Stern. The particular day-to-day scheduling will depend upon the crotonase requirements and the preparation facilities of the individual laboratory.

Properties

Physical Properties. Sedimentation equilibrium analyses show that liver crotonase contains 6 subunits, which are combined in the native enzyme solely through noncovalent interactions.[12] Analyses in the presence of 6 M guanidine hydrochloride, with or without 0.02 M dithiothreitol, provide a subunit (weight average) molecular weight of 28,000 ± 500, in good agreement with the results of chromatography on a guanidine hydrochloride–agarose column[12] and of acrylamide electrophoresis in the presence of sodium dodecyl sulfate.[26] Sedimentation analyses under nondenaturing conditions give a native (weight average) molecular weight of 164,000 ± 3,000. Although this value is lower than the 210,000 reported earlier[1,2,4] from light scattering measurements, it and the inferred hexameric subunit structure are supported by a variety of additional determinations including amino acid analysis,[12] quantitative amino and carboxyl terminal determinations,[12] and analysis of the number of acetoacetyl coenzyme A binding sites.[16] Upon extrapolation to zero protein concentration, the sedimentation coefficient, corrected to standard conditions ($s_{20,w}^{\circ}$), is 8.42 S, corresponding to $f/f_0 = 1.2$ for a native molecular weight of 164,000.[12] It is of interest that crotonase purified from *Clostri-*

[26] R. M. Waterson and R. L. Hill, unpublished observations (1971).

dium acetobutylicum has a comparable native molecular weight (158,000), but contains only 4 subunit polypeptide chains.[27]

The purity of the crystalline protein is demonstrated (1) by its homogeneity in sedimentation velocity and sedimentation equilibrium analyses[12]; (2) by its homogeneity upon electrophoresis on cellulose acetate, at a variety of pH values near neutrality, and in acrylamide gels, both in the presence and absence of 6 M urea[12]; and (3) by a unique amino terminal sequence of 20 residues, determined by automated Edman degradation of the S-carboxymethylated protein.[28] Peptide mapping of the trypsin-digested, carboxymethylated enzyme is consistent with a single primary structure for all 6 subunits.[12] Each subunit possesses a single binding site for acetoacetyl-CoA, and thus most certainly a single substrate binding site.[16]

The absorption spectrum of crotonase is that expected for a tryptophan-containing protein with no light-absorbing cofactors.[2,4,29] For the native enzyme the absorbance maximum is at 280 nm ($E_{1 cm}^{0.1\%} = 0.576$),[12] and $A_{280}/A_{260} = 1.71 \pm 0.02$. Based upon electrophoresis on cellulose acetate strips, the isoelectric point in 0.01 M potassium phosphate, 0.003 M EDTA is approximately pH 7.5.[29]

The subunit polypeptide chain of crotonase commences with amino terminal serine and is composed of approximately 260 amino acid residues including a single residue of histidine and 1 or 2 residues of tryptophan.[12] The average tryptophan content by three different methods is 1.6 ± 0.1 residues per subunit, indicating either anomalous spectral and chemical behavior of the tryptophan residue(s), or microheterogeneity in the near-identical primary structures of the subunits. The total half-cystine content of 5 residues/subunit, determined by amino acid analysis after cysteic acid oxidation, can be accounted for entirely by the total sulfhydryl groups titratable with DTNB in 6 M guanidine hydrochloride. Crotonase is thus devoid of both inter- and intrasubunit disulfide bonds.[12]

Catalytic Properties. The fundamental enzymology of liver crotonase has been reviewed earlier by Stern,[1,3,4] and here only selected basic facts are repeated along with information resulting from more recent studies. While most hydratases require either a metal or a pyridoxal type of cofactor for hydration, neither crotonase nor fumarase possesses any cofactor requirement.[30]

[27] R. M. Waterson, F. J. Castellino, G. M. Hass, and R. L. Hill, *J. Biol. Chem.* **247**, 5266 (1972).

[28] W. F. Moo-Penn, H. M. Steinman, and R. L. Hill, unpublished observations (1972).

[29] G. M. Hass, Ph.D. Thesis, Duke University, Durham, North Carolina (1969).

[30] R. L. Hill and J. W. Teipel, *in* "The Enzymes" (P. D. Boyer, ed.), 3rd ed., Vol. 5, p. 539. Academic Press, New York, 1971.

Much investigation has been done upon the crotonase-catalyzed hydration of CoA thio esters of α,β-unsaturated acids. Crotonyl-CoA is the most active crotonase substrate reported with a turnover number of 350,000–400,000 moles/mole/minute and a K_m value of 20 μM, at pH 7.4, 25°.[1-4,16] At equilibrium at 25° the ratio, β-hydroxybutyryl-CoA–crotonyl-CoA, is 3.5.[3] The pH dependence of crotonyl-CoA hydration has been studied over the range from 5.3 to 10.4.[3] The maximum turnover number, at pH 9.4, is approximately twice the value at pH 7.4 and 4 times that at pH 5.8. Ionization of phosphate groups on the substrate, as well as ionization of amino acid side chains, most certainly contributes to the observed pH-dependent behavior.

The action of crotonase has been reported, upon a series of CoA thio esters of *trans*-α,β-unsaturated, straight chain monocarboxylic acids, of an even number of carbon atoms.[16] From a maximum value for crotonyl-CoA, the turnover number at pH 7.4 decreases continuously with increasing chain length of the acyl group, reaching a minimum of 2300 for *trans*-hexadecenoyl-CoA, representing the longest acyl group investigated. Within this same series, the K_m value at pH 7.4 increases only modestly, from 20 μM for crotonyl-CoA to 500 μM for hexadecenoyl-CoA. Crotonase hydration of these substrates is stereospecific, forming the L(+)-β-hydroxyacyl-CoA derivative, and conforms to simple Michaelis–Menten kinetics. In contrast to the liver enzyme, crotonase from *Clostridium acetobutylicum* shows pronounced substrate inhibition by crotonyl-CoA, while obeying the Michaelis–Menten scheme with hexenoyl-CoA.[27] In addition, the bacterial enzyme does not hydrate the C_8–C_{16} enoyl-CoA thio esters of the above series.

Liver crotonase has little specificity regarding the nature of the acyl group of CoA substrates. A variety of carboxylic acids, which are structural analogues of crotonic acid in possessing a trans-α,β double bond, are crotonase substrates when thio esterified with coenzyme A, including those formed with 3-methyl crotonic acid,[3,4] 2-chloro-, 3-chloro-, and 4-bromocrotonic acid,[17] acrylic[4,31] and 2-methyl acrylic acid,[3,4] sorbic acid,[3,4] cyclohexene-1-carboxylic acid, and cinnamic acid.[17] Thus, the substrate binding site of crotonase possesses flexibility to accommodate an assortment of uncharged substituents (methyl, halogen, and phenyl groups). Anionic carboxyl substituents can apparently be accommodated if displaced from the α,β double bond as the CoA thio ester of 2-methyl-propene-1,3-dicarboxylic acid, 3-β-methyl glutaconic acid, is hydrated.[4] Carboxyl substituents at the double bond, however, are not accommodated (CoA thio esters of 2-propene-1,2-dicarboxylic acid and 2-methyl fumaric acid, itaconic acid and mesaconic acid[3,4]).

[31] G. Rendina and M. J. Coon. *J. Biol. Chem.* **225**, 523 (1957).

This lack of specificity extends to the double bond geometry; *cis*-crotonyl-CoA is hydrated to D(−)-β-hydroxybutyryl-CoA.[32,33] Crotonase also hydrates *cis*-2-hexenoyl-CoA, and *cis*-2-methyl crotonyl-CoA.[3,4] Positional isomers with the double bond in the β,γ location, 3-butenoyl (vinylacetyl) and *trans*-3-hexenoyl-CoA,[3,4] also serve as substrates. It was first recognized by Stern that in acting upon the former β,γ-unsaturated ester, crotonase is serving as a positional isomerase in an *ex officio* fashion.[4] Thus, 3-butenoyl-CoA is hydrated to L(+)-β-hydroxybutyryl-CoA, which in turn may be dehydrated to 2-butenoyl (crotonyl) CoA, thus affecting translocation of the double bond from the β,γ to the α,β position.

Liver crotonase is much more selective regarding the donor of the thiol group of the ester bond. For the crotonyl acyl group, maximal activity is found with the thio ester of coenzyme A; pantethiene is the next most active thiol donor, whose thio ester is hydrated at only about 0.15% the rate of crotonyl-CoA.[3,4,25] No hydration activity has been observed with the crotonyl thio esters of glutathione, mercaptoacetic acid, mercaptoethylamine, N-acetyl mercaptoethylamine, or N,N-diethyl mercaptoethylamine, nor with free crotonic acid.[3,4]

In addition to the positional isomerase activity, a number of nonhydration activities of crotonase have been reported earlier by Stern,[4] only one of which has been the subject of further, more recent investigation.[25] Crotonyl pantethiene has a V_{max} about 0.015% that of crotonyl-CoA. A marked increase in hydration activity is observed upon addition of free CoA to a mixture of crotonase and crotonyl pantethiene. This activation originally was attributed to a transferase activity of the enzyme, which was proposed to catalyze the migration of the crotonyl group from pantethiene to CoA, thus generating a substrate of much higher turnover number.[4] Reinvestigation has shown that such a transfer does not occur and that the activation is related to purely noncovalent interactions.[25] The phenomenon appears to be one of the substrate complementation class, in which the K_m of crotonyl pantethiene is lowered through a synergistic binding of coenzyme A, analogous to the activation of tryptic hydrolysis of N-acetyl glycine methyl ester by alkyl guanidinium and alkyl ammonium salts.[34]

Regulation and Inhibition of Crotonase Action. Among the enzymes involved in β-oxidation of fatty acids, crotonase is distinctive in two regards: (1) the striking dependence of V_{max} upon the chain length

[32] S. J. Wakil, *Biochim. Biophys. Acta* **19**, 497 (1956).
[33] S. J. Wakil, *in* "Lipid Metabolism" (S. J. Wakil, ed.), p. 1. Academic Press, New York, 1970.
[34] T. Inagami and S. S. York, *Biochemistry* **7**, 4045 (1968).

of the acyl group (*vide supra*) and (2) the strong competitive inhibition by acetoacetyl-CoA ($K_i = 20$ μM), an end product of mitochondrial β-oxidation.[16] These two facts have suggested that crotonase may act physiologically in the regulation of β-oxidation. The proposed scheme has been supported by studies of fatty acid oxidation in mitochondrial preparations.[16] In the absence of acetoacetyl-CoA, the rate of the crotonase-catalyzed hydration of CoA substrates, of all chain lengths, is greater than or equal to the catalytic rate of the other enzymes in the β-oxidation pathway. In the presence of excess acetoacetyl-CoA, the rate of crotonase action upon all substrates is depressed. However, only the rate for the long-chain thio esters (C_{14} and C_{16}) is reduced below that of the other β-oxidation enzymes, which are not inhibited by acetoacetyl-CoA. Thus, acetoacetyl-CoA may serve as a means of regulating the overall rate of β-oxidation through rendering as rate limiting the crotonase hydration of C_{14} and C_{16} CoA esters entering the pathway.

Earlier studies upon the inhibition of crotonase by sulfhydryl reagents[3,4] have been extended through chemical and catalytic characterization of the enzyme treated with a variety of thiol-reactive compounds. The results with these compounds, representing a rather broad spectrum of size, charge, hydrophobicity, and mechanism of reaction with sulfhydryl groups, provide a consistent description of the reactivity of crotonase cysteinyl residues and their apparent catalytic involvement.[17,25]

1. A single, apparently unique sulfhydryl group, of the five per subunit, is hyperreactive toward all compounds tested. This side chain is, however, most certainly not required for catalytic hydration, and its modification, even by bulky reagents (DTNB or 1-fluoro-2,4-dinitrobenzene) does not substantially diminish crotonase activity.

2. Another, apparently unique, cysteinyl residue is located in the vicinity of the substrate binding site. Its modification may inhibit the enzyme, apparently by steric blocking of the binding site by the reagent moiety covalently bound. However, if the reagent moiety is sufficiently small, inhibition is minimal. This thiol group thus does not appear to be catalytically crucial.

3. Following modification of the hyper-reactive thiol, progressive reaction of the remaining 4 cysteinyl residues per subunit results in inactivation if the reagent is bulky, but permits substantial retention of activity if the reagent is small. Disruption of the structural integrity of the hexamer thus seems to be responsible for the observed activity loss.

4. None of these studies support the existence of a unique catalytic –SH group in crotonase.

Models have been formulated from mitochondrial β-oxidation, in which the failure to detect intermediates, once fatty acids are committed to the pathway, is explained by transient esterification of fatty acyl groups to sulfhydryl functions on β-oxidation enzymes.[35,36] Support for such models has, in part, been sought in previous reports of a crotonyl-transferase activity of crotonase, possibly involving a catalytic sulfhydryl group of the enzyme.[4] In light of the recent investigations of crotonase catalytic function[16,25] and cysteinyl involvement[17,25] such models warrant reevaluation.

[35] P. B. Garland, D. Shepherd, and D. W. Yates, *Biochem. J.* **97,** 587 (1965).
[36] G. D. Greville and P. K. Tubbs, *Essays Biochem.* **4,** 155 (1968).

Section III

HMG-CoA Enzymes

[19] Mitochondrial 3-Hydroxy-3-methylglutaryl-CoA Synthase from Chicken Liver

By W. DOUGLAS REED and M. DANIEL LANE

$$\text{Acetyl-CoA} + \text{acetoacetyl-CoA} \xrightleftharpoons{+\text{HOH}}$$
$$\text{3-hydroxy-3-methylglutaryl-CoA} + \text{CoA}$$

The thermodynamically favorable condensation of acetyl-CoA and acetoacetyl-CoA to form 3-hydroxy-3-methylglutaryl-CoA is catalyzed in chicken liver mitochondria by 3-hydroxy-3-methylglutaryl-CoA synthase (HMG-CoA synthase) according to the equation above. This enzyme is localized in the mitochondrial matrix[1] and occupies a potentially favorable position for the regulation of hepatic ketogenesis. The distinctly different properties of this enzyme from those of the cytoplasmic HMG-CoA synthase (see Chapter 20 in this volume and Clinkenbeard et al.[2]) is consistent with independent regulation of ketogenesis and cholesterogenesis.

Assay Method

Principle. 3-Hydroxy-3-methylglutaryl-CoA synthase activity can be conveniently determined by a radiochemical assay which monitors the incorporation of $[1-^{14}C]$acetyl-CoA (volatile ^{14}C activity when taken to dryness at 95° in 6 N HCl) into HMG-CoA (nonvolatile ^{14}C activity when taken to dryness at 95° in 6 N HCl). Alternatively, a spectrophotometric assay modified from that developed by Ferguson and Rudney[3] can be used which measures the acetyl-CoA–dependent disappearance of acetoacetyl-CoA absorbance at 300 nm.

Radiochemical Assay

Reagents

Tris (Cl⁻) buffer, 0.4 M, containing 0.4 mM EDTA, pH 8.0
$[1-^{14}C]$Acetyl CoA, 2.0 mM, pH 4.5 (approximately 3×10^6 cpm/μmole; the specific activity must be accurately determined).
Acetoacetyl-CoA, 1.0 mM, pH 4.5
Liquid scintillator, 22.0 g of 2,5-diphenyloxazole (PPO), 0.4 g

---

[1] Reed, W. D., and Lane, M. D., in preparation.
[2] Clinkenbeard, K. D., Sugiyama, T., and Lane, M. D., in preparation.
[3] Ferguson, J. J., and Rudney, H., *J. Biol. Chem.* **234**, 1072 (1959).

1,4-bis-2-(4-methyl-5-phenyloxazolyl)-benzene (dimethyl POPOP), 2668 ml toluene and 1332 ml Triton X-100.

Procedure. The complete reaction mixture contains the following components (in micromoles): Tris (Cl⁻), 20; EDTA, 0.02; [1-¹⁴C]acetyl-CoA, 0.04; acetoacetyl-CoA, 0.01; and 0.15–1.0 milliunits of HMG-CoA synthase in a total volume of 0.2 ml. The reaction is initiated by addition of [1-¹⁴C]acetyl-CoA to the assay mixture following a 2-minute preliminary incubation at 30°. At appropriate times 0.04 ml aliquots are removed and pipetted into 15 × 45 mm glass vials (flat-bottomed shell vials) containing 0.2 ml of 6 N HCl, and the acidified aliquot taken to dryness at 95° in a forced draft oven. Water (0.3 ml) is added to each vial followed by 3 ml of liquid scintillator; the vials are capped and inserted into clear acrylic holders (Packard, catalog No. 6000144). The nonvolatile ¹⁴C activity (as [¹⁴C]HMG-CoA) is determined with a liquid scintillation spectrometer. Under the conditions described, a linear increase in nonvolatile ¹⁴C activity with time is observed until at least 80% of the acetoacetyl-CoA present is utilized.

Spectrophotometric Assay

Reagents

Tris (Cl⁻) buffer, 0.4 M, containing 0.4 mM EDTA, pH 8.2
Acetyl-CoA, 10.0 mM, pH 4.5
Acetoacetyl-CoA, 1.0 mM, pH 4.5
MgCl₂, 1.0 M

Procedure. The complete assay mixture contains (in micromoles): Tris (Cl⁻) pH 8.2, 100; EDTA, 0.1; acetyl-CoA, 0.2, acetoacetyl-CoA, 0.05; MgCl₂, 20; and HMG-CoA synthase in a total volume of 1.0 ml. Following a 2-minute preliminary incubation of HMG-CoA synthase at 30° in the above assay mixture lacking acetyl-CoA to establish a base line absorbance at 300 nm, the reaction is started by the addition of acetyl-CoA. The rate of HMG-CoA synthesis is monitored by following the disappearance of the acetoacetyl-CoA–Mg²⁺ complex absorbance at 300 nm ($\epsilon_{300\,nm}^{1\,cm} = 13.6 \times 10^3\ M^{-1}$) dependent on acetyl-CoA (if Mg²⁺ is omitted, $\epsilon_{300\,nm}^{1\,cm} = 3.6 \times 10^3\ M^{-1}$). Under these conditions, the decrease in $A_{300\,nm}$ is linear with time until at least 80% of the acetoacetyl-CoA has reacted and is proportional to enzyme concentration within the limits specified. The $\epsilon_{300\,nm}$ is sensitive to changes in pH, EDTA, and/or Tris concentration.

Definition of Units and Specific Activity. One unit of HMG–CoA synthase is defined as the amount of enzyme which catalyzes the formation

TABLE I
PURIFICATION OF HMG-CoA SYNTHASE

Stage of purification	Total activity[a] (units)	Total protein (mg)	Specific activity (units/mg protein)	Yield (%)
1. Original homogenate (160 g supernatant)	3333	194,000	0.017	
2. Mitochondria (pressed)	1070	37,900	0.028	100
3. Matrix fraction	342	12,400	0.033	32
4. DEAE-cellulose chromatography	249	3,038	0.082	23
5. Phosphocellulose chromatography	114	97	1.17	11
6. Sephadex G-150 chromatography	43	42	1.0	4

[a] The spectrophotometric assay for HMG-CoA synthase was employed in this purification.

of 1 μmole of HMG–CoA per minute under the assay conditions described. Both the radiochemical and the spectrophotometric assays yield identical results. Specific activity is expressed as units per milligram of protein.

The radiochemical assay cannot be used with preparations which contain HMG-CoA lyase, i.e., prior to step 5 of Table I. The activity of HMG-CoA synthase determined by the spectrophotometric assay on preparations containing a $\geqq$ number of units of acetoacetyl-CoA thiolase per unit of HMG-CoA synthase is divided by two in order to take into account the CoA-dependent disappearance of acetoacetyl-CoA by acetoacetyl-CoA thiolase; the extent of the thiolase reaction is limited by the quantity of CoA generated in the synthase-catalyzed reaction.

The choice of assay for the HMG-CoA synthase is dictated by the activity and purity of the sample, i.e., whether HMG-CoA lyase is present. If low enzymatic activities per aliquot are involved (<3 nmoles/minute) the radioactive assay was employed since it facilitates the use of longer reaction times for each of a number of samples.

Purification Procedure

The purification procedure described here is carried out at 4°. A summary of this purification appears in Table I. The HMG-CoA synthase is unstable in the crude state and this contributes to the large initial losses of activity and therefore to the low total yield.

Initial Homogenate. Fresh liver from laying hens is ground in a Universal No. 1 meat chopper and then homogenized in a loose-fitting Potter-Elvehjem tissue homogenizer with a Teflon pestle. One kilogram of liver is homogenized in 2 liters of media containing 0.25 M sucrose, 0.1 mM EDTA, and 2 mM N-2-hydroxyethylpiperazine-N-2-ethanesulfonic acid (HEPES) brought to pH 7.2 with potassium hydroxide. This homogenate is diluted with an equal volume of the original buffer and centrifuged at 160 g for 15 minutes. The pellet from this centrifugation containing nuclei, cellular debris, and unbroken cells is discarded while the supernatant is centrifuged at 9000 g for 15 minutes. The pellet from this spin is resuspended in a volume of 20 mM potassium phosphate, pH 7.0; 0.1 mM EDTA; and 0.1 mM dithiothreitol (DTT) which is equal to the original number of grams of tissue, i.e., 1000 ml. This suspension is centrifuged at 9000 g for 15 minutes and the pellet resuspended again in 1000 ml of the same potassium phosphate buffer. The mitochondria in this suspension are then ruptured by three passes through a Manton-Gaulin laboratory homogenizer (M-G Manufacturing Co., Inc., Everett, Mass.) at 6000 psi, which has been previously cooled to about 2° with the cold potassium phosphate buffer. The effluent from the press is centrifuged at 105,000 g for 60 minutes, and the supernatant containing the matrix components is filtered through four layers of cheesecloth.

DEAE-Cellulose Chromatography. A DEAE-cellulose column (8 $\times$ 80 cm, Chromaflex by Kontes) is poured to approximately 3.5 liters with standard diethyl-amino ethyl cellulose (exchange capacity 0.9 mEq/g, Carl Schliecher and Schuell Co.) and equilibrated with 0.5 M potassium phosphate buffer pH 7.0; 0.1 mM DTT; and 0.1 mM EDTA. After equilibration, the column is washed with 10 column volumes of 10 mM potassium phosphate buffer, pH 7.0, containing 0.1 mM DTT and 0.1 mM EDTA. The matrix fraction (from the 105,000 g centrifugation) containing about 15 g of protein is applied to this column and eluted from the same 10 mM potassium phosphate buffer. The effluent is continually monitored for protein (OD at 280 nm) and HMG-CoA synthase activity (using the spectrophotometric assay.) The enzyme is eluted immediately after the "breakthrough" volume. The fractions exhibiting the peak activity which comprise approximately 70% of the total applied are pooled. The DEAE pooled fractions contain approximately 3 g of protein.

Phosphocellulose Chromatography. A phosphocellulose column (5 $\times$ 95 cm, exchange capacity 0.88 mEq/g, Carl Schliecher and Schuell Co.) was poured to approximately 1500 ml. The phosphocellulose is prepared by sequential washing with 0.1 N NaOH, water until neutral, 0.1 N HCl, water until neutral, and 50 mM potassium phosphate buffer, pH 7.0. Finally, the phosphocellulose column is washed with 10 column volumes of 10 mM potassium phosphate buffer, pH 7.0, containing 0.1 mM DTT and 0.1 mM EDTA. The pooled active fractions from the DEAE-cellulose column are immediately (within 2 hours after chromatography) applied to the phosphocellulose column and followed by one column volume of the phosphate buffer. The enzyme is eluted with a 6-liter linear potassium phosphate gradient of 5–100 mM (a constant concentration of 0.1 mM of DTT and EDTA is maintained during the elution) as a single HMG-CoA synthase activity when the concentration of phosphate in the eluate reaches about 50 mM. The enzyme is virtually free of HMG-CoA thiolase at this stage of purification (<0.02% of mitochondrial thiolase). The fractions which are most active in HMG-CoA synthase are pooled, and the enzyme is precipitated by inward dialysis of ammonium sulfate to a final concentration of 55% saturation (in 20 mM potassium phosphate buffer, pH 7.0). After centrifugation of the enzyme suspension the sedimented protein is dissolved in 20 mM potassium phosphate buffer, pH 7.0, containing 0.1 mM DDT and EDTA, and dialyzed to equilibrium against at least a 500:1 v/v excess of the 20 mM potassium phosphate buffer, pH 7.0, containing 0.1 mM DTT and EDTA.

Gel Filtration. The dialyzed enzyme solution from the preceding step is applied to a column (2.5 × 90 cm, 500 ml, Pharmacia Fine Chemicals, Inc.) of Sephadex G-150. The enzyme is eluted with the same buffer with which it was previously equilibrated, i.e., 20 mM potassium phosphate buffer containing the usual 0.1 mM DTT and EDTA. The HMG-CoA synthase activity is eluted as a single peak coincident with a protein peak at a V_e/V_o of 1.90. The fractions containing the majority of the activity are pooled and brought to 60% saturation with ammonium sulfate (in 20 mM potassium phosphate buffer, pH 7.0) by dialysis of the ammonium sulfate into the protein solution. The precipitated enzyme is recovered by centrifugation and then dialyzed against 10 mM potassium phosphate in 0.1 mM DTT and EDTA at least 500:1 v/v ratio. The enzyme is relatively stable in this condition for 4–10 days. The enzyme may also be stored in a 30% glycerin solution at −20° for 2–3 months with no loss of activity.

Homogeneity and Molecular Properties. HMG-CoA synthase prepared as described herein is homogeneous in the analytical ultracentrifuge and by disc electrophoresis in polyacrylamide gel.[1] The homogeneous synthase has a sedimentation coefficient ($s_{20,w}$) of 5.7 S and a molecular weight, determined by equilibrium sedimentation, of 105,000 ± 2850. This enzyme is composed of subunits with molecular weights determined by sodium dodecyl sulfate–polyacrylamide gel electrophoresis of 53,000. The relation between absorbance at 280 nm and refractometrically determined protein concentration is given by $C = 0.88 (A_{280\,nm}^{1\,cm})$, where C is protein concentration in milligrams per milliliter and A is absorbance at 280 nm with a 1-cm light path. The absorbancy ratio A_{280}/A_{260} of the pure enzyme is 2.14. To convert protein concentration as determined by the method of Lowry et al.,[4] with bovine serum albumin as standard, to the refractometrically determined protein concentration, the former is multiplied by 0.775.

Kinetic Parameters. The synthase is markedly inhibited by acetoacetyl-CoA above 20 μM at acetyl-CoA concentrations between 50 and 400 μM. The K_m value for acetyl-CoA at 50 μM acetoacetyl-CoA is about 500 μM. As the concentration of acetoacetyl-CoA is lowered, the K_m for acetyl-CoA decreases markedly; at 5 μM acetoacetyl-CoA, the K_m for acetyl-CoA is approximately 50 μM.

[4] Lowry, O. H., Rosebrough, N. J., Farr, A. L., and Randall, R. N., *J. Biol. Chem.* **193**, 265 (1951).

[20] Cytosolic 3-Hydroxy-3-methylglutaryl-CoA Synthase from Chicken Liver

By KENNETH D. CLINKENBEARD, TATSUO SUGIYAMA, and
M. DANIEL LANE

Acetyl-CoA + acetoacetyl-CoA + $H_2O \rightleftharpoons$
3-hydroxy-3-methylglutaryl-CoA + CoA

Cytosolic 3-hydroxy-3-methylglutaryl-CoA (HMG-CoA) synthase catalyzes the essentially thermodynamically irreversible condensation of acetyl-CoA with acetoacetyl-CoA as illustrated in the above equation. Two distinct synthases, one cytosolic and the other mitochondrial, are present in chicken liver.[1,2] The cytosolic synthase, for which the purification and properties are outlined here, catalyzes the second step of cholesterogenesis from acetyl-CoA.[1,3-6] The preparation of the mitochondrial

[1] K. D. Clinkenbeard, T. Sugiyama, and M. D. Lane, in preparation.
[2] W. D. Reed and M. D. Lane, this volume [19].
[3] T. Sugiyama, K. Clinkenbeard, J. Moss, and M. D. Lane, *Fed. Proc., Fed. Amer. Soc. Exp. Biol.* **31**, 475 (1972).
[4] T. Sugiyama, K. Clinkenbeard, J. Moss, and M. D. Lane, *Biochem. Biophys. Res. Commun.* **48**, 255 (1972).
[5] K. D. Clinkenbeard, T. Sugiyama, J. Moss, W. D. Reed, and M. D. Lane, *J. Biol. Chem.* **248**, 2275 (1973).
[6] C. A. Barth, H. J. Hackenschmidt, E. E. Weis, and K. F. A. Decker, *J. Biol. Chem.* **248**, 738 (1973).

HMG-CoA synthase from chicken liver which functions in ketogenesis is described by Reed and Lane.[2]

Assay Methods

Principle. HMG-CoA synthase activity can be conveniently determined by a radiochemical assay which monitors the incorporation of [^{14}C]acetyl-CoA (volatile ^{14}C activity when taken to dryness at 95° in 6 N HCl) into HMG-CoA (nonvolatile ^{14}C activity when taken to dryness at 95° in 6 N HCl). Alternatively, a spectrophotometric assay modified from that of Ferguson and Rudney[7] can be used which measures the acetyl-CoA–dependent disappearance of acetoacetyl-CoA at 300 nm.

Radiochemical Assay

Reagents

Tris(Cl⁻) buffer, 0.4 M containing 0.4 mM EDTA, pH 8.0, at 30°
[1-^{14}C]Acetyl-CoA, 2.0 mM, pH 4.5 (approximately 3×10^6 cpm/μmole; the specific activity must be accurately determined).
Acetoacetyl-CoA, 1.0 mM, pH 4.5
Liquid scintillator, 22.0 g of 2,5-diphenyloxazole (PPO) 0.4 g/liter, 4-bis-2-(4-methyl-5-phenyloxazolyl)-benzene (dimethyl POPOP), 2668 ml toluene and 1332 ml Triton X-100.

Procedure. The complete reaction mixture contains the following components (in micromoles): Tris(Cl⁻), 20; EDTA, 0.02; [1-^{14}C]acetyl-CoA, 0.04; acetoacetyl-CoA 0.01; and 0.1–1.0 milliunit of HMG-CoA synthase in a total volume of 0.2 ml. The reaction is initiated by addition of [1-^{14}C]acetyl-CoA to the otherwise complete reaction mixture after a preliminary incubation at 30° for 2 minutes. At 2, 4, 6, and 8 minutes after addition of acetyl-CoA, 0.04 ml aliquots are removed and pipetted into 15 × 45 mm glass vials (flat-bottomed shell vials) containing 0.2 ml of 6 N HCl, and the acidified aliquot taken to dryness at 95° in a forced draft oven. Water (0.3 ml) is added to each vial followed by 3 ml of liquid scintillator; the vials are capped and inserted into clear acrylic holders (Packard, catalog No. 6000144). The nonvolatile ^{14}C activity (as [^{14}C]-3-hydroxy-3-methylglutaryl-CoA) is determined with a liquid scintillation spectrometer. Under the conditions described, a linear increase in nonvolatile ^{14}C activity with time is observed until at least 70% of the acetoacetyl-CoA present is consumed. It should be noted

[7] J. J. Ferguson, Jr. and H. Rudney, *J. Biol. Chem.* **234**, 1070 (1959).

that the radiochemical assay cannot be employed for crude enzyme preparations containing HMG-CoA lyase.

Spectrophotometric Assay

Reagents

Tris(Cl$^-$) buffer, 0.2 M containing 0.2 mM EDTA, pH 8.0, at 30°
Acetyl-CoA, 10.0 mM, pH 4.5
Acetoacetyl-CoA, 1.0 mM, pH 4.5
MgCl$_2$, 1.0 M

Procedure. The complete assay mixture contains (in micromoles): Tris (Cl$^-$), 100; EDTA, 0.1; acetyl-CoA, 0.2; acetoacetyl-CoA, 0.05; MgCl$_2$, 20; and HMG-CoA synthase in a total volume of 1.0 ml. Following a 2-minute preliminary incubation (30°) of 0.5–5.0 milliunits of the enzyme in assay mixture lacking acetyl-CoA to establish a base line rate of absorbance change at 300 nm, the reaction is initiated with acetyl-CoA. The rate of HMG-CoA synthesis is monitored by following the disappearance of the acetoacetyl-CoA at 300 nm ($\epsilon_{300\,nm}^{1\,cm} = 16.1 \times 10^3$ M^{-1} [1]) dependent on acetyl-CoA. Under these conditions the decrease in $A_{300\,nm}$ is linear with time until approximately 70% of the acetoacetyl-CoA has reacted and is proportional to enzyme concentration within the limits specified. HMG-CoA synthase activity is equal to one-half the rate of acetoacetyl-CoA consumption for preparations containing acetoacetyl-CoA thiolase activity $\geq$ HMG-CoA synthase activity. This correction is made in order to account for the CoA-dependent consumption of acetoacetyl-CoA catalyzed by thiolase—the extent of which is governed by the amount of CoA generated by HMG-CoA synthesis.

Definition of Unit and Specific Activity. One unit of HMG-CoA synthase is defined as the amount of enzyme which catalyzes the formation of 1 μmole of HMG-CoA per minute under the assay conditions described. Both the radiochemical and the spectrophotometric assays yield identical results. Specific activity is expressed as units per milligram of protein.

Purification Procedure

The purification procedure is based on that of Clinkenbeard *et al.*[1] The results of a typical purification of cytosolic HMG-CoA synthase from the cytoplasmic cell fraction of chicken liver are summarized in Table I. All operations are carried out at 4°.

Preparation of the Cytoplasmic Fraction. One kilogram of fresh chicken liver (approximately 20-month-old laying hens) obtained at a

TABLE I

PURIFICATION OF CYTOSOLIC HMG-CoA SYNTHASE FROM CHICKEN LIVER

Step	Total activity[b] (units)	Protein (mg)	Specific activity (units/mg protein)	Yield (%)
1. Cell-free extract[a]	58.4	140,000[c]	0.0004	100
2. Cytoplasmic fraction	53.8	54,000[c]	0.001	92
3. Calcium phosphate gel fractionation	44.1	1,550[c]	0.029	76
4. Phosphocellulose chromatography	24.7	82[d]	0.300	42
5. DEAE-cellulose chromatography	19.5	27[d]	0.72	33
6. Sephadex G-200 gel filtration	8.0	16[d]	0.50	14

[a] From 1.0 kg of fresh chicken liver.
[b] Measured by the spectrophotometric assay.
[c] Determined by the biuret method.[8]
[d] Determined by the method of Warburg and Christian.[9]

poultry processing plant is washed with ice-cold homogenizing medium containing 0.25 M sucrose, 0.1 mM EDTA, and 2.0 mM N-2-hydroxyethylpiperazine-N-2-ethane sulfonic acid (HEPES), pH 7.2, and sliced into cubes. These are suspended in 2.0 liters of the above medium and homogenized batchwise (250 ml/batch) with a loose fitting Potter-Elvehjem tissue homogenizer by two passes of the motor-driven Teflon pestle. After dilution to 4 liters with homogenizing medium, the homogenate is passed through two layers of cheesecloth, and cellular debris, nuclei and mitochondria removed by centrifugation at 20,000 g for 30 minutes. It is possible to isolate the mitochondrial HMG-CoA synthase from this loosely packed pellet.[2] After the addition of 50 ml of 40 mM toluene sulfonyl fluoride in 95% ethanol[10] to the supernatant fraction, a second centrifugation at 20,000 g for 60 minutes is conducted to remove additional debris.

Ammonium Sulfate (30–45% Saturated) Fractionation. To the supernatant fraction is added 1.0 M potassium phosphate, pH 7.0, to produce a final concentration of 50 mM potassium phosphate. This solution is

[8] E. Layne, Vol. III, p. 450.
[9] O. Warburg and W. Christian, *Biochem. Z.* **310**, 384 (1942).
[10] Toluene sulfonyl fluoride is employed in the early purification steps to minimize proteolysis.

brought to 45% saturation with solid ammonium sulfate (277 g/liter), and after standing for 30 minutes, the precipitated protein is collected by centrifugation. The pellet is resuspended in 2.5 liters of 30% saturated ammonium sulfate containing 50 mM potassium phosphate, pH 7.0, 0.1 mM EDTA, and 1 mM toluene sulfonyl fluoride, and after standing for 30 minutes the precipitate is removed by centrifugation. The supernatant solution (30–45% saturated ammonium sulfate fraction) is then brought to 65% saturation by the addition of solid ammonium sulfate (209 g/liter), and the precipitated protein is collected by centrifugation.

Calcium Phosphate Gel Fractionation. Calcium phosphate gel (29 mg dry weight/ml) is prepared by the method of Keilin and Hartree.[11] The pellet from the preceding step is resuspended in 1.0 liter of 10 mM potassium phosphate, pH 7.0, containing 0.1 mM EDTA and 1 mM toluene sulfonyl fluoride, and is dialyzed against 50 liters of this same buffer but with toluene sulfonyl fluoride omitted. To the dialyzed enzyme are added 1470 ml of gel suspension (2.9 mg of gel dry weight/mg of protein) and the concentrations of potassium phosphate, pH 7.0, and toluene sulfonyl fluoride brought to 10 mM and 1.0 mM, respectively. After stirring for 30 minutes the gel is recovered by centrifugation and the supernatant solution (10 mM potassium phosphate) saved. The gel pellet is resuspended in 2.0 liters of 50 mM potassium phosphate, pH 7.0, and the supernate collected. The gel is eluted again with 50 mM potassium phosphate buffer in an identical manner. The original supernate (10 mM phosphate) and the 50 mM phosphate eluates are pooled, brought to 0.5 mM toluene fluoride by the addition of 100 mM toluene sulfonyl fluoride in 95% ethanol, and the protein precipitated by the addition of solid ammonium sulfate (309 g/liter).

Phosphocellulose Chromatography. Phosphocellulose ion exchanger purchased from Schleicher and Schuell Company is prepared for enzyme purification by successive washings with 0.1 N NaOH, water (until neutral), 0.1 N HCl, water (until neutral), 50 mM potassium phosphate, pH 7.0, and, finally, 20 mM potassium phosphate, pH 7.0. The enzyme precipitate from the previous step is recovered by centrifugation, dissolved in 200 ml of 20 mM potassium phosphate, pH 7.0, 0.5 mM EGTA and 2 mM toluene sulfonyl fluoride, and then dialyzed overnight against 12 liters of 20 mM potassium phosphate, pH·7.0. After removal of insoluble material by centrifugation, the enzyme solution is applied to a column (5 × 85 cm) of phosphocellulose, and the column washed with 1.2 liters of 20 mM potassium phosphate, pH 7.0, prior to initiation of a 5-liter linear gradient of potassium phosphate, pH 7.0 (20–400 mM); the eluate

[11] D. Keilin and E. F. Hartree, *Proc. Roy. Soc., Ser. B* **124,** 397 (1938).

is collected fractionally. Greater than 98% of the HMG-CoA synthase activity eluted from the column appears in the eluate as a peak when the phosphate concentration reaches approximately 120 mM. The most active fractions are pooled, the enzyme precipitated by the addition of solid ammonium sulfate (492 g/liter), and the precipitate collected by centrifugation. Approximately 2% of the HMG-CoA synthase activity from phosphocellulose chromatography is present in the "breakthrough" volume and is not carried further in the purification scheme. Some of the molecular and catalytic properties of this activity (HMG-CoA synthase II) have been reported.[4]

DEAE-Cellulose Chromatography. Standard DEAE-cellulose ion exchanger was purchased from Schleicher and Schuell Company and prepared for enzyme purification by washing with 0.5 M potassium phosphate, pH 7.0, followed by 20 mM potassium phosphate, pH 7.0. The enzyme pellet recovered from phosphocellulose chromatography is dissolved in 20 ml of 20 mM potassium phosphate, pH 7.0, containing 0.1 mM EDTA, and dialyzed overnight against 4 liters of the same buffer. The dialyzed enzyme is applied to a column (1.6 × 78 cm) of DEAE-cellulose, and the column washed with 100 ml of 20 mM potassium phosphate, pH 7.0, containing 0.1 mM EDTA. A 1-liter linear gradient of potassium phosphate, pH 7.0 (20–250 mM), is applied to the column and the eluate collected fractionally. The majority of the HMG-CoA synthase activity (approximately 95%) is eluted from the column at approximately 120 mM potassium phosphate, while a small amount of activity (approximately 5%) was present in the breakthrough volume. The HMG-CoA synthase present in the breakthrough is referred to as synthase I in a previous communication.[4] Fractions containing the bulk of the synthase activity (eluting at 120 mM potassium phosphate) are pooled and precipitated by overnight dialysis against a solution containing 50 mM potassium phosphate, pH 7.0, 0.1 mM EDTA and sufficient ammonium sulfate to reach 70% saturation at equilibrium. The precipitated enzyme is collected by centrifugation.

Gel Filtration with Sephadex G-200. Enzyme from the previous step is dissolved in 1.5 ml of 50 mM potassium phosphate, pH 7.0, containing 50 mM potassium chloride and applied to a column (1.6 × 78 cm) of Sephadex G-200 equilibrated with the same buffer. The enzyme is eluted with this buffer as a single coincident activity and protein peak. The enzyme after the gel filtration step is stable for 2–4 months at 4° in solution (50 mM potassium phosphate, pH 7.0; 3–5 mg enzyme/ml) or as a suspension under 60% saturated ammonium sulfate. Any loss of activity during this period can be regained by treatment with dithiothreitol.

Previously, we reported the purification of multiple forms of HMG-

CoA synthase (synthases I and II) from frozen chicken liver.[4] These forms are distinctly different from the major HMG-CoA synthase purified from the cytoplasmic fraction (outlined here) or from the mitochondrial fraction[2] of fresh chicken liver. Minor activities of synthase I and II are present in the cytoplasmic fraction of fresh chicken liver but not in the mitochondrial fraction.[2] While the relationship between synthases I and II and the major cytoplasmic HMG-CoA synthase of chicken liver is not understood, it appears that synthases I and II may have arisen via proteolysis of the major cytoplasmic species of HMG-CoA synthase.

Molecular and Catalytic Properties. As illustrated in Table I, cytosolic HMG-CoA synthase is purified approximately 1200-fold in good yield by the procedure outlined. The enzyme preparation is free of acetoacetyl-CoA thiolase following the DEAE-cellulose chromatography (step 5). The enzyme purified as described is homogeneous as judged by its behavior in the analytical ultracentrifuge and by standard and dodecyl sulfate–polyacrylamide disc gel electrophoresis.[1] An $s_{20,w}^0$ of 6.3 S and a molecular weight of 100,000 (by sedimentation equilibrium; partial specific volume of 0.73) were obtained for the homogeneous enzyme.[1] Electrophoresis of the disassociated enzyme on dodecyl sulfate–disc gels gives rise to a single stained protein band corresponding to a subunit weight of 55,000, indicating that the native enzyme is a dimer of identical or similar polypeptide chains.[1]

The pH optimum for the cytosolic synthase is pH 9.4, while the pH range over which the enzyme exhibits >50% of peak activity extends from pH 8.0 to 9.6. The apparent K_m for acetoacetyl-CoA is <2.5 μM, and this substrate is a potent inhibitor of HMG-CoA synthesis at concentrations >30 μM. An apparent K_m for acetyl-CoA of 0.3 mM was determined with an acetoacetyl-CoA concentration of 0.1 mM. The chicken liver enzyme like the yeast enzyme[12] is sensitive to inhibition by thiol group reagents such as p-chloromercuribenzoate and Ellman's reagent, 5,5′-dithiobis-(2-nitrobenzoate).

In addition to the synthesis of HMG-CoA, the cytosolic synthase catalyzes the transacetylation of acetyl-CoA with dephospho-CoA yielding acetyl-dephospho-CoA.[13] This reaction is quite rapid in comparison to the overall reaction, and hence supports the suggestions of Steward and Rudney[14] and Middleton[15] than an acetyl-enzyme intermediate is involved in catalysis. Acetyl-CoA and acetoacetyl-CoA hydrolase activ-

[12] B. Middleton and P. K. Tubbs, *Biochem. J.* **126**, 27 (1972).

[13] K. Clinkenbeard and M. D. Lane, *Fed. Proc., Fed. Amer. Soc. Exp. Biol.* **32**, 627 (1973).

[14] P. R. Stewart and H. Rudney, *J. Biol. Chem.* **241**, 1222 (1965).

[15] B. Middleton, *Biochem. J.* **126**, 35 (1972).

ities are also associated with the purified enzyme; however, these activities are weak in comparison to the rate of the synthetic reaction catalyzed by HMG-CoA synthase.[1,13]

The actions of highly purified cytosolic acetoacetyl-CoA thiolase and HMG-CoA synthase were found to rapidly convert acetyl-CoA to HMG-CoA in a thermodynamically favorable manner. The apparent equilibrium constant for the reaction shown in Eq. (1),

$$3 \text{ Acetyl-CoA} + H_2O \rightleftharpoons \text{HMG-CoA} + 2 \text{ CoA} \tag{1}$$

at pH 8.0 and 30° is 1.33.[5]

[21] Cytosolic Acetoacetyl-CoA Thiolase from Chicken Liver

By KENNETH D. CLINKENBEARD, TATSUO SUGIYAMA, and M. DANIEL LANE

$$\text{Acetoacetyl-CoA} + \text{CoA} \rightleftharpoons 2 \text{ acetyl-CoA}$$

Cytosolic acetoacetyl-CoA thiolase catalyzes the reversible thiolytic cleavage of acetoacetyl-CoA by CoA to yield two molecules of acetyl-CoA as illustrated in the above equation. Recently, it has been demonstrated that the 3-hydroxy-3-methylglutaryl-CoA necessary for hepatic cholesterogenesis is synthesized in the cytoplasmic cell compartment from acetyl-CoA by the actions of two cytosolic enzymes: acetoacetyl-CoA thiolase and 3-hydroxy-3-methylglutaryl-CoA synthase.[1-6]

Assay Methods

Principle. The equilibrium of the reaction shown in the equation above lies far to the right ($K_{app} = 5.0 \times 10^4$ at pH 8.1, 25°,[7]). Thiolase activity is most conveniently determined in the direction of thiolytic cleavage by

[1] K. D. Clinkenbeard, T. Sugiyama, J. Moss, W. D. Reed, and M. D. Lane, *J. Biol. Chem.* **248**, 2275 (1973).

[2] T. Sugiyama, K. Clinkenbeard, J. Moss, and M. D. Lane, *Fed. Proc., Fed. Amer. Soc. Exp. Biol.* **31**, 475 (1972).

[3] T. Sugiyama, K. Clinkenbeard, J. Moss, and M. D. Lane, *Biochem. Biophys. Res. Commun.* **48**, 255 (1972).

[4] C. A. Barth, H. J. Hackenschmidt, E. E. Weis, and K. F. A. Decker, *J. Biol. Chem.* **248**, 738 (1973).

[5] K. D. Clinkenbeard, T. Sugiyama, and M. D. Lane, in preparation.

[6] K. D. Clinkenbeard, T. Sugiyama, and M. D. Lane, this volume [20].

[7] J. R. Stern, M. J. Coon, and A. del Campillo, *J. Amer. Chem. Soc.* **75**, 1517 (1953).

following the CoA-dependent disappearance of acetoacetyl-CoA using a modification of the assay developed by Stern.[8] Thiolase-catalyzed formation of acetoacetyl-CoA from acetyl-CoA can be measured spectrophotometrically at 303 nm in the presence of divalent magnesium ion at pH 8.8. The latter mentioned factors promote enolization of acetoacetyl-CoA[9] thereby enhancing its extinction coefficient ($\epsilon_{303\,nm}^{1\,cm} = 20.7 \times 10^3$ M^{-1}) and shifting the otherwise unfavorable equilibrium more toward condensation.

Thiolytic Cleavage Assay

Reagents

Tris (Cl$^-$) buffer, 0.2 M, containing 0.2 mM EDTA pH 8.2 at 30°
Acetoacetyl-CoA, 4.0 mM, pH 4.5
CoA, 4.5 mM, pH 4.5

Procedure. The complete assay mixture contains the following components (in micromoles): Tris(Cl$^-$), 100; EDTA, 0.1; acetoacetyl-CoA, 0.12; CoA, 0.09; and 5–50 milliunits of thiolase in a total volume of 1.0 ml. After a 3-minute preliminary incubation of thiolase in the above reaction mixture lacking CoA to establish the rate of CoA independent disappearance of acetoacetyl-CoA, the reaction is initiated by the addition of 90 nmoles of CoA. The rate of thiolysis is followed at 30° by the disappearance of acetoacetyl-CoA absorbance at 300 nm ($\epsilon_{300\,nm}^{1\,cm} = 3.6 \times 10^3\ M^{-1}$). The reaction is linear with respect to time until at least 60% of the available CoA is utilized.

Condensation Assay

Reagents

Tris(glycine) buffer, 0.2 M, pH 8.8
Acetyl-CoA, 8.2 mM, pH 4.5
MgCl$_2$, 1.0 M

Procedure. The complete assay mixture contains the following components (in micromoles): Tris(glycine), 100; acetyl-CoA, 0.82; MgCl$_2$, 50; and enzyme in a final volume of 1.0 ml. After incubation of the complete assay mixture lacking enzyme for 3 minutes, the reaction is initiated by the addition of 3–30 milliunits of thiolase; the synthesis of acetoacetyl-

[8] J. R. Stern, Vol. I, p. 573.
[9] G. Hartmann and F. Lynen, *in* "The Enzymes" (P. D. Boyer, H. Lardy, and K. Myrbäck, eds.), 2nd ed., Vol. 5, pp. 381–386. Academic Press, New York, 1961.

CoA is followed spectrophotometrically at 303 nm ($\epsilon_{303\,nm}^{1\,cm} = 20.7\ M^{-1}$) and 30°. A linear increase in $A_{303\,nm}$ is observed for at least 3 minutes, and the initial rate is proportional to enzyme concentration under these conditions.

Definition of Unit and Specific Activity. One unit of thiolase catalyzes the CoA-dependent cleavage of 1 μmole of acetoacetyl-CoA per minute under the conditions described. Specific activity is expressed in units per milligram of protein.

Purification Procedure

Although the fractionation method outlined below does not include an initial subcellular fractionation, this procedure is selective for and leads to homogeneous preparations of the cytosolic acetoacetyl-CoA thiolase.[1] The results of each step of a typical purification beginning with 1.2 kg of liver are summarized in Table I. All operations are carried out

TABLE I
PURIFICATION OF THE CYTOSOLIC THIOLASE FROM CHICKEN LIVER

Step	Total activity[a] (units)	Protein (mg)	Specific activity (units/mg protein)	Yield (%)
1. Cell-free extract[b]	41,300	234,180[c]	0.18	100
2. Supernate	39,890	142,450[c]	0.28	97
3. First (30–45% saturated) ammonium sulfate fractionation	22,180	34,840[c]	0.64	53
4. Calcium phosphate gel fractionation	10,040	7,660[c]	1.31	24
5. Phosphocellulose chromatography	8,630	520[d]	16.5	21
6. Second (30–45% saturated) ammonium sulfate fractionation	6,980	230[d]	30.3	17
7. DEAE-cellulose chromatography	6,720	83[d]	81.0	16
8. Hydroxylapatite chromatography	3,020	24[d]	126	7
9. Sephadex G-200 gel filtration	1,708	8.4[d]	203	4

[a] Acetoacetyl-CoA cleavage assay.
[b] From 1.2 kg of chicken liver.
[c] Determined by the method of Lowry *et al.*[10]
[d] Determined by the method of Warburg and Christian.[11]

at 4° and all solutions in contact with the enzyme contain 0.1 mM EDTA and 5 mM β-mercaptoethanol unless otherwise specified. The purification outlined here is that described by Clinkenbeard et al.[1]

Extraction and First (30–45% Saturated) Ammonium Sulfate Fractionation. Fresh livers from laying hens obtained at a poultry processing plant are quickly chilled on ice and stored at −20° for up to 2 months. Thawed liver (1.2 kg total) is cut into small cubes, suspended in 4 volumes of buffer containing 0.085 M K$_2$HPO$_4$, 0.009 M KH$_2$PO$_4$, 0.070 M KHCO$_3$, and 0.1 mM EDTA and then homogenized using a Sorvall Ominimixer at maximum speed for three 30-second periods. Following centrifugation of the homogenate at 13,000 g for 45 minutes, the supernate is collected, brought to 50% saturation with solid ammonium sulfate (313 g/liter), and after standing for 30 minutes, the precipitate is collected by centrifugation. The pellet is resuspended in 2 liters of 45% saturated ammonium sulfate containing 0.05 M potassium phosphate, pH 7.0. After standing for 30 minutes the precipitate is collected by centrifugation and the extraction of the pellet with 45% saturated ammonium sulfate repeated. The pellet is extracted twice in a similar manner with 2 liters of 30% saturated ammonium sulfate containing 50 mM potassium phosphate, pH 7.0; the supernate, after centrifugation, is retained and brought to 70% saturation by addition of solid ammonium sulfate (273 g/liter).

Calcium Phosphate Gel Fractionation. Calcium phosphate gel (29 mg dry weight/ml) is prepared by the method of Keilin and Hartree.[12] The pellet from the preceding step (30–45% saturated ammonium sulfate cut) is dissolved in 1 liter of 10 mM potassium phosphate, pH 7.0, and dialyzed against 50 liters of the same buffer overnight. To the dialyzed enzyme (38.4 g of protein) are added 2.7 liters of gravity-packed calcium phosphate gel prepared as described above. The suspension is brought to a final concentration of 5.0 mM potassium phosphate, pH 7.0, stirred for 30 minutes and the gel recovered by centrifugation at room temperature. After washing the gel twice with 2 liters of 10 mM potassium phosphate, pH 7.0, as described above, the washes (5 mM and two 10 mM potassium phosphate) are combined, brought to 70% saturation with solid ammonium sulfate, and the precipitate collected by centrifugation.

Phosphocellulose Chromatography. Phosphocellulose ion exchanger purchased from Schleicher and Schuell Company, is prepared for enzyme purification by successively washing with 0.1 N NaOH, water (until neu-

[10] O. H. Lowry, N. J. Rosenbrough, A. L. Farr, and R. J. Randall, *J. Biol. Chem.* **193**, 265 (1951).

[11] O. Warburg and W. Christian, *Biochem. Z.* **310**, 384 (1942).

[12] D. Keilin and E. F. Hartree, *Proc. Roy. Soc., Ser. B* **124**, 397 (1938).

tral), 0.1 N HCl, water (until neutral), 0.05 M potassium phosphate, pH 7.0, and finally, 20 mM potassium phosphate, pH 7.0, containing 0.1 mM EDTA and 5.0 mM β-mercaptoethanol. The enzyme from the previous step is dissolved in and dialyzed against 20 mM potassium phosphate, pH 7.0, overnight. After dialysis and removal of any insoluble material by centrifugation, 176 ml of enzyme solution (approximately 7.5 g of protein) are applied to a phosphocellulose column (9 $\times$ 67 cm) previously equilibrated as described above. After washing the column with 2 liters of 20 mM potassium phosphate, pH 7.0, the enzyme is eluted with an 8-liter linear potassium phosphate gradient (20 mM to 0.45 M) as a single thiolase activity peak at a phosphate concentration of about 150 mM. The most active fractions are pooled and the enzyme precipitated with solid ammonium sulfate (472 g/liter).

Second (30–45% Saturated) Ammonium Sulfate Fractionation. The protein precipitate from the preceding step is suspended in 10 ml of 45% saturated ammonium sulfate containing 50 mM potassium phosphate, pH 7.0, and allowed to stand for 15 minutes; the precipitate collected by centrifugation is extracted as described above with 10 ml of 30% saturated ammonium sulfate containing 50 mM potassium phosphate, pH 7.0. The precipitate, after centrifugation, is discarded, solid ammonium sulfate added (84 g/liter) to the supernate and the precipitate recovered. The pellet is dissolved in 10 ml of 20 mM potassium phosphate, pH 7.0, and dialyzed overnight against 8 liters of the same buffer.

DEAE-Cellulose Chromatography. Schleicher and Schuell standard DEAE-cellulose ion exchanger is prepared for enzyme purification by washing with 0.5 M potassium phosphate, pH 7.0, followed by 20 mM potassium phosphate, pH 7.0, containing 0.1 mM EDTA and 5 mM β-mercaptoethanol. The dialyzed enzyme (about 323 mg protein) is applied to a column (2.2 $\times$ 30 cm) of DEAE-cellulose equilibrated as described above and the column washed with 20 mM potassium phosphate, pH 7.0. Thiolase is eluted almost quantitatively in the "breakthrough" fraction. The most active fractions are pooled, protein precipitated with solid ammonium sulfate (472 g/liter), and the precipitate collected by centrifugation.

Hydroxylapatite Chromatography. After dialysis against 20 mM potassium phosphate, pH 7.0, the DEAE-cellulose chromatographically purified enzyme (about 75 mg of protein) is applied to a hydroxylapatite (Bio-Gel HT, Bio-Rad Laboratories) column (2.4 $\times$ 26 cm) previously equilibrated with the same buffer. Stepwise elution includes 185 ml of 20 mM, followed by 600 ml of 70 mM potassium phosphate buffer, pH 7.0. Thiolase activity is eluted as a single peak with the 70 mM phosphate buffer; no additional thiolase activity is eluted by 500 mM phosphate buffer. The pooled active fractions are brought to 70% saturation with solid ammonium sulfate and the precipitated enzyme recovered by centrifugation.

Gel Filtration with Sephadex G-200. Thiolase (approximately 24 mg of protein) from the preceding step is dialyzed against 20 mM potassium phosphate, pH 7.0, and applied in a volume of 1.3 ml to a Sephadex G-200 column (1.5 × 86 cm) previously equilibrated with the same buffer. A single thiolase activity peak is eluted with the first protein peak at a V_e/V_o of 1.5 by 20 mM potassium phosphate buffer, pH 7.0. The cytosolic acetoacetyl-CoA thiolase, purified 1100-fold through the gel filtration step (Table I), is virtually free of 3-hydroxy-3-methylglutaryl-CoA synthase (<0.3 milliunit synthase/unit of thiolase).[1] At any stage in the purification after phosphocellulose chromatography (step 5, Table I), the enzyme pellet after precipitation with ammonium sulfate, centrifugation, and removal of supernatant solution is stable for up to 12 months when stored tightly stoppered at −90°.

Molecular and Catalytic Properties. The molecular properties of the purified chicken liver cytosolic acetoacetyl-CoA thiolase are quite similar to those of pig heart acetoacetyl-CoA thiolase.[13] The chicken liver enzyme purified as described is homogeneous as judged by its behavior in the analytical ultracentrifuge and on sodium dodecyl sulfate and polyacrylamide disc gel electrophoresis.[1] The purified enzyme has a sedimentation coefficient ($s_{20,w}$) of 7.94 S.[1] Gel filtration on a calibrated column of Sephadex G-200 yields an estimated molecular weight of 188,000, while the more reliable sedimentation equilibrium method gave a molecular weight of 169,000.[1] A subunit weight of 41,000 was determined by sodium dodecyl sulfate gel electrophoresis indicating that the native enzyme is a tetramer of identical or similar polypeptide chains.[1]

The purified cytosolic acetoacetyl-CoA thiolase exhibits a pI (isoionic point) of 7.1, which is exactly the pI of the sole species of acetoacetyl-CoA thiolase present in the cytoplasmic cell fraction of chicken liver. Intracellular distribution studies indicate that the cytosolic acetoacetyl-CoA thiolase accounts for approximately 70% of the acetoacetyl-CoA thiolase activity in the laying hen.[1] It has also been shown that the cytoplasmic thiolase activity of chicken liver is subject to depression by cholesterol feeding[1]; this and its cytoplasmic localization indicate involvement in hepatic cholesterogenesis.

The kinetic parameters of the cytosolic acetoacetyl-CoA thiolase are summarized in Table II. Maximal activity in the cleavage direction is obtained at pH 8.4, although the pH optimum range, within which greater than 85% of the peak activity is observed, extends from pH 8.0 to 9.0.

[13] U. Gehring, C. Riepertinger, and F. Lynen, *Eur. J. Biochem.* **6**, 264 (1968).

TABLE II
KINETIC DATA FOR CYTOSOLIC THIOLASE FROM CHICKEN LIVER

Parameter	
Cleavage direction	
pH optimum[a]	8.4
$K_m(AcAcCoA)$[b]	38.5 μM
$K_m(CoA)$[c]	6.4 μM
V_m	1.35 μmoles/minute/unit thiolase
Molecular activity	54,000 moles/minute/mole thiolase
Condensation direction	
$K_m{}^d$	270 μM
$V_m{}^d$	55.4 nmoles/minute/unit thiolase
Molecular activity	1,770 moles/minute/mole thiolase

[a] Initial velocity measured in the cleavage direction as described except that pH was varied between 7.0 and 9.2 using 0.1 M potassium phosphate, Tris (Cl⁻) or glycine (Na⁺) buffers. Initial velocity of thiolysis was calculated using appropriate $\epsilon_{300\ nm}^{1\ cm}$ determined experimentally for acetoacetyl-CoA at each pH value employed.

[b] Initial velocity was measured for thiolysis as described except that acetoacetyl-CoA concentration was varied between 20 and 200 μM.

[c] Initial velocity was measured for thiolysis as described except that CoA concentration was varied between 4.35 and 435 μM.

[d] Initial velocity was measured in the condensation direction as described except that acetyl-CoA concentrations were varied between 0.082 and 0.82 mM and the final concentration of $MgCl_2$ was 50 mM.

CoA is inhibitory in the cleavage direction at concentrations >30 μM in the presence of saturating acetoacetyl-CoA concentrations (120 μM).[1]

[1] J. J. Ferguson and H. Rudney, *J. Biol. Chem.* **234**, 1072 (1959).

[22] 3-Hydroxy-3-methylglutaryl-CoA Synthase from Bakers' Yeast

By B. MIDDLETON and P. K. TUBBS

$CH_3CO—SCoA + CH_3COCH_2CO—SCoA + H_2O$

$$\rightarrow CH_3—\underset{\underset{CH_2CO_2H}{|}}{\overset{\overset{CH_2CO—SCoA}{|}}{C}}—OH \quad + CoA$$

3-Hydroxy-3-methylglutaryl-CoA (HMG-CoA)synthase (EC 4.1.3.5) was discovered in bakers' yeast by Ferguson and Rudney[1] who demonstrated the occurrence and stoichiometry of the above reaction.

Assay Method

Principle. The spectrophotometric assay is based on the absorption of the enol form of acetoacetyl-CoA at 303 nm.[2] The synthase activity is measured by observing the acetyl-CoA stimulated decrease in this absorption, Mg^{2+} is routinely included in the assay in order to increase the apparent extinction coefficient of the acetoacetyl-CoA.[3] Mg^{2+} has no direct effect on the enzyme activity.

Reagents

Tris-HCl, pH 8.2, 50 mM containing 20 mM MgCl$_2$
Acetoacetyl-CoA, prepared by reaction of CoA with diketen,[4] 6.4 mM
Acetyl-CoA,[5] 8.6 mM

Procedure. In the standard assay a 1-cm light path silica cuvette contains (at 30°), 50 mM tris-HCl, pH 8.2, 20 mM MgCl$_2$, 16 μM acetoacetyl-CoA and enzyme in a total volume of 1.99 ml. The enzyme is added last and the rate of acetoacetyl-CoA hydrolysis (if any) is measured. Then acetyl-CoA (10 μl of 8.6 mM) is added and the increase in rate of acetoacetyl-CoA disappearance at 303 nm measures the HMG-CoA synthase activity. Under these conditions the apparent extinction coefficient of the acetoacetyl-CoA is 20×10^3 liter mole^{-1} at 303 nm. It should be noted that the presence of acetoacetyl-CoA thiolase in the HMG-CoA synthase preparation can give spuriously high rates of apparent synthase activity,[1] particularly if there is a significant amount of free CoA in the acetyl-CoA. It is, therefore, important to keep the free CoA content of acetyl- and acetoacetyl-CoA as low as possible.

Protein Determination. The biuret method of Gornall *et al.*[6] is used during the first three steps of the purification. After the DEAE-cellulose chromatography the direct spectrophotometric method[7] can be used.

Units. A unit of enzymatic activity is defined as the amount of enzyme necessary to transform 1 μmole of acetoacetyl-CoA into product per minute under the conditions described. Specific activity is expressed in units per milligram of protein.

[2] F. Lynen, *Fed. Proc., Fed. Amer. Soc. Exp. Biol.* **12**, 683 (1953).
[3] J. R. Stern, *J. Biol. Chem.* **221**, 33 (1956).
[4] T. Wieland and L. Rueff, *Angew. Chem.* **65**, 186 (1953).
[5] E. R. Stadtman, Vol. 3 [137].
[6] A. G. Gornall, C. J. Bardawill, and M. M. David, *J. Biol. Chem.* **177**, 751 (1949).
[7] E. Layne, Vol. 3 [73].

Purification Procedure[8]

Autolysis. Fresh bakers' yeast is crumbled into a stainless steel beaker and toluene added to 110 ml/kg of yeast. The mixture is heated in a water bath at 55° until the yeast has warmed to 38°. The yeast is then allowed to autolyze at 38° for 8–12 hours. Within these time limits the yield and specific activity of the HMG-CoA synthase remain approximately constant; further incubation causes losses. At the end of this period an equal volume of cold deionized water (900 ml/kg of yeast) is added with stirring and the cell debris removed by centrifugation at 15,000 g for 20 minutes.

Ethanol Fractionation. The clear, light yellow extract is allowed to warm to 18–20° and cold (−5°) 96% (v/v) ethanol is added slowly, with stirring, to a final concentration of 25% (v/v). The preparation is then stirred at room temperature for 2 hours. The copious white precipitate is then removed by centrifugation (2000 g for 20 minutes) at 18° and the clear supernatant cooled to −3° and allowed to stand at this temperature for 40 minutes. The reddish brown sticky precipitate is collected by centrifugation (2000 g for 20 minutes) at −3°. It is dissolved in 10 mM potassium phosphate, pH 7.8, containing 1 mM dithiothreitol to give a final volume of 100 ml/kg of yeast. All subsequent steps are performed at 4°.

Precipitation at pH 5.6. Acetic acid (1 M) is added slowly, with stirring, until the pH is 5.6. The precipitate is collected by centrifugation and dissolved in 50 mM potassium phosphate pH 7.8.

DEAE-Cellulose Chromatography. The pH 5.6 fraction is diluted with 10 mM potassium phosphate, pH 7.8, to give a protein concentration of 30 mg/ml. It is then applied to a column of DEAE-cellulose previously equilibrated with 10 mM potassium phosphate, pH 7.8, containing 0.5 mM dithiothreitol. The protein load is 12.5 mg/ml of packed exchanger. The column is washed with one column volume of the above buffer and the synthase then eluted with 75 mM potassium chloride contained in the above buffer.

Calcium Phosphate-Gel Adsorption. The DEAE-cellulose eluate is immediately adjusted to pH 6.8 with 1 M acetic acid and calcium phosphate gel (1.3 mg/mg of protein) is added. After 5 minutes the gel is collected by centrifugation and the supernatant discarded. The enzyme is immediately eluted from the gel with two washes of 200 mM potassium phosphate pH 8.3. It is important that the elution step be carried out swiftly since the enzymatic activity recoverable from the gel decreases

[8] B. Middleton and P. K. Tubbs, *Biochem. J.* **126,** 27 (1972).

TABLE I

PURIFICATION OF HMG-CoA SYNTHASE FROM YEAST[a]

Procedure	Volume (ml)	Total activity (unit)	Protein concn. (mg/ml)	Specific activity (unit/mg)	Yield (%)
Centrifuged autolysate	12110	2900	37	0.007	(100)
Ethanol fractionation	950	2460	65	0.04	85
pH 5.6 precipitation	210	2000	60	0.16	70
DEAE-cellulose column	2300	810	0.52	0.47	28
Calcium phosphate gel	40	470	16.5	0.7	16
Sephadex G-150 (after concentration)	18	350	9.2	2.1	12

[a] From Middleton and Tubbs.[8]

rapidly with time. After this step the enzyme can be stored for some weeks at $-20°$ in the presence of 30% (v/v) glycerol without activity loss.

Gel Filtration on Sephadex G-150. The enzyme is first freed from glycerol by passage through a column of Sephadex G-25 at $20°$ and then concentrated by rotary evaporation at a bath temperature of $28°$. Batches of 10 ml of enzyme are then applied to a column of Sephadex G-150 (64×3.7 cm) equilibrated with 20 mM tris-HCl, pH 7.5, containing 0.5 mM dithiothreitol and 10% (v/v) glycerol. The step is performed at $4°$. Fractions (10 ml) are collected and those containing HMG-CoA synthase are pooled. The pooled eluate is concentrated by rotary evaporation, as before, until the glycerol concentration is about 40–50% (v/v). The enzyme is stored at $-20°$. The complete purification from 9.1 kg of fresh bakers' yeast is summarized in Table I.

Properties

Stability. At all stages of the above purification the enzyme is labile to freezing and thawing. The purified enzyme also loses activity at $4°$ in the absence of glycerol but when stored at $-20°$ in the presence of dithiothreitol and at glycerol concentrations of 30–50% (v/v) the purified synthase retains complete activity for periods of up to 3 months.

Purity. HMG-CoA synthase purified by the above method is not homogeneous but has very low activities of acetoacetyl-CoA thiolase and other enzymes capable of interfering with the synthase reaction. Thus, a synthase preparation with specific activity of 2.1 units/mg would have an acetoacetyl-CoA thiolase activity of less than 0.004 unit/mg and an acetoacetyl-CoA hydrolase activity of 0.015 unit/mg.

Molecular Weight. By gel filtration on Sephadex G-200, HMG-CoA synthase has a molecular weight of 130,000.[8]

Isoelectric Point. Using isoelectric focusing, Kornblatt and Rudney[9] have shown that the isoelectric point of yeast HMG-CoA synthase is 5.6.

Substrate Specificity. The enzyme is absolutely specific for the acyl-moiety of both its substrates. Thus neither propionyl- nor butyryl-CoA can replace acetyl-CoA, and 3-ketohexanoyl-CoA cannot replace aceto-acetyl-CoA.[10] HMG-CoA synthase is not absolutely specific for the CoA moiety of its substrates. Thus acetyl-3'-dephospho-CoA,[11] acetylpan-tetheine,[12] and acetylglutathione[12] can all act as acetyl-CoA analogues in the reaction but only acetoacetyl-ACP[12,13] has been found to replace acetoacetyl-CoA as a substrate.

Kinetic Properties.[10] Yeast HMG-CoA synthase is strongly inhibited by acetoacetyl-CoA. This substrate inhibition is competitive with respect to acetyl-CoA binding and is affected by the pH. It is less marked at pH values above 8.0. At pH 8.0 the observed kinetic parameters (at infinite concentration of the invariant substrate) are K_m(acetyl-CoA) = 14 μM, K_m (acetoacetyl-CoA) = $\ll 0.4$ μM, and K_i (acetoacetyl-CoA) = 8 μM. At pH 8.9 these become: K_m(acetyl-CoA) = 18 μM, K_m (acetoacetyl-CoA) = 3 μM, and K_i (acetoacetyl-CoA) = 20 μM. Free CoA and all acyl-CoA derivatives that are simple acyl analogues of substrates or products of the enzyme act as reversible inhibitors.

Irreversible Inhibitors. HMG-CoA synthase is very susceptible to inhibition by low concentrations of certain thiol group reagents[8] such as N-ethylmaleimide, p-chloromercuribenzoate, Cd^{2+}, and various arsenicals. Iodoacetate and iodoacetamide[8,12] are less effective. Preincubation with acetyl-CoA protects against irreversible loss of activity by thiol group reagents[8].

Mechanism of Reaction. Data from kinetic studies[10] and studies with irreversible inhibitors[8] suggest a two-step reaction mechanism in which initial binding of acetyl-CoA yields an acetyl-enzyme covalent intermediate. This is followed by the transfer of the acetyl group from the enzyme to acetoacetyl-CoA yielding HMG-CoA. Both the formation of acetyl enzyme[11,14] and its reaction with acetoacetyl-CoA[11] have been studied.

[9] J. A. Kornblatt and H. Rudney, *J. Biol. Chem.* **246**, 447 (1971).
[10] B. Middleton, *Biochem. J.* **126**, 35 (1972).
[11] B. Middleton and P. K. Tubbs, *Biochem. J.* **137**, 15 (1974).
[12] P. R. Stewart and H. Rudney, *J. Biol. Chem.* **241**, 1212 (1966).
[13] H. Rudney, P. R. Stewart, P. W. Majerus, and P. R. Vagelos, *J. Biol. Chem.* **241**, 1226 (1966).
[14] P. R. Stewart and H. Rudney, *J. Biol. Chem.* **241**, 1222 (1966).

Section IV

Hydrolases

[23] Hormone-Sensitive Triglyceride Lipase from Rat Adipose Tissue

By John C. Khoo and Daniel Steinberg

Adipose tissue contains at least two distinct triglyceride lipases. Lipoprotein lipase (LPL) is responsible for the degradation of the triglyceride moiety of circulating lipoproteins and thus controls the uptake of triglyceride fatty acids from plasma into adipose tissue. This enzyme is characterized by its requirement for a serum lipoprotein activator [or for a specific apoprotein (apoLp-Glu) responsible for the activating effect of serum lipoproteins].[1] Hormone-sensitive lipase (HSL) is responsible for the degradation of stored triglycerides in response to fat-mobilizing hormones (catecholamines, glucagon, ACTH, etc.). Although these two lipases are clearly distinct enzymes and have many properties that differentiate one from the other, they are sufficiently similar such that assays carried out on mixtures of the two will probably reflect some of each under most conditions; for example, the pH optimum for LPL (8.4) is significantly higher than that for HSL (7.0), but assays carried out at pH 8.4 will include some of the activity of HSL and vice versa. One can use the differential effects of activators and inhibitors to help distinguish the two activities in a mixture, but it is difficult if not impossible to obtain complete dissociation except by physical resolution of the two enzymes.

Adipose tissue also contains high levels of activity against di- and monoglycerides. Evidence has been presented to show that at least some of the monoglyceride lipase activity is distinct from HSL. Whether or not there is a separate diglyceride lipase is less clear.[2] In the case of highly purified pancreatic lipase[3] and LPL of milk,[4] it is clear that the single enzyme protein has activity against tri-, di-, and monoglyceride. Whether the same is true of HSL remains uncertain since even the most highly purified enzyme, as discussed below, retains activities against di- and monoglyceride that actually exceed the activity against triglyceride. Consequently, the rate-limiting step in the release of free fatty acid (FFA) is the hydrolysis of the first ester bond, whether assaying crude

[1] W. V. Brown, R. I. Levy, and D. S. Fredickson, *Biochim. Biophys. Acta* **200**, 573 (1970).

[2] R. A. Heller and D. Steinberg, *Biochim. Biophys. Acta* **270**, 65 (1972).

[3] P. Desnuelle, *in* "The Enzymes" (P. D. Boyer, ed.), 3rd ed., Vol. 7, p. 161. Academic Press, New York, 1972.

[4] T. Egelrud and T. Olivecrona, *Biochim Biophys. Acta* **306**, 115 (1973).

extracts or purified HSL. This is verified by the absence of any significant accumulation of lower glycerides during the assay, with parallel release of glycerol and FFA. However, under special circumstances, some accumulation of lower glycerides has been observed during fat mobilization.[5] With any given preparation, particularly with purified fractions, it is therefore advisable to verify that glycerol and FFA release are indeed parallel before accepting the rate of release of FFA as a measure of triglyceride lipase activity.

Finally, it should be noted that adipose tissue also contains activity against short-chain fatty acid esters such as tributyrin.[6] It has become conventional to refer to activity against short-chain glyceryl esters as well as activity against simple monoesters as "esterase activity" as opposed to "lipase activity," which is used to refer to activity against long-chain glycerides. While the terminology is somewhat inexact, the distinction is important since "esterases," although not fully characterized, have properties and significance that differ markedly from those of HSL and other long-chain glyceride hydrolases.

Assay Method

Principle. Free [^{14}C]oleic acid produced by hydrolysis of [^{14}C]triolein (labeled equally in all three acyl groups) is extracted and adsorbed onto an anion exchange resin. The adsorbed fatty acid is then displaced by strong base and the radioactivity is determined in a liquid scintillation counter (Method 1). Alternatively, the incubation mixture at the end of the assay period is subjected to a liquid–liquid partition to separate [^{14}C]oleic acid from unhydrolyzed [^{14}C]triolein and an aliquot of the upper aqueous phase containing the FFA is taken for scintillation counting (Method 2).

Method 1

Reagents

Amberlite (Mallinckrodt, IRA-400 c.p., in $RN(CH_3)_3^+$ Cl^- form, 20–50 mesh)
NaOH, 2 N
Hexane

[5] R. O. Scow, F. A. Stricker, T. Y. Pick, and T. R. Clary, *Ann. N.Y. Acad. Sci.* **131**, 288 (1965).
[6] L. R. Crum, R. C. Harbecke, J. J. Lech, and D. N. Calvert, *Biochim. Biophys. Acta* **198**, 229 (1970).

Triton X-100
NCS (Nuclear-Chicago, quaternary ammonium solubilizer)
Benzene
[^{14}C]Triolein
Unlabeled triolein, 0.02 M, in benzene
Gum arabic (5% in water)
Bovine serum albumin, 10%, in 0.2 M phosphate buffer, pH 6.8
EDTA, 0.3 M, pH 6.8
Dole's extraction mixture (isopropyl alcohol–heptane–sulfuric acid, 40:10:1)[7]
Liquid scintillator: 4 g 2',5'-diphenyloxazole (PPO) and 0.1 g 1,4-bis-2(4-methyl-5-phenyloxazole)-benzene(dimethyl-POPOP) in 1 liter of toluene.

Preparation of Dehydrated Hydroxyl-Charged Ion Exchange Resin.[8] Amberlite (about 800 g) is stirred with water and the fine particles are removed by decantation. The resin is then packed into a 5.5 × 60 cm column and converted to the hydroxyl form by washing with 2 liters of 2 N NaOH. The column is then washed with water until the effluent is neutral. Dehydration is achieved by first washing with 2 liters of isopropyl alcohol and then 2 liters of hexane. The dry Amberlite is stored in hexane in sealed containers.

Preparation of Triolein Emulsion. To obtain a low background, it is necessary to treat a benzene solution of the [^{14}C]triolein (diluted appropriately with unlabeled triolein) with basic Amberlite to remove contaminating FFA. Triolein stored for any length of time, even at very low temperatures, shows a progressive increase in [^{14}C]FFA level. After this treatment, the stock solution of triolein in benzene can be sealed under nitrogen and stored at 4° and can be used for periods up to 2 weeks without important increase in [^{14}C]FFA background. From this stock solution, substrate is prepared fresh each day for lipase assays. The appropriate aliquot of stock solution is transferred to a 12-ml conical tube and evaporated to dryness under nitrogen at 65°. A 5% solution of gum arabic at 4° is added and sonicated with the microtip attachment of a Branson cell disruptor, model 140 D, at near maximum energy level for about 30 seconds or until no oil droplets are visible at the surface. Prolonged sonication can cause unacceptably high background values. The final emulsion contains 2 μmoles of triolein and approximately 200,000 cpm in 0.1 ml.

Lipase Assay. Lipase assay is generally carried out at 30° for 1 hour

[7] V. P. Dole, *J. Clin. Invest.* **35**, 150 (1956).
[8] T. F. Kelley, *J. Lipid Res.* **9**, 799 (1968).

in a final volume of 2 ml containing enzyme, 2 μmoles of [^{14}C]triolein (approximately 200,000 cpm), 60 μmoles EDTA, 40 mg albumin, and 80 μmoles phosphate buffer, pH 6.8. The reaction is stopped by adding 5 ml of Dole's extraction mixture. The tube is shaken vigorously and 4 ml of heptane (containing 0.4 μmole of nonradioactive oleic acid as carrier) and 1 ml of water are added. The mixture is shaken vigorously and after 15 minutes' standing at room temperature to allow complete phase separation, 4 ml of the upper phase is transferred to a 20-ml scintillation vial containing 0.5 g Amberlite immersed in 1 ml of 0.05 M nonradioactive triolein (technical grade) in heptane. The scintillation vial is swirled for at least 1 minute using a Vortex mixer fitted with a custommade adapter to hold scintillation vials. Thorough mixing at this step is essential to insure complete adsorption of FFA onto the resin. The heptane is then suctioned off into a radioactive trap using a Pasteur pipette. The Amberlite is washed 4 more times with 5 ml aliquots of hexane, carefully rinsing down the walls of the vial from the very upper rim each time. After the first washing, the mixing time is not critical. After the fifth washing, the hexane is aspirated and 1 ml of NCS is added and the vial is heated in a 65° water bath for 15 minutes. This heating step is essential to displace completely the [^{14}C]oleic acid adsorbed onto the resin. Radioactivity is assayed by adding 10 ml of the scintillation mixture to the vial; removal of the resin from the bottom of the vial is not necessary. Blank assays yield background values that range between 80 and 200 cpm; thus, the amount of enzyme used should cause release of at least 1000–5000 cpm of [^{14}C]oleic acid. Under these conditions, duplicate assays yield highly reproducible results, almost always agreeing within less than 5%. In a representative series of duplicate assays, the mean difference between duplicates (14 pairs) was 1.8 ± 0.2%.

Method 2

An alternative and mechanically simpler method utilizes a liquid–liquid partition system for isolation of [^{14}C]oleic acid released,[9] a modification of the method of Belfrage and Vaughan.[10] Assay is carried out in a 13 × 100 mm disposable glass test tube. To the enzyme in 0.2 ml is added 0.6 ml of a substrate mixture containing 1 μmole [^{14}C]triolein, 30 μmoles EDTA, 20 mg albumin, and 40 μmoles phosphate buffer, pH 6.8. The reaction is stopped by adding 3 ml of the fatty acid extraction mixture (chloroform–methanol–benzene, 2:2.4:1) containing 0.3 μmole of

[9] R. C. Pittman, unpublished data (1973).
[10] P. Belfrage and M. Vaughan, *J. Lipid Res.* **10**, 341 (1969).

nonradioactive oleic acid, followed by addition of 0.1 ml of 1 N NaOH (final pH 11–11.5). The mixture is shaken vigorously in a Vortex mixer for at least 15 seconds and then centrifuged at 2500 rpm in an International centrifuge at room temperature for 10 minutes. An aliquot of the upper phase (1.8 ml) is transferred to a scintillation vial and 10 ml of scintillation fluid–Triton X-100 mixture (v/v, 2:1) are added. Radioactivity is assayed in a liquid scintillation counter; quenching (approximately 30%) is generally constant but should be checked. Recovery of added [^{14}C]oleic acid is approximately 80%. Reproducibility is comparable to that obtained with Method 1 (duplicates agree within less than 5%). The major advantage of this method is that it is more readily applicable to large numbers of samples.

To determine the rate of release of glycerol during incubation, triolein labeled in the glycerol moiety can be used. The labeled glycerol released is isolated by the trichloroacetic acid method of Schotz and Garfinkel.[11]

Purification of Hormone-Sensitive Lipase from Rat Adipose Tissue. Sprague-Dawley rats (200–300 g) are fasted for 48 hours in order to reduce the tissue content of lipoprotein lipase. The rats are killed by decapitation, and epididymal fat pads from 50 rats are then incubated for 2 hours in Krebs-Ringer bicarbonate buffer, pH 7.4 (containing 4% albumin) under 95% O_2–5% CO_2 (two fat pads per flask containing 4 ml of buffer). This incubation further reduces the lipoprotein lipase content of the tissue and also allows the activated form of HSL to revert to the nonactivated form. After incubation, the pooled fat pads are homogenized in 2 volumes of ice-cold buffer containing 0.25 M sucrose, 1 mM EDTA, and 10 mM Tris, pH 7.4, in a small Waring blender for 60 seconds. The homogenate is centrifuged at low speed to remove the floating fat cake and the fat-free infranatant fraction is then centrifuged at 78,000 g for 1 hour in a Beckman-Spinco model L centrifuge using a No. 30 rotor.[12] The supernatant fraction from the centrifugation is placed in an ice bath and carefully adjusted to pH 5.2 by addition of 0.1 N acetic acid. After 15 minutes, the precipitate formed is collected by centrifugation at 1000 g for 10 minutes. The precipitate is taken up

---

[11] M. C. Schotz and A. S. Garfinkel, *J. Lipid Res.* **13**, 824 (1972).

[12] HSL is predominantly a soluble enzyme. When fat cells are homogenized in an iso-osmotic medium, approximately one-third of the total enzymatic activity is tightly bound to the fat cake. Of the enzyme in the fat-free infranatant fraction, 80% is recovered in a 160,000 g supernatant fraction. Cell fractionation studies show that less than 2% is in the plasma membrane fraction [J. C. Khoo, L. Jarett, S. E. Mayer, and D. Steinberg, *J. Biol. Chem.* **247**, 4812 (1972)]. Crum and Calvert suggested that this enzyme was primarily associated with the plasma membrane. This was based on results using tributyrin as substrate for assaying fat cell ghosts [L. R. Crum and D. N. Calvert, *Biochim. Biophys. Acta* **225**, 161 (1971)].

TABLE I
PURIFICATION OF TRIGLYCERIDE LIPASE FROM RAT ADIPOSE TISSUE[a]

Fraction	Total activity (μEq FFA/ hour)	Total protein (mg)	Specific activity (μEq FFA/mg protein/ hour)	Relative specific activity	Yield (%)
78,000 g supernatant	250	435	0.58	1	100
pH 5.2 precipitate	235	91	2.6	4.5	94
d < 1.12 fraction	105	15.7	6.7	11.5	42
d 1.06–1.12 fraction	58	8.3	7.0	12.1	23
Agarose fraction	48	3.0	16.0	27.6	19
87,000 g supernatant	38	0.61	62.0	106	15

[a] Data from Huttunen et al.[13]

in 10 mM Tris, pH 7.4, containing 1 mM EDTA (volume one-tenth that of the original extract). The pH of this solution, which is generally around 6–6.5, is adjusted to 7.4 by addition of 0.2 M Tris, pH 8.0.

Concentrated sucrose is then added to bring the solution density to 1.12 and the mixture is centrifuged for 48 hours at 40,000 rpm in a 40.3 Beckman-Spinco rotor at 4°. The opalescent top layer (approximately one-fourth of the total volume) is collected by slicing the tube and this fraction is then adjusted to density 1.06 and centrifuged for another 36 hours. The top fraction from this centrifugation, which contains any lipid droplets that may have been carried through the fractionation, is discarded and the bottom fraction, about one-third of the total volume, is subjected to chromatography on 6% agarose. Sample (8–10 ml) is loaded on a 2.5 × 50 cm column and eluted at a flow rate of 30 ml/hour using 0.25 M sucrose containing 1.0 mM EGTA, 0.5 mM dithiothreitol, and 20 mM Tris-HCl, pH 7.4. Most of the lipase activity emerges from the column in the void volume. Finally, the void-volume fraction is centrifuged at 87,000 g for 30 minutes, which removes sedimenting membrane-like material. Approximately 77% of the lipase activity remains in the supernatant fraction at this stage. This procedure yields approximately a 100-fold purification over the original 78,000 g supernatant fraction with recovery of about 15%. The results of a representative purification are shown in Table I.

Analysis of HSL purified in this way shows that it is almost 50%

[13] J. K. Huttunen, A. A. Aquino, and D. Steinberg, Biochim. Biophys. Acta 224, 295 (1970).

by weight lipid, mostly phospholipid, consistent with its low density (mean density in sucrose density gradients 1.08). Because of the difficulty in preparing large quantities of enzyme, physical studies have been limited. Analytical ultracentrifuge studies utilizing ultraviolet scanning show $s_{20,w}$ values between 30 and 32; sedimentation equilibrium studies by the method of Yphantis yield a molecular weight of 7.2 million.[13] Electron micrographs show fairly uniform spherical particles, without evident substructure, ranging in diameter from 180 to 220 Å. Attempts to purify the enzyme further (by delipidation using various techniques, isoelectric focusing, and agarose gel electrophoresis) have not been successful. Thus, the most highly purified HSL available appears to have the properties, not of a single pure enzyme protein, but rather of a large, lipid-rich complex. The possibility that it represents an enzyme complex containing separate enzymes for hydrolysis of lower glycerides (and possibly other hydrolases) has not been ruled out.

Activation of Hormone-Sensitive Lipase by Cyclic AMP–Dependent Protein Kinase. It has now been established that the increase in rate of adipose tissue lipolysis effected by the lipolytic hormones (epinephrine, norepinephrine, glucagon, and ACTH) is mediated by an increase in the activity of cyclic AMP–dependent protein kinase.[14-16] Partially purified lipase prepared from tissues not previously exposed to hormones can be activated rapidly by 50–100% in a reaction requiring cyclic AMP, cyclic AMP–dependent protein kinase, ATP and Mg^{2+}, and this activation can be completely prevented by addition of protein kinase inhibitor purified from rabbit skeletal muscle.[17] Activation in cruder fractions, still containing endogenous protein kinase[7] may be just as effective in the absence of added exogenous protein kinase.[15,18,19] The correspondence between this *in vitro* activation system and the process occurring in response to lipolytic hormones is supported by the fact that enzyme prepared from tissues incubated with lipolytic hormones before homogenization shows a much smaller degree of activation.[16]

Activation is carried out at 30° for 10 minutes in a final volume of 0.2 ml containing 0.1 ml of enzyme solution, 0.01 mM cyclic AMP, 15

[14] T. R. Soderling, J. D. Corbin, and C. R. Park, *J. Biol. Chem.* **248**, 1822 (1973).

[15] J. K. Huttunen, D. Steinberg, and S. E. Mayer, *Biochem. Biophys. Res. Commun.* **41**, 1350 (1970).

[16] J. C. Khoo, D. Steinberg, B. Thompson, and S. E. Mayer, *J. Biol. Chem.* **248**, 3823 (1973).

[17] D. A. Walsh, C. D. Ashby, C. Gonzales, D. Calkins, E. H. Fischer, and E. G. Krebs, *J. Biol. Chem.* **246**, 1977 (1971).

[18] J. D. Corbin, E. M. Reiman, D. A. Walsh, and E. G. Krebs, *J. Biol. Chem.* **245**, 4849 (1970).

[19] S-C. Tsai, P. Belfrage, and M. Vaughan, *J. Lipid Res.* **11**, 466 (1970).

μg/ml of cyclic AMP–dependent protein kinase purified from skeletal muscle through the first DEAE-cellulose chromatography step,[20] 0.5 mM ATP, 5 mM Mg^{2+}, 1 mM theophylline, 1 mM dithiothreitol 0.5mM EGTA, and 50 mM Tris-HCl, pH 8.0. ATP, cyclic AMP, or both are omitted from control tubes. As indicated above, exogenous protein kinase is usually not needed when assaying cruder preparations but can be added to insure maximum rates of activation in preparations where the levels of endogenous protein kinase are not known. The cofactor concentrations listed are chosen to maximize the rate of activation. For the rat enzyme, the apparent K_m values for cyclic AMP and ATP are 1.1×10^{-4} and 5×10^{-3} mM, respectively. For studies in which the initial rate of activation is to be measured, the protein kinase and cofactor concentrations must be reduced to slow the reaction to a more readily measurable rate. At the end of the activation period, the appropriate aliquot (or the entire 0.2 ml) of the reaction mixture is added to the substrate for assay as described above in a final volume of 2 ml. Activation is arrested by the high EDTA concentration in the assay mixture as well as by the dilution.

Phosphorylation of Hormone-Sensitive Lipase. Conditions are as described for activation except that [γ-^{32}P]ATP is substituted for unlabeled ATP. The phosphorylation is terminated by adding 0.1 ml of 0.1 M EDTA, pH 6.8, containing 2.5 mg albumin to serve as protein carrier, followed immediately by addition of 5 ml of 5% trichloroacetic acid. The tube is shaken in a Vortex mixer and held in an ice bath for 10 minutes. The precipitate is centrifuged down and redissolved in 1 ml of 0.1 N NaOH. As soon as the precipitate has dissolved, another 5 ml of 5% trichloroacetic acid is added and this process is repeated twice in order to remove any trapped ATP or inorganic phosphate radioactivity. Finally, the precipitate is redissolved in 0.5 ml of 0.1 N NaOH which is transferred to a scintillation vial. The tube is rinsed with another 0.5 ml of 0.1 N NaOH and radioactivity is determined in a liquid scintillation counter after addition of 10 ml of liquid scintillator–Triton X-100 (v/v, 2:1). The time course of lipase activation has been shown to parallel closely the time course for phosphorylation.[21] In three experiments using 100-fold purified hormone-sensitive lipase, the phosphate incorporation ranged between 2 and 4 moles per 10^6 g of enzyme protein. The enzyme-bound ^{32}P is stable in the presence of 0.25 N HCl but over 80% is released by incubation in 0.25 N NaOH at 37° for 5 hours. As discussed above, the purified lipase preparation is known to contain monoglyc-

[20] D. A. Walsh, J. P. Perkins, and E. G. Krebs, *J. Biol. Chem.* **243**, 3763 (1968).
[21] J. K. Huttunen and D. Steinberg, *Biochim. Biophys. Acta* **239**, 411 (1971).

eridase and diglyceridase activities and it is not certain that all of the incorporated radioactivity is associated with hormone-sensitive triglyceride lipase. Since, however, the activities against monoglycerides and diglycerides are unaffected by activation with cyclic AMP–dependent protein kinase, it is unlikely that they account for any of the ^{32}P incorporation.

Inhibitors. Most organic solvents (acetone, butanol, ethanol, methanol, isopropyl alcohol, and chloroform) and many detergents (Tergitol, Cetytrimethylammonium bromide, deoxycholate, and Triton X-100) inhibit rat HSL activity. Free fatty acids, the end product of the lipase reaction, inhibit but this inhibition is prevented by the inclusion of sufficient albumin to bind FFA as released. Other inhibitors include diisopropylfluorophosphate, Paraoxon, fluoride, and calcium. Tsai *et al.*,[22] have described an irreversible inactivation that requires ATP-Mg^{2+} and ascorbic acid but the nature of this reaction remains unclear.

Species Differences. Basal levels of hormone-sensitive lipase activity and responsiveness to the various lipolytic hormones differ widely from species to species.[23] The activation system shown to be effective in rat tissue has been shown to be equally effective in the activation of HSL from mouse, hamster,[24] and human adipose tissue[25] and even more so with HSL from chicken and turkey adipose tissue.[24] With preparations from the latter two species, activation is more striking than it is with enzyme from rat or human tissue (from 200% to as high as 700%). On the other hand, crude extracts of adipose tissue from swine, cow, rabbit, guinea pig, and dog have yielded variable or negative results. The reasons for these variable results are not clear but no systematic attempt has been made as yet to explore alternative conditions for activation in these species. While the possibility remains that the activation process in some species may differ qualitatively from that originally elucidated in rat adipose tissue, the close similarity in the activation systems in birds, rats, and man suggests that this may be a universal mechanism in higher animals.

[22] S-C. Tsai, H. M. Fales, and M. Vaughan, *J. Biol. Chem.* **248**, 5278 (1973).
[23] D. Rudman and M. Di Girolamo, *Advan. Lipid Res.* **5**, 35 (1967).
[24] J. C. Khoo and D. Steinberg, unpublished results (1973).
[25] J. C. Khoo, A. A. Aquino, and D. Steinberg, *J. Clin. Invest.* **53**, 1124 (1974).

[24] Carboxylesterases from Pig and Ox Liver

By Nicholas P. B. Dudman and Burt Zerner

Carboxylesterases are remarkably efficient enzymes,[1,2] which catalyze the hydrolysis of a wide range of alkyl and aryl esters of aliphatic and aromatic carboxylic acids. The hydrolysis of anilides (e.g., acetanilide, phenacetin, and xylocaine) and of thiol esters (e.g., ethyl thiolacetate, phenyl thiolacetate) is also catalyzed.[3,4] In addition, carboxylesterases catalyze the transfer of the acyl moiety of substrate molecules to nucleophiles such as alcohols[4] and amines.[5,6] Fruton[5] and others[6] have demonstrated that amino acid esters act as acyl acceptors, and that ox liver carboxylesterase will, for example, catalyze the synthesis of phenylalanyl-phenylalanine methyl ester and of the dipeptide itself, starting with phenylalanine methyl ester as substrate.[5]

The purifications of pig and ox liver carboxylesterases described here have been developed within a broadly based program concerned with the mechanism of action of hydrolytic enzymes. Other carboxylesterases which have been highly purified in this program are those from chicken liver, horse liver, sheep liver, and shark liver; three of these purifications are detailed by Scott and Zerner.[6a] Considerable variations exist in chemical composition, physical properties, and catalytic effects between these enzymes, which are all designated EC 3.1.1.1 under Enzyme Commission nomenclature. Since the original entry EC 3.1.1.1 was based on the properties of a preparation from horse liver with a high activity towards ethyl butyrate as substrate,[7] we suggest that the horse enzyme be given a five-number classification such as EC 3.1.1.1.1, and that distinct carboxylesterases from other sources (such as the ox or pig) should be distinguished by changing the fifth number of the classification (*viz.*, 3.1.1.1.2, etc.).

[1] D. J. Horgan, E. C. Webb, and B. Zerner, *Biochem. Biophys. Res. Commun.* **23,** 23 (1966).

[2] J. K. Stoops, D. J. Horgan, M. T. C. Runnegar, J. de Jersey, E. C. Webb, and B. Zerner, *Biochemistry* **8,** 2026 (1969).

[3] K. Krisch, *Biochem. Z.* **337,** 546 (1963).

[4] P. Greenzaid and W. P. Jencks, *Biochemistry* **10,** 1210 (1971).

[5] T. A. Krenitsky and J. S. Fruton, *J. Biol. Chem.* **241,** 3347 (1966); M. I. Goldberg and J. S. Fruton, *Biochemistry* **8,** 86 (1969).

[6] H. C. Benöhr and K. Krisch, *Hoppe-Seyler's Z. Physiol. Chem.* **348,** 1102 (1967).

[6a] This volume [25].

[7] D. J. Horgan, J. K. Stoops, E. C. Webb, and B. Zerner, *Biochemistry* **8,** 2000 (1969).

Assay Methods

Assay of Carboxylesterase Activity. Carboxylesterases catalyze the hydrolysis of ethyl butyrate, and the rate of this reaction, determined with a pH-stat, is a measure of enzyme activity.

Incubate stoppered tubes (2.5×5.7 cm) containing 10 ml of 0.0125 M ethyl butyrate at $38 \pm 0.1°$ for 15 minutes. Start the reaction by adding a 100-μl aliquot of enzyme to one tube, and adjust the pH to 7.5. Maintain this pH by the addition of 0.0100 N NaOH contained in a 0.5-ml Agla micrometer syringe burette. The burette is operated by an automatic titrator (e.g., Radiometer TTT1c) equipped with a recorder (e.g., Radiometer SBR2c). Under the assay conditions, no correction for spontaneous hydrolysis or for the absorption of CO_2 is required. Further, butyric acid ($pK_a' = 4.8$) is completely ionized.

Pig liver carboxylesterase will give linear reaction traces regardless of the purity of the enzyme, but ox liver carboxylesterase yields slightly curved traces owing to substrate inhibition. However, the initial portions of ox esterase reaction traces, which are essentially linear, yield entirely reproducible assays of esterase activity. One unit of carboxylesterase activity is defined as the amount of enzyme required to catalyze the hydrolysis of 1 μmole of ethyl butyrate per minute, under the assay conditions described above.[8]

Protein Estimation. Assay protein concentration by measuring the absorbance at 280 nm (A_{280}) in a 1-cm cuvette. In all stages of the purification of pig and ox liver carboxylesterases, except the final step, the concentration of protein in an enzyme solution with $A_{280} = 1$ is taken to be 1 mg/ml. Follow the purification of pig liver or ox liver carboxylesterase by measuring: (i) specific activity = activity (in units per milliliter/A_{280}), which compares the amount of enzyme present with the total amount of protein; (ii) A_{280}/A_{260}, which gives a measure of nucleic acid contamination[9]; and (iii) A_{280}/A_{410}, which measures a contaminating heme pigment extracted from liver powders.

Active Site Titration of Pig and Ox Liver Carboxylesterase. The active site normality of pig liver carboxylesterase is obtained by measuring the p-nitrophenol released when the esterase reacts with p-nitrophenyl dimethylcarbamate (NPDMC).

Incubate 3 ml of 0.15 M Tris buffer, pH 8.16, at $25 \pm 0.1°$ in a

---

[8] In terms of the new unit of enzymatic activity, the katal, adopted by the International Union of Biochemistry,[8a] one unit of carboxylesterase activity is equal to 16.67 nanokatals (nkt).

[8a] "Enzyme Nomenclature," p. 27. Elsevier, Amsterdam, 1973.

[9] O. Warburg and W. Christian, *Biochem. Z.* **310**, 384 (1941).

cuvette, the top of which is sealed to avoid evaporation, in the sample compartment of a spectrophotometer fitted with a 0–0.1 absorbance recorder. Add a 50-μl aliquot of stock NPDMC (50 mM) in acetonitrile, and commence recording the absorbance at 400 nm to provide a base line for the measurement. Add an aliquot (50–100 μl) of enzyme and follow the liberation of p-nitrophenolate ion at 400 nm. There will be an initial rapid release, or "burst," of p-nitrophenol, followed by a much slower zero-order hydrolysis of NPDMC, and this slow reaction is measured for periods of up to 1 hour until the reaction trace can be accurately extrapolated back to zero time. From the extrapolated A_{400}, calculate the concentration of p-nitrophenol released in the burst ($[P_1]_{burst}$), from which he normality, $[E]_0$, of the enzyme solution is calculated using the equation of Ouellet and Stewart[10]:

$$[P_1]_{burst} = \left[\frac{k_{+2}}{k_{+2} + k_{+3}} \cdot \frac{[S]_0}{[S]_0 + K_m} \right]^2 \cdot [E]_0$$

Under the conditions of assay the rate constant for carbamoylation of the esterase, k_{+2}, is at least 1000-fold larger than the decarbamoylation constant, k_{+3}. The concentration of NPDMC ($[S]_0 = 8 \times 10^{-4}$ M) is much larger than K_m ($\sim 2.5 \times 10^{-6}$ M), and under these conditions $[P_1]_{burst} = [E]_0$ (very nearly).

Ox liver carboxylesterase may be titrated in the same way with o-nitrophenyl dimethylcarbamate, and both pig and ox liver carboxylesterases may be titrated using p-nitrophenyl diethyl phosphate. Carboxylesterases inhibited by dimethylcarbamate esters are essentially fully reactivated by dialysis against 0.15 M Tris, pH 8.2, containing 0.25 M hydroxylamine, and 5 mM cysteine (to scavenge metal ions, especially Cu^{2+}) at 25°.

Purification Procedure for Pig Liver Carboxylesterase[11]

General. Pig liver carboxylesterase is slowly denatured at pH 5 and below, and therefore exposure of the enzyme to low pH should be brief.

Regeneration of Media. Regenerate the ion-exchange media by emptying the columns and washing the resins thoroughly with 0.1 M trisodium phosphate (pH 12.5) to remove residual protein. Then wash the resins with water and column buffer as described below. Clean Sephadex

[10] L. Ouellet and J. A. Stewart, *Can. J. Chem.* **37**, 737 (1959).
[11] This procedure is an adaption by P. A. Inkerman and B. Zerner of the procedure of Horgan *et al.*[7]

columns by passing an excess of column buffer through them, until the eluent shows negligible absorbance at 280 and 260 nm.

Step. 1. Preparation of Chloroform–Acetone Powder. Homogenize 1 kg of fresh pig liver mince with 5 liters of redistilled $CHCl_3$ (-30 to $-10°$) in an explosion-proof Waring blender. Filter the $CHCl_3$ through a large Büchner funnel (24 cm, Whatman No. 542 paper). Homogenize the pasty residue with 5 liters of redistilled acetone (-30 to $-10°$) and filter as before. Wash the filter cake with acetone (-30 to $-10°$) until the filtrate is colorless, and remove excess acetone by compressing the filter cake with a rubber dam fitted over the Büchner funnel. Divide the filter cake finely and dry *in vacuo* ($10°$) for about 20 hours, condensing the residual acetone in two traps at $-80°$. Powder the material by grinding in a Waring blender, and remove final traces of acetone by evacuating the powder over concentrated H_2SO_4 ($10°$). The thoroughly dried powder (200–250 g) should be stored in an air-tight jar and is stable for 6 months at $4°$.

Step 2. Extraction of Chloroform–Acetone Powder. Extract a total of 2.4 kg powder in 400-g lots, with 10 volumes of 0.1 M citrate buffer (pH 3.9) at $4°$. Add the powder slowly to the buffer which is stirred continuously, and measure the pH with a glass/saturated calomel electrode. As the protein is extracted the pH will rise and should be maintained between pH 4.3 and 4.5 by dropwise addition of 2 N HCl. The pH will become steady after about 5 minutes and the suspension is then stirred for a further 30 minutes. After centrifuging at 3000 g ($2°$) for 40 minutes, raise the pH of the supernatant extract to 8.0 with 2 N NaOH. This causes the precipitation of flocculent inactive protein which will be removed in step 3. The extract should contain about 340 g of protein, including 2.7×10^6 units (3.5 g) of carboxylesterase. However, fresh powders may vary by more than a factor of 2 in enzyme content. Nonetheless, the specific activities of carboxylesterase preparations from powders of different potency are about the same after $(NH_4)_2SO_4$ fractionation (step 3).

Step 3. Ammonium Sulfate Fractionation. Perform this and all subsequent operations at 0 to $4°$. Add solid $(NH_4)_2SO_4$ (B. D. H. AnalaR or Mann Research Laboratories Special Enzyme Grade; 278 g/liter) slowly to the stirred extract. When all the salt has dissolved, stand the mixture for 30 minutes, and remove the precipitate by centrifuging at 3000 g for 45 minutes. Add solid $(NH_4)_2SO_4$ (169 g/liter) to the supernatant fluid as above, and collect the precipitate by centrifuging for 45 minutes at 3000 g. The precipitate is dissolved in, and dialyzed against, 0.05 M acetate buffer, pH 5.5. After dialysis, remove any turbidity or precipitate by centrifugation at 10,000 g for 40 minutes. The dialysate now contains about 66 g of protein including 2×10^6 units of carboxylesterase.

Step 4. Chromatography on CM-Cellulose. Prepare CM-cellulose for chromatography by washing it extensively with distilled water. Stir the suspension for a short time, allow it to settle, and decant the supernatant containing the fines. This is repeated up to a dozen times. Treat the yellowish white suspension with 0.01 N NaOH for 30 minutes to decolorize the resin, and again wash it with distilled water until the pH is about 7. Then equilibrate the resin with buffer (0.05 M acetate, pH 5.5), and pack it into a column (6 × 65 cm). Pass buffer (2 liters) through the column, and insure complete equilibration by comparing the pH and A_{280} of the eluent and feed buffers.

Add approximately 100 ml of solution from step 3 carefully to the column which is then washed with buffer. The enzyme will pass through the column without binding. Collect the eluent from this and other chromatography columns in an automatic fraction collector and assay for protein and enzymatic activity. Combine fractions from this column with specific activities greater than 50, and immediately add sufficient 2 M Tris buffer, pH 8.3, to raise the pH to about 8.0. Precipitate the protein by adding solid $(NH_4)_2SO_4$ (560 g/liter), and centrifuge as in step 3. Dialyze the precipitate once against 100 volumes of 0.05 M Tris buffer, pH 7.5. Step 4 is performed 8 times to process all the extract prepared in step 2. The yield of carboxylesterase is about 1.7×10^6 units in 26 g of protein.

Time is saved in this step and in step 5 by running parallel columns of CM-cellulose or Sephadex G-200, respectively.

Step 5. Gel Filtration through Sephadex G-200. Prepare Sephadex G-200 by adding the dry powder slowly to an excess of distilled water with a magnetic stirrer. Wash the gel by decantation several times and allow it to swell in water overnight. Equilibrate the gel with 0.05 M Tris buffer, pH 7.5, and pack into a column (5 × 95 cm). Pass the dialysate from step 4 through the column washed with the same buffer. This step is performed 6 times; between samples the column is washed exhaustively with buffer until the eluent and feed buffer have the same A_{280}. The carboxylesterase solution from this step contains about 7.7 g of protein, including 1.6×10^6 units of enzyme.

Step 6. Chromatography on DEAE-Cellulose. Prepare DEAE-cellulose in a similar way to CM-cellulose, and equilibrate in 0.05 M Tris buffer, pH 7.5. Apply fractions from gel filtration with specific activities above 50 directly to a column (6 × 60 cm) of the DEAE-cellulose. The enzyme binds to the resin. Wash the column free of extraneous protein with the starting buffer, and then elute the enzyme with a linear salt gradient. The mixing vessel for the gradient contains 1.25 liters of buffer, and the other vessel contains 1.25 liters of 0.3 M NaCl dissolved in buffer. Enzyme with a specific activity above 360 is precipitated with $(NH_4)_2SO_4$ (560 g/liter) and collected by centrifugation.

The yield is about 1.68 g of protein containing 7.0×10^5 units of carboxylesterase.

Step 7. Rechromatography on Sephadex G-200. Dialyze the protein precipitate from step 6 against 0.05 M Tris buffer, pH 7.5, and pass the resulting solution through Sephadex G-200 as described in step 5. Precipitate fractions with specific activity greater than 400 with $(NH_4)_2SO_4$ as described in step 5, and dialyze the protein precipitate against 0.1 M acetate buffer, pH 5.0. This step yields about 6.8×10^5 units of carboxylesterase in 1540 A_{280} units.

Step 8. Chromatography on CM-Sephadex. Prepare CM-Sephadex by washing with dilute alkali and dilute acid as recommended by the manufacturers (Pharmacia, Sweden). Then wash it with distilled water, and equilibrate with the starting buffer (0.1 M acetate, pH 5.0). Apply the dialysate from step 7 to a column (2.3 × 80 cm) of CM-Sephadex; the carboxylesterase is bound. Wash the column with buffer, and when protein ceases to appear in the eluent, elute the enzyme with a combined pH-salt gradient. Establish the gradient by placing 250 ml of starting buffer in the mixing vessel, and 250 ml of starting buffer containing 0.5 M $(NH_4)_2SO_4$ in the other vessel of a linear gradient system. The flow rate of this column will drop markedly when the protein is bound to the resin, but will increase (from about 7 to 60 ml/hour) during gradient elution. The protein and carboxylesterase activity will be eluted from the column together in single, nearly coincidental peaks, as seen in Fig. 1. Carboxylesterase from the trailing part of the peak is of high and consistent specific activity, and fractions from this region with specific activities above 570 are combined and precipitated with $(NH_4)_2SO_4$ as in step 4. Dialyze the precipitate against 0.05 M Tris, pH 7.5. This step yields about 2.6×10^5 units (330 mg) of carboxylesterase containing negligible nucleic acid ($A_{280}/A_{260} = 1.74$)[9] or red pigment ($A_{280}/A_{410} = 123$).

Step 9. Rechromatography on Sephadex G-200. Pass enzyme from step 8 with specific activity from 530 to 570 through a column (2.3 × 90 cm) of Sephadex G-200 equilibrated with 0.05 M Tris buffer, pH 7.5, containing 0.15 M KCl. Pool fractions with specific activities over 570 to obtain a further 1.1×10^5 units (140 mg) of highly purified carboxylesterase ($A_{280}/A_{260} = 1.72$, $A_{280}/A_{410} = 280$).

Table I provides a summary of the purification of pig liver carboxylesterase.

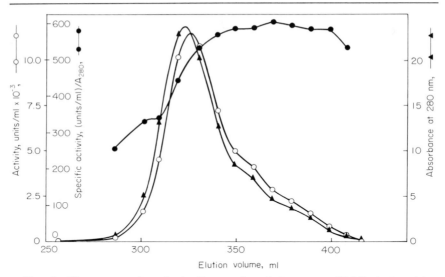

Fig. 1. Chromatography of pig liver carboxylesterase on CM-Sephadex (step 8). Experimental conditions: column, 2.3 × 80 cm; sample, 6.8 × 10⁵ units of enzyme, 1540 A_{280} units; buffer, 0.1 M acetate, pH 5.0; linear gradient, 250 ml buffer and 250 ml buffer containing 0.5 M (NH₄)₂SO₄.

Crystallization of the Enzyme. Enzyme from steps 8 or 9 can be crystallized readily in the following way. Add solid (NH₄)₂SO₄ to a solution almost saturated with respect to the enzyme until the first turbidity ap-

TABLE I
PURIFICATION OF PIG LIVER CARBOXYLESTERASE

Procedure	A_{280}/A_{410}	Specific activity [(units/ml)/ A_{280}]	Yield (%)
Extraction[a,b]	~10	~8	100
Ammonium sulfate fractionation	12–17	30	72
CM-Cellulose	30	54–83	64
Sephadex G-200	80	150–286	60
DEAE-Cellulose	49	415	25.3
Sephadex G-200	106	440	24.6
CM-Sephadex	123	582	9.6[c]
Sephadex G-200	280	580	4.2[c]

[a] 2.75 × 10⁶ units are extracted from 2.4 kg CHCl₃–acetone powder.

[b] A_{280}/A_{260} of extract is ~1.

[c] Total yield of enzyme with specific activity 580 and above is 13.8%.

pears. This is removed by centrifugation, and the supernatant fluid is evaporated slowly under an inverted beaker in a refrigerator at 4°. The enzyme crystallizes as large thin plates which are very fragile.

Properties of Pig Liver Carboxylesterase

Purity of the Enzyme. The elution profile from CM-Sephadex (Fig. 1), from which 330 mg of enzyme of fairly constant specific activity (580) and constant A_{280}/A_{260} ratio (1.73–1.75) was isolated, indicates that the enzyme is relatively homogeneous. Electrophoresis on polyacrylamide gel (pH 8.9) shows a single band of protein, when up to 37 μg of enzyme are loaded. Gels stained with the far more sensitive esterase stain reveal the presence of a small quantity of faster moving enzyme. This material may be an artifact caused by the dissociation of the enzyme (see below).

Stability. Pig liver carboxylesterase has been stored at −15° in 0.02 M potassium phosphate, pH 7.4, for 1½ years without losing significant activity.[4] Enzyme stored at 5° under similar conditions loses only ∼6% of its activity over 7 months.[7] At pH 5.0, 7% of the carboxylesterase activity is lost in 16 days,[7] and at pH 4.5 the enzyme rapidly dissociates to monomeric units (see below) which lose about 30% of activity in 8 hours at 4°.[12] At pH 2.3 all activity is lost in 9 hours at 20°.[12]

Physical Properties. The A_{280} of a solution of pig liver carboxylesterase (1 mg/ml) in 0.15 M Tris, pH 8.16, is 1.34_1.[13]

The enzyme reversibly polymerizes at neutral pH when the concentration is raised (0.15 mg/ml and above).[14] The polymers dissociate to monomeric units, rapidly at pH 4.2,[12] and slowly at pH 7.5 when the enzyme concentration is low (e.g., 5 μg/ml).[14] The presence of salt (0.5 M NaCl or LiBr) has been found to increase dissociation. Although the molecular weight of the carboxylesterase has been the subject of many investigations, recent estimates from different laboratories of the monomer molecular weight still vary widely, most values lying between 55,000 and 85,000 daltons.[12,14-17] Recent results from this laboratory, in which pig liver carboxylesterase was examined in the ultracentrifuge over a wide concentration range and also by frontal analysis of gel filtration

[12] D. L. Barker and W. P. Jencks, *Biochemistry* **8**, 3879 (1969).

[13] R. L. Blakeley and B. Zerner, this volume [26].

[14] P. A. Inkerman, D. J. Winzor, and B. Zerner, unpublished results (1972).

[15] D. J. Horgan, J. R. Dunstone, J. K. Stoops, E. C. Webb, and B. Zerner, *Biochemistry* **8**, 2006 (1969).

[16] R. C. Augusteyn, J. de Jersey, E. C. Webb, and B. Zerner, *Biochim. Biophys. Acta* **171**, 128 (1969).

[17] E. Heymann, W. Junge, and K. Krisch, *FEBS Lett.* **12**, 189 (1971).

TABLE II
PHYSICAL PROPERTIES OF PIG LIVER CARBOXYLESTERASE

Sedimentation coefficient, $s_{20,w}^0$ (S)	8.12,[a] 8.48,[b] 8.3,[c] 8.2[d]
Dependence of $s_{20,w}$ on concentration, $k(\times 10^2$, liter/g)	0.94,[a] 1.1,[b] 0.85,[c] 1.95[d]
Diffusion coefficient, $D_{20,w}$ ($\times 10^7$, cm^{2}sec^{-1})	4.12,[a] 4.57,[b] 4.3,[c] 4.6[d]
Partial specific volume, $\bar{v}$ (ml/g)	0.740,[a] 0.74$_0$,[b] 0.738[e]
Isoelectric point, pI	5.0[a,c]
Absorbance at 280 nm, $A_{1\,cm}^{0.1\%}$	1.38,[a] 1.34$_1$[c]
Apparent molecular weight ($\times 10^{-5}$)	1.68,[a] 2.06,[b] 1.63,[c] 1.75[d]

[a] From Barker and Jencks.[12]
[b] From Inkerman et al.[14]
[c] From Horgan et al.[15]
[d] From Boguth et al.[19]
[e] From Klapp et al.[22]

show that the apparent molecular weight at 5 μg/ml (pH 7.5) is $\sim$66,000, and at >5 mg/ml is 206,000 $\pm$ 8,000.[14] Several values of the apparent molecular weight of more concentrated enzyme have been reported, generally from 160,000 to 180,000 daltons.[12,15,18,19] The equivalent weight of pig liver carboxylesterase, determined by titration of enzyme of specific activity 580, using p-nitrophenyl diethyl phosphate, is 66,800 $\pm$ 600.[20] This value, taken with the molecular weight study mentioned above, shows that the enzyme possesses one active site per monomer of $\sim$66,000 daltons. The apparent molecular weight of the enzyme in more concentrated solutions indicates that the enzyme contains a number of polymers with more than two monomeric units in these solutions. Further, electron microscopy supports the existence of both dimers and trimers in concentrated pig liver carboxylesterase solutions.[21]

The isoelectric point of the enzyme, determined by starch-gel electrophoresis in acetate buffers ($\mu = 0.025$) is 5.0.[7]

Values for some physical characteristics of pig liver carboxylesterase are set out in Table II.[12,14,15,19,22] The sedimentation coefficient, $s_{20,w}^0$, the constant k describing the dependence of $s_{20,w}$ on enzyme concentration, c, in milligrams per milliliter, where $s_{20,w} = s_{20,w}^0(1 - kc)$, the diffusion coefficient, and the apparent molecular weight have all been determined at enzyme concentrations at and above 1 mg/ml, at neutral pH, and therefore refer to the polymerized state of the enzyme.

[18] S. A. Kibardin, *Biokhimiya* **27**, 82 (1962).
[19] W. Boguth, K. Krisch, and H. Niemann, *Biochem. Z.* **341**, 149 (1965).
[20] S. E. Hamilton and B. Zerner, unpublished results (1972).
[21] B. Krisch and K. Krisch, *FEBS Lett.* **24**, 146 (1972).
[22] B. Klapp, K. Krisch, and K. Borner, *Hoppe-Seyler's Z. Physiol. Chem.* **351**, 81 (1970).

Chemical Properties. Highly purified pig liver carboxylesterase has been phosphorylated with [^{32}P]di-isopropyl phosphorofluoridate, and the sequence of amino acids around the single radioactively labeled serine per monomer has been determined in two laboratories.[16,23] The sequence has been reported as

$$\overset{*}{\text{Gly--Glu--Ser--Ala--Gly--Gly--Glu--Ser}}$$

by Augusteyn and colleagues,[16] and as

$$\text{Gly--Glu--}\overset{*}{\text{Ser}}\text{--Ala--Gly--}\overset{\text{Asp}}{\underset{\text{Glu}}{}}\text{--Gly--Ser}$$

by Heymann *et al.*[23] The sequence reported by Augusteyn *et al.*[16] is analogous to the sequence reported by them for ox:

$$\overset{*}{\text{Gly--Glu--Ser--Ala--Gly--Ala--Glu--Ser}}$$

which in turn is homologous with the sequence for horse published earlier by Jansz *et al.*[24]:

$$\overset{*}{\text{Gly--Glu--Ser--Ala--Gly--Gly--(Glu,Ser)}}$$

In view of these similarities, and because the sequence of Heymann *et al.*[23] contains interchanged residues of $\underset{\text{Glu}}{\overset{\text{Asp}}{}}$ and Gly, the sequence of Augusteyn *et al.*[16] is preferred.

Klapp *et al.* have compared the tryptic peptide fingerprint maps of pig and ox liver carboxylesterases, which are clearly different, reflecting differences in the amino acid compositions of the two proteins.[12,22] The N-terminal amino acid of pig liver carboxylesterase is glycine, and of the ox liver enzyme is leucine.[17]

There is no evidence that the enzyme is activated or inhibited significantly by metal ions normally found in metalloenzymes.[3,15] The carboxylesterase does not have a highly reactive --SH group necessary for catalysis or for the stability of the enzyme.

Catalysis. As indicated earlier, pig liver carboxylesterase catalyzes a wide range of transfer reactions, and enjoys very high catalytic efficiency. Table III illustrates the consistently high values of k_{cat} for the pig liver carboxylesterase-catalyzed hydrolysis of a number of butyrate esters at 25°. Table IV is a similar collection of values of k_{cat} for the

[23] E. Heymann, K. Krisch, and E. Pahlich, *Hoppe-Seyler's Z. Physiol. Chem.* **351,** 931 (1970).

[24] H. S. Jansz, C. H. Posthumus, and J. A. Cohen, *Biochim. Biophys. Acta* **33,** 396 (1959).

TABLE III
HYDROLYSIS OF BUTYRIC ACID ESTERS BY PIG LIVER CARBOXYLESTERASE AT 25°[a]

Butyrate	k_{cat} (sec^{-1})	K_m (10^5 M)
Phenyl	519	10
p-Nitrophenyl	367	14
o-Nitrophenyl	348	17
m-Nitrophenyl	345	25
2,4-Dinitrophenyl	233	11
Ethyl	551	50

[a] For experimental details, see Stoops et al.[2]

TABLE IV
HYDROLYSIS OF PHENYL ESTERS BY PIG LIVER CARBOXYLESTERASE AT 25°[a]

Substrate	k_{cat} (sec^{-1})
Phenyl formate	660
Phenyl acetate	380
Phenyl butyrate	519
Phenyl valerate	860

[a] For experimental details, see Stoops et al.[2]

hydrolysis of phenyl esters of aliphatic acids. From these tables it is clear that pig liver carboxylesterase exhibits a high reactivity and fairly low specificity in its catalytic function. In its reactions with optically active substrates the enzyme again exhibits low, although distinct, stereospecificity, as illustrated in its catalysis of the hydrolysis of D- and L-tyrosine ethyl ester, and N^α-benzyloxycarbonyl-D- and N^α-benzyloxycarbonyl-L-tyrosine p-nitrophenyl ester. (See Table V; values of k_{cat} for the α-chymotrypsin-catalyzed hydrolyses of these esters are inserted for comparison.)

The spectrophotometric observation of a burst in the titration of the ox and pig liver carboxylesterase with nitrophenyl dimethylcarbamates[1,25] demonstrates that a dimethylcarbamoyl-enzyme intermediate is formed. The formation of analogous acyl-enzyme intermediates in other reactions catalyzed by carboxylesterases seems likely, especially in view of the formation of acyl-chymotrypsin in the α-chymotrypsin-

[25] M. T. C. Runnegar, K. Scott, E. C. Webb, and B. Zerner, Biochemistry 8, 2013 (1969).

TABLE V
STEREOSELECTIVE HYDROLYSES CATALYZED BY PIG LIVER
CARBOXYLESTERASE AND BY α-CHYMOTRYPSIN AT 25°[a]

Substrate	Pig liver carboxylesterase k_{cat} (sec^{-1})	α-Chymotrypsin k_{cat} (sec^{-1})
L-Tyrosine ethyl ester	71	39
D-Tyrosine ethyl ester	71	Very low
Cbz-L-tyrosine p-nitrophenyl ester	24	128
Cbz-D-tyrosine p-nitrophenyl ester	0.9	0.2

[a] For experimental details, see Stoops et al.[2]

catalyzed hydrolysis of activated acetate esters and of "specific" substrates, and in view of the similarities in chemistry between α-chymotrypsin and the esterases.[1,2] The existence of an acyl-enzyme intermediate for the carboxylesterase-catalyzed reactions is supported by the kinetics of hydrolysis of butyrate esters,[2] and by the partitioning of the acyl group between the various nucleophiles in solution.[4-6] However, the *direct* demonstration of such acyl-enzyme intermediates has so far not been accomplished. This could be because of the high values of k_{cat} for good substrates (Tables III and IV; hydrolysis of acetyl-carboxylesterases would be faster than acetyl-α-chymotrypsin by a factor of $\sim 10^5$), coupled with the instability of carboxylesterases at low pH, conditions under which other acyl-enzymes are most stable and have been isolated.[26,27]

Pig liver carboxylesterase behaves in a complex way kinetically. Assays of the enzyme at high substrate concentrations exhibit substrate activation; for example, the enzyme is about twice as reactive toward phenyl butyrate at high concentrations as expected from rates at low levels of the ester.[2] The enzyme may also be activated by low concentrations of organic solvents such as benzene, acetone, and CH_3CN.[2,28] Activation by organic solvent molecules tends to blanket substrate activation,[2,28] and since many of the kinetic data reported for pig and other carboxylesterases have been obtained in the presence of CH_3CN, these results should be interpreted accordingly. The kinetics of activation by

[26] M. L. Bender, E. T. Kaiser, and B. Zerner, *J. Amer. Chem. Soc.* **83**, 4656 (1961).
[27] M. L. Bender, G. R. Schonbaum, and B. Zerner, *J. Amer. Chem. Soc.* **84**, 2540 (1962); M. L. Bender and E. T. Kaiser, ibid. p. 2556.
[28] D. L. Barker and W. P. Jencks, *Biochemistry* **8**, 3890 (1969).

substrate and by organic solvents indicate that each catalytically active enzyme unit has, in addition to its catalytically active site, a second site which binds organic molecules such as substrate, or benzene, acetone, or CH_3CN, which increase catalytic efficiency. Although the enzyme polymerizes readily at neutral pH, very dilute enzyme that is exclusively monomeric after gel filtration is catalytically active.[15] Also, gel filtration of the very dilute enzyme in a solution containing benzene at a concentration ($\sim 0.01\ M$) which fully activates the enzyme has shown that the enzyme remains completely monomeric.[2] Therefore, it is clear that both the catalytic site and the activator binding site are on the same monomeric unit and that binding of activator molecules does not induce polymerization.

Pig liver carboxylesterase when phosphorylated with bis(p-nitrophenyl) phosphate does not follow simple first- or second-order kinetics but reacts in a biphasic manner.[29] No satisfactory explanation of this observation has yet been advanced. Similarly, acylation of this enzyme during titration with NPDMC is not accurately described by a simple carbamoylation–decarbamoylation scheme, although the scheme gives a good approximation for the rate of acylation.[15] While constants for the titration system, which have been determined from the simple scheme, are thus approximations, titration of the pure enzyme by NPDMC and p-nitrophenyl diethyl phosphate give identical values for the equivalent weight,[1] and assay of the enzyme concentration by titration yields reproducible kinetic values on a wide range of substrates.[2] Thus these titrations are valid measures of the concentration of carboxylesterases.

Purification Procedure for Ox Liver Carboxylesterase[30]

General Methods. Spectrophotometric protein determination, the pH-stat rate assay of carboxylesterase activity toward ethyl butyrate, ammonium sulfate precipitation, and the preparation and regeneration of chromatographic media are described above in the purification procedure for pig liver carboxylesterase.

Step 1. Preparation of Ox Liver Acetone Powder. Homogenize 1 kg of fresh ox liver mince with 5 liters of redistilled acetone (-30 to $-10°$) in an explosion-proof Waring blender. Filter the acetone through a large Büchner funnel (24 cm, Whatman No. 542 paper). Wash the filter cake and dry it in exactly the same way that the pig liver powder was treated. Store the thoroughly dried powder (200–250 g) in an air-tight jar at $4°$.

---

[29] E. Heymann and K. Krisch, *Hoppe-Seyler's Z. Physiol. Chem.* **348**, 609 (1967).
[30] This procedure is that of Runnegar *et al.*[25]

Step 2. Extraction of Carboxylesterase from Acetone Powders. Extract 1.2 kg of acetone powder in 400 g lots with 9 volumes of 0.1 M citrate buffer, pH 4.0, at 3° for 45 minutes with constant stirring. Centrifuge the suspension at 3000 g (2°) for 45 minutes. Raise the pH of the reddish brown clear supernatant from 4.5–4.7 to 7.5 by the dropwise addition of 2 N NaOH. A turbidity appears which is not readily removed by centrifugation. The extract should contain 255 g of protein and about 4×10^5 units (3.9 g) of carboxylesterase. However, fresh powders may vary considerably in enzyme content, which we have found to be particularly low in time of drought.

Extracts with an esterase content of about 70 units/ml yield enzyme of maximum specific activity after step 6 of this procedure, while extracts with a lower enzyme content (e.g., 25 units/ml) yield enzyme of lower specific activity. Thus extracts containing high levels of esterase should be used in this procedure to produce enzyme of the highest purity.

Step 3. Ammonium Sulfate Fractionation. Perform this and all subsequent operations at 0–4°. By the addition of solid $(NH_4)_2SO_4$, fractionate the extract in the same way, and with the same $(NH_4)_2SO_4$ concentrations as pig liver carboxylesterase. Redissolve the ox liver carboxylesterase in 0.025 M citrate buffer, pH 5.6, and dialyze it against the same buffer. The yield of protein is about 39 g, and of enzyme, 2.6×10^5 units.

Step 4. Chromatography on CM-Cellulose. Pass the dialysate from step 3 through a column (5×90 cm) of CM-cellulose in 0.025 M citrate buffer, pH 5.6. The carboxylesterase does not bind. This step is performed 4 times. Combine the enzyme fractions with highest specific activity, precipitate with $(NH_4)_2SO_4$ (560 g/liter) and dialyze the precipitated protein against 0.05 M acetate buffer, pH 5.30. Approximately 7 g of protein containing 2.3×10^5 units of enzyme are obtained from this step.

Step 5. Chromatography on CM-Sephadex. Apply the dialysate from step 4 to a CM-Sephadex column (5×90 cm) in 0.05 M acetate buffer, pH 5.30, in which the carboxylesterase binds to the resin. After thorough washing with the starting buffer, elute the bound enzyme with a linear gradient consisting of 3 liters of starting buffer in the mixing vessel and 3 liters of 0.2 M sodium acetate in the other vessel. The enzyme is eluted over the region 4.4–5.0 liters. Figure 2 shows an elution profile for the column. Runnegar *et al.*[25] have found by gel electrophoresis that ox liver carboxylesterase consists of three closely spaced bands. Esterase from region 1 of Fig. 2 is largely the anodic band, with a little of the middle band. Esterase from region 2 contains all three bands, and esterase from region 3 is almost entirely the most cathodic band.

Step 6. Gel Filtration through Sephadex G-100. The three pooled por-

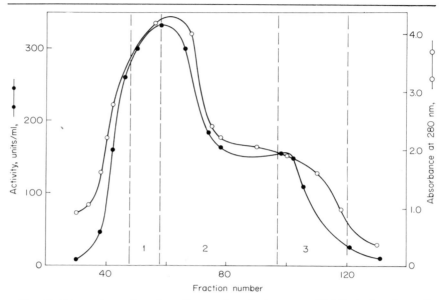

Fɪɢ. 2. Chromatography of ox liver carboxylesterase on CM-Sephadex (step 5). Experimental conditions: column, 5×90 cm; sample, 7 g of protein containing 2.3×10^5 units of enzyme; buffer, 0.05 M acetate, pH 5.30; linear gradient, 3 liters of starting buffer and 3 liters of 0.2 M sodium acetate; each fraction contains 6.2 ml, and the enzyme is eluted over the region 4.4–5.0 liters of the gradient; fractions from regions 1, 2, and 3 are pooled separately. (See text).

tions (regions 1, 2, and 3 of Fig. 2) are purified individually in this step. Precipitate the enzyme with $(NH_4)_2SO_4$ (560 g/liter), take up the precipitate in a minimum of 0.05 M Tris, pH 7.6, and dialyze the solution against this buffer. Centrifuge the solution to remove any precipitate or turbidity and pass the clear solution through a column $(3.3 \times 70$ cm) of Sephadex G-100 in the Tris buffer. This step yields about 2.1×10^4 units (200 mg) of carboxylesterase from region 1, 5.5×10^4 units (530 mg) from region 2, and 1.1×10^4 units (110 mg) from region 3.

Table VI provides a summary of the purification of the components of ox liver carboxylesterase.

Properties of Ox Liver Carboxylesterase

Purity of the Enzyme. Each component, after gel filtration on Sephadex G-100, has been shown to give a single symmetrical peak in the analytical ultracentrifuge. No inert protein bands could be detected on starch or polyacrylamide electrophoresis. Refiltration on Sephadex G-100

TABLE VI
PURIFICATION OF OX LIVER CARBOXYLESTERASE

Procedure	Total protein[a] (A_{280} × liters) (g)	A_{280}/A_{410}	Specific activity [(units/ ml)/A_{280}]	Yield (%)
Extraction[b,c]	255	4	1.5	100
Ammonium sulfate fraction	39	6	6.8	67
CM-Cellulose	7	62	32.6	58
CM-Sephadex				
Region 1	0.412	190	73	7.6
Region 2	0.954	120	67	16.8
Region 3	0.245	150	68	4.2
				28.6
Sephadex G-100				
Region 1	0.277	>200	77	5.4
Region 2	0.692	>200	78	14
Region 3	0.153	>200	73	2.8
				22.2

[a] For the purpose of this column, a solution having $A_{280} = 1$ is defined as containing 1 mg/ml of protein.

[b] 3.95×10^5 units are extracted from 1200 g of powder in three extractions.

[c] A_{280}/A_{260} of extract is ∼1.

or G-200 did not increase the specific activity. Specific activities in the same range have been obtained reproducibly in other large-scale preparations. There is no appreciable absorbance at 410 nm and samples of concentrated protein ($A_{280} = 10$–30) appear colorless.

It is possible to produce electrophoretically pure forms of the most anodic and most cathodic variants by vicious cutting of CM-Sephadex fractions in regions 1 and 3, respectively. However, as described below, the kinetic behavior and physical properties of the electrophoretic variants appear to be very similar. The problem, therefore is likely to be of less significance than originally anticipated.

Stability. A sample of ox liver carboxylesterase which was not divided into components 1, 2, and 3 after CM-Sephadex chromatography was passed through Sephadex G-100 in step 6 and then precipitated with $(NH_4)_2SO_4$ (560 g/liter). The enzyme, of specific activity 78, was stored as a suspension in $(NH_4)_2SO_4$ solution for 8 months at 4°, and retained full specific activity.

Another preparation which had been frozen and thawed in neutral buffers a number of times over a period of 2 years appeared to be modified. It consisted (originally) of the two most anodic electrophoretic bands. The presteady state reaction with o-nitrophenyl dimethylcarbamate was slowed considerably. There was also a 20% decrease in specific activity toward ethyl butyrate. Gel filtration of the modified enzyme in 5 mM Tris base, 1 mM EDTA, pH ~8.6, showed that the enzyme dissociated more readily than the native enzyme (see below), although at higher concentrations the polymerized state was observed.

Physical Properties. Ox liver carboxylesterase is similar to the pig enzyme in existing in a polymerized state at neutral pH at high enzyme concentrations.[31,32] The ox enzyme dissociates to monomeric units when the concentration is lowered, and starts to dissociate at higher concentrations (~1 mg/ml)[33] than does the pig enzyme.[6] Several studies of the molecular weight of ox liver carboxylesterase have been carried out, resulting in values for the molecular weight of the monomeric unit lying in the range 52,000 to 86,000 daltons.[17,31,34-36] The apparent molecular weight of the polymeric enzyme has been estimated at 167,000 daltons.[31] Gel filtration and ultracentrifuge studies have shown that the electrophoretically distinct components isolated by CM-Sephadex chromatography (step 5) have the same molecular weight and do not constitute various degrees of association of the enzyme.[32]

The s_{20} for ox liver carboxylesterase in 0.05 M Tris buffer, pH 7.0, obeys the relationship

$$s_{20} = 8.3_0 (1-0.011_5\, c)$$

where c is the enzyme concentration in milligrams per milliliter. This relationship holds at enzyme concentrations between 3 and 10 mg/ml.[33] At protein concentrations lower than 3 mg/ml the observed s_{20} value does not increase as expected, and below 1 mg/ml it decreases, probably caused by the tendency of the enzyme to dissociate.[33]

The partial specific volume of the enzyme has been calculated to be 0.742 ml/g.[22]

For highly purified enzyme, $A^{0.1\%}_{280,1\,cm} = 1.34$.[25]

The equivalent weight of ox liver carboxylesterase containing all three electrophoretic variants is 68,000, derived from three independent titra-

[31] H. C. Benöhr and K. Krisch, *Hoppe-Seyler's Z. Physiol. Chem.* **348**, 1115 (1967).
[32] M. T. C. Runnegar, E. C. Webb, and B. Zerner, *Biochemistry* **8**, 2018 (1969).
[33] M. T. C. Runnegar, Ph.D. Thesis, p. 114. University of Queensland, Brisbane (1968).
[34] D. J. Ecobichon, *Can. J. Biochem.* **47**, 799 (1969).
[35] D. J. Ecobichon, *Can. J. Biochem.* **50**, 9 (1972).
[36] S. Bauminger and L. Levine, *Biochim. Biophys. Acta* **236**, 639 (1971).

tions (o- and p-nitrophenyl dimethylcarbamates, and p-nitrophenyl diethyl phosphate) of enzyme of specific activity 74, adjusted to a specific activity of 78.[25] Titration of apparently less pure samples of specific activity 65–68 has been reported to give equivalent weights of 54,000 to 59,000 when correction was made to maximum specific activity of 78.[25] This anomaly has been traced to the fact that the ox liver enzyme is subject to inhibition by substrate; thus, the figures which are presumably in error are the specific activity figures. Had the titration data not been corrected, the equivalent weights determined would be 67,500 (region 1), 64,800 (region 2), and 64,500 (region 3).[14]

Chemical Properties. These have been discussed in conjunction with those of the pig carboxylesterase (see above).

Catalysis. The catalytic properties of ox liver carboxylesterase resemble those of the pig enzyme fairly closely. The ox enzyme hydrolyzes a wide range of substrates, although substrate specificity is clearly different from that of the pig esterase. Table VII lists rate constants and values of K_m for the ox esterase-catalyzed hydrolysis of a series of butyrate esters. As the table illustrates, any differences in the catalytic properties between the different electrophoretic variants of ox liver carboxylesterase are minor.

A significant difference between pig and ox liver carboxylesterases lies in their relative rates of reaction with *ortho*- and *para*-nitrophenyl butyrates, (cf. Table III with Table VII), and with o- and p-nitrophenyl dimethylcarbamates. Pig liver carboxylesterase reacts with o- and p-nitrophenyl dimethylcarbamates at about the same rate. However, the *ortho*-substituted ester reacts more rapidly than the *para*- with the ox enzyme by a factor of 10. Similarly, k_{cat} values for the pig liver carboxylesterase-catalyzed hydrolysis of o- and p-nitrophenyl butyrates are

TABLE VII

HYDROLYSIS OF BUTYRIC ACID ESTERS BY OX LIVER CARBOXYLESTERASE AT $25°$[a]

	k_{cat} (sec^{-1})		K_m (10^5 M)	
	Ox 1[b]	Ox 3[b]	Ox 1[b]	Ox 3[b]
Phenyl	160	200	17	25
p-Nitrophenyl	110	110	22	34
o-Nitrophenyl	210	240	6	8
m-Nitrophenyl	60	65	16	19

[a] For experimental details, see Stoops *et al.*[2]

[b] Ox 1 and ox 3 refer to the enzyme from region 1 and region 3, respectively, of the CM-Sephadex chromatogram (Fig. 2).

virtually identical. With ox liver carboxylesterase, k_{cat} for the *ortho* ester is twice that for the *para*. The factor of two in k_{cat} values probably represents a much larger factor in the acylation of the enzyme which is presumed to occur during catalysis. Therefore, the *para*-nitro group on the phenyl ring appears to cause considerable steric hindrance to acylation of the ox enzyme but not to acylation of the pig enzyme. It seems likely that the methyl side chain of the alanine present in the active site peptide of ox but not of pig is the source of the observed steric hindrance. This suggestion is supported by the specificity of sheep liver carboxylesterase which resembles that of the pig enzyme; the active site sequence of the sheep enzyme is the same as that of the pig.[16]

The rate of acylation of ox liver carboxylesterase during titration is not exactly described by the usual simple kinetic scheme, a characteristic shown also by the pig enzyme. Intensive investigation of the ox system has shown that the simple kinetic scheme constitutes a very good approximation to the observed rate, however.[25] It is possible that the kinetics of acylation of the ox esterase are complicated by a relatively slow dissociation of the enzyme during acylation.

Values of k_{cat} for the ox liver carboxylesterase-catalyzed hydrolysis of butyric acid esters (Table VII) and of phenyl acetate ($k_{cat} = 300$ sec^{-1} at 25°) and phenyl trimethylacetate ($k_{cat} = 6.5$ sec^{-1} at 25°)[2] illustrate the high reactivity and relatively low specificity of the enzyme. In addition to the hydrolysis of esters, ox liver carboxylesterase has been found to catalyze the hydrolysis of anilides[3] and the transfer of the acyl groups of substrates to nucleophiles such as amines (*e.g.*, aliphatic esters of amino acids) in solution.[5,6]

[25] Carboxylesterases from Chicken, Sheep, and Horse Liver

By KEITH SCOTT and BURT ZERNER

In the preceding paper[1] the purification and properties of pig and ox liver carboxylesterases are described. This paper extends the description to include the corresponding enzymes from chicken, sheep, and horse livers. The same basic techniques are used for all of the purification pro-

---

[1] N. P. B. Dudman and B. Zerner, this volume [24].

cedures which therefore differ one from the other only in small, though probably crucial, experimental details.

General Methods

The esterase rate assay, spectrophotometric protein determination, the preparation of chloroform–acetone and acetone powders, the extraction of esterase activity from these powders, ammonium sulfate fractionation, the sources, preparation, and regeneration of chromatographic media, and active-site titration are described in the preceding paper.[1]

One unit of esterase activity is defined as the amount of enzyme required to catalyze the hydrolysis of 1 μmole of ethyl butyrate per minute, under the assay conditions described previously.[1] Specific activity is activity (in units/ml) /A_{280}.

Both chicken and sheep liver carboxylesterases are subject to substrate inhibition with ethyl butyrate at concentrations greater than 4.0×10^{-3} M.[2] For this reason, the rate assays for these enzymes produce curved traces. However, the assays are completely reliable if initial rates are measured. The horse liver enzyme produces linear traces.

General Properties

A comparison of the kinetic properties of chicken, horse, sheep, ox, and pig liver carboxylesterases has been made recently by Stoops et al.[2]

The amino acid compositions of these five esterases have also been reported recently[3] and are shown in Table I.

A discussion of the molecular weight of the pig liver enzyme has recently been published in papers reporting the gel-chromatographic behavior and ultracentrifugation of the esterase.[4,5] Similar studies on the chicken liver enzyme are reported in the same papers.

Purification Procedure for Chicken Liver Carboxylesterase[6]

Step 1. Extraction of Chloroform–Acetone Powders. Chloroform–acetone powder (2 kg) is extracted in 400-g batches, each with

[2] J. K. Stoops, S. E. Hamilton, and B. Zerner, unpublished results.
[3] K. Scott and B. Zerner, unpublished results.
[4] P. A. Inkerman, Ph.D. Thesis, Univ. of Queensland, 1974.
[5] P. A. Inkerman, D. J. Winzor, and B. Zerner, unpublished results.
[6] P. A. Inkerman, K. Scott, M. T. C. Runnegar, S. E. Hamilton, E. A. Bennett, and B. Zerner, unpublished results.

TABLE I
AMINO ACID COMPOSITION OF LIVER CARBOXYLESTERASES[a]

Amino acid	Chicken	Ox	Sheep	Horse	Pig
Asp	49.46	56.07	56.21	46.75	52.87
Thr	24.52	29.93	30.29	32.14	34.87
Ser	37.60	42.75	39.62	41.67	40.50
Glu	78.08	55.39	55.58	64.80	60.52
Pro	29.12	46.41	47.71	43.19	45.67
Gly	45.40	46.10	51.16	53.56	55.34
Ala	55.39	47.58	47.21	51.06	48.82
Cys[b]	7.26	5.98	5.21	6.07	4.27
Val	50.54	45.12	48.13	49.65	50.62
Met	15.14	16.09	14.81	19.16	13.50
Ile	28.90	24.14	22.76	26.10	20.02
Leu	48.56	64.84	67.15	55.07	61.64
Tyr	21.39	16.08	18.30	14.33	15.97
Phe	32.34	32.74	31.22	34.62	32.62
Lys	39.24	37.58	35.09	42.30	37.12
His	11.07	16.35	15.19	16.50	14.17
Arg	26.07	20.77	20.57	13.99	20.25
Trp	9.20	11.20	11.32	10.30	12.00

[a] Residues per 68,000 daltons.
[b] Half-cystine + cysteine.

4 liters of 0.1 M phosphate buffer, pH 7.4, at ~25° for 30 minutes while stirring constantly. The suspensions are centrifuged at 4° for 1 hour at 3000 g to produce clear, reddish brown solutions with a pH of 7.1–7.2.

The specific activity of various chloroform–acetone powder extracts has, in our laboratory, varied from 0.40 to 1.62 (units/ml)/A_{280}, presumably owing to variation in the esterase content of the livers. Despite this, the procedure has proved uniformly successful.

Step 2. Ammonium Sulfate Fractionation. A 50–70% saturated ammonium sulfate cut is taken on each of the five supernatants from step 1 (pH ~7.1). The precipitates, collected by centrifugation at 3000 g for 1 hour at 4°, are taken up in and dialyzed against 0.075 M acetate buffer, pH 4.7. Prior to column chromatography, the dialysates are centrifuged at 3000 g for 45 minutes at 4°.

Ammonium sulfate fractionation produces a 20-fold increase in specific activity, thereby reducing the amount of protein to such an extent that the five separate extractions can be chromatographed, after fractionation, on one CM-cellulose column.

Step 3. Chromatography on CM-Cellulose. CM-Cellulose chromatography is carried out on a column (5.5 × 85 cm) equilibrated with 0.075 M acetate buffer, pH 4.7. The 5 dialysates from step 2 are applied individually to the one column at a flow rate of <1 ml/minute. After each dialysate is applied, the column is washed with 2 liters of starting buffer (0.075 M acetate, pH 4.7). When all the dialysates have been loaded, the column is washed with ~25 liters of starting buffer to remove unbound protein. Under these conditions the enzyme binds, but only if the procedure is accurately followed. The esterase is eluted with a combined pH-salt gradient consisting of 1.5 liters of starting buffer in the mixing vessel of a linear gradient system and 1.5 liters of starting buffer containing 0.3 M sodium acetate in the other vessel. Esterase activity elutes over the range 2.4–2.5 liters of the gradient. Fractions with a specific activity greater than 150 are pooled and concentrated by adding solid $(NH_4)_2SO_4$ to 80% saturation. The precipitate is dialyzed against 0.02 M phosphate buffer, pH 8.0, and the dialysate is centrifuged at ~10,000 g for ~40 minutes at 4° to remove any insoluble material prior to DEAE-cellulose chromatography.

Chicken liver carboxylesterase is somewhat unstable at low pH; therefore, this chromatographic step and the preceding dialysis are carried out expeditiously.

Step 4. Chromatography on DEAE-Cellulose. DEAE-Cellulose chromatography is carried out on a column (4.5 × 75 cm) equilibrated with 0.02 M phosphate buffer, pH 8.0. The dialysate from step 3 is loaded onto the column at a flow rate of <1 ml/minute. The column is washed exhaustively with the starting buffer (0.02 M phosphate, pH 8.0) to remove unbound protein. The esterase is eluted over the range 2.9–4.0 liters of a linear salt gradient. The mixing vessel contains 2 liters of starting buffer, and the other vessel contains 2 liters of 0.2 M NaCl in starting buffer. The specific activity is nearly constant at ~240 throughout the elution of the major part of the esterase activity (Fig. 1).

The purification of chicken liver carboxylesterase is summarized in Table II. The A_{280}/A_{260} and A_{280}/A_{410} values indicate the successful removal of nucleic acid and heme pigment, respectively. The complete purification takes about 3–4 weeks, and reproducibly yields ~1 g of purified enzyme from 2 kg of chloroform–acetone powder.

Crystallization of the Enzyme. Concentrated samples of purified enzyme (>20 mg/ml) can be crystallized as follows. Solid $(NH_4)_2SO_4$ is added slowly from a glass rod while stirring constantly until the first signs of turbidity appear. The solution is then centrifuged (e.g., at 30,000 g for 15 minutes at 4°) to remove the amorphous protein, and the supernatant is allowed to evaporate at 4° until crystallization occurs. With

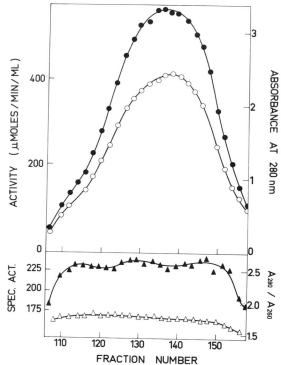

Fig. 1. Chromatography of chicken liver carboxylesterase on DEAE-cellulose (step 4). Each fraction contained 26 ml. (○) A_{280}, (●) activity, (△) A_{280}/A_{260}, and (▲) specific activity.

TABLE II

PURIFICATION OF CHICKEN LIVER CARBOXYLESTERASE

Step	Total protein[a] (g)	A_{280}/A_{260}	A_{280}/A_{410}	Specific activity	Yield (%)
1. Extraction[b]	1552	0.80–0.84	23–38	0.4–0.93	100
2. 50–70% saturated ammonium sulfate cut[c]	51	1.13–1.16	23–34	16.6–17.1	79
3. CM-Cellulose	3.14	1.79	79	175	51
4. DEAE-Cellulose	1.29[d]	1.81	>500	240	29

[a] For the present purposes, a solution having an A_{280} of 1 is defined as containing 1 mg/ml of protein.

[b] Esterase activity (1.07×10^6 units) was extracted from 2 kg of chloroform–acetone powder in five extractions.

[c] A separate $(NH_4)_2SO_4$ cut was taken of each extract.

[d] Equivalent to 1.06 g, based on $A_{1\,cm}^{1\%} = 12.15$ at 280 nm for pure enzyme.

$(NH_4)_2SO_4$ at 54% saturation, crystallization begins within a few hours and is $\sim$75% complete after 24 hours. At pH 7.6 (0.1 M phosphate), crystals are clearly defined needles, whereas at pH 4.7 (0.075 M acetate) they form large clusters.

Properties of Chicken Liver Carboxylesterase

Purity of the Enzyme. The enzyme prepared by the procedure described is significantly pure, as indicated by the constancy of the specific activity throughout elution from DEAE-cellulose and by the A_{280}/A_{260} and A_{280}/A_{410} values. When the purified enzyme is chromatographed on Sephadex G-200 or rechromatographed on CM- and DEAE-celluloses the specific activity shows no increase. Further, polyacrylamide gel electrophoresis (6% gels, 0.03 M Tris-glycine, pH 8.9) shows no protein bands which do not have esterase activity.[6] However, four closely migrating, but distinctly separated, bands of esterase activity are observed, of which two comprise over 95% of the total activity.

Stability. A solution of chicken liver carboxylesterase (20 mg/ml) in 0.05 M Tris buffer (pH 7.5, 0.15 M KCl) has been stored for 2 years at 4° without any change in specific activity or equivalent weight of the enzyme. The esterase is similarly stable at pH 8.0 (4°) for at least 2 months, but prolonged storage at pH 4.7 results in loss of activity. Enzyme held at 50° (pH 8) loses about half of its activity in 100 hours.[6]

Physical Properties. The molecular weight of chicken liver carboxylesterase has been estimated to be 65,000 $\pm$ 1500 by sedimentation equilibrium (pH 7.5).[4] The diffusion coefficient has been measured by low-speed synthetic boundary spreading, by the van Holde method, and by zonal gel filtration. The three methods give essentially the same value ($D_{20,w} = 6.7 \times 10^{-7}$ cm^2 sec^{-1}). A molecular weight of 58,000 to 64,000 daltons is derived from the Svedberg equation using this value for the diffusion coefficient, the measured sedimentation coefficient, and $\bar{v}$ calculated from the amino acid composition.[4] The enzyme is eluted from Sephadex G-200 at a volume which is constant throughout a concentration range of $\sim$0.002–5.0 mg/ml. This is in contrast to the pig enzyme, which associates in the microgram per milliliter to the milligram per milliliter concentration range to form a trimer.[4,5] There is no evidence for polymerization of the chicken enzyme, and this has been attributed to the fact that it contains a significantly greater percentage of (acidic + basic) residues than the pig enzyme.[3]

The equivalent weight of chicken liver carboxylesterase has been determined to be 67,000 $\pm$ 1000.[6] This was calculated from the normality found by titration with Paraoxon. The $A_{1\ cm}^{1\%}$ at 280 nm is 12.1$_5$ (0.02 M

TABLE III
PHYSICAL PROPERTIES OF CHICKEN LIVER CARBOXYLESTERASE

Property	Value	Ref.
Sedimentation coefficient, $s_{20,w}$ (S)	4.2–4.6	4
Diffusion coefficient, $D_{20,w} \times 10^7$ (cm² sec⁻¹)	6.7	4
Partial specific volume, $\bar{v}$ (ml/g)	0.738, 0.734	3, 4
Isoelectric point	4.9, 5.1	4
$A_{1\,cm}^{1\%}$ at 280 nm	12.1$_5$	6

phosphate, pH 8.0, 25°).[6] An essentially identical equivalent weight can be obtained from the stoichiometry of the reaction of the enzyme with [³²P]DFP.[6,7] Titration with bis-(p-nitrophenyl) phosphate gives a biphasic burst, which is quantitatively the same as that given by paraoxon.[6] The pig enzyme also gives a biphasic burst with this titrant, as discussed in the preceding paper.[1] The equivalent weight estimations demonstrate that the enzyme has one active site per molecule (∼65,000 daltons).

The isoelectric point of the chicken esterase was found to be 5.1 at an ionic strength of 0.025 (starch gel electrophoresis) and 4.9 at an ionic strength of 0.1 (moving boundary electrophoresis).[4]

Some of the physical properties of chicken liver carboxylesterase are shown in Table III.

Purification Procedure for Sheep Liver Carboxylesterase[6]

Step 1. Extraction of Chloroform–Acetone Powders. Chloroform–acetone powder (400 g) is extracted with 3.6 liters of 0.1 M citrate buffer, pH 4.0, at 4° for 1 hour with constant stirring. The suspension is centrifuged at 2° for 45 minutes at 3000 g. The pH of the clear, reddish brown supernatant is raised to 7.5 by the dropwise addition of 2 N NaOH to the magnetically stirred solution. A flocculent precipitate is formed, but it is not necessary to remove it at this stage since it is easily removed in the first stage of the ammonium sulfate fractionation.

Step 2. Ammonium Sulfate Fractionation. A 40–60% saturated $(NH_4)_2SO_4$ cut is taken on the supernatant (pH 7.5) from step 1. The precipitate is taken up in and dialyzed against 0.05 M acetate buffer, pH 5.4. The dialysate is centrifuged at ∼10,000 g for ∼40 minutes at 4° to remove any insoluble material before chromatography.

[7] R. C. Augusteyn, J. de Jersey, E. C. Webb, and B. Zerner, *Biochim. Biophys. Acta* **171**, 128 (1969).

Step 3. Chromatography on CM-Cellulose. CM-Cellulose chromatography is carried out on a column (90 × 5 cm) equilibrated with 0.05 *M* acetate buffer, pH 5.4. After the dialysate is applied, the column is washed with the equilibrating buffer to remove unbound protein. During this washing, about 20% of the esterase activity is eluted. This is not used in the subsequent purification procedure. The bound esterase is eluted with 0.05 *M* acetate buffer, pH 5.6. Fractions with a specific activity greater than ∼100 are pooled and concentrated by adding solid $(NH_4)_2SO_4$ to 65% saturation. The precipitate is dialyzed against 0.05 *M* acetate buffer, pH 5.25, and the dialysate is centrifuged at ∼10,000 *g* for ∼40 minutes at 4° to remove insoluble material prior to CM-Sephadex chromatography.

CM-Cellulose chromatography can be carried out at pH 5.6, where none of the esterase binds to the resin. This avoids the reduction in yield caused by discarding the esterase which does not bind at pH 5.4. However, the purification achieved in this step at pH 5.4 is ∼2½ times better than that at pH 5.6.

Step 4. Chromatography on CM-Sephadex. CM-Sephadex chromatography is carried out on a column (85 × 4.5 cm) equilibrated with 0.05 *M* acetate buffer, pH 5.25. The dialysate from step 3 is loaded onto the column, which is then washed with starting buffer to remove unbound material. The bound esterase is eluted over the region 3.9–4.4 liters of a pH-salt gradient, consisting of 3 liters of starting buffer in the mixing vessel of a linear gradient system, and 3 liters of 0.15 *M* sodium acetate in the other vessel. Fractions with a specific activity greater than ∼200 are pooled and concentrated by adding solid $(NH_4)_2SO_4$ to 65% saturation. The precipitate is collected by centrifugation and dialyzed against 0.05 *M* phosphate buffer, pH 7.55. The dialysate is centrifuged at ∼10,000 *g* for ∼40 minutes at 4° to remove insoluble material prior to gel filtration.

Step 5. Chromatography on Sephadex G-200. The dialysate from step 4 is loaded onto a Sephadex G-200 column (91 × 2.2 cm), previously equilibrated in 0.05 *M* phosphate buffer, pH 7.55. Fractions are collected, pooled, and concentrated by adding solid $(NH_4)_2SO_4$ to 65% saturation. The precipitate is collected by centrifugation and dialyzed against 0.05 *M* phosphate buffer, pH ∼7.5. In some cases, this concentration step is necessary to achieve the specific activity of 270. The elution profile from Sephadex G-200 is shown in Fig. 2. There is no significant change in the specific activity throughout the elution of ∼80% of the esterase activity.

The purification of sheep liver carboxylesterase is summarized in Table IV. The procedure yields ∼200 mg of purified enzyme from 400 g of chloroform–acetone powder.

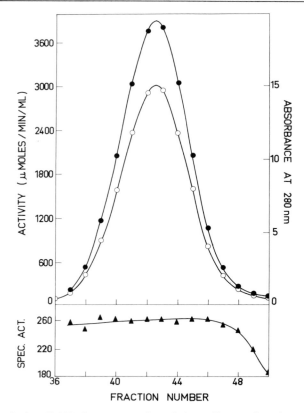

Fig. 2. Sephadex G-200 chromatography of sheep liver carboxylesterase (step 5). Each fraction contained ~4 ml. (◯) A_{280}, (●) activity, and (▲) specific activity.

An alternative procedure, starting with acetone powder and chromatographing on CM-cellulose at pH 5.65 rather than 5.4, has also produced enzyme of specific activity ~270. The yield given by this alternative procedure is higher than that given by the procedure described (30 vs. 9%) but the A_{280}/A_{410} is much lower (73 vs. 384).

Properties of Sheep Liver Carboxylesterase

Purity of the Enzyme. The elution profile from Sephadex G-200 (Fig. 2) suggests a high degree of purity in the eluted enzyme. Starch gel electrophoresis of the purified enzyme shows one major band of esterase activity with a minor, faster migrating component. Staining of the gel for protein shows only one band, corresponding to the major band of esterase activity.[6]

TABLE IV
PURIFICATION OF SHEEP LIVER CARBOXYLESTERASE

Step	Total protein[a] (g)	A_{280}/A_{260}	A_{280}/A_{410}	Specific activity	Yield (%)
1. Extraction[b]	67	0.92	13.5	11.7	100
2. 40–60% saturated ammonium sulfate cut	18	1.37	9.0	24	55
3. CM-Cellulose	1.1	1.63	19.8	148	20
4. CM-Sephadex	0.37	1.75	231	255	12
5. Sephadex G-200	0.25	1.79	384	274	9

[a] For the present purposes, a solution having an A_{280} of 1 is defined as containing 1 mg/ml of protein.

[b] A total of 7.89×10^5 units of activity are extracted from 350 g of chloroform–acetone powder.

Physical Properties. Very little work has been done on the physical properties of sheep liver carboxylesterase. The concentration dependence of the sedimentation coefficient has been studied, producing the relationship $s_{20} = 8.1_0 (1-0.075_3 c)$ over the concentration range 0.15–1.0 g/100 ml, where c is the concentration in gram per 100 ml and s is in Svedberg units (S).[6] A molecular weight of ~150,000 daltons is indicated by the value of 8.1 S for $s_{20,\text{buffer}}^0$. Ecobichon[8] has reported a molecular weight of $65,000 \pm 3000$ for the "cytoplasmic carboxylesterase" present in a sheep liver homogenate, determined by gel filtration on Sephadex G-200. The enzyme was presumably at very low concentration in that experiment. In our laboratory, an apparent molecular weight of ~65,000 was obtained for sheep liver carboxylesterase at pH 4.7, using gel filtration on Sephadex G-100.[9] The equivalent weight was found to be ~69,500 by titration with *p*-nitrophenyl dimethylcarbamate.[6] Therefore, there is evidence for a species, existing at low concentration and at low pH, with a molecular weight of ~65,000 to 70,000 daltons and bearing one active site. The evidence from velocity sedimentation for a species of higher molecular weight suggests a concentration-dependent polymerization. That it is also pH-dependent is indicated by the fact that the enzyme is eluted from Sephadex G-100 at the void volume at pH 7.6,[6] whereas a molecular weight of ~65,000 daltons is deduced from the elution volume at pH 4.7. A value of 13.7_7 was found for $A_{1\,\text{cm}}^{1\%}$ at 280 nm (0.1 M Tris buffer, pH 7.5).

[8] D. J. Ecobichon, *Can. J. Biochem.* **50**, 9 (1972).

[9] P. A. Inkerman and B. Zerner, unpublished results.

Purification Procedure for Horse Liver Carboxylesterase[6]

Step 1. Extraction of Acetone Powder. Acetone powder is extracted in two 400-g batches, each with 3.6 liters of 0.1 M acetate buffer, pH 4.5, at 4° for 1 hour with constant stirring. The suspensions are centrifuged at 2° for 1 hour at 3000 g. Fractionation with $(NH_4)_2SO_4$ is carried out on the clear, reddish brown supernatants without pH adjustment (pH ~4.5).

Step 2. Ammonium Sulfate Fractionation. A 50–60% saturated ammonium sulfate cut is taken on each supernatant separately. The precipitates are taken up in and dialyzed against 0.05 M acetate buffer, pH 5.0. The dialysates are centrifuged at ~10,000 g for ~40 minutes to remove any insoluble material before chromatography.

Step 3. Chromatography on Sephadex G-100. The two dialysates are chromatographed separately, and two columns are used for each dialysate. Chromatography is carried out on Sephadex G-100 columns (~75 × 5 cm) equilibrated with 0.05 M acetate buffer, pH 5.0. A volume of ~50–60 ml is applied to each column. For this reason, the precipitates in step 2 must be taken up in a minimum volume if columns of this size are used. Fractions with a specific activity greater than ~40 are pooled and concentrated by adding solid $(NH_4)_2SO_4$ to 85% saturation. The precipitates are collected by centrifugation and dialyzed against 0.01 M acetate buffer, pH 5.5. The dialysates are centrifuged at ~10,000 g for ~40 minutes at 4° to remove insoluble material prior to DEAE-cellulose chromatography.

Step 4. Chromatography on DEAE-Cellulose. DEAE-Cellulose chromatography is carried out on a column (43 × 4.5 cm) equilibrated with 0.01 M acetate buffer (pH 5.5). The four dialysates from step 3 are pooled and loaded onto the DEAE-cellulose column, which is then washed with the equilibrating buffer to remove unbound material. During this washing, a large peak of protein is eluted which contains a small amount of esterase activity (~3%). The bound esterase is eluted over the region 1.9–2.4 liters of a pH gradient, produced from a two-reservoir linear gradient system containing 2 liters of starting buffer in the mixing reservoir and 2 liters of 0.02 M acetic acid in the other reservoir. Fractions with a specific activity of greater than ~130 are pooled and concentrated as in step 3.

Step 5. Chromatography on CM-Cellulose. The dialysate from step 4 is chromatographed on a CM-cellulose column (79 × 3 cm), previously equilibrated in 0.05 M acetate buffer, pH 5.0. The column is thoroughly washed with the equilibrating buffer to remove unbound protein. The bound esterase is eluted using a two-reservoir linear gradient system with

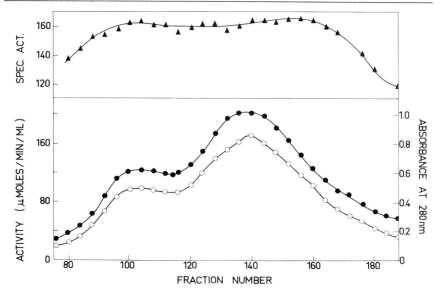

Fig. 3. CM-Cellulose chromatography of horse liver carboxylesterase. Each fraction contained ~6 ml. (○) Activity, (●) A_{280}, and (▲) specific activity.

2 liters of starting buffer in the mixing reservoir and 2 liters of 0.1 M sodium acetate in the other. The esterase elutes over the region 1.7–2.5 liters. The specific activity remains virtually constant (~160) throughout the elution of ~65% of the esterase activity (Fig. 3). However, the elution profile shows a bimodal boundary in the esterase peak. Samples taken from various fractions across the peak have been electrophoresed on starch and polyacrylamide gels, and subjected to sedimentation velocity experiments, but no evidence for more than one component has been found.[6] The most likely explanation of the bimodal boundary is an interaction between protein and buffer.[10]

The purification of horse liver carboxylesterase is summarized in Table V. The procedure yields ~230 mg of purified enzyme from 800 g of acetone powder.

Properties of Horse Liver Carboxylesterase

Purity of the Enzyme. Although the elution profile from the final step suggests some inhomogeneity, the constancy of the specific activity argues aganst this. Further, starch and polyacrylamide gel electrophoresis of the purified enzyme indicate homogeneity. Both types of gel electrophoresis

[10] J. R. Cann, Vol. 25 [11].

TABLE V

Purification of Horse Liver Carboxylesterase

Step	Total protein[a] (g)	A_{280}/A_{260}	A_{280}/A_{410}	Specific activity	Yield (%)
1. Extraction[b] (i)	72	1.05	5.2	1.5	100
(ii)	70	0.97	5.1	1.5	100
2. 50–60% saturated (i)	6.9	1.58	6.4	15.1	96
ammonium sulfate cut (ii)	4.8	1.53	8.8	15.7	71
3. Sephadex G-100 (i)	0.959	1.55	3.5	88	78
(ii)	0.816	1.57	2.4	76	58
4. DEAE-cellulose	0.666	1.58	4.1	135	42[c]
5. CM-cellulose	0.277	1.79	39	157	20[c]

[a] For the present purposes, a solution having an A_{280} of 1 is defined as containing 1 mg/ml protein.

[b] A total of 2.15×10^5 units of activity are extracted from 800 g of acetone powder in two extractions, (i) and (ii).

[c] Calculated on the basis of the combined activity of (i) and (ii).

show one major band of esterase activity with a minor, faster migrating band. When the gel is stained for protein, only one band is seen, which has the same mobility as the major esterase band.[6]

Physical Properties. Ecobichon[8] found that the "cytoplasmic carboxylesterase" from horse liver gave one peak on Sephadex G-100 with a molecular weight estimated at $70,000 \pm 3000$. Chromatography on Sephadex G-100 of enzyme that had been frozen and thawed, or concentrated, gave two peaks in each case. The molecular weights were estimated on Sephadex G-200 to be 65,000 and 96,000. In sedimentation velocity experiments, we have measured a value of ~ 8 S for $s_{20,buffer}$, at a concentration of 0.56 mg/ml,[6] suggesting that the molecular weight at this concentration is much higher than that found at a presumably lower concentration by Ecobichon.

The equivalent weight of horse liver carboxylesterase was first reported by Boursnell and Webb.[11] They obtained a value of $\sim 96,000$ using [^{32}P]DFP. However, the enzyme they used, which was prepared by the method of Burch,[12] was probably significantly more impure than the 20% for which they allowed.[6] An equivalent weight of $\sim 70,000$ has been obtained by titration with *p*-nitrophenyl dimethylcarbamate.[6] The purified enzyme has an $A_{1\,cm}^{1\%}$ of 12.0_1 at 280 nm (0.1 M Tris buffer, pH 7.05).

The evidence described above suggests that horse liver carboxyl-

[11] J. C. Boursnell and E. C. Webb, *Nature (London)* **164**, 875 (1949).
[12] J. Burch, *Biochem. J.* **58**, 415 (1954).

esterase undergoes association, and that the subunit has a molecular weight of ~70,000 daltons and bears one active site. In this respect it is similar to the sheep, pig, and ox liver enzymes.[1,4,13]

[13] M. T. C. Runnegar, K. Scott, E. C. Webb, and B. Zerner, *Biochemistry* **8**, 2013 (1969).

[26] Appendix: An Accurate Gravimetric Determination of the Concentration of Pure Proteins and the Specific Activity of Enzymes

By Robert L. Blakeley and Burt Zerner

In studies of subunit structure, equivalent weight and mechanism of action of enzymes, an accurate measure of the weight-concentration of the highly purified protein in solution is required. Concentration is usually measured spectrophotometrically, utilizing chromophores associated with the protein itself or chromophores produced in the Folin-Lowry[1] or biuret[2] procedures. In any spectrophotometric assay of protein concentration, it is necessary to know the relationship between the weight-concentration of the protein and the absorbance measured. Described below is an absolute gravimetric method for determining the weight-concentration of a protein solution. This method is useful in standardizing the measurement of concentration of a pure protein by *any* spectrophotometric method.[3] The need for a reliable and accurate method is highlighted by the range of values of $A_{1\,cm}^{1\%}$[4] reported in the literature[5] for specific highly purified proteins.

[1] O. H. Lowry, N. J. Rosebrough, A. L. Farr, and R. J. Randall, *J. Biol. Chem.* **193**, 265 (1951); J. Eichberg and L. C. Mokrasch, *Anal. Biochem.* **30**, 386 (1969).
[2] R. F. Itzhaki and D. M. Gill, *Anal. Biochem.* **9**, 401 (1969).
[3] The absorbance produced by a given weight-concentration of a protein ("dry weight") in the Folin-Lowry[1] and biuret[2] procedures is a function of its amino acid composition, and hence an absolute standardization is necessary if these procedures are to be used in quantitative work with pure proteins.
[4] Abbreviations: A_{280}, etc., absorbance at 280 nm etc.; $A_{1\,cm}^{1\%}$, absorbance of a solute (which obeys Beer's law) at a concentration of 10 mg/ml of solution using an optical path of 1 cm; kat (katal) [or μkat (microkatal)], i.e., that amount of enzyme which catalyzes the transformation of 1 mole (or 1 μmole) of substrate per second under defined conditions.
[5] D. M. Kirschenbaum, *in* "Handbook of Biochemistry" (H. Sober, ed.), 2nd ed., p. C-71. Chem. Rubber Publ. Co., Cleveland, Ohio, 1970; D. M. Kirschenbaum, *Int. J. Protein Res.* **3**, 109, 157, 237, and 329 (1971); **4**, 63 and 125 (1972).

One of the most important physical characteristics of an enzyme is its specific activity. Because of the convenience of measuring the concentration of protein by means of its A_{280}, it is suggested below that, where possible, the specific activity of an enzyme be expressed in terms of the empirical A_{280} rather than in terms of milligram per milliliter or milligram of protein.

Materials and Methods

Preparation of the Drying Vessel. The heavy-walled glass vessel shown in Fig. 1 weighs about 40 g and has a 10-ml working volume (to the dotted line). The shoulder above the oval end rests in the groove of the pan in a Mettler model B6 balance. The Quickfit B7 stopper provides an air-tight seal without grease. The B24 cone connects to the vacuum line through a glass tube (2 cm diam) containing a coarse fritted glass disk.

Clean the vessel and stopper in chromic acid, rinse with fresh 5% HF for 20 seconds, boil in distilled water, and dry at 160°. Cool the vessel and stopper in a clean desiccator in the presence of P_2O_5, stopper the vessel and allow it to equilibrate for a standard interval (say, 30 minutes) in a desiccator containing saturated aqueous magnesium acetate. Touch the vessel *only* with fingers freshly cleaned with low-boiling petroleum ether. Equilibrate a balance case with saturated magnesium acetate and calibrate the balance's optical scale. Zero the balance, weigh the vessel rapidly to 0.02 mg, and recheck the zero. Rinse the vessel with distilled water, wipe it dry with paper tissues, and wipe it 4 times with tissues soaked with petroleum ether; rotate the vessel with continuous pressure while moving the tissues from top to bottom. No residue should be visible on the glass. Reequilibrate the vessel in the desiccator and reweigh it. If the vessel is clean and dry, its weight may be read immediately upon

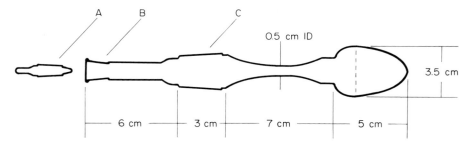

FIG. 1. Drying vessel: A, B7 cone; B, B7 socket; and C, B24 cone. Broken line indicates top of 10 ml working volume.

placing it on the balance. A drifting weight means that equilibration with moisture has not been adequate. A nonreproducible weight means that the washing and wiping of the vessel have not been done with sufficient care.

Dry Air Inlet System. The vacuum line consists of a mercury diffusion pump, a liquid nitrogen trap, and a small manifold. Air is admitted to the manifold through a fine capillary which feeds through a series of 16 loops (4.5 × 25 cm) of 0.5 cm glass tubing *half*-immersed in Dry Ice–acetone. The loop traps are connected to the manifold by means of surgical rubber tubing previously dried in vacuum over P_2O_5. The inlet system is thoroughly dried by pulling many volumes of air slowly through it prior to drying the protein; this system dries air efficiently since moisture is visible only in the first one or two loops at the end of the experiment.

Procedure. 1. Dialyze the protein solution exhaustively at about 5 mg/ml in prewashed casing against buffered 1 mM EDTA and then against distilled water.[6]

2. Centrifuge the protein solution and record the ultraviolet spectrum of an aliquot in buffer on a spectrophotometer (such as a Cary 14) previously calibrated with alkaline potassium chromate[7] and with benzene vapor.

3. Pipette an aliquot of the protein solution containing 10–30 mg into the preweighed drying vessel and wash it with distilled water.

4. Freeze the solution by swirling the vessel in Dry Ice–acetone, degas it by thawing under gentle vacuum, *evaporate* it at ∼30°, and, finally, heat it at 120–140° in diethyl phthalate for 2 hours under high vacuum (10^{-5} mm Hg).

5. Cool the sample and then admit air carefully through the predried air inlet system.

6. Stopper the vessel tightly immediately upon removal from the vacuum line, then degrease it and weigh it according to the standard procedure. The ungreased Quickfit B7 stopper prevents moisture from being reabsorbed by the protein for at least several hours if it has been inserted firmly.

---

[6] The effectiveness of exhaustive dialysis in removing all but minor contamination by metal ions is attested by the low total content of metal (<0.1% by weight) found in jack beam urease [C. Gazzola, R. L. Blakeley, and B. Zerner, *Proc. Aust. Biochem. Soc.* **5**, 11 (1972)]. For sulfhydryl proteins, the water should be boiled and then cooled and used under oxygen-free nitrogen. For euglobulins, a volatile buffer such as ammonium acetate could presumably be used in place of distilled water.

[7] G. W. Haupt, *J. Res. Nat. Bur. Stand.* **48**, 414 (1952).

7. After constant weight is established, heat the vessel for another hour under high vacuum and then reweigh it. The extra period of drying should not alter the weight of the vessel and contents.

Comments

The vessel can be reproducibly weighed to 0.1 mg or better, so that the weight of a 10–30 mg sample of protein may be obtained correct to $\pm 1\%$ or better. The reproducibility of weighing and the effectiveness of drying are thoroughly checked in each determination. This procedure is absolute and is suitable for proteins with nonprotein chromophores. No special equipment is required apart from a simple vacuum line.

Comparison with Other Methods

Pig Liver Carboxylesterase. The value of $A_{1\,cm}^{1\%}$ at 280 nm for pure pig liver carboxylesterase was determined by the present method to be 13.4_1.[8] The value 13.05 determined earlier in these laboratories[9] was based on removal of water under high vacuum *without heating*, and clearly reflects about 2.7% water which is tightly bound and is removed even in *high* vacuum only at high temperature. (The necessity of heating proteins in *moderate* vacuum to remove the last traces of water has been previously noted.[10,11]) The value 13.8 ± 0.3 based on a refractometric comparison with bovine serum albumin[12] is as close to 13.4_1 as can be expected considering that the refractive index of proteins is a function of their amino acid composition.[13] Refractometric comparisons are even less reliable if a heme group[13] or presumably any other chromophore is present.

Jack Bean Urease. The value of $A_{1\,cm}^{1\%}$ for jack bean urease was determined by the present method to be 6.01 at the absorption maximum at 278.5 nm and 5.89 at 280 nm.[14] Ultraviolet spectra of pure urease, even in buffers containing 2-mercaptoethanol, show an absorbance at 360 nm of 2–3% of the total absorbance at 278.5 nm.[14] The spectra also show a gentle and almost linear increase in absorbance as the wavelength is decreased across the region 360–320 nm.[14] Since the chromophores inherent in pro-

[8] P. A. Inkerman and B. Zerner, unpublished results (1972).

[9] D. J. Horgan, J. R. Dunstone, J. K. Stoops, E. C. Webb, and B. Zerner, *Biochemistry* **8**, 2006 (1969).

[10] M. J. Hunter, *J. Phys. Chem.* **70**, 3285 (1966).

[11] R. Goodrich and F. J. Reithel, *Anal. Biochem.* **34**, 538 (1970).

[12] D. L. Barker and W. P. Jencks, *Biochemistry* **8**, 3879 (1969).

[13] J. Babul and E. Stellwagen, *Anal. Biochem.* **28**, 216 (1969); T. L. McMeekin, M. Wilensky, and M. L. Groves, *Biochem. Biophys. Res. Commun.* **7**, 151 (1962).

[14] R. L. Blakeley, E. C. Webb, and B. Zerner, *Biochemistry* **8**, 1984 (1969).

teins do not absorb significantly in this region,[15] we have routinely corrected the A_{280} (or $A_{278.5}$) for the difference between the protein spectrum extrapolated from this region and the solvent base line similarly extrapolated.[14] This correction amounts to 6–9% of the total absorbance at 278.5 nm. If this correction were not made, the value of $A_{1\,cm}^{1\%}$ at 278.5 nm would be about 6.4 for urease by the present technique. This may be compared with values of 7.71,[16] 6.40,[17] 7.54,[17] and 7.71[11,18] which have presumably not involved the base line correction and which are based on various procedures for preparing the enzyme and measuring its "dry weight." For jack bean urease (molecular weight, 483,000 daltons[19]), the number of subunits is six based on the molecular weight in the presence of denaturing agents,[20,21] and the number of essential sulfhydryl groups is eight as determined by titration with silver ion.[22] The value of $A_{1\,cm}^{1\%}$ and the definition of experimental conditions for measuring absorbance of the protein are clearly critical in titrations of essential amino acid residues in urease, since, for example, the difference between an $A_{1\,cm}^{1\%}$ of 6.4 and 7.7 corresponds to about one extra group titrated per species of MW 483,000.

The Specific Activity of Enzymes

The concept of specific activity of an enzyme serves two useful functions. One is as a guide to purification in successive steps, and the other is as a characterizing property of a pure chemical. Both these uses have their lifeblood in convenience. In practice, a given enzyme solution is assayed for its concentration of enzymatically active material (in microkatal per liter)[23] and for its total concentration of material (most conveniently as A_{280}). We therefore support the practice of defining specific activity as (microkatal per liter)/A_{280}.[24] This definition of specific activity is dimen-

[15] H. Edelhoch, *Biochemistry* **6**, 1948 (1967), and references cited therein.

[16] G. Gorin, E. Fuchs, L. G. Butler, S. L. Chopra, and R. J. Hersh, *Biochemistry* **1**, 911 (1962).

[17] G. Gorin and C.-C. Chin, *Anal. Biochem.* **17**, 49 (1966).

[18] F. J. Reithel and J. E. Robbins, *Arch. Biochem. Biophys.* **120**, 158 (1967).

[19] J. B. Sumner, N. Gralén, and I.-B. Eriksson-Quensel, *J. Biol. Chem.* **125**, 37 (1938).

[20] F. J. Reithel, J. E. Robbins, and G. Gorin, *Arch. Biochem. Biophys.* **108**, 409 (1964).

[21] C. J. Bailey and D. Boulter, *Biochem. J.* **113**, 669 (1969).

[22] G. Gorin and C.-C. Chin, *Biochim. Biophys. Acta* **99**, 418 (1965).

[23] "Enzyme Nomenclature," p. 27. Elsevier, Amsterdam, 1973. It is a pleasure to acknowledge the benefit of discussion with Professor E. C. Webb.

[24] The former definition[14] of specific activity as (IU/ml)/A_{280} is equivalent to (16.67 μkat/liter)/A_{280}.

sionally equivalent to microkatal per gram and in effect makes the working approximation that a solution with an A_{280} of 1 contains 1 mg/ml of protein. While this approximation is valid only to within a factor of three,[5] it is nonetheless a convenient and useful approximation at all stages of purification. Adjustment of the value of A_{280} by a factor related to the A_{280}/A_{260} ratio[25] is a needless complication during the purification of proteins. The spectrophotometric determination of material is ideal in that fractionation with respect to nucleic acids (A_{280}/A_{260})[25] and heme proteins (A_{280}/A_{410}) is instantly indicated. Those proteins which are insoluble except in the presence of lipids or detergents are not amenable to this technique unless a stage of purity is reached at which the lipid or detergent becomes unnecessary for solubility.

It has been noted that A_{224},[26] A_{215},[27] and A_{194}[28] may be used for determining protein concentration. In certain situations it is necessary to use the low wavelength absorption of peptides and proteins,[29] but its use is dangerous in protein purification because many impurities and buffer components absorb appreciably in this region.

[25] O. Warburg and W. Christian, *Biochem. Z.* **310**, 384 (1941); H. M. Kalckar, *J. Biol. Chem.* **167**, 461 (1947).
[26] W. E. Groves, F. C. Davis, Jr., and B. H. Sells, *Anal. Biochem.* **22**, 195 (1968).
[27] W. J. Waddell, *J. Lab. Clin. Med.* **48**, 311 (1956).
[28] M. M. Mayer and J. A. Miller, *Anal. Biochem.* **36**, 91 (1970).
[29] J. R. Spies, D. C. Chambers, and E. J. Coulson, *Arch. Biochem. Biophys.* **84**, 286 (1959); J. B. Murphy and M. W. Kies, *Biochim. Biophys. Acta* **45**, 382 (1960).

[27] Phospholipase D from Peanut Seeds

EC 3.1.4.4 Phosphatidylcholine phosphatidohydrolase

By Michael Heller, Nava Mozes, and Eddie Maes

Phosphatidylcholine + R–OH $\rightleftharpoons$ phosphatidyl—OR + choline

Peanut seeds contain an enzyme catalyzing the transfer of a phosphatidyl moiety from glycerophosphatides to a primary alcohol or water. With water as acceptor, hydrolysis occurs with formation of phosphatidic acid, whereas with primary alcohols (e.g., methanol, ethanol, propanol, and glycerol) a transfer reaction occurs resulting in the formation of a phosphatidyl alcohol. The product retains its L-α configuration in the phosphatidyl residue. If an acceptor containing an asymmetric carbon (e.g., glycerol) is used, a DL mixture is formed,[1] e.g.:

L-α-Phosphatidylcholine + glycerol
$$\rightleftharpoons \text{L-α-phosphatidyl DL-α-glycerol + choline}$$

Assay Method[2]

Principle. The activity of phospholipase D can be determined by the liberation of choline from phosphatidylcholine in the presence of an acceptor containing a primary hydroxyl group (either water or an alcohol). The products are separated by partitioning between a chloroform-rich phase and an aqueous–methanol phase.[3] The reaction may be monitored by measuring the amount of aqueous-methanol soluble radioactivity arising from the [^{3}H]methyl groups of the choline portion of the substrate. Alternatively, the choline formed may be assayed by any conventional technique for choline determination.[4]

Reagents. [^{3}H]Choline-labeled lecithin (phosphatidylcholine) may be obtained from tissue cultures (BSC$_1$ line of monkey kidney cells[5]) grown in the presence of [^{3}H]methyl-choline. Alternatively, a synthetic medium containing the tritiated choline may be innoculated with fresh yeast.[6] Ovolecithin has been isolated from egg yolk according to Pangborn,[7] and the product obtained used directly. Although small amounts of lysolecithin, sphingomyelin, and phosphatidylethanolamine were present, they did not affect the enzyme's assay. The ovolecithin thus obtained could also be purified chromatographically on a column of aluminum oxide[8] yielding a product containing 99% ovolecithin. The labeled lecithin is diluted with the nonradioactive ovolecithin in chloroform to give a 0.1 M solution with a specific radioactivity of 1–3 μCi/μmole of lecithin (based on phosphate content).

Acetate buffer, 0.2 M, pH 5.6
CaCl$_2$, 0.5 M
Sodium dodecyl sulfate (SDS), 25 mM

---

[1] S. F. Yang, S. Freer, and A. A. Benson, *J. Biol. Chem.* **242**, 477 (1967).
[2] R. Tzur and B. Shapiro, *Biochim. Biophys. Acta* **280**, 290 (1972).
[3] J. Folch, M. Lees, and G. H. Sloane-Stanley, *J. Biol. Chem.* **226**, 497 (1957).
[4] J. C. Dittmer and M. A. Wells, Vol. 14, Sect. 53, p. 482.
[5] Y. Ascher, M. Heller, and Y. Becker, *J. Gen. Virol.* **4**, 65 (1969).
[6] A. D. Bangham and R. M. C. Dawson, *Biochem. J.,* **75**, 133 (1960).
[7] M. C. Pangborn, *J. Biol. Chem.* **188**, 471 (1951).
[8] M. A. Wells and D. J. Hanahan, Vol. 14, Sect. 33, p. 178.

Chloroform–methanol, 2:1 (by volume)
Insta-gel[9]

Procedure. From the chloroform solution of the tritiated lecithin, 0.05 ml is pipetted into a 25-ml test tube, the solvent evaporated to dryness under a stream of nitrogen, and 0.1 ml of SDS and 0.25 ml of acetate buffer added. The tube is shaken vigorously on a mixer (Vortex). A homogeneous suspension should be obtained without flakes. Then 0.1 ml of $CaCl_2$ is added and the reaction initiated by addition of a suitable amount of enzyme (5–30 milliunits or 0.5–5 μg of protein). The final volume is made with water to 1 ml. The enzyme should always be added *after* the addition of $CaCl_2$.

After incubation at 30° for 10 minutes with constant shaking, the reaction is terminated by the addition of 4 ml of chloroform–methanol (2:1); the tube is vigorously mixed (on a Vortex mixer) and kept in ice until it is centrifuged for 5–10 minutes at 500 g. An aliquot of 0.5 ml of the aqueous–methanol upper phase is withdrawn directly into a small glass counting vial; 3.5 ml of Insta-gel scintillation mixture is added and the radioactivity determined in a Packard Tri-Carb liquid scintillation counter or other suitable detection systems.

Units. Activity is expressed in units which represent the amount of enzyme that catalyzes the release of 1 μmole of choline per minute under the conditions specified above. The amount of choline formed is proportional to the amount of protein added only up to 30 milliunits.[10] Specific activity is defined as enzyme units per milligram of protein. The concentrations of protein are determined by the method of Lowry *et al.*,[11] with bovine serum albumin as standard, or by determining the absorption at 280 nm, assuming that 1 mg/ml of protein has an absorbance of 1.0.

Enzyme Purification[12]

Dry peanut seeds (*Arachis hypogea* var. *virginia*) were obtained from a local seeds selection institute following the summer harvest. The seeds

[9] A tradename given by Packard Instruments Inc. to a scintillation fluor containing detergents employed to overcome the presence of water in samples for radioactivity determinations by scintillation.

[10] The most purified enzyme preparations (following the last step) exhibited a yet-unexplained activation which is obtained by the increase of the enzyme concentrations beyond 30 milliunits. The specific activity at this stage may therefore vary depending on the enzyme concentrations.

[11] O. H. Lowry, N. J. Rosebrough, A. L. Farr, and R. J. Randall, *J. Biol. Chem.* **193**, 265 (1951).

[12] M. Heller, N. Mozes, E. Maes, and I. Abramovitz, *Biochim. Biophys. Acta* (in press).

may be stored in a cool place for several months without losing the enzymatic activity. All operations of the purification scheme described below have been carried out at temperatures below 4° unless otherwise noted.

Step 1. Soluble Extracts of Peanuts. Two-hundred grams of dry seeds are surface sterilized with a commercial detergent containing approximately 4% (w/v) SDS, rinsed thoroughly with water and soaked overnight in water at 26° in a thermostatic incubator. The brown seed cover (testa) is removed and the cotyledons homogenized with 600 ml of a solution containing 0.25 M sucrose, 50 mM Tris, 2 mM EDTA, and 3 mM 2-mercaptoethanol, pH 7.4, in a chilled Waring blender at top speed for 1 minute. The homogenate is filtered through cheesecloth and centrifuged for 15 minutes at 30,000 g in a Sorvall centrifuge. The floating fat and the debris are discarded, and the clear supernatant centrifuged again for 60 minutes at 105,000 g in a Spinco centrifuge using rotor No. 30. The resulting yellowish supernatant is aspirated and the pH adjusted to 7.4.

Step 2. Ammonium Sulfate Fractionation. Solid $(NH_4)_2SO_4$ is added to a final concentration of 20% (w/v), care being taken to maintain the pH at 7.4 or somewhat higher. Following 60 minutes of stirring the mixture is centrifuged at 30,000 g for 15 minutes (Sorvall centrifuge). The precipitate is suspended in a minimal volume of a solution containing 50 mM Tris, 5 mM EDTA, and 3 mM 2-mercaptoethanol, pH 8.0 (abbreviated TEM) and dialyzed against several changes of 50–100 volumes of TEM. The preparation at this stage may either be used immediately for further purification or lyophilized and stored at −20° without appreciable loss of activity. [Note: in some batches of peanuts variations in the distribution of protein during the $(NH_4)_2SO_4$ fractionation may cause a lower yield of active enzyme which subsequently affects the chromatographic separation pattern.]

Step 3. Chromatography on DEAE-Cellulose. The lyophilized powder from step 2 is dissolved in a minimum volume of TEM, exhaustively dialyzed against TEM, and applied to a 2.5 × 50 cm column containing approximately 20 g of dry ion exchanger packed and equilibrated in TEM (if all of the lyophilized powder does not dissolve, the preparation should be centrifuged for 10 minutes at 30,000 g and the precipitate discarded). Approximately 300 ml of TEM are first applied to the column. A linear salt gradient consisting of 1 liter of TEM buffer and 1 liter 1 M KCl in the same buffer is then applied and fractions of 15 ml each are collected at a rate of 45 ml/hour. The enzymatic activity is eluted at concentrations of 0.25 M to 0.4 M KCl. The active fractions are pooled and concentrated by ultrafiltration through PM-30 Diaflo membranes (Amicon). The excess KCl is removed by several washes with TEM on the Diaflo membrane. At this stage, concentrated enzyme solutions may be lyophilized and stored either at −20 or 4°. Alternatively, it may be dialyzed immediately against several changes of 50 mM Tris, 1 mM EDTA, 0.25 mM dithiothreitol (DTT), and 3.5% (w/v) glycerol pH 7.9 (abbreviated buffer G) for the next step.

Step 4. Sepharose 6B Chromatography. The appropriately dialyzed preparation obtained in step 3 is applied to a 2.5 × 80 cm Sepharose 6B column operated with upward flow and eluted at a rate of 30–50 ml/hour. Fractions of 7 ml are collected during the elution by buffer G. The enzyme was eluted at approximately $V_e = 2V_0$. The handling of the active fractions proceeded similarly to that of the previous step.

Step 5. Preparative Acrylamide Disc Gel Electrophoresis.[13,14] The concentrated protein solution of the previous step is dialyzed against 50 mM Tris-glycine buffer, pH 8.9, and made dense by addition of several grains of solid sucrose or a few drops of glycerol. It is then layered on top of the 3.5% acrylamide spacer gel, polymerized on a 7.5 or 10% acrylamide which is used as the separating gel. The instructions of the manufacturer (Canalco[13]) should be followed for the preparation of the gels as well as the setting up of the column and the conditions to be used for elution. The latter is regulated by a pump to give a flow rate of 15–20 ml/hour and fractions of 3–5 ml are collected. Runs usually last about 8–10 hours, using a constant current of 10 mA and a voltage of 300–500 V. The activity is monitored and the enzymatically active fractions pooled dialyzed against the same buffer and lyophilized. The amount of material eluted at this stage is small and may have low absorbance values before concentration. It does, however, show one major band on analytical disc gel electrophoresis.

Using the above procedure, phospholipase D preparations were obtained with specific activities exceeding 200 units/mg of protein.

Properties

Activators and Inhibitors. With lecithin as substrate, no activity was detected in the absence of Ca^{2+}. This cation, at optimal concentrations of 40–60 mM, gives maximal hydrolysis rates. It cannot be replaced by Mg^{2+}.[12,15] Coarse, aqueous dispersions of most phospholipids are either not affected or are hydrolyzed at a very slow rate.[15,16]

Ultrasonic irradiation for about 10 minutes at 2° under N_2 forms leci-

[13] Prep. Disc. Instruction manual, Canal, Industrial Corporation, Rockville.
[14] L. Shuster, Vol. 22, Sect. 34, p. 412.
[15] M. Heller, E. Aladjem, and B. Shapiro, *Bull. Soc. Chim. Biol.* **50**, 1395 (1968).
[16] M. Heller and R. Arad, *Biochim. Biophys. Acta* **210**, 276 (1970).

thin dispersions which are susceptible to the enzyme.[16,17] Organic solvents are capable of activating the hydrolysis of phospholipids. The most useful were diethyl ether or acetone (but not chloroform or petroleum-ether) which exert maximal effect at about 25–50% (v/v). Primary alcohols, e.g., methanol, ethanol, or glycerol, are excellent activators of the reaction catalyzed by phospholipase D. However, by virtue of their ability to replace water as acceptor of the phosphatidyl moiety, the alcohols are, in fact, acting both as substrates and as activators (probably affecting the physical state of the phospholipids).[15,16] Detergents such as SDS, sodium taurocholate (STC), or deoxycholate (DOC) activate the enzymatic hydrolysis, whereas Triton X-100 or cetyltrimethyl ammonium bromide have only very weak effects.[15,18] The negatively charged SDS was found superior to the others as an activator, and its maximal effect was obtained at a "molar" ratio of lecithin to SDS of about 2:1.[2,18] The molar ratios are calculated on the basis of formula weights. A similar activation by STC was obtained with a "molar" ratio of 10:8.[18] The products of the hydrolytic or transphosphatidyl reactions (e.g., calcium salts of phosphatidic acid, phosphatidylglycerol, or phosphatidylmethanol) increase the rate of the lecithin hydrolysis or lecithin–methanol transfer reactions.[16] Vesicles of biologic membranes (e.g., rat liver microsomes) also activate the hydrolysis of lecithin.[16]

β-Lipoprotein from bovine or rat serum, at a range of concentrations of 1–5 mg of protein per milliliter, were found to be potent inhibitors.[16] Similarly, bovine serum albumin inhibited the hydrolysis of lecithin at a range of concentration of 0.5–2 mg of albumin per milliliter.[2]

Specificity and Some Practical Applications. Phospholipase D has a broad specificity and with aqueous dispersions, and in the presence of the appropriate activators, it catalyzed hydrolysis or transfer reactions of all phospholipids tested, with the exception of sphingomyelin[2,15–16,19,20]: lecithin (activation by ether, acetone, SDS, STC, DOC, ultrasonic irradiation, or when membrane-bound), phosphatidylethanolamine (activation when membrane-bound), phosphatidylserine (activation when membrane-bound), phosphatidylglycerol (activation by ether), diphosphatidylglycerol (cardiolipin; activation by STC, DOC, or even without activator). The phospholipids bound to natural membranes, e.g., erythrocyte plasma membranes (ghosts)[19,20] or rat liver endoplasmic reticulum (microsomes),[2,16] are readily hydrolyzed even in the absence of an acti-

[17] M. Heller and I. Abramovitz, unpublished observations (1973).
[18] E. Aladjem, M.Sc. Thesis, Hebrew University, Jerusalem (1969).
[19] M. Heller, B. Roelofsen, R. F. A. Zwaal, C. Woodward, and L. L. M. van Deenen, unpublished observations (1971).
[20] B. Roelofsen and L. L. M. van Deenen, *Eur. J. Biochem.* **40,** 245 (1973).

vator. The phospholipids of serum β-lipoproteins are resistant to the enzyme.[16]

The hydrolytic as well as the transphosphatidyl activities of phospholipase D seem to be associated with the same protein. The ratio of their activities remain constant throughout purification.[2] The transfer reactions to alcohols yield 95% conversion of lecithin to phosphatidylmethanol with 20 to 25% (v/v) of methanol as acceptor. Similar concentrations of ethanol were optimal for the transfer, while higher concentrations were inhibitory. Transfer of the phosphatidyl moiety from lecithin to glycerol requires the presence of SDS and about 4 M of the acceptor. Total conversion was about 50%, of which one-third to one-half was phosphatidylglycerol, the rest being phosphatidic acid.

Large-scale preparations of a variety of phospholipids is thus possible with any starting phospholipid, except sphingomyelin, having the desired fatty acid composition.[15,16]

pH Optimum and K_m. Phospholipase D has a pH optimum of 5.5–6.0 with SDS or diethyl ether as activators, and 5.45 for ultrasonically irradiated lecithin.[16–18] These values were obtained with the following buffers: acetate, maleate, pyridine, collidine, and dimethylglutarate.[15,17,18] The apparent K_m values for the hydrolysis of lecithin (*ex ovo*) were as follows: with ether as activator; 1.25 × 10⁻² M; with ultrasonically irradiated substrate: 3.38 × 10⁻³ M.[15,17]

Stability. A. STORAGE.[2,12,18] Solutions of the enzyme at various stages of purification are more stable at 4 than at —20°. Dilute enzyme solutions (approximately 0.1 mg of protein per milliliter) could be kept on the laboratory shelf for a few days, provided the pH was maintained above 7.4, in the presence of DTT. Addition of glycerol to a final concentration of 0.5 M or, better yet, higher, extended the stability period up to about 3 weeks.

Dry powders (after dialysis and lyophilization) are stable at 4 or —20° for long periods, provided the freeze-drying was done on rather concentrated enzyme solutions. Enzyme suspensions in 3 M or 4 M $(NH_4)_2SO_4$ are very useful, and may be stored at 4°.

B. TEMPERATURE.[18] Enzyme solutions at pH above 7.4 could be warmed for 10 minutes at 45° without losing activity, but 50 and 100% losses were obtained at 55 and 65°, respectively.

C. ACID.[12] Isoelectric focusing and pH gradient elution on DEAE-cellulose revealed an isoelectric point (pI) at 4.65. At this pH or at the optimal pH of around 5.6 or any other acid pH for that matter, the enzyme is unstable and undergoes irreversible loss of activity.

Section V

Miscellaneous Enzymes

[28] Succinyl-CoA: Propionate CoA-Transferase from *Propionibacterium shermanii*[1]

EC 2.8.3.6 Succinyl-CoA:propionate CoA-transferase

By Marvin Schulman and Harland G. Wood

$$\text{Succinyl-CoA} + \text{propionate} \rightleftarrows \text{propionyl-CoA} + \text{succinate} \quad (1)$$
$$\text{Succinyl-CoA} + \text{acetate} \rightleftarrows \text{acetyl-CoA} + \text{succinate} \quad (2)$$

Assay Method

CoA-transferase from *P. shermanii* catalyzes reactions (1) and (2) equally well. It is most conveniently assayed spectrophotometrically at 340 nm by measuring the formation of acetyl-CoA by coupling with malate dehydrogenase and citrate synthase as shown below.

$$\text{Malate} + \text{NAD}^+ \xrightarrow{\text{malate dehydrogenase}} \text{oxalacetate} + \text{NADH} + \text{H}^+ \quad (3)$$
$$\text{Acetyl-CoA} + \text{oxalacetate} \xrightarrow{\text{citrate synthase}} \text{citrate} + \text{CoA} \quad (4)$$

Acetyl-CoA formed in reaction (2) is condensed with oxalacetate [formed in reaction (3)] to yield citrate and the rate of formation of NADH equals the rate of formation of acetyl-CoA.

Reagents

Trizma (Tris base reagent grade, Sigma)

Sodium DL malate (Sigma)

β-DPN (NAD) (Sigma)

Malate dehydrogenase (5 mg of protein per milliliter, 1100 IU/mg of protein, Boehringer)

Citrate synthase (2 mg of protein per milliliter, 70 IU/mg of protein, Boehringer)

Succinyl-CoA prepared by the method of Simon and Shemin[2] as modified by Swick and Wood[3] using succinic anhydride (John L. Baker), which is recrystallized, and CoA (P. L. Biochemicals). An approximate determination of succinyl-CoA is made by the hydroxamate method of Lipman and Tuttle[4] at the time of preparation and a final determination by spectrophotometric assay via

[1] This work was supported in part by USPHS, NIH Grant GM11839.

[2] E. J. Simon and D. Shemin, *J. Amer. Chem. Soc.* **75**, 2520 (1953).

[3] R. W. Swick and H. G. Wood, *Proc. Nat. Acad. Sci. U.S.* **42**, 28 (1960).

[4] F. Lipmann and L. C. Tuttle, *J. Biol. Chem.* **159**, 21 (1945).

reactions (2), (3), and (4). The assay mixture in micromoles is as follows: Tris-Cl (pH 8.0), 25; sodium malate, 22.5; NAD, 0.25; sodium acetate, 15 and in IU citrate synthase, 0.3; malate dehydrogenase, 2.1; CoA-transferase, 0.5; and approximately 0.02 μmole succinyl-CoA in a final volume of 0.25 ml in a 0.5-ml cuvette. The values of the enzymatic determination are about 70% of those of the hydroxamate method and all concentrations given below are based upon the enzymatic assay.

Enzyme: Dilutions of CoA-transferase are made in 0.1 M phosphate buffer (pH 6.8) containing 0.1 mM sodium ethylene-diaminetetraacetic acid (EDTA). EDTA stabilizes the enzyme in dilute solution and is routinely added to all buffers.

Spectrophotometric Assay

The assay mixture in micromoles per milliliter is as follows: Tris-Cl (pH 8.0), 100; sodium malate, 2; NAD, 0.25; sodium acetate, 15; succinyl-CoA, 0.15–0.20; and in IU per milliliter of citrate synthase, 0.3; and malate dehydrogenase, 2. For convenience, Mixture 1 containing reagents at 40 times the required concentrations is made up as follows: 1.0 M Tris-Cl (pH 8.0), 1 ml; 0.4 M sodium malate, 0.01 ml; 0.01 M NAD, 1.0 ml; and water to 4 ml. Mixture 2 contains 100 times the required concentrations of enzymes and is prepared as follows: 220 units malate dehydrogenase (0.04 ml of suspension containing 5500 IU/ml) and 35 units citrate synthase (0.25 ml of suspension containing 140 IU/ml) are made up to 1 ml with 0.1 M phosphate buffer (pH 6.8). The assays are conducted in 0.5 ml cuvettes with a 10-mm light path and 2 mm width containing the following: Mixture 1, 0.1 ml; 1.5 M sodium acetate, 0.01 ml; Mixture 2, 0.01; succinyl-CoA, 0.01 ml ($\sim$0.15 μmole); CoA-transferase ($\sim$0.002 IU) and water to give 0.25 ml. The reaction is initiated by addition of the CoA-transferase and is conducted at 25°. The formation of NADH is linear with time for 3–5 minutes.

Units. Units of enzymes are expressed as micromoles NAD reduced per minute and specific activity as units per milligram of protein. Protein is measured spectrophotometrically.[5]

Purification of CoA-Transferase

Reagents and Equipment

Pyrex beads 100 μ in diameter (Minnesota Mining and Manufacturing Corp.) washed with concentrated HCl, distilled water, and dried.

---

[5] See Vol. 3 [73]. The factors 1.45 $\times$ A_{260}–0.74 A_{260} are used.

DEAE-cellulose (Type 40 capacity 0.9 mEq/g, Brown Co.) washed successively with 0.1 N H_2SO_4, water, 0.1 N HCl, water, and finally 50 mM phosphate buffer (pH 6.8).

Cellulose phosphate (reagent grade, capacity 0.8 mEq/g, Brown Co.) washed as was the DEAE-cellulose.

Barnstead purity meter or Radiometer conductivity meter, Model CDM3. The conductivity of 10, 20, 30%, etc. saturated solutions of $(NH_4)_2SO_4$ at 1:50,000 dilutions are determined with the Barnstead purity meter or 1:100 dilutions with the Radiometer conductivity meter and plotted against concentration to establish a standard linear relationship. To determine the salt concentration of an unknown, the unknown is diluted as above and its conductivity determined. The concentration is calculated as percent saturation of $(NH_4)_2SO_4$ as read from the standard curve.

Eppenbach Mill (Gilford-Wood Co., Hudson, New York)

$(NH_4)_2SO_4$ (special enzyme grade, Mann Research Laboratories)

Sephadex G-200 (Pharmacia)

Magnetic stirrer

Sorvall refrigerated centrifuge, RC-2 or RC-2B

Whatman No. 4 filter paper

Büchner funnel, 15 cm in diameter, and suction flask

Sodium EDTA

Potassium phosphate

TEAE-cellulose (reagent grade, 0.93 mEq/g, Brown Co.) washed as was the DEAE-cellulose.

Fraction collector and test tubes

Dialysis tubing, 2 cm

Columns, 7×25 cm and 3×60 cm for cellulose ion-exchange columns and a 5×90 cm Pharmacia column fitted with adopter plungers for gel filtration.

Growth of Propionibacterium shermanii. The medium[6] contains Na_2CO_3, 80 mM; KH_2PO_4, 90 mM; glycerol, 0.35 M; and the following in milligrams per liter: autolyzed yeast extract powder, 3500 (Yeast Products Inc., Paterson, N.J.); $Co(NO_3)_2 \cdot 6H_2O$, 10; calcium pantothenate 1; thiamine hydrochloride, 1; and biotin, 1. The carbonate and phosphate solutions are sterilized separately and the remaining components in combination (1 hour at 120°). The three solutions are mixed just before inoculation. The fermentation is at 30° usually with 18 liters of medium in 20-liter carboys with cotton plugs, and the flasks are shaken by hand once each day for a few minutes. A liter of a vigorously fermenting culture of *P. shermanii* (52W) growing in the same medium is used as the

[6] See Vol. 13 [36].

inoculum. It is necessary to inoculate with bacteria that have been transferred 4 or 5 times in this medium to obtain a culture which grows rapidly. The pH of the medium is about 7.8 at the beginning and drops to about 5.8 after prolonged fermentation. The cells are harvested in a Sharples centrifuge after about 7 days and the yield of cells is 3–5 g/liter.

Step 1. Preparation of the Cell-Free Extract. Unless otherwise stated all operations are carried out at 0–4°. The extract is routinely prepared from a large quantity of cells in the Eppenbach mill. The following is a typical example: 300 g of Pyrex beads are added to 300 g of packed cells (as obtained from the Sharples centrifuge) together with 150 ml of 0.2 M phosphate (pH 6.8) containing 0.4 mM cysteine. The mixture is ground at top speed for 30 minutes and the mill is cooled by circulation of glycol antifreeze at −5°. If the temperature rises above 15°, grinding is interrupted until the mill cools to 5°. The resulting mixture is centrifuged for 30 minutes at 16,000 g (GSA rotor at 10,000 rpm). The sedimented beads and cellular material are resuspended in 300 ml 0.2 M phosphate (pH 6.8) and the grinding is repeated. The resulting suspension is centrifuged at 16,000 g. The sedimented beads and cellular material are suspended in 200 ml cold distilled water and centrifuged at 16,000 g for 20 minutes. The three supernatant solutions are combined and centrifuged at 43,000 g (SS-34 rotor at 19,000 rpm) for 30 minutes. Approximately 600 ml of extract containing 15–20 g of protein and 2000–3000 units of CoA-transferase is obtained.

Step 2. Adsorption on DEAE-Cellulose and Batch Elution. CoA-transferase is adsorbed by DEAE-cellulose from 50 mM phosphate buffer (pH 6.8) and is eluted with 0.35 M phosphate. The 600-ml of extract is diluted 6-fold with cold distilled water and about 2 kg of moist filtered DEAE-cellulose is added. The slurry is stirred for 1 hour; a portion is removed, centrifuged at 20,000 g (GSA rotor at 13,000 rpm) for 15 minutes, and the CoA-transferase activity of the supernatant is determined. Most of the CoA-transferase (95%) is adsorbed; if it is not, more DEAE-cellulose is added and the process is repeated. When the enzyme is adsorbed, the slurry is filtered on a Büchner funnel using Whatman No. 4 filter paper.

The DEAE-cellulose is then suspended in 3 liters of 0.1 M phosphate buffer (pH 6.8), stirred for 20 minutes, and filtered as above. The filtrate does not contain CoA-transferase. The CoA-transferase is eluted by two washes with 0.35 M phosphate; in the first, the DEAE-cellulose is suspended in 3500 ml, stirred 1 hour, and filtered as above; in the second elution, 2500 ml of 0.35 M phosphate is used and the stirring is for 30 minutes. Approximately 90% of the CoA-transferase is recovered in the combined filtrate and the specific activity is around 0.4 unit/mg of protein. The combined filtrate is brought to 90% saturation by addition of solid $(NH_4)_2SO_4$, stirred 1 hour, and centrifuged at 16,000 g for 30 minutes at $0°$. The precipitate contains all the CoA-transferase and can be stored indefinitely at $-15°$; it also contains transcarboxylase,[6] malate dehydrogenase,[7] P-enolpyruvate carboxytransphosphorylase,[8] methylmalonyl-CoA mutase,[9] methylmalonyl-CoA racemase,[10] and pyruvate, phosphate dikinase.[11]

Step 3. Chromatography on Cellulose Phosphate. A 7×20 cm column of cellulose phosphate equilibrated with 50 mM phosphate (pH 6.8) containing 1 mM β-mercaptoethanol is prepared by packing successive 2 cm layers under 1 psi air pressure. The column is washed with 500 ml of 50 mM phosphate (pH 6.8). Enzyme (5–10 g) from Step 2 is dialyzed for 6 hours in 2 cm tubing against 2 liters of 0.1 M phosphate buffer (pH 6.8). The buffer is changed at 2-hour intervals and the conductivity of the dialysate is determined with either a Barnstead purity meter or a Radiometer conductivity meter. If necessary, the dialysate is diluted with cold distilled water to a salt concentration equivalent to or lower than 50 mM $(NH_4)_2SO_4$. The enzyme solution (300 ml containing 1800–2700 units of enzyme) is placed on the column which is then washed with 50 mM phosphate (pH 6.8). Most of the protein including CoA-transferase, malate dehydrogenase,[7] P-enolpyruvate carboxytransphosphorylase,[8] methylmalonyl-CoA mutase,[9] and pyruvate, phosphate dikinase[11] does not adhere to the column and is recovered in the initial effluent and the 50 mM phosphate wash. Transcarboxylase[6] and methylmalonyl-CoA racemase[10] are adsorbed on cellulose phosphate and elute later. CoA-transferase is precipitated from the combined initial effluent and 50 mM phosphate wash by addition of solid $(NH_4)_2SO_4$ to 90% saturation. The suspension is stirred for 1 hour, and the precipitate is collected by centrifugation at 16,000 g for 20 minutes.

Step 4. Fractionation with Ammonium Sulfate. The protein of Step 3 (1.0–2.6 g) is dissolved in 50 mM phosphate buffer (pH 6.8) to give a concentration of approximately 20 mg of protein per milliliter. The salt concentration of the solution is determined using either a Barnstead or a Radiometer conductivity meter and assuming the salt is ammonium sulfate (carried in the protein precipitate), a further addition of solid $(NH_4)_2SO_4$ is made to bring the solution to 35% saturation. After stirring for 1 hour, the suspension is centrifuged at 16,000 g for 20 minutes. The supernatant solution is brought to 50% saturation with $(NH_4)_2SO_4$,

[7] S. H. G. Allen, R. W. Stjernholm, and H. G. Wood, *J. Bacteriol.* **87,** 171 (1964).
[8] See Vol. 13 [47].
[9] See Vol. 13 [35].
[10] See Vol. 13 [33].
[11] H. J. Evans and H. G. Wood, *Biochemistry* **10,** 721 (1971).

stirred for another hour, and centrifuged as above. The precipitate contains P-enolpyruvate carboxytransphosphorylase.[8] The supernatant solution is brought to 65% saturation with $(NH_4)_2SO_4$ and after stirring for 1 hour is centrifuged as above. The precipitate contains pyruvate, phosphate dikinase.[11] The supernatant solution is brought to 90% saturation with $(NH_4)_2SO_4$, stirred for 3 hours, and centrifuged at 16,000 g for 20 minutes. The precipitate, which contains CoA-transferase, malate dehydrogenase,[7] and methylmalonyl-CoA mutase,[9] is dissolved in a minimum of 50 mM phosphate buffer (pH 6.8) and 0.1 mM EDTA. It is dialyzed against 20 volumes of the same buffer. The salt concentration of the dialysate is determined using a Barnstead or a Radiometer conductivity meter, and if the conductivity is above the equivalent of 50 mM $(NH_4)_2SO_4$, the enzyme solution is diluted to the equivalent of 30 mM.

Step 5. Chromatography on TEAE-Cellulose. TEAE-cellulose suspended in 50 mM phosphate (pH 6.8) and 0.1 mM EDTA is tightly packed in successive 3–5 cm layers each under 4 psi air pressure to give a column 3 × 50 cm which is then equilibrated with 50 mM phosphate (pH 6.8) containing 0.1 mM EDTA. The enzyme solution of Step 4 (1500–2000 units and 0.4–0.7 g of protein) is passed into the column which is then washed with 2 liters of 0.05 M followed by 2 liters of 0.1 M phosphate buffer (pH 6.8) each containing 0.1 mM EDTA. Enzyme is then eluted with a 4-liter linear gradient from 0.1 M to 0.5 M phosphate (pH 6.8) in 0.1 mM EDTA. Methylmalonyl-CoA isomerase elutes in the 0.1 M phosphate wash, and the vast majority of malate dehydrogenase elutes just ahead of CoA-transferase. The fractions are assayed for CoA-transferase, protein is estimated spectrophotometrically,[5] and the fractions with high specific activity are pooled and precipitated by addition of $(NH_4)_2SO_4$ to 90% saturation. The suspension is stirred for 1 hour, kept in the cold for 6–12 hours, centrifuged at 43,000 g for 30 minutes, and the precipitate is dissolved in a minimum of 0.1 M phosphate buffer (pH 6.8) in 0.1 mM EDTA to give about 20 mg of protein per milliliter.

Step 6. Gel Filtration with Sephadex-G 200. The enzyme from Step 5 is further purified by ascending chromatography on a column of Sephadex G-200. Three to five milliliters (1000 units) of the concentrated enzyme is applied to a 2-liter (5 × 90 cm) Pharmacia column of Sephadex G-200 equilibrated with 0.1 M phosphate buffer (pH 6.8) in 0.1 mM EDTA and is eluted with the same buffer. CoA-transferase elutes in a symmetrical peak just after the void volume of the column. The fractions with highest specific activity (10–35 units/mg of protein) are pooled, precipitated at 90% saturation with $(NH_4)_2SO_4$, and dissolved in a minimum of 0.1 M phosphate (pH 6.8) in 0.1 mM EDTA.

Results of the Purification Procedure. The range of results of typical

TABLE I

Purification of CoA-Transferase[a]

Step	Total protein (mg)	Specific activity (unit/mg protein)	Recovery of enzymatic activity (%)
1. Crude extract	15,000–20,000	0.1–0.160	100
2. DEAE-cellulose batch elution 0.35 M	4,500–9,700	0.3–0.45	~90
3. Cellulose phosphate eluant	1,000–26,000	0.8–1.5	~70
4. Ammonium sulfate 65–90%	400–700	2–3	~50
5. TEAE-cellulose eluant	50–150	8–13	~40
6. Sephadex G-200 eluant	15–60	15–38	~35

[a] Based upon preparations from 300 g of bacteria.

purifications are summarized in Table I and show a 160–380-fold increase in specific activity with a 35% recovery of enzymatic activity from the crude extract. The enzyme of Step 6, although giving a symmetrical peak on Sephadex G-200 and Biogel P-200, contains a small amount of malate dehydrogenase which is not removed by repeated chromatography. The contamination by malate dehydrogenase is calculated to be less than 5% assuming a specific activity of 1000 for pure malate dehydrogenase from *P. shermanii*.[7]

Properties of the Enzyme

Stability. CoA-transferase is quite stable at protein concentrations greater than 1 mg/ml in 0.1 M phosphate (pH 6.8) containing 0.1 mM EDTA when kept frozen at −20°. Approximately, a 30% loss of enzymatic activity occurs during 1 year of storage. The enzyme is routinely stored in small portions to avoid repeated thawing and refreezing of unused portions.

Sedimentation Pattern.[7] The enzyme has an $s_{20,w}$ value of 10.5 S.

pH Optimum. The enzyme has a broad optimum between pH 6.5 and 8 and has 50% activity at pH 5.5; there is no enzymatic activity at pH 8.5.

K_m Values.[7] The apparent K_m for succinyl-CoA in the transfer to propionate [reaction (1)] is 6.8×10^{-5} M and for propionate in this reaction is 6.25×10^{-5} M. The apparent K_m for succinyl-CoA in the transfer to acetate [reaction (2)] is 1.3×10^{-4} M, and for acetate in the same reaction is 7.0×10^{-3} M.

Substrate Specificity.[7] The enzyme catalyzes a reversible transfer of CoA between succinyl-CoA and either acetate or propionate at equal rates. The enzyme does not catalyze transfer from acetyl-CoA to methyl-malonate, from β-hydroxy-β-methylglutaryl-CoA to acetate or from ace-toacetyl-CoA to succinate.

[29] Acetyl Coenzyme A:Long-Chain Base Acetyltransferase from the Microsomes of *Hansenula ciferri*[1,2]

By YECHEZKEL BARENHOLZ and SHIMON GATT

$$RNH_2 + CH_3CO \cdot SCoA \rightarrow RNH \cdot COCH_3 + CoASH \quad (1)$$
$$R \cdot CHOH \cdot CHNH_2 \cdot CH_2OH$$
$$+ CH_3CO \cdot SCoA \rightarrow RCHOAc \cdot CHNHAc \cdot CH_2OAc + CoASH \quad (2)$$

This enzyme transfers the acetyl portion of acetyl coenzyme A to the amino groups of primary amines of 6–18 carbon atoms and to the amino and hydroxyl groups of sphingosine base. The enzyme differs from the arylamine acetyltransferase of the soluble fraction of *H. ciferri*[3,4] or of mammalian and avian liver.[5-7]

Assay Methods

Two assay methods were used. In one, radioactively labeled acetyl-CoA is used and the radioactivity of the acetylated base is determined. In the second, the reaction is run in the presence of DTNB [5,5-dithiobis-(2-nitrobenzoic acid)] and the color resulting from the coenzyme A released is read.

Method 1

Principle. The base and ^{3}H-labeled acetyl-CoA are incubated with the enzyme. Chloroform, methanol, and water are added and the radio-

[1] Y. Barenholz and S. Gatt, *J. Biol. Chem.* **247**, 6827 (1972).
[2] Y. Barenholz and S. Gatt, *J. Biol. Chem.* **247**, 6834 (1972).
[3] Y. Barenholz and S. Gatt, *Biochim. Biophys. Acta* (1974) (in press).
[4] Y. Barenholz and S. Gatt, this volume [30].
[5] T. C. Chou and F. Lipmann, *J. Biol. Chem.* **196**, 89 (1952).
[6] H. Weissbach, B. G. Redfield, and J. Axelrod, *Biochim. Biophys. Acta* **54**, 190 (1961).
[7] W. W. Weber and S. N. Cohen, *Mol. Pharmacol.* **3**, 266 (1967).

activity in the lower chloroform phase is measured (Method 1a). This determines the total acetylation (i.e., acetylation of both the amino and hydroxyl groups of the sphingosine bases). When the N-acetylation (i.e., that of only the amino group of the sphingosine base) is measured, the reaction mixture is hydrolyzed with methanolic KOH. Chloroform and water are added and the radioactivity in the lower phase is measured (Method 1b). DTNB is present in the reaction mixtures to prevent an inhibitory effect of the coenzyme A released during the reaction.

Method 1a

Reagents

[2-³H₃]acetyl-CoA,[8] 5 mM
Long-chain base in chloroform–methanol, 2:1, 10 mM
Potassium phosphate, pH 7.5, 1 M
Egg phospholipids,[9] 0.2 mg of phosphorus per ml
Chloroform–methanol, 1:1
Pure solvents upper phase,[10] a mixture of methanol–water–chloroform 94:96:6
DTNB, 10 mM in 0.2 M potassium phosphate, pH 7.5

Procedure. The solution of the base is pipetted into a test tube of about 1 cm diameter and the organic solvent is evaporated under a stream of nitrogen in a 50–60° water bath. Water is added, followed by 0.02 ml of acetyl-CoA, 0.02 ml of buffer, 0.02 ml of DTNB and enzyme (50–1000 μg). The final volume is 0.2 ml. After 1–2 hours at 37°, with shaking, 1.6 ml of chloroform–methanol 1:1 is added, followed by 0.6 ml of water. The tube is mixed, centrifuged, and the upper phase is aspirated off. The lower phase is washed 3 times successively with 0.8 ml each of pure solvents upper phase. The chloroform phase is transferred to a counting vial; 0.025 ml of the solution of egg phospholipids and a carborundum chip are added. The solvent is evaporated on a thermostatic bath, adjusted to 65°, 1 ml of toluene is added to the still-warm vial, followed by 10 ml of toluene scintillator. The vial is counted in a scintillation counter.

---

[8] The tritium-labeled acetyl-CoA was synthesized using acetic anhydride and coenzyme A by the method of S. Smith, Ph.D. thesis, University of Birmingham, Birmingham, England, 1965.

[9] An egg yolk is blended with 50 ml of acetone. The acetone is decanted and the phospholipids are further treated, 3 times each, with acetone. The residue is dissolved in chloroform–methanol, 2:1, to a final concentration of 0.2 mg of phosphorus per milliliter.

[10] J. Folch, M. Lees, and G. H. Sloane-Stanley, *J. Biol. Chem.* **226**, 497 (1957).

Method 1b

Reagents

[2-^{3}H$_3$]acetyl-CoA, 5 mM
Long-chain base in chloroform–methanol, 2:1, 10 mM
Potassium phosphate, pH 7.5, 1 M
N KOH in methanol
DTNB, 10 mM in 0.2 M potassium phosphate, pH 7.5
Pure solvents upper phase,[10] methanol–water–chloroform 94:6:6
Egg phospholipids,[9] 0.2 mg of phosphorus per ml
Chloroform

Procedure. The incubation mixture is the same as in Method 1a. After 1–2 hours at 37°, with shaking, 0.5 ml of methanolic KOH is added, followed by 0.3 ml of water. The tube is heated for 10 minutes at 50–60° to hydrolyze the ester bonds, and 0.5 ml of chloroform is added. The tube is mixed and centrifuged and the upper phase is aspirated off. The lower phase is washed 3 times successively with 0.5 ml each of pure solvents upper phase. The chloroform phase is transferred to a counting vial; 0.025 ml of the solution of egg phospholipids and a carborundum chip are added. The solvent is evaporated in a thermostatic bath, adjusted to 65°, 1 ml of toluene is added to the still-warm tube followed by 10 ml of toluene scintillation. The vial is counted in a scintillation counter.

Method 2

Principle. DTNB produces a yellow color when reacted with free coenzyme. This color is read at 413 nm.[11]

Reagents

Acetyl-CoA, 5 mM
Amine, 10 mM in chloroform–methanol, 2:1
Potassium phosphate, pH 7.5, 1 M
DTNB, 10 mM in 0.2 potassium phosphate, pH 7.5
Chloroform–methanol, 1:1

Procedure. Incubation and termination are the same as in Method 1a. The upper phase is transferred to a cuvette with a light path of 1

[11] G. L. Ellman, *Arch. Biochem. Biophys.* **74**, 443 (1958).

cm and the color intensity is read at 413 nm; 1 nmole reads 0.009 OD unit.

Preparation of Enzyme

Step 1. Growth of Yeast. Hansenula ciferrii, NRRL, Y-1031, F 60-10, may be obtained from the American type collection. The lyophilized culture is transferred to 100 ml erlenmeyers containing 40 ml of a yeast maintenance medium consisting of 0.3% malt extract, 0.3% yeast extract, 0.5% peptone, and 1% glucose. The sterilization of the medium is done without the glucose, a sterile solution of the latter compound is added separately to the growth medium. The erlenmeyer is incubated at 26° using a New Brunswick shaking incubator at 120 rpm. Alternatively, the culture can be grown without agitation, but under an efficient stream of filtered air (at least 50 ml/minute), using a 100-ml erlenmeyer. After 24 hours, 0.5 ml is transferred to a similar growth medium which is again incubated as above. After 24 hours the culture should be examined for bacterial contamination, using a phase-contrast microscope. Five milliliters of this medium are transferred to a 2-liter erlenmeyer containing 800 ml of a growth medium similar to the above except that the glucose content is 2%. After 24 hours' growth with agitation or aeration (aeration rate, 1 liter of air per minute), aliquots of this medium are used for large-scale growth of the yeast.

Two alternate procedures were used: In the first, 400 ml were transferred to a New Brunswick fermentor containing 5 liters of the growth medium (having 2% glucose). Antifoam was added and the culture was grown at 26° and 220 rpm, with aeration, using 0.5 volume of air per minute.[12] In the second procedure, 160 ml each, of the culture was transferred to 5 liter erlenmeyers containing 2 liters of the growth medium. Five erlenmeyers were aerated, at 2.5 liters/minute through tubes with a sintered glass filter for efficient aeration. After 40 hours at 26°, the yeast was harvested using a Sharpless centrifuge, or in a MSE-Mystral refrigerated centrifuge for 30 minutes at 2500 *g*. In either case, the supernatant is discarded.[13] The yeast paste (200–250 g) is used for the next step.

Step 2. Disruption of the Yeast Cells. The yeast paste is suspended

---

[12] L. J. Wickerham and F. H. Stodola, *J. Bacteriol.* **80**, 484 (1960).

[13] The supernatant may be utilized for the isolation of large quantities of sphingosine bases, mostly phytosphingosine.[1,12] Better yields of these bases may be obtained if the growth is done in the presence of 5–7% glucose for 4 days and the medium, after growth, is stored for 3 more days at 4°.

in 4 volumes of cold water, centrifuged at 4°, and the sediment is suspended in 2 volumes of 0.05 M potassium phosphate, pH 7.0, containing 20% glycerol by volume (PG).[14] The yeast cells are disrupted using a French press, precooled to 4°, at 20,000 psi. The juice is collected into an erlenmeyer on cracked ice.

Step 3. Preparation of Microsomes. The extract is centrifuged for 10 minutes at 1200 g or for 4 hours at 100,000 g. The supernatant is discarded and the sediment is suspended in PG, 0.05 ml/g of packed yeast cells. The suspension is stored at −20°.

Properties

Stability. The enzyme, a suspension in PG, retained full activity for at least 3 years at −20°. If the microsomes were suspended in phosphate buffer, devoid of glycerol, the enzymatic activity was lost within 1 week or less at −20°.

General Properties of the Reaction.[1,2] The transfer reaction is directly proportional to enzyme concentration (up to at least 1 mg of microsomal protein) and to the reaction time (for at least 2 hours). The optimal pH of the reaction depends on the type of buffer used; this is related to the charge imparted to the micelles of the base. The curves which describe the rate of acetyl transfer as a function of the concentration of acetyl-CoA (while maintaining the base at a fixed concentration) are hyperbolic. Similar curves in which acetyl-CoA is the fixed and dihydrosphingosine or hexadecylamine are the variable substrates have a nonsymmetric sigmoidal shape. Serum albumin increases the reaction rates, and at an optimal molar ratio of albumin to base of about 0.5 the V/S curves are hyperbolic.

Specificities of the Substrate and Enzyme. Sphingosine, dihydrosphingosine, phytospingosine, and the N-acetylated derivatives of these bases are substrates for the acetyl transfer reaction. The products are bases which are acetylated both at the amino and the hydroxyl groups. The N-butyryl, N-octanoyl, and N-palmitoyl derivatives are not substrates, neither is 1-O-galactosyl sphingosine. Normal amines of C_6–C_{18} carbon atoms can be used for the acetyl transfer reaction. Acetyl coenzyme A is the sole donor of the acyl group; butyryl-CoA is not utilized. The enzyme catalyzes both an N and O transfer of the acetyl group. These two activities could not be separated from each other.[2] The above contrasts with the specificity of the acetyl-CoA:amine acetyltransferase

---

[14] If the cells cannot be disrupted immediately, they may be stored at −20° in the phosphate–glycerol medium for periods up to at least 1 week.

from the soluble fraction of *H. ciferri*,[3,4] which utilizes normal amines of 6–16 carbon atoms and water-soluble amines, but not sphingosine bases.

Inhibitors. The reaction is inhibited by coenzyme A, a product of the reaction, and by detergents such as Triton X-100, bile salts, and sodium dodecyl sulfate.

[30] Acetyl Coenzyme A:Amine Acetyltransferase from the Soluble Fraction of *Hansenula ciferri*[1,2]

By YECHEZKEL BARENHOLZ and SHIMON GATT

$$RNH_2 + CH_3CO \cdot SCoA \rightarrow RNH \cdot COCH_3 + CoASH$$

This enzyme transfers the acetyl group of acetyl coenzyme A to an amine acceptor. Similar enzymes from mammalian and avian tissue were described by other investigators.[3–5] In all these cases, water-soluble alkyl or arylamines were employed. The acetyltransferase from the soluble fraction of the yeast *H. ciferri* utilizes water-soluble amines but also acetylates normal amines up to hexadecylamine.

Assay Methods

Three assay methods are used. In two (one for the water-soluble and the second for the long-chain amines) radioactively labeled acetyl-CoA is used and the radioactivity of the acetylated amino is determined. In the third, the reaction is terminated, DTNB [5,5-dithiobis(2-nitrobenzoic acid)] is added and the color resulting from the coenzyme A released is read.

Method 1

Principle. A water-soluble amine and ^{3}H-labeled acetyl-CoA are incubated with the enzyme. Borate buffer, pH 10, is added to hydrolyze the residual acetyl-CoA. The acetylated amine is then extracted into isoamyl

[1] Y. Barenholz, I. Edelman, and S. Gatt, *Biochim. Biophys. Acta.* (1974) (in press).

[2] Y. Barenholz, I. Edelman, and S. Gatt, *Isr. J. Chem.* 9, 24 (1971).

[3] T. C. Chou and F. Lipmann, *J. Biol. Chem.* 196, 89 (1952).

[4] H. Weissbach, B. G. Redfield, and J. Axelrod, *Biochim. Biophys. Acta* 54, 190 (1961).

[5] W. W. Weber and S. N. Cohen, *Mol. Pharmacol.* 3, 266 (1967).

alcohol in the presence of NaCl (45% saturation), and the radioactivity in the upper alcoholic phase is determined. This method was used with the following substrates: histamine, tryptamine, serotonin, tyramine, and hydroxytyramine.

Reagents

[2-³H₃]acetyl-CoA,[6] 5 mM
Amine, aqueous solution,[7] 10 mM
Potassium phosphate, pH 7.5, 1 M
Potassium borate, pH 10, 0.5 M in saturated NaCl
Isoamyl alcohol

Procedure. The reagents are pipetted into a test tube of about 1 cm diam in the following order: acetyl-CoA, 0.02 ml; amine, 0.02 ml; water (to a final volume of 0.2 ml); buffer, 0.02; and enzyme. After 15 minutes at 37°, 0.17 ml of borate–NaCl buffer is added, followed by 1 ml of isoamyl alcohol. The tube is centrifuged, the upper isoamyl alcohol layer is transferred, using a Pasteur pipette to a test tube containing 0.3 ml of the borate–NaCl buffer. The tube is mixed and centrifuged and the upper phase is transferred to a counting vial. One milliliter of ethanol is added to the counting vial followed by 10 ml of toluene scintillator (5 g of PPO and 130 mg of dimethyl POPOP), and the vial is counted in a liquid scintillation counter.

Method 2

Principle. A long-chain primary normal amine (C₆–C₁₆) and ³H-labeled acetyl-CoA are incubated with the enzyme. Chloroform, methanol, and water are added and the radioactivity in the lower chloroform phase is determined.

Reagents

[2-³H₃]acetyl-CoA, 5 mM
Amine in chloroform–methanol 2:1,[7] 10 mM
Potassium phosphate, pH 7.5, 1 M
Chloroform–methanol, 1:1

---

[6] The tritium-labeled acetyl-CoA was synthesized, using acetic anhydride and coenzyme A by the method of S. Smith, Ph.D. thesis, University of Birmingham, Birmingham, England, 1965.

[7] With amines of C₈ or less, aqueous solutions can be pipetted directly into the test tube. With amines having a longer chain, solutions in chloroform–methanol are used.

"Pure solvents upper phase"[8]: a mixture of methanol–water–chloroform, 94:96:6

Egg phospholipids,[9] 0.2 mg of phosphorus per milliliter

Procedure. The amino solution[10] is pipetted into a test tube of about 1 cm diam and the organic solvent is evaporated under a stream of nitrogen in a 50–60° water bath. Water is added, followed by 0.02 ml of actyl-CoA, 0.02 ml of buffer, and enzyme. The final volume is 0.2 ml. After 15 minutes at 37°, with shaking, 1.6 ml of chloroform–methanol, 1:1, is added, followed by 0.6 ml of water. The tube is mixed, centrifuged, and the upper phase is aspirated off. The lower phase is washed 3 times successively with 0.8 ml, each of pure solvents upper phase. The chloroform phase is transferred to a counting vial; 0.025 ml of the solution of egg phospholipids and a carborundum chip are added. The solvent is evaporated on a thermostatic plate, adjusted to 65°, 1 ml of toluene is added to the still-warm vial, followed by 10 ml of toluene scintillator. The vial is counted in a scintillation counter.

Method 3

Principle. DTNB produces a yellow color when reacted with free coenzyme A.[11] After incubation, this reagent is added to the reaction mixture and the color is read at 413 nm.

Reagents

Acetyl-CoA, 5 mM
Amine, 10 mM in water or in chloroform–methanol, 2:1
Potassuum phosphate, pH 7.5, 1 M
DTNB, 10 mM in 0.2 M potassium phosphate, pH 7.5
Ethanol–ether, 2:1 (v/v)

Procedure. The amine solution is added to the incubation tube. If a chloroform–methanol solution of a long-chain amine is used, the organic solvent is evaporated under a stream of nitrogen. Water (to give a final

[8] J. Folch, M. Lees, and G. H. Sloane-Stanley, *J. Biol. Chem.* **226**, 497 (1957).
[9] An egg yolk is blended with 50 ml of acetone. The acetone is decanted and the phospholipids are further treated, 3 times each, with acetone. The residue is dissolved in chloroform–methanol, 2:1, to a final concentration of 0.2 mg of phosphorus per milliliter.
[10] The concentration of the amine depends on the chain length. With amines of C_8 or less, it can reach 2 mM; with C_{12} 1 mM, with C_{14} 0.5 mM, and with C_{16} only 0.2 mM.
[11] G. L. Ellman, *Arch. Biochem. Biophys.* **74**, 443 (1958).

volume of 0.2 ml), 0.02 ml of acetyl-CoA, 0.02 ml of buffer and enzyme are added and the tube is incubated, with shaking for 15 minutes at 37°; 0.2 ml of ethanol–ether, 2:1, is added, followed by 0.03 ml of the DTNB solution. The color is read in a cuvette with a light path of 1 cm and a volume of 0.3 ml at 413 nm. One nanomole reads 0.027 OD unit. This method should be used only for reactions which release up to 30 nmoles of CoA. For reaction releasing greater quantities of coenzyme A, larger volumes of ethanol–ether must be used.

Method 4

Principle. This method enables determination of both the coenzyme A released and the [³H]acetyl bound to the amine; it is applicable only for C_6–C_{16} amines.

Reagents

[2-³H₃]Acetyl-CoA, 5 mM
Amine, 10 mM
Potassium phosphate, pH 7.5, 1 M
DTNB, 10 mM in potassium phosphate, pH 7.8
Potassium phosphate, pH 7.8, 0.3 M
Chloroform–methanol, 1:1

Procedure. The procedure is similar to that of Method 2, except that, after the addition of the mixture of chloroform and methanol, 0.6 ml of potassium phosphate, pH 7.8, and 0.03 ml of DTNB are added. The phases are separated and the color in the upper phase is read at 413 nm.

Units. One unit of enzyme is defined as the amount which releases 1 nmole of coenzyme A or transfers 1 nmole of acetyl group of 1 minute. Protein is determined by the method of Lowry *et al.*[12]

Purification Procedure[1,13]

Step 1. Growth of Yeast. Hansenula ciferii, NRRL, Y-1031, F 60-10, may be obtained from the American type collection. The lyophilized culture is transferred to 100 ml erlenmeyers containing 40 ml of a yeast maintenance medium consisting of 0.3% malt extract, 0.3% yeast extract, 0.5% peptone, and 1% glucose. The sterilization of the medium is done

---

[12] O. H. Lowry, N. J. Rosebrough, A. L. Farr, and R. J. Randall, *J. Biol. Chem.* **226**, 497 (1957).

[13] Y. Barenholz and S. Gatt, *J. Biol. Chem.* **247**, 6827 (1972).

without the glucose, a sterile solution of the latter compound is added separately to the growth medium. The erlenmeyer is incubated at 26° using a New Brunswick shaking incubator at 120 rpm. Alternatively, the culture can be grown without agitation, but under an efficient stream of filtered air (at least 50 ml/minute), using a 100-ml erlenmeyer. After 24 hours, 0.5 ml is transferred to a similar growth medium which is again incubated as above. After 24 hours the culture should be examined for bacterial contamination, using a phase-contrast microscope. Five milli-liters of this medium are transferred to a 2-liter erlenmeyer containing 800 ml of a growth medium similar to the above except that the glucose content is 2%. After 24 hours' growth with agitation or aeration (aeration rate about 1 liter of air per minute), aliquots of this medium are used for large-scale growth of the yeast.

Two alternate procedures were used: In the first, 400 ml were trans-ferred to a New Brunswick fermentor containing 5 liters of the growth medium (having 2% glucose). Antifoam was added and the culture was grown at 26° and 220 rpm, with aeration, using 0.5 volume of air per minute.[14] In the second procedure, 160 ml each of the culture was trans-ferred to 5 liter erlenmeyers containing 2 liters of the growth medium. Five erlenmeyers were aerated at 2.5 liter/minute through tubes with a sintered glass filter for efficient aeration. After 40 hours at 26°, the yeast was harvested using a Sharpless centrifuge, or in a MSE-Mystral refrigerated centrifuge for 30 minutes at 2500 g. In either case the super-natant is discarded.[15] The yeast paste (200–250 g) is used for the next step.

Step 2. Disruption of the Yeast Cells. The yeast paste is suspended in 4 volumes of cold water, centrifuged at 4°, and the sediment is sus-pended in 2 volumes of 0.05 M potassium phosphate, pH 7.0, containing 10% glycerol by volume (PG).[16] The yeast cells are disrupted using a French press, precooled to 4°, at 20,000 psi. The juice is collected into an erlenmeyer on cracked ice.

Step 3. Preparation of High-Speed Supernatant. The extract is cen-trifuged for 10 minutes at 1,200 g, the supernatant is collected and cen-trifuged for 70 minutes at 250,000 g or for 4 hours at 100,000 g. The supernatant is filtered through glass wool to remove floating fat. This

[14] L. J. Wickerham and P. H. Stodola, *J. Bacteriol.* **80**, 484 (1960).

[15] The supernatant may be utilized for the isolation of large quantities of sphingosine bases, mostly phytosphingosine (see refs. 13 and 14). Better yields of these bases may be obtained if the growth is done in the presence of 5–7% glucose for 4 days and the medium, after growth, is stored for 3 more days at 4°.

[16] If the cells cannot be disrupted immediately, they may be stored at −20° in the phosphate–glycerol medium for periods up to at least 1 week.

supernatant (about 450 ml) may be frozen and stored at −20° till further used; 150 ml are thawed, taking care to maintain the temperature at no more than 10° and filtered through a Diaflo PM-10 or UM20E membrane (63 mm diam) under a nitrogen pressure of 25 psi. The filtration is interrupted when the volume reaches about 10 ml. At this stage this solution should be further processed immediately, or if necessary after one night's storage at 4°.

Step 4. Gel Filtration. The residual solution on the Diaflo filter is transferred onto a column (125 × 3 cm) of Sephadex G-50, fine, previously swelled, loaded onto the column and washed in PG. This solution is also used to elute the protein. Six milliliter fractions are collected at 10° using a refrigerated fraction collector. The rate of elution is 20 ml/hour; about 100 fractions are collected. Aliquots of the fractions are taken for enzymatic assay and protein determination. The enzyme is eluted around fraction 45–55. The fractions containing enzymatic activity are pooled and stored at −20°. The specific activity of the enzyme is about 60–70 μ/mg with dodecylamine as substrate.

Properties[1]

Proportionality. The purified enzyme is directly proportional to enzyme concentration up to at least 12 μg of protein and incubation time of 20–30 min. The linearity with time is improved by increasing the concentration of glycerol. Using the high-speed supernatant, the curves describing the rate as a function of enzyme concentration are not linear, and the specific activities are not constant, but increase upon dilution of the enzyme with PG.

Stability. The high-speed supernatant, in PG, retained full activity for at least 3 years at −20° and the Sephadex effluent for at least 2 years at the same temperature. Repeated freezing and rethawing should be avoided. Glycerol, 10% or even more, must be present throughout the complete purification procedure. If the yeast cells are disrupted in phosphate buffer, free of glycerol, all activity is lost within 72 hours at −20°. The enzyme is very sensitive to heating and loses its activity after 1–2 minutes at 40°. Protection against heat is afforded by either high concentrations of glycerol (30% or more) or by addition of acetyl-CoA (0.5–1 mM).[2]

General Properties of the Reaction. No acetyl transfer reaction is obtained below pH 6 or above pH 11, irrespective of the type of buffer or substrate. The optimal pH depends on the type of buffer used. It is 7.5 in potassium phosphate and about 9.5 in Tris-ethanolamine or cyclohexylaminopropane sulfonic acid (CAPS). Tricine and Tris buffer are inhibitory. The curves which describe the rate as a function of substrate concentration (V/S) are hyperbolic at a fixed concentration of the amine and varying concentrations of acetyl-CoA. They are also hyperbolic when acetyl-CoA is the fixed substrate and the varying substrate is a water-soluble amine or a normal, primary amine of 6–10 carbon atoms. With amines of 12–16 carbon atoms, substrate inhibition ensues at concentrations greater than the critical micellar concentration of these substrates (above about 0.5 mM with dodecylamine, above 0.3 mM with tetradecylamine, and 0.2 with hexadecylamine). Values of K_m and V_{max} are calculated from kinetic curves of the type employed for a bisubstrate reaction.[17]

Substrate Specificity. The enzyme transfers the acetyl group of acetyl-CoA only. Butyryl-CoA and palmitoyl-CoA are not utilized; neither is acetyl-AMP or acetyl pantetheine. The following amines are acceptors of the acetyl group of acetyl-CoA: normal amines of 6–16 carbon atoms; glucosamine, histamine, tryptamine, hydroxytyramine, serotonin, norepinephrine, and *p*-nitro aniline. The following amines are not substrates: normal amines of 1–4 carbon atoms, polyamines of the spermidine group, amino acids, secondary or tertiary amines, or sphingosine bases. This contrasts with the microsomal transferase which utilizes sphingosine bases and long-chain amines, but not water-soluble amines.[13,18]

Inhibitors. The reaction is inhibited by coenzyme A, a product of the reaction. Other inhibitors are butyryl- or palmitoyl-CoA, and SH inhibitors.

[17] W. W. Cleland, *Biochim. Biophys. Acta* **67**, 104, 173, and 188 (1963).
[18] Y. Barenholz and S. Gatt, this volume [29].

[31] Desaturation of Long-Chain Fatty Acids by Animal Liver

By P. W. HOLLOWAY

Fatty acyl-CoA + NADH
$$+ \ O_2 \rightarrow \text{unsaturated fatty acyl-CoA} + NAD^+ + H_2O$$

Olefinic bonds may be introduced into saturated or unsaturated fatty acid chains by a process of desaturation. The enzyme system catalyzing this reaction, the fatty acyl coenzyme A desaturase, is found associated with the microsomal fraction of homogenates of liver, adipose tissue, adrenals, and testicles.[1] This report will be concerned only with the liver

[1] R. R. Brenner, *Lipids* **6**, 567 (1971).

microsomal desaturase and, primarily, with the desaturation of stearyl-CoA by hen liver microsomes.

Assay Method

Principle. Stearyl-CoA desaturase activity is estimated from the conversion of [1-14C]stearyl-CoA into [1-14C]oleate.[2] The incubation mixture is saponified and the liberated [1-14C]stearate and [1-14C]oleate are separated by argentation thin-layer chromatography. An alternative assay based upon formation of ^3H_2O from [9,10³H₂]stearyl-CoA has been described.[3,4]

Reagents

Potassium phosphate buffer, 1 M, pH 7.2 at 37°

NADH, 5 mM in 10 mM potassium phosphate buffer, pH 7.2 at 0°. The NADH solution is prepared fresh daily.

[1-14C]Stearyl-CoA, 0.5 mM in 10 mM potassium phosphate buffer pH 6.0 at 0°. The [1-14C]stearyl-CoA of high specific activity is prepared by the method of Kornberg and Pricer[5] or may be purchased from New England Nuclear, Boston, Mass. The radioactive material is diluted to approximately 3 μCi/μmole with non-labeled stearyl-CoA obtained from PL Biochemicals, Inc., Milwaukee, Wis.

Ethanolic KOH, 10%

Aqueous KOH, saturated

H_2SO_4, 2 M

Pentane, redistilled or pesticide grade

Chloroform, reagent grade

Hexane, reagent grade

Diethyl ether, reagent grade

Methanol, reagent grade

Etherial diazomethane prepared fresh daily. A diazomethane generator is easily made from a 50-ml round bottom flask. The neck of the flask is extended to 15 cm by a piece of 18 mm glass tubing

[2] When microsomal preparations are used the radioactivity is found distributed among many lipid products. The amount of oleyl-CoA formed is determined after conversion of all fatty acyl residues to their corresponding methyl esters. Hence, it is convenient to refer to the product as [1-14C]oleate although at the end of the incubation period, this radioactivity is distributed among oleyl-CoA, oleic acid, and other lipids.

[3] B. R. Talamo and K. Bloch, *Anal. Biochem.* **29**, 300 (1969).

[4] A. R. Johnson and M. I. Gurr, *Lipids* **6**, 78 (1971).

[5] A. Kornberg and W. E. Pricer, Jr., *J. Biol. Chem.* **204**, 329 (1953).

and a 25-cm length of 8 mm diameter glass tubing is attached to
the neck at a 45° angle (pointing up) 5 cm above the flask. This
side arm is now bent down through a 45° angle 5 cm from the flask
in order to be parallel to the neck of the flask, but pointing down.
This generator is free from ground glass joints which can explo-
sively decompose diazomethane. Into the flask is placed 2 g of
N,N'-dimethyl-N,N'-dinitrosoterephthalamide, 20 ml of diethyl
ether, and 5 ml of methanol. The flask is cooled in ice and when
cold 2 ml of saturated aqueous KOH is added. After 10 minutes
(or longer, the mixture may be kept in ice for several hours) the
side arm of the flask is placed in an 18-mm test tube, cooled in
ice, and the flask is stoppered with a rubber stopper. Upon warming the flask with water at 50° the diazomethane distills over and
collects in the 18-mm test tube. It is used within 30 minutes or
is returned to the generator and redistilled.

AgNO$_3$ impregnated silica gel H thin-layer plates 0.5 mm thick. Four
20 × 20 cm plates may be prepared from 30 g of silica gel H and
75 ml 4% aqueous AgNO$_3$. The slurry is spread with a plastic
spreader (metal corrodes). The plates are dried at 110° for 2 hours
and are stored, if necessary, in a desiccator in the dark. If stored,
they are reactivated at 110° for 1 hour.

Standard thin-layer chromatography (TLC) mixture containing 25
mg of methyl oleate and 60 mg of methyl stearate per milliliter
of chloroform

2,7-Dichlorofluoresceine, 0.2% in 95% ethanol

Procedure. Incubations are performed in open 16 × 150 mm test
tubes, and the following reagents are added to the test tubes in ice: water
to 0.5 ml, 30 μl phosphate buffer, 20 μl [1-^{14}C]stearyl-CoA, 10 μl NADH,
and enzyme protein 100–500 μg. The reaction is started by transferring
the test tubes to a water bath at 37° and gently shaking them for 15
seconds. After 15 minutes at 37° the test tubes are placed in ice and
to each tube is rapidly added 0.2 ml of ethanolic KOH. The tubes are
transferred to a boiling water bath and are capped with glass marbles.
After 20 minutes the tubes are removed and cooled to room temperature.
If necessary the incubation tubes may be stored at −20° until next day.
The reaction mixture is acidified with 0.2 ml 2 M H$_2$SO$_4$, 0.1 ml of methanol is added, and the fatty acids are extracted with three 3-ml portions
of pentane. It may be necessary to centrifuge the tubes to separate the
pentane and aqueous layers. The combined pentane extracts, in a 15-ml
glass conical centrifuge tube, are evaporated to dryness in a stream of
nitrogen at room temperature. The residue is methylated with 2–3 drops

of etherial diazomethane (an excess, yellow color). After 5 minutes at room temperature the excess diazomethane is removed in a stream of nitrogen.

The fatty acid methyl esters are dissolved in 50 μl of chloroform and 5 μl are removed for radioactive assay to check on the efficiency of extraction (usually 95%). A 20-μl portion is applied as a single spot to a AgNO$_3$ impregnated silica gel H plate. Either 4 or 5 samples may be run on a single 20 $\times$ 20 cm plate together with two 5-μl aliquots of the standard TLC mixture applied at both sides of the plate. The plate is developed with hexane–diethyl ether (9:1 v/v). After development the plate is air-dried and the edges of the plate, where the standards are located, are sprayed lightly with dichlorofluorescein. The positions of the standard methyl oleate and stearate (R_f 0.35 and 0.65, respectively) are located under ultraviolet light, and the silica gel from the areas corresponding to methyl stearate, methyl oleate, and origin from each assay spot is scraped from the thin-layer chromatogram into scintillation vials containing a toluene solution of scintillator. The radioactivity in the three areas derived from each assay spot is determined. The percentage desaturation in an incubation is calculated from the ratio of radioactivity in the oleate region compared to the total radioactivity recovered from the thin-layer chromatogram of that assay spot. With conversions of 20% or less the reaction is usually linear with time; however, for precise determination of desaturase activity several protein concentrations and incubation times should be used. From the percentage desaturation the amount of oleate formed may be calculated. A tube lacking protein is used as a control. It should be noted that there is seldom any radioactive material remaining at the origin of the thin-layer chromatogram, any radioactivity in this position could be attributed to hydroxy esters but in this laboratory has always indicated faulty saponification or methylation.

Units. One unit of desaturase is defined as the amount of protein catalyzing the formation of 1 μmole of oleate per minute under the conditions of the assay. Specific activity is expressed in milliunits per milligram of protein.

Resolution of Desaturase

Isolation of Hen Liver Microsomes. Livers from old laying hens are collected in ice at the slaughterhouse. The livers may be kept on ice overnight or processed the same day. Three-hundred grams of liver are blended for 30 seconds with 450 ml of 0.1 M potassium phosphate buffer (pH 7.2 at 0°) in a Waring blender at 5°. The products of two such blendings are combined and centrifuged in a Sorvall GSA rotor at 9,000 rpm (13,200 g max) for 20 minutes. The supernatant is poured off through four layers of cheesecloth and centrifuged in two Beckman No. 30 rotors at 30,000 rpm (78,000 g average) for 120 minutes. The supernatant is poured off and the microsomal pellets are removed (avoiding the central dark area), and homogenized in a Potter-Elvehjem Teflon homogenizer in 0.25 M sucrose. Yield from 300 g of hen liver 3.5 g of crude microsomal protein. Specific activity 0.15–0.50 milliunits per milligram of protein. The microsomes may be kept several weeks at −20° with no loss of desaturase activity.

Preparation of Particulate Fraction (P_2). A "solubilizing solution" is prepared from 240 ml of glycerol, 80 ml 1 M KCl, 80 ml 1 M potassium citrate (pH 7.7 at 0°), and 40 ml 10% sodium deoxycholate (pH 8.5 at 20°). To 42 ml of solubilizing solution at 0° is added 0.3 ml of 0.2 M dithiothreitol and 30 ml of microsomal suspension (60 mg of protein/ml). The mixture is stirred gently for 20 minutes at 5° and centrifuged at 105,000 g (average) for 60 minutes. The supernatant fluid is filtered through glass wool and diluted, with stirring, with 0.75 volume of 0.1 mM dithiothreitol at 0°. The mixture is stirred for 20 minutes at 5° and centrifuged at 105,000 g (average) for 60 minutes. The supernatant fluid is isolated and diluted with 1.25 volumes of 0.1 mM dithiothreitol at 0°. The mixture, after stirring for 20 minutes at 5°, is centrifuged at 105,000 g (average) for 60 minutes and the resulting pellets are suspended in 10 mM potassium phosphate (pH 7.2 at 0°) containing 0.1 mM dithiothreitol. This particulate fraction is designated "P_2 fraction" and may be stored at −20° for several weeks with no loss of desaturase activity.

Sephadex G-200 Chromatography. The P_2 fraction is diluted to 20 ml with 0.25 M sucrose at 0°. The suspension is homogenized and to it is added 1 ml of sodium Tricine buffer (pH 8.0 at 0°), 0.1 ml of 0.2 M dithiothreitol, and 1 ml of 10% sodium deoxycholate. The mixture is rehomogenized, sonicated for 2 seconds at 0° with a Biosonik II sonicator (Bronwill Scientific, Rochester, N.Y.) using the 4-mm tip at 20% output and applied to a column (40 × 430 mm) of Sephadex G-200 stabilized by a 5-mm layer of Sephadex G-75 on the top. The column is equilibrated at 5° with 20 mM sodium bicarbonate (pH 7.7 at 0°) containing 0.1 mM dithiothreitol and 0.2% sodium deoxycholate. The column is eluted overnight at 5° with the same buffer and 16 ml (approximately 20 minutes) fractions are collected. This procedure completely resolves cytochrome b_5 from the majority of the protein applied to the column. The fractions which contain cytochrome b_5 are located by measurement of reduced minus oxidized difference spectra. A 1-ml aliquot is removed from each fraction, 7.4 μg of purified rat liver microsomal NADH-cytochrome b_5 reductase[6,7] are added, and the mixture is diluted to 2.5 ml with 10 mM potassium phosphate buffer (pH 7.2 at 37°). The mixture is divided equally between two cuvettes and a base line of equal light absorbance is established at room temperature in a Cary 14 recording spectrophotometer with 0.1 OD full-scale sensitivity. To the sample cuvette is added 0.1 μmole NADH and the NADH-reduced minus oxidized spectrum is recorded. The difference in absorbance between 424 and 410 nm is a measure of cytochrome b_5. Cytochrome b_5 is usually eluted in tubes 16–23.

The material eluted from the Sephadex G-200 column immediately after the void volume (tubes 10–15) is applied to a column (40 × 180 mm) of Sephadex G-25 equilibrated at 5° with 10 mM potassium phosphate buffer (pH 7.2 at 0°) containing 0.1 mM dithiothreitol. This procedure removes the deoxycholate. The cloudy eluate is poured into an equal volume of stirred saturated ammonium sulfate solution at 5° and the precipitate is collected by centrifugation. The pellets are suspended in a small volume of 10 mM potassium phosphate buffer (pH 7.2 at 0°) containing 0.1 mM dithiothreitol. This fraction is designated the "P_3 fraction" and may be kept for up to 2 weeks at −20° with no loss of activity. Before it is assayed for desaturase activity the thawed P_3 fraction is clarified by a 2-second sonication with a Biosonik II using the 4-mm tip at 20% output.

The P_3 fraction when assayed for desaturase activity is inactive; desaturase activity can however be restored to a high level by addition of NADH-cytochrome b_5 reductase, cytochrome b_5 isolated by a detergent procedure,[8,9] lipid dispersion, and sodium deoxycholate (Table I). The recovery of desaturase activity is shown in Table II.

The Sephadex G-200 column retains some protein and lipid material with the result that the top of the column gradually becomes impermeable to the solubilized P_2 fraction. It is advisable to remove the Sephadex G-75 and repack the column, using the same Sephadex G-200, after four separations have been performed. After eight separations have been performed on one batch of Sephadex G-200 the column should be repacked with fresh Sephadex G-200.

Preparation of Lipid Dispersion. In 3 ml of benzene are dissolved 17.5 mg of purified egg yolk phosphatidylcholine[10] and 4.7 mg of oleic

[6] S. Takesue and T. Omura, *J. Biochem.* (*Tokyo*) **67**, 267 (1970).
[7] P. W. Holloway, *Biochemistry* **10**, 1556 (1971).
[8] L. Spatz and P. Strittmatter, *Proc. Nat. Acad. Sci. U.S.* **68**, 1042 (1971).
[9] P. W. Holloway and J. T. Katz, *Biochemistry* **11**, 3689 (1972).
[10] B. J. Litman, *Biochemistry* **12**, 2545 (1973).

TABLE I

REGENERATION OF DESATURASE ACTIVITY

Components[a]	Stearyl-CoA desaturase (nmoles oleate formed)
P_3	<0.1
$P_3 + F_p$	<0.1
$P_3 + F_p$ + deoxycholate	<0.1
$P_3 + F_p$ + cyt b_5	1.0
$P_3 + F_p$ + cyt b_5 + deoxycholate	1.4
$P_3 + F_p$ + cyt b_5 + lipid	1.2
$P_3 + F_p$ + cyt b_5 + deoxycholate + lipid	2.0
F_p + cyt b_5 + deoxycholate + lipid	<0.1

[a] The standard assay mixture contained the additional components indicated in the following amounts:
    0.4 mg of P_3 protein, 3.7 μg of NADH-cytochrome b_5 reductase protein (F_p), 1.7 mM sodium deoxycholate, 8 μg cytochrome b_5 protein (cyt b_5) (isolated by detergent solubilization), or 0.12 mg of lipid.

acid. The benzene is removed by lyophilization and the residue is taken up in 3.7 ml of 20 mM Tris acetate (pH 8.2 at 0°) containing 1 mM EDTA. The mixture is sonicated under nitrogen with a Biosonik II sonicator, operated at 50% output with the 12.5-mm tip. The sonication vessel is cooled in ice while argon is blown onto the surface of the mixture. After six 30-second periods of sonication the lipid suspension will have cleared to an opalescent liquid. The lipid dispersion is centrifuged at 40,000 rpm in a Beckman No. 40 rotor for 15 minutes, and the supernatant lipid dispersion is removed from the dark pellet. The lipid dispersion is stored at 4° under argon in a small bottle closed with a silicon rubber septum held in place with a Teflon-lined cap (5 ml Microflex tubes are suitable, Konte Glass Co., Vineland, N.J.). The lipid dispersion is re-

TABLE II

RECOVERY OF DESATURASE ACTIVITY

Fraction	Total protein (mg)	Total activity (milliunits)	Specific activity (milliunits/mg)
Microsomes	1800	270	0.15
P_2 fraction	200	116	0.58
P_3 fraction	55	30[a]	0.54[a]

[a] Assayed in the presence of optimal levels of NADH-cytochrome b_5 reductase, cytochrome b_5, lipid dispersion, and sodium deoxycholate as described in Table I.

moved as needed with a syringe, and the bottle is flushed periodically with argon. No peroxidization or hydrolysis can be detected after 2 months of storage, and the lipid retains its ability to restore desaturase activity for at least this length of time.

Properties

Specificity. Hen liver microsomes will desaturate saturated fatty acids of chain lengths 12 through 22 to the corresponding *cis*-Δ^9-monoenoic acid.[11] Maximum desaturation occurs with C_{14} and C_{18} fatty acids.[11] The hydrogens abstracted during the conversion of stearic to oleic acid are *cis* and are both of the D configuration.[12]

Unsaturated fatty acids may be further desaturated by microsomes. Olefinic bonds are introduced between C-5 and C-6 or between C-6 and C-7 and perhaps in other positions.[1] It has been suggested that the same desaturase is not used for desaturation of saturated and unsaturated fatty acids.[1]

The desaturation of palmityl-, stearyl-, or oleyl-CoA requires oxygen and NADH or NADPH.[13] NADH is the preferred electron donor for stearyl-CoA desaturation.[14,15]

With stearyl-CoA as substrate oleyl-CoA is the first product[16,17] although, because of the presence of fatty acyl CoA thioesterases and other enzymes which utilize CoA esters, the substrate and initial product are rapidly converted into other lipids when hen liver microsomes are used.

Inhibitors. The stearyl-CoA desaturase activity of intact microsomes, or the resolved and reconstituted desaturase system described in Table I, is inhibited by cyanide but not by carbon monoxide. Cyanide (0.1–1.0 mM) produces approximately 50% inhibition of the desaturase activity of both the intact microsomes or the reconstituted system.[9,14,15,18] Desaturation of stearyl-, palmityl-, and oleyl-CoA is inhibited by N-ethyl-maleimide and p-hydroxymercuribenzoate.[13,14,19] Stearyl-CoA desatura-

[11] A. R. Johnson, A. C. Fogerty, J. A. Pearson, F. S. Shenstone, and A. M. Bersten, *Lipids* **4**, 265 (1969).

[12] L. J. Morris, *Biochem. J.* **118**, 681 (1970).

[13] P. W. Holloway, R. Peluffo, and S. J. Wakil, *Biochem. Biophys. Res. Commun.* **12**, 300 (1963).

[14] N. Oshino, Y. Imai, and R. Sato, *Biochim. Biophys. Acta* **128**, 13 (1966).

[15] P. D. Jones, P. W. Holloway, R. O. Peluffo, and S. J. Wakil, *J. Biol. Chem.* **244**, 744 (1969).

[16] I. K. Vijay and P. K. Stumpf, *J. Biol. Chem.* **246**, 2910 (1971).

[17] C. T. Holloway and P. W. Holloway, *Lipids* **9**, 196 (1974).

[18] T. Shimakata, K. Mihara, and R. Sato, *J. Biochem. (Tokyo)* **72**, 1163 (1972).

[19] P. W. Holloway and S. J. Wakil, *J. Biol. Chem.* **245**, 1862 (1970).

tion is inhibited by oleyl-CoA[14] in agreement with the proposal that oleyl-CoA is the first product of this reaction. The inhibition of the desaturation of saturated fatty acids by sterculic acid (a C_{19} fatty acid with a cyclopropene ring at the 9,10 position) is thought to result from the reaction of this material with a sulfydryl group on the desaturase.[11]

Other Properties. The pH optimum of the stearyl-CoA desaturase in intact microsomes and a reconstituted system is at pH 7.2.[18] The apparent Michaelis constants found are for NADH 20 μM, NADPH 12 μM, and stearyl-CoA 17 μM.[14] The value for stearyl-CoA is undoubtedly influenced by the presence of other enzymes which utilize the stearyl-CoA as well as the micellar nature of stearyl-CoA.

Mechanism of Stearyl-CoA Desaturation

The evidence for the involvement of the NADH-specific electron transport chain of microsomes in stearyl-CoA desaturation, as shown in Scheme I, may be summarized.

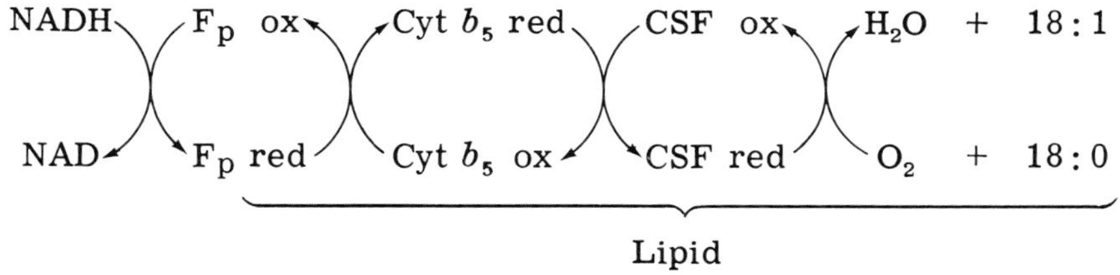

SCHEME I

In Scheme I F_p is the NADH-cytochrome b_5 reductase, cyt b_5 is cytochrome b_5, and CSF is an unknown component the "cyanide sensitive factor."[14,18]

(i) The reaction requires O_2 and NADH (or NADPH).

(ii) Cyanide inhibits, carbon monoxide does not.[14]

(iii) Destruction of NADH-cytochrome b_5 reductase by N-ethylmaleimide causes loss of desaturase activity which can be restored by addition of purified NADH-cytochrome b_5 reductase.[19]

(iv) Desaturation is absolutely dependent upon the presence of cytochrome b_5 and NADH-cytochrome b_5 reductase.[9,18]

(v) The desaturase has been resolved into three protein components by detergent treatment: NADH-cytochrome b_5 reductase, cytochrome b_5 and an uncharacterized fraction, presumably containing the cyanide sensitive factor of Scheme I.[9,18] All three fractions are required, together with lipid dispersions, for restoration of desaturase activity.[9]

(vi) Lipid is required in both the intact microsomal system[15] and the reconstituted system.[9] It is suggested that lipid is required at two distinct points of the electron transport chain.[9]

Acknowledgments

The egg yolk phosphatidylcholine was a generous gift of Dr. E. A. Dawidowicz. This work was supported by Grants P-559, BC-71A and BC-71B from the American Cancer Society and Grant GM 18406 from the U.S. Public Health Service. This work was carried out during the tenure of an Established Investigatorship of the American Heart Association.

[32] Purification of Phospholipid Exchange Proteins from Beef Heart

By D. B. ZILVERSMIT and L. W. JOHNSON

The soluble fraction of a variety of animal and plant tissues contains proteins which stimulate the exchange of phospholipids between intracellular membrane fractions (mitochondria and microsomes)[1-6] or between sonicated phospholipid vesicles,[7,8] chylomicrons,[7] or fat droplets in fat emulsions.[7] Phospholipid exchange proteins have been purified from beef liver[9,10] and from beef heart[11,12] cytosol. In the assays described below phosphatidylcholine was used to measure exchange activity.

Assay Method

Principle of Assay. Transfer of ^{32}P-labeled phosphatidylcholine from artificial vesicles to mitochondria is measured with and without added

[1] K. W. A. Wirtz and D. B. Zilversmit, *J. Biol. Chem.* **243**, 3596 (1968).

[2] W. C. McMurray and R. M. C. Dawson, *Biochem. J.* **112**, 91 (1969).

[3] M. Akiyama and T. Sakagami, *Biochim. Biophys. Acta* **187**, 105 (1969).

[4] K. W. A. Wirtz and D. B. Zilversmit, *Biochim. Biophys. Acta* **193**, 105 (1969).

[5] L. Wojtczak, J. Baranska, J. Zborowski, and Z. Drahota, *Biochim. Biophys. Acta* **249**, 41 (1971).

[6] D. B. Zilversmit, *J. Biol. Chem.* **246**, 2645 (1971).

[7] C. Ehnholm and D. B. Zilversmit, *Biochim. Biophys. Acta* **274**, 652 (1972).

[8] A. B. Abdelkader and P. Mazliak, *Eur. J. Biochem.* **15**, 250 (1970).

[9] K. W. A. Wirtz, H. H. Kamp, and L. L. M. van Deenen, *Biochim. Biophys. Acta* **274**, 606 (1972).

[10] H. H. Kamp, K. W. A. Wirtz, and L. L. M. van Deenen, *Biochim. Biophys. Acta* **318**, 313 (1973).

[11] K. W. A. Wirtz and D. B. Zilversmit, *FEBS Lett.* **7**, 44 (1970).

[12] C. Ehnholm and D. B. Zilversmit, *J. Biol. Chem.* **248**, 1719 (1973).

exchange protein. After a 40-minute period for the exchange of phospholipids, the particles are separated by sedimentation of the mitochondria; exchange is measured as the appearance of radioactivity in the mitochondria or as its loss from the vesicles. In practice the loss of ^{32}P from the vesicles is the more convenient measure of exchange activity.

Radioactive Lipids. Labeled phosphatidylcholine is prepared by injecting a rat intraperitoneally 16 hours before sacrifice with 800 μCi ^{32}P$_i$. The minced liver is extracted with 20 volumes of chloroform–methanol, 2:1 (v/v), and the extract is washed by the procedure of Folch *et al.*[13] to remove nonlipid material. The solvent is evaporated under nitrogen, and lipids are redissolved in chloroform for chromatography. They are applied to silica gel H thin layer plates containing Ultraphor,[14] a fluorescent marker for lipids. The lipids are chromatographed with chloroform–methanol–acetic acid–water, 25:15:4:2 (v/v) and the plates are dried under a stream of nitrogen. The phosphatidylcholine band is identified by comparison with an appropriate standard and is scraped and eluted with chloroform–methanol–water, 80:35:5 (v/v). For larger preparations of phosphatidylcholine, thin-layer chromatography can be replaced by chromatography on an alumina column by a modification of the procedure of Singleton *et al.*[15] The chloroform solution of lipids is applied to a 1.5 $\times$ 15 cm column containing neutral alumina (activity I, EM Laboratories) equilibrated with chloroform. Neutral lipids are eluted with chloroform–methanol 9:1 (120 ml) and phospholipids with the same solvent plus 0.5% concentrated ammonium hydroxide. Fractions near the phospholipid front (100 ml) contain only phosphatidylcholine; these can be pooled and stored in the dark under nitrogen at 4° for at least one half-life of ^{32}P (i.e., 2 weeks). Dry lipids should not be exposed to air.

[^{14}C]Triolein is purified by thin-layer chromatography on silica gel H with the solvent system hexane–diethyl ether–acetic acid, 60:40:1 (v/v). It is eluted with chloroform and stored at 4°. Rechromatography is required periodically to check for the presence of oleic acid and partial glycerides.

Liposomes. [^{32}P]Phosphatidylcholine with 10% (w/w) butylated hydroxytoluene as antioxidant and a trace amount of [^{14}C]triolein are dried under a stream of nitrogen from the stock solutions. The lipid is resuspended in 5 ml of diethyl ether and redried under nitrogen to a thin film in a test tube. Buffer containing 0.25 M sucrose, 1 mM EDTA, 0.05

[13] J. Folch, M. Lees, and G. H. Sloane-Stanley, *J. Biol. Chem.* **226**, 497 (1957).

[14] J. W. Copius Peereboom, *J. Chromatogr.* **4**, 323 (1960).

[15] W. S. Singleton, M. S. Gray, M. L. Brown, and J. L. White, *J. Amer. Oil Chem. Soc.* **42**, 53 (1965).

M Tris Cl, pH 7.4, is added to give a lipid concentration of 1.67 mg/ml. The capped tube, after flushing with nitrogen, is vigorously agitated for 10 minutes to disperse the lipid which is then allowed to swell for 1 hour. The test tube is suspended in a sonicating bath (Branson HD-50) and sonicated for 30 minutes at 25° during which time the turbidity of the liquid decreases markedly. The liposomes are centrifuged for 20 minutes at 15,000 *g* to remove large particles.

Mitochondria. Mitochondria may be obtained from heart or liver, but heart mitochondria, prepared according to Green *et al.*,[16] have desirable physical characteristics. They are obtained by grinding bovine heart and adding a solution of 0.25 *M* sucrose, 0.01 *M* potassium phosphate, and 20 m*M* EDTA in the amount of 1.5 liters/kg tissue. An equal volume of the sucrose buffer solution containing 4 ml 6 *N* KOH is added, and the mixture is homogenized for 30 seconds. The final pH should be 7.4. The homogenate is centrifuged 20 minutes at 1400 *g* and the resulting supernatant is separated in a Sharples continuous flow centrifuge operated at 50,000 rpm. The mitochondria are suspended in 0.25 *M* sucrose, 0.05 *M* Tris Cl, and 1 m*M* EDTA and are washed twice in the same medium. They may be stored at −20°, but they should be rewashed with sucrose–EDTA–Tris Cl after thawing.

Incubation. Exchange protein activity is measured by incubation of liposomes (10 μg of phospholipid phosphorus) with mitochondria (12.5 mg of mitochondrial protein) and exchange protein in 4 ml of 0.25 *M* sucrose, 0.05 *M* Tris Cl, and 1 m*M* EDTA for 40 minutes at 37°. Constant agitation assures homogeneity of the mixture. The exchange of phospholipid is terminated by chilling and centrifuging the incubation mixture at 15,000 *g* for 15 minutes. Aliquots of the supernatants are counted in appropriate scintillation cocktails (e.g., that of Gordon and Wolfe[17]). The fraction of [^{32}P]phosphatidylcholine transferred from the liposomes to the mitochondria during incubation is measured by the decrease in the ratio of ^{32}P to ^{14}C of the liposomes. This decrease serves as the measure of phospholipid exchange activity. [^{14}C]Triolein serves as a nonexchangeable marker for liposomes. The small amount of [^{32}P]phosphatidylcholine which exchanges in the absence of exchange protein is subtracted as background.

Purification Procedure

Acid pH Step. The postmitochondrial supernatant, containing heart muscle cytosol and microsomes (recovered during the preparation of

[16] D. E. Green, R. L. Lester, and D. M. Ziegler, *Biochim. Biophys. Acta* **23**, 516 (1957).
[17] C. F. Gordon and A. L. Wolfe, *Anal. Chem.* **32**, 574 (1960).

mitochondria), is used as a source of exchange protein. About 900 ml
of the ice-cold solution is adjusted to pH 5.1 with 3 N HCl. After 1 hour
of flocculation the precipitate is removed by centrifugation for 15 minutes
at 10,000 g. The pH is returned to 7.4 by titration with 18 N NaOH.

Ammonium Sulfate Step. Solid ammonium sulfate is added to 0.90
saturation at 4° and the protein is allowed to precipitate overnight. The
precipitate is collected by centrifugation and dialyzed against 10 liters
of 0.15 M potassium phosphate buffer, pH 7.4, for 16–20 hours at 4°.

Sephadex Step. The dialyzed protein (about 5 g) is loaded onto a
Sephadex G-75 column (5 × 70 cm) which has previously been equilibrated with 0.15 M potassium phosphate buffer. The column is eluted
at 2 ml/cm²/hour with the same buffer at 4°. A typical elution profile
is shown in Fig. 1. The fractions containing exchange activity are pooled
and concentrated to 15 ml with an Amicon ultrafiltration cell and a
PM-10 membrane. The concentrate is then rechromatographed on Sephadex G-75 in the same manner (Fig. 2) and again concentrated by
ultrafiltration.

Isoelectric Focusing Step. The protein concentrate is dialyzed overnight at 4° against 1% glycine, pH 6.4, and then subjected to isoelectric
focusing at 15° with a pH gradient from 4 to 6. The activity profile from
this step is shown in Fig. 3. Two peaks of exchange activity are found,

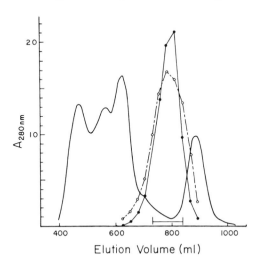

Fig. 1. Elution profile for Sephadex G-75 column. About 5 g of dialyzed protein
are chromatographed on a 5 × 70 cm Sephadex G-75 column equilibrated with
0.15 M potassium phosphate, pH 7.4. Flow rate is 33 ml/hr. (———) Absorbance
at 280 nm, (○) phospholipid exchange activity, and (●) specific activity in arbitrary units. From Ehnholm and Zilversmit.[12]

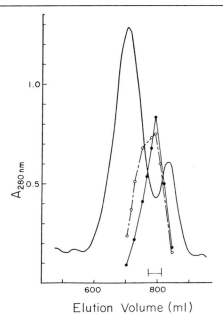

FIG. 2. Elution profile for the second Sephadex G-75 column. The pooled, concentrated protein from the first Sephadex column are reapplied to the same column. Symbols as in Fig. 1. From Ehnholm and Zilversmit.[12]

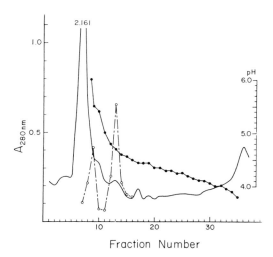

FIG. 3. Activity profile for isoelectric focusing. Protein from Sephadex G-75 was focused with 600 V for 48 hours at 15° over a pH range of 4–6. (——) Absorbance at 280 nm, (○) exchange of activity, and (●) pH. From Ehnholm and Zilversmit.[12]

TABLE I
PURIFICATION OF PHOSPHOLIPID EXCHANGE PROTEIN[a]

Step	Volume (ml)	Protein (mg)	Specific activity	Recovery (%)	Purification factor
pH 5.1	900	7470	7.0	100	
(NH$_4$)$_2$SO$_4$	50	4900	9.1	85	1.3
Sephadex G-75	32	123	82.2	20	11.7
Sephadex G-75	50	20.5	416	16	59
Isoelectric focusing					
Fraction 9	3	0.32	1252	0.8	179
Fraction 13	3	0.48	2068	2	295

[a] From Ehnholm and Zilversmit.[12]

one at pH 4.7 and the other at 5.5. The sucrose and ampholine may be removed by dialysis against Tris or phosphate buffers, pH 7.4, followed by gel filtration on Sephadex G-25. The degree of purification at each step is shown in Table I.

Carboxymethyl Cellulose Step.[18] An alternative to isoelectric focusing is the use of carboxymethyl cellulose. This avoids the necessity of removing ampholines and gives about the same degree of purification. To expedite this step, the preceding Sephadex G-75 columns are preequilibrated and run with 0.042 M Tris acetate, pH 7.4. Two passes through the Sephadex G-75 column can be avoided if two Sephadex G-75 columns are connected in series. The active fractions from the Sephadex columns are pooled, concentrated to 15 ml by ultrafiltration, and titrated with acetic acid to pH 6.0. The protein is applied to a 3.8 × 45 cm carboxymethyl cellulose column, which has been equilibrated with 0.042 M Tris acetate at pH 6.0. The column is eluted with the same buffer to give a profile as shown in Fig. 4. The active fractions are pooled, titrated to pH 7.4 with Tris base, and concentrated by ultrafiltration. In this form the protein retains 70–80% of its activity for 3 weeks at 4°. The inclusion of 5 mM β-mercaptoethanol in all buffers may improve the stability of the protein, but it is not essential.

Purity of Isolated Proteins. Fractions 9 and 13 from the isoelectric focusing column have been subjected to disc gel electrophoresis[19] on 7% polyacrylamide gels followed by staining with Coomassie blue. The protein with the highest specific activity (fraction 13) had an isoelectric point of 4.7 and showed one main band with 90% of the total absorbance in a gel loaded with 50 μg of protein. One minor band with slightly higher mobility was also observed.

[18] L. W. Johnson and D. B. Zilversmit, unpublished observation.
[19] B. J. Davis, *Ann. N.Y. Acad. Sci.* **121**, 404 (1964).

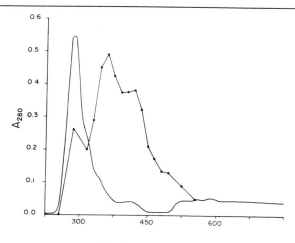

Elution   Volume. (ml)

Fig. 4. Elution profile for carboxymethyl cellulose column. Protein from Sephadex G-75 in 0.042 M Tris acetate, pH 6.0, is chromatographed on a 3.8 × 45 cm carboxymethyl cellulose column with the same buffer as eluant. (——) Absorbance at 280 nm and (●) exchange activity.

Properties of Phospholipid Exchange Proteins

As previously noted, these proteins promote the exchange of phosphatidylcholine between a wide variety of membranes, lipoproteins, emulsions, and even monolayers.[20] The crude fractions show exchange activity for phosphatidylethanolamine and phosphatidylinositol as well as for phosphatidylcholine. The purified fractions appear to be specific for phosphatidylcholine and do not accelerate exchange of phosphatidylethanolamine, phosphatidylinositol, or cholesterol.[10,12] The purified protein from beef heart may transport some sphingomyelin,[12] but the liver protein shows no such activity.[10] Phosphatidylcholines with varying degrees of fatty acid unsaturation appear to be exchanged indiscriminately.[21] In addition to exchanging phospholipids the proteins have been shown to accelerate one-way transport of phosphatidylcholine[22] and of sphingomyelin.[12]

Two phospholipid exchange proteins have been isolated from beef heart[12]; their isoelectric points are 4.7 and 5.5. One protein has been

[20] R. A. Demel, K. W. A. Wirtz, H. H. Kamp, W. S. M. Geurts van Kessel, and L. L. M. van Deenen, *Nature (London), New Biol.* **246**, 102 (1973).
[21] K. W. A. Wirtz, L. M. G. van Golde, and L. L. M. van Deenen, *Biochim. Biophys. Acta* **218**, 176 (1970).
[22] Y. Kagawa, L. W. Johnson, and E. Racker, *Biochem. Biophys. Res. Commun.* **50**, 245 (1973).

found in beef liver and has an isoelectric point of 5.8.[10] The molecular weights, estimated from polyacrylamide gel electrophoresis in sodium dodecyl sulfate, are **21,000** and **25,900** for the two proteins from beef heart[12] and **22,000** for the liver protein.[10] Amino acid analysis of the liver protein indicates a high average hydrophobicity.[10] The purified liver protein binds one mole of phosphatidylcholine per mole of protein.[10] It contains one cystine residue per molecule and is inactivated by the reagent N-ethyl maleimide.[10,23]

Acknowledgments

This research was supported in part by Public Health Service Research Grant HL 10940 from the National Heart and Lung Institute, U.S. Public Health Service, in part by a Career Investigator grant of the American Heart Association, and in part by funds provided by the State University of New York.

[23] D. R. Illingworth and O. W. Portman, *Biochim. Biophys. Acta* **280**, 281 (1972).

Section VI

General Analytical Methods

[33] Enzymatic Determination of Long-Chain Fatty Acyl-CoA

By D. Veloso and R. L. Veech

The method consists of two main steps: enzymatic hydrolysis of long-chain acyl-CoA (fatty acyl-CoA + H_2O → CoA + fatty acid) and subsequent determination of the CoA produced. Several enzymes have been reported to release CoA from long-chain acyl-CoA; *viz.*, pigeon[1] and rat[2] liver fatty acid synthase, pig brain palmityl-CoA deacylase,[3] *Escherichia coli* palmityl thioesterases,[4,5] rat liver microsome acyl-CoA hydrolase,[6] and pancreatic lipase.[7] Rat liver fatty acid synthase can be prepared easily in a highly pure form[2] and offers a convenient basis for the assay of long-chain acyl-CoA. Fatty acid synthase hydrolyzes palmityl- and stearyl-[8] and oleyl-CoA.[9] Its action on other long-chain acyl-CoA is not known. The concentration of long-chain acyl-CoA in the samples to be hydrolyzed should be below 5 μM to avoid errors resulting from micelle formation.[1,6] This low concentration of long-chain CoA necessitates the use of a cycling technique in order to measure the concentration conveniently in a spectrophotometer. The CoA released from long-chain acyl-CoA is measured with an enzymatic cycling procedure using the reactions catalyzed by carnitine acetyltransferase [acetyl-CoA:carnitine O-acetyltransferase, EC 2.3.1.7] (CAT) and citrate synthase [EC 4.1.3.7] (CS).

The oxalacetate used in the citrate synthase reaction is generated by the malate dehydrogenase reaction [L-malate:NAD oxidoreductase, EC

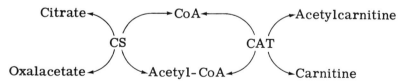

[1] S. Kumar, J. A. Dorsey, R. A. Muesing, and J. W. Porter, *J. Biol. Chem.* **245**, 4732 (1970).
[2] S. Kumar, G. T. Phillips, and J. W. Porter, *Int. J. Biochem.* **3**, 15 (1972).
[3] P. A. Srere, W. Seubert, and F. Lynen, *Biochim. Biophys. Acta* **33**, 313 (1959).
[4] E. M. Barnes, Jr., and S. J. Wakil, *J. Biol. Chem.* **243**, 2955 (1968).
[5] E. M. Barnes, Jr., A. C. Swindell, and S. J. Wakil, *J. Biol. Chem.* **245**, 3122 (1970).
[6] R. E. Barden and W. W. Cleland, *J. Biol. Chem.* **244**, 3677 (1969).
[7] E. D. Barber and W. E. M. Lands, *Biochim. Biophys. Acta* **250**, 361 (1971).
[8] G. T. Phillips, J. E. Nixon, J. A. Dorsey, P. H. W. Butterworth, C. J. Chesterton, and J. W. Porter, *Arch. Biochem. Biophys.* **138**, 380 (1970).
[9] D. Veloso and R. L. Veech, in preparation.

1.1.1.37] establishing a connection between the enzymatic cycling reactions and a NADH formation system:

CoA + acetylcarnitine $\rightleftharpoons$ acetyl-CoA + carnitine
Acetyl-CoA + H_2O + oxalacetate^{2-} $\rightleftharpoons$ CoA + citrate^{3-} + H^+
Malate^{2-} + NAD$^+$ $\rightleftharpoons$ oxalacetate^{2-} + NADH + H^+

For the CoA recycling to proceed, the formation of more oxalacetate is necessary and consequently more NADH is formed. The rate of formation of NADH concentration can be followed in a spectrophotometer (340 nm) or fluorometer. Similar to the enzymatic cycling procedure for measurement of pyridine nucleotides,[10] the concentrations of CoA and acetyl-CoA must be below their K_m in order that the reaction rates be dependent on their concentrations. Coenzyme A is the standard for the cycling procedure because no reliable long-chain acyl-CoA preparations are commercially available.

Reagents

Fatty Acid Synthase. The enzyme is partially purified from rat liver[11] and is suitable for use after dialysis of the second ammonium sulfate precipitate. Fatty acid synthase activity is determined according to Nepokroeff *et al.*[11] and expressed as 1 unit = 1 μmole malonyl-CoA used per minute per milligram of protein. Before use, the enzyme is activated by mixing it with an equal volume of a mixture of 0.8 M potassium phosphate buffer, pH 7.0, 0.02 M dithiothreitol, and 1.2 mM EDTA and incubating the mixture at 30° for 30 minutes. The diluted activated enzyme can be stored at −70° and reactivated once by immersion in a water bath (30°) for 30 minutes.

Coenzyme A Standard. Prepare a 3 mM solution of CoA in 10 mM dithiothreitol, standardize, and store in small portions at −20°. Each tube should not be thawed more than 3 times. The concentration of CoA in the standard is determined by the method of Tubbs and Garland,[12] with 3-acetyl pyridine adenine nucleotide as coenzyme. Prior to use, dilute the enzyme 300-fold (spectrophotometry) or 750-fold (fluorometry) with a 1 mM dithiothreitol solution.

Hydrolysis Mixture. This mixture, prepared according to Kumar *et al.*[1] contains 10 ml of 1 M potassium phosphate buffer, pH 6.8, 1 ml 0.05 M EDTA, 1 ml 0.1 M dithiothreitol, and 38 ml of distilled water. The

[10] O. H. Lowry, J. V. Passonneau, D. W. Schulz, and M. K. Rock, *J. Biol. Chem.* **236**, 2746 (1961).
[11] C. M. Nepokroeff, M. R. Lakshmanan, and J. W. Porter, this volume [6].
[12] P. K. Tubbs and P. B. Garland, Vol. 13 [72].

concentration of the reagents in this mixture must be altered according to the volume of sample to be hydrolyzed. The final reaction mixture after addition of sample contains the following concentration: 0.2 M phosphate, 1 mM EDTA, and 2 mM dithiothreitol.

CoA Cycling Mixture I (Spectrophotometric Assay). The cycling mixture consists of 16 ml 0.1 M potassium phosphate buffer, pH 7.2, 1 ml 0.1 M L-malate, 0.2 ml 0.1 M NAD⁺, 0.1 ml 0.05 M EDTA, 0.5 ml 0.02 M acetyl-DL-carnitine, 12.3 ml of distilled water, and 25 μl of malate dehydrogenase (1100 units/mg, 5 mg/ml). This mixture is prepared 10 minutes before the cycling assay and left at room temperature in order to equilibrate the malate dehydrogenase reaction. The reagents for the cycling assay must be neutralized with KOH and stored at −20°.

CoA Cycling Mixture II (Fluorometric Assay). For the fluorometric cycling assay, the spectrophotometric cycling mixture is modified as follows: 10 ml potassium phosphate buffer, 0.1 ml malate, 0.1 ml NAD⁺, 0.05 ml EDTA, 0.25 ml acetyl-DL-carnitine, 19.25 ml distilled water, and 20 μl of malate dehydrogenase.

Enzyme Mixture. Prior to use, mix gently 75 μl of citrate synthase (70 units/mg, 2 mg/ml), 75 μl of carnitine acetyltransferase (80 units/mg, 5 mg/ml), and 170 μl of distilled water.

Procedure (Spectrophotometric Assay)

Hydrolyses of the Long-Chain Acyl-CoA. If the samples to be assayed are solutions that do not contain CoA or short-chain CoA derivatives or enzymes that interfere with the long-chain acyl-CoA determination (e.g., solutions of commercial long-chain acyl-CoA), incubate sample aliquots (2–5 nmoles of long-chain acyl-CoA) in 1 ml of the hydrolysis mixture and 0.04 unit of activated fatty acid synthase at 30° for 40 minutes in a shaking bath. If the concentration of long-chain acyl-CoA in the sample is very low and large volumes of sample are used for hydrolyses, a smaller volume of a more concentrated hydrolysis mixture should be used. After incubation, place the tubes in ice, add 0.1 ml of 2 N HCl into each tube, and neutralize with N KOH. Blanks are handled in the same way except that the addition of fatty acid synthase is omitted. The use of a blank has two aims: (1) to detect any CoA or CoA derivative (e.g., acetyl-CoA) not hydrolyzed by fatty acid synthase that could react during the CoA cycling procedure, and (2) to allow a mixture similar to that used in the assay mixture to be added to the CoA standards in order to correct for any activator or inhibitor of the CoA cycling reactions present in the samples.

If the material to be analyzed contains CoA or short acyl-CoA derivatives (less than twelve carbons), a separation of these compounds from long-chain acyl-CoA is possible with perchloric acid.[12] If rat tissues are going to be assayed, homogenize pulverized frozen tissues in preweighed centrifuge tubes (1 g of brain of fed or starved rat, 1 g of liver of fed rat, and 0.4 g of 48 hour starved rat) with 8 ml of 0.85 M perchloric acid. The frozen tissues can be stored at $-70°$ without any loss of long-chain acyl-CoA. Centrifuge the homogenates at 35,000 g and $0°$, for 15 minutes. Discard the supernatant and treat the pellet twice more with perchloric acid in order to remove adherent CoA or short-chain CoA. Weigh the tube plus the pellet and calculate the amount of liquid retained in the pellet: weight of tube plus pellet minus tube weight minus dry weight of pellet equals liquid in pellet. Add to the pellet 10 ml of the hydrolysis mixture, homogenize, and neutralize with KOH. For optimum results the amount of long-chain CoA should be from 2 to 5 μM in this solution. For calculations V_1 is defined as milliliters of volume retained in the pellet plus milliliters of hydrolysis mixture plus milliliters of KOH added. Mix very thoroughly and pipette 1 ml of the suspension (V_2) into another centrifuge tube. Add fatty acid synthase (0.04 unit), mix, and incubate at $30°$ for 40 minutes. After the incubation, place the samples in ice, treat with 0.1 ml of 2 N HCl, and centrifuge at 35,000 g at $0°$ for 15 minutes ($V_3 = V_2 + $ HCl added). Pipette an aliquot sample of the supernatant (V_4) and neutralize with N KOH ($V_5 = V_4 + $ KOH added). The sample is ready for cycling but can be stored at $-70°$. Since the sample volumes are too small to be neutralized using a common pH meter, the amount of KOH necessary for neutralization is calculated from the preparation of tissue blank. For the preparation of the tissue blank, the remaining suspension (*ca.* 9.5 ml) left after pipetting 1 ml for hydrolysis of long-chain acyl-CoA is treated with one-tenth of its volume of 2 N HCl and centrifuged at 35,000 g at $0°$ for 15 minutes. Neutralize a measured volume of the supernatant with N KOH. The volume of KOH used is noted and from it the volume of KOH necessary to neutralize small volumes of samples can be calculated. The tissue blank is then ready for cycling.

Measurement of the CoA Formed. Pipette 2.5 ml of the CoA cycling mixture I into a series of 1 cm light path spectrophotometer cuvettes: 1 (zero reference); 5 (CoA standard curve) and the number of cuvettes necessary for the samples and the tissue blank of each sample. For the CoA standard curve, pipette 5, 10, 15, 20, and 25 μl ($Std_1, Std_2 \ldots Std_5$) of the diluted CoA standard and 50 μl of a blank of tissue and mix. If tissues with very different composition (such as protein content) are assayed, a CoA standard curve with the tissue blank of each type of tissue should be used. Add 50 μl of each tissue blank ($B_1, B_2, \ldots B_n$) or sample

(S_1, S_2, . . . S_n) (for calculations, defined as V_6) to the group of cuvettes and mix. Read the cuvettes at 340 nm. Cycling reactions are initiated at timed intervals by the addition of 25 μl of the enzyme mixture (CS and CAT). Leave the cuvettes in such a way that all of them have the same temperature. In our experiments we have used a 10-cell cuvette holder, adapted for readings with the spectrophotometer Zeiss PMQ II. The cuvettes are read 30, 45, 60 minutes, etc., after initiation of the cycling reaction. For calculations $\Delta E_1 = \Delta E$ of sample minus ΔE of respective tissue blank; $\Delta E_2 = \Delta E$ of standard whose reading is the closest to the sample reading minus ΔE of the blank used. Figure 1 presents

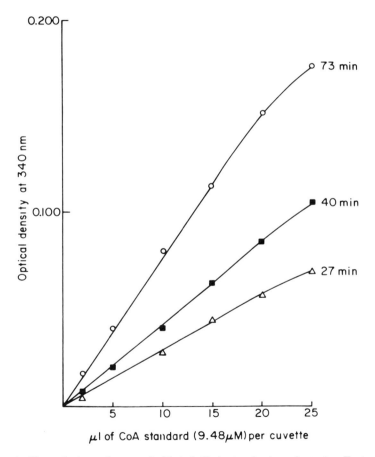

Fig. 1. Plot of the volumes of diluted CoA standard against the E_2 (spectrophotometry) obtained for each volume of standard. For conditions of experience see text. The time of readings are indicated.

a CoA standard curve used for determination of long-chain acyl-CoA in liver of fed rats. The concentration of long-chain acyl-CoA, expressed as micromoles per gram wet weight tissue, is equal to:

[diluted standard CoA] $\times$ volume standard (Std_1, Std_2, Std_3, Std_4, or Std_5)

$$\times \frac{\Delta E_1}{\Delta E_2 \times V_6} \times \frac{\text{total volume in cuvette (sample)}}{\text{total volume in cuvette (standard)}}$$

$$\times \frac{V_5}{V_4} \times \frac{V_3}{V_2} \times \frac{V_1}{\text{tissue wt g}}$$

For calculations all the volumes are expressed as milliliters.

Procedure (Fluorometric Assay)

For measurement of smaller concentrations of long-chain acyl-CoA the fluorometric method is preferred. The standard curve was obtained in the following way: 1 ml of cycling mixture II was pipetted into fluorometric tubes and 1, 5, 10, 15, 20, and 25 μl of the diluted CoA standard (4.25 μM) were added. After equilibration of the temperature, the tubes were read using a ratio fluorometer supplied by the Farrand Optical Company. Cycling reactions were initiated at timed intervals by the addition of 5 μl of the enzyme mixture (CS and CAT).

Remarks

The method allows the estimation of a wide range of amounts of long-chain acyl-CoA. There are three steps in the method that allow this flexibility:

1. In the first step by allowing various amounts of long-chain acyl-CoA solution or varying amounts of tissue to be used. In our experiments similar results were obtained when amounts of a pooled rat liver powder from 0.4 to 2 g were treated with 8 ml of perchloric acid.

2. In the second step by using various volumes of hydrolysis mixture for diluting the solution of long-chain acyl-CoA or making the suspension of a tissue. For hydrolyses, concentrations of long-chain acyl-CoA between 2 and 5 μM were used. Although concentrations higher than 5 μM cannot be used because of micelle formation, in our studies concentrations smaller than 2 μM were not assayed.

3. In the third step by using spectrophotometric or fluorometric methods.

[34] Measurement of Rates of Lipogenesis with Deuterated and Tritiated Water

By J. M. LOWENSTEIN, H. BRUNENGRABER, and M. WADKE

I. Introduction

Deuterium oxide was the first labeled precursor to be used for demonstrating the *de novo* synthesis of fatty acids.[1-3] It was shown that deuterium from D_2O is incorporated into long-chain fatty acids, but this finding was not developed into a quantitative method for measuring rates of fatty acid synthesis. This application has come into use only recently, except that the use of tritium oxide is now usually preferred to deuterium oxide.[4-10] In the interim, most assessments of the rate of fatty acid synthesis *in vivo,* in perfused organs, and in tissue slices, were made by measuring the amount of [14C]acetate incorporated into long-chain fatty acids. The unreliability of this method has been discussed elsewhere.[11] Briefly, in nonruminant mammals the pathway of fatty acid synthesis utilizes little or no free acetate, and the size and the rate of turnover

[1] D. Rittenberg and R. Schoenheimer, *J. Biol. Chem.* **121**, 235 (1937).
[2] H. Waelsch, W. M. Sperry, and V. A. Stoyanoff, *J. Biol. Chem.* **135**, 291 (1940).
[3] D. Stetten, Jr. and G. E. Boxer, *J. Biol. Chem.* **157**, 271 (1944).
[4] J. N. Fain, R. O. Scow, E. J. Urgoiti, and S. S. Chernick, *Endocrinology* **77**, 237 (1965).
[5] H. G. Windmueller and A. E. Spaeth, *J. Biol. Chem.* **241**, 2891 (1966).
[6] H. G. Windmueller and A. E. Spaeth, *Arch. Biochem. Biophys.* **122**, 362 (1967).
[7] R. L. Jungas, *Biochemistry* **7**, 3708 (1968).
[8] J. M. Lowenstein, *J. Biol. Chem.* **246**, 629 (1971).
[9] H. Brunengraber, M. Boutry, and J. M. Lowenstein, *J. Biol. Chem.* **248**, 2656 (1973).
[10] M. Wadke, H. Brunengraber, J. M. Lowenstein, J. J. Dolhun, and G. P. Arsenault, *Biochemistry* **12**, 2619 (1973).
[11] J. M. Lowenstein *in* "Handbook of Physiology-Endocrinology" (N. Freinkel and D. F. Steiner, eds.), Vol. 1, p. 415. American Physiological Society, Washington, D.C., 1972.

of the free acetate pool may vary depending on the nutritional and hormonal status of the animal. Similar problems are encountered when [¹⁴C]glucose is used as precursor; for example, in the perfused liver added glucose contributes from 3 to 33% of the carbon used for fatty acid synthesis. The fraction utilized depends on the concentration of added glucose and on other conditions.[9]

The use of hydrogen-labeled water does not suffer from such disadvantages. The conversion of carbohydrate precursors into fatty acids can be expected to yield a constant ratio of hydrogen incorporated to fatty acid synthesized. The conversion of certain other precursors, such as acetate and ethanol, into fatty acids can be expected to yield a somewhat different ratio of labeled hydrogen incorporated to fatty acid synthesized, but this can be allowed for by making appropriate corrections.

In studies of the control mechanisms that affect the rate of fatty acid synthesis it is necessary to measure rates before and after imposing a change. The variability in the rate of fatty acid synthesis in individual animals kept on carefully controlled, schedule-fed diets is such that five or six animals must be used for each rate measurement. Similar considerations apply to liver perfusion experiments. However, each individual animal or preparation may show a reproducible percentage change in response to an identical change in conditions. Therefore, if rates of fatty acid synthesis can be measured on the same animal or liver, before and after imposing a change, the number of experiments that needs to be performed is greatly reduced. The availability of D_2O and ^3H_2O makes possible two such consecutive measurements. The methodology for counting ^3H-labeled fatty acids is simple and straightforward. Mass spectrometry of fatty acid methyl ester is used for assays of deuterium content. If the deuterium content of the water used in the perfusion is high, the newly synthesized fatty acids contain sufficient deuterium so that they can be separated from the preexisting, unlabeled fatty acids by gas chromatography. In this case simple measurements of the peak areas suffice to obtain rates of fatty acid synthesis.

II. Tritium Method

The method to be described is applied to rat liver, but it can be used for most other tissues without modification. At the end of the experiment the liver is frozen rapidly by being pressed between blocks of aluminum which have been cooled in liquid nitrogen. The livers are stored in a liquid nitrogen refrigerator. Tissue treated in this manner can be used for metabolite analyses as well as measurements of fatty acid synthesis. The frozen liver is powered in the frozen state using a stainless steel pestle and mortar cooled with Dry Ice or liquid nitrogen. The resulting powder is weighed, and is dispersed in 19 volumes chloroform–methanol (2:1, by volume), and the suspension is shaken gently for 24 hours.

Alternatively, a portion of fresh liver obtained at the end of the experiment can be weighed and dropped directly into 19 ml chloroform–methanol (2:1, by volume) contained in a Sorvall Omni-Mixer. (This instrument is an overhead blender-homogenizer.) The amount of reagents are quoted for 1 g of fresh liver. The liver is then homogenized. Shaking of this suspension is not necessary.

In either case the chloroform–methanol extract is allowed to stand until the insoluble material has settled, and the clear supernatant is decanted. Alternatively, the extract can be filtered. It is then evaporated to dryness in a stream of nitrogen, and the residue is saponified in 2 ml 5 N NaOH at 80° for 3 hours. The solution is diluted with 5 ml of water and is extracted 3 times with 8 ml petroleum ether (boiling point range 60–80°). The pooled petroleum ether extracts are evaporated to dryness. The residue is dissolved in 5 ml acetone–ethanol (1:1, by volume), one drop of 10% acetic acid is added, and this is followed by 2.0 ml of 0.5% digitonin in 50% ethanol.[12] The resulting mixture is allowed to stand overnight, and the digitonide precipitate is harvested by centrifugation. The digitonide precipitate is washed with 6 ml of acetone–ether (1:2, by volume) followed by 6 ml of ether. The precipitate contains 3β-hydroxysterols. It is dissolved in 1 ml methanol, and may be counted in this form.

The aqueous residue from the alkaline extraction is acidified with 1.25 ml of 10 N H_2SO_4, and is then extracted 3 times with 8 ml of petroleum ether. The extract is evaporated at 40° in a stream of nitrogen. The residue contains the long-chain fatty acids. It is dissolved in a known volume of petroleum ether or other suitable organic solvent. An aliquot of this solution is evaporated and counted in toluene scintillation fluid. Other aliquots can be used for the preparation of methyl esters for gas chromatography and other procedures. If prolonged storage of the fatty acids is necessary, this is best achieved in petroleum ether solutions in tightly stoppered glass bottles, in an atmosphere of nitrogen. Counting efficiencies of all samples are determined by recounting each sample in the presence of internal standards of either [³H]toluene or [¹⁴C]toluene or both. Rates of fatty acid and 3β-hydroxysterol synthesis can be expressed as micromoles hydrogen from ³H₂O incorporated per gram dry weight per hour. These values are converted to micromoles of acetyl group incorporated

¹² W. M. Sperry and M. Webb, *J. Biol. Chem.* **187**, 97 (1950).

into fatty acids by multiplying by 1.15 (refs. 7 and 10) and to micromoles of acetyl group incorporated into 3β-hydroxysterols by multiplying by 1.31.[13]

The specific radioactivity of the ^3H_2O used in the experiment is determined by withdrawing a sample in the middle of the period during which ^3H_2O is used. The sample is diluted and counted using naphthalene–dioxane scintillation fluid.[14]

If it is desired to determine the radioactivity or the masses of the individual fatty acids, an aliquot of the petroleum ether extract containing the free fatty acids (20–25 mg of fatty acids) is evaporated to dryness under N_2 at 40°. The fatty acids are then esterified with 25 ml of methanol containing 1% by volume concentrated sulfuric acid by refluxing for 4 hours. The resulting solution is poured into an equal volume of water and the methyl esters are extracted with three 10 ml portions of petroleum ether (reagent grade, b.p. 40–60°). The combined petroleum ether extracts are washed with an aqueous solution of sodium bicarbonate and finally with water. The petroleum ether extract containing the methyl esters is then dried over anhydrous sodium sulfate for a minimum of 1 hour. The petroleum ether solution is evaporated under N_2 at 40°. The methyl esters of the fatty acids are dissolved in a small volume of petroleum ether and are stored under nitrogen.

The mixture of fatty acid methyl esters is separated on a gas chromatography apparatus equipped with a thermal conductivity detector.[15] Stainless steel columns (10 ft $\times$ $\frac{3}{16}$ in. i.d.), packed with 10% EGSS-X as the liquid phase on Gas-Chrom P (100–120 mesh) as the supporting phase, may be used. The columns are operated at 190–200° with helium gas at a flow rate of 40 ml/minute and an inlet pressure of 18 psi.[16] Identification of the peaks is achieved by comparing the retention time with that of standard fatty acid methyl esters.

Peaks corresponding to the fatty acids (16:0, 18:0, 18:1, etc.) are collected individually in U-shaped Pyrex glass tubes which are attached directly to the outlet of the heated exit port of the thermal conductivity detector. Collection of the peaks is started a short time before the peaks emerge from the column to avoid possible loss of polydeuterated peaks, which tend to elute ahead of nondeuterated peaks. However, under the operating conditions of our semipreparative columns, fractionation of deuterated from nondeuterated fatty acid methyl esters is not observed.

[13] H. Brunengraber, J. Sabine, M. Boutry, and J. M. Lowenstein, Arch. Biochem. Biophys. **150**, 392 (1972).

[14] G. A. Bray, Anal. Biochem. **1**, 273 (1960).

[15] R. G. Ackman, Vol. 14 [49].

[16] The detector, packing material, and other conditions are suitable but not unique (see also ref. 15).

(Such fractionation is observed on more efficient columns; see below.)
The collection tubes are packed loosely with Pyrex glass wool and may
be immersed in a mixture of Dry Ice and ethanol to facilitate sample
collection. Recovery is about 75%. The methyl esters are rinsed from
the tubes with 15 ml of petroleum ether. The solvent is evaporated to
dryness in a conical tube under N_2 at 30°. The residue is dissolved in
a known amount of solvent and is divided into three equal parts (A, B,
and C).

Portion A is used for mass spectrometric analyses as described below.
Portion B is used for radioactivity determinations at 12° in a liquid scintillation counter. The scintillating solution consists of 0.4% 2,5-diphenyl-
oxazole and 0.005% 1,4-bis[2(5-phenyloxazolyl)]-benzene in toluene.
Counting efficiencies are determined by the use of appropriate internal
standards. Portion C is used for the quantitative determination of the
fatty acid methyl ester. A known amount of internal standard (methyl
pentadecanoate) is added to the sample, the contents are mixed, and an
aliquot is analyzed on a gas chromatography apparatus equipped with
a flame ionization detector. Glass columns [6 ft × $\frac{1}{13}$ in. (2 mm) i.d.]
packed with 10% EGSS-X on Gas Chrom P (100–120 mesh) are operated
at 190°. Nitrogen is used as carrier gas at a flow rate of 40 ml/minute
and an inlet pressure of 25 psi. The area under each peak is determined
gravimetrically. (It was found to be directly proportional to the weight
of the methyl esters represented by the peak.) The amount in each sample
peak is calculated by comparing the area of the sample peaks (A_x)
with the area of the internal standard (A_{st}) as follows: $x = (A_x/A_{st}) \times$
(amount of internal standard). Specific radioactivity (dpm/μmole) is
calculated by dividing total disintegrations per minute determined on
sample B, by the amount of fatty acid methyl ester determined on
sample C.

III. Deuterium Incorporation into Fatty Acids
Measured by Mass Spectrometry

The methyl esters of long-chain fatty acids are obtained as described
in the preceding paragraph, and their deuterium content is determined
by mass spectrometry. The isotope ratios can be measured from the recorded mass spectra with a 12-in. Gerber variable scale, Model TP
007100B (The Gerber Scientific Instrument Co., Hartford, Conn.).

Originally, for each sample the isotope ratios were measured on ten
partial mass spectra and then averaged.[10] Subsequently, we have found
it adequate to measure two partial mass spectra. An outline of the calculation is described in this paragraph for a sample of methyl palmitate.
The same calculation is used for fatty acid methyl esters isolated from

experiments in which 10 or 100% D_2O is used. Typical isotope ratios obtained for authentic, unlabeled methyl palmitate follow: m/e 270, relative intensity 100.0; 271, 19.3; 272, 2.40; 273–275 inclusive, ≤ 0.2. The peaks for the unlabeled controls with m/e 273–275 are neglected in the calculations. For a perfusion experiment with 10% D_2O a typical average would be m/e 270 relative intensity 100.0; 271, 24.6; 272, 7.25; 273, 3.00; 274, 1.27; 275, 0.57. From these relative intensities the distribution in mole percent is calculated following an established procedure.[17] It affords the following answer when the numerical values for the 10% D_2O perfusion given above are used in the calculation: 88.97 mole-% unlabeled species, 4.71 mole-% monodeuterated species, 3.41 mole-% dideuterated species, 1.89 mole-% trideuterated species, 0.69 mole-% tetradeuterated species, and 0.33 mole-% pentadeuterated species. The percentage deuterium in various molecules is calculated by dividing the actual number of deuterium atoms in a molecule by the number of deuterium atoms in the hypothetical, fully deuterated molecule, times 100, and amounts to 0% D for undeuterated species, 2.94% D for monodeuterated species, 5.88% D for dideuterated species, 8.82% D for trideuterated species, 11.8% D for tetradeuterated species, and 14.7% D for pentadeuterated species. The percent deuterium·content of the sample is obtained by multiplying the percent deuterium in a species by the mole fraction of that species in the sample and summing the products obtained for all deuterated species. This amounts to 0.64% deuterium for the specific case discussed as an example.

To calculate the percentage of deuterium incorporated in 90–95% D_2O experiments as a result of chain elongation rather than *de novo* synthesis, the total percentage of deuterium incorporated is first calculated as described in the previous paragraph. In one case this amounted to 7.3% deuterium for methyl stearate. The sum of the contributions to this percentage caused by the presence of mono- through tetradeuterated species in the sample of fatty acid methyl ester is then calculated. (In all cases examined there was no contribution to the total percentage deuterium incorporated in the fatty acid methyl ester from species with more than four deuterium atoms and fewer than fourteen.) In the case under consideration the sum of the contributions of mono- through tetradeuterated species for methyl stearate amounted to 0.25%. This is divided by the total deuterium content of the sample. The resulting quotient times 100 gives the percent deuterium incorporated during the experiment as a result of chain elongation. It amounted to 3% for methyl stearate in the example under discussion.

[17] K. Biemann, Mass Spectrometry-Organic Chemical Applications, McGraw-Hill, New York, 1972, p. 204.

IV. Deuterium Incorporation into Fatty Acids Measured by Gas Chromatography

The average number of deuterium atoms incorporated into fatty acids synthesized by rat liver perfused in 100% D_2O has been calculated to be 22.3 and 24.9 for palmitate and stearate, respectively.[10] Compared to the unlabeled compounds, this represents an increase in mass of 8.5%. Gas chromatography linked to mass spectrometry showed that the deuterium-labeled fatty acids emerge on the leading edges of the individual peaks. Columns of relatively low efficiency were used in these experiments.[18] We concluded that more efficient columns may separate the newly synthesized, deuterated species from the preexisting, unlabeled fatty acids. This conclusion has been confirmed in practice.

Method

The column is a stainless steel capillary[15] [150 ft long × 0.01 in. i.d. (46 m long × 0.25 mm i.d.)]. The wall of the column is coated with EGSS-X.[19] We use a Perkin-Elmer model 900 gas chromatograph equipped with a flame ionization detector.[19] The column is operated isothermally between 150° and 180°. Injection port and detector temperatures are set at 250°. Nitrogen is used as carrier gas at an inlet pressure of 20 psi.

The sample capacity of the column is very small, and the sample is introduced indirectly using a sample injection splitter (Perkin-Elmer part number 009-0598) which is installed at the injection port of the instrument. A split ratio of 1:250 is employed routinely by using restrictor number 2 (a part supplied with the splitting system) in the sample splitter. Normally, 1 μl of a petroleum ether solution of the mixture of fatty acids is injected. Identification of the unlabeled peaks is achieved by comparing their retention times with those of standard fatty acid methyl esters. The deuterium-labeled fatty acid methyl esters emerge before the corresponding unlabeled compounds. A typical chromatogram obtained from a perfusion experiment in which a rat liver was perfused with 98% D_2O is shown in Fig. 1.

The area under each peak is determined either gravimetrically, or by measuring the peak height multiplied by peak width at half height, or by using an integrator. The area of the peak is proportional to the weight of the compound in the peak. To obtain the mole fraction, the

---

[18] We wish to thank Dr. G. Arsenault and Dr. J. J. Dolhun for running the combined gas chromatography–mass spectrometry analyses.

[19] Purchased from The Perkin-Elmer Corp., Norwalk, Conn.

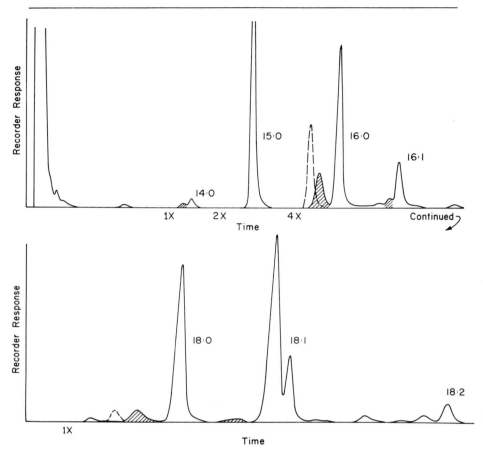

Fig. 1. Chromatogram of mixture of methyl esters of fatty acids from a rat liver that had been perfused with 98% D_2O for 2 hours. EGSS-X coated capillary column, 150 ft long × 0.01 in. i.d. Temperature 170°; nitrogen carrier gas at 20 psi; flame ionization detector. Attenuations as marked. The perfusion method is described elsewhere.[20] The deuterated peaks are cross-hatched; these peaks are absent when the perfusion is carried out in H_2O instead of D_2O. The broken line shows the elution position of perdeutero palmitate and perdeutero stearate.[21]

weight of the peak is divided by the molecular weight. (As has already been indicated, in rat liver perfused with 100% D_2O the average molecular weight of the methyl esters of newly synthesized palmitate and stearate is 292.7 and 322.4, respectively.)

[20] This volume [56].
[21] We wish to thank Dr. J. McCloskey for a generous gift of these compounds.[22]
[22] This volume [43].

V. Applicability

The ^3H_2O method can be applied in most situations both *in vivo* and *in vitro*. One of the advantages of the method is that it can be used in the presence of ^{14}C-labeled substances. The only constraint placed on such double-labeling experiments is that the ratio of ^3H and ^{14}C specific activities used in the experiment must be adjusted so that the two labels appear in the end products in ratios that make possible the counting of ^3H in the presence of ^{14}C.

The concentration of D_2O required for the D_2O method depends on the sensitivity and resolution of the analytical method and on the amount of fatty acid synthesized *de novo* in relation to the total amount present initially. When mass spectrometry is used (see Section III), D_2O concentrations of 5–10% are sufficient, except in cases where the rate of fatty acid synthesis is low. When high resolution gas chromatography is used (see Section IV), D_2O concentrations of 50–100% are necessary. This applies to conditions and columns with the resolution described in Fig. 1; if columns with a higher resolution are used, the D_2O concentration can be lowered accordingly.

Acknowledgment

This article is based on work supported by the John A. Hartford Foundation.

[35] Immunology of Prostaglandins[1]

By Rose Mary Gutierrez-Cernosek,[2] Lawrence Levine,[3] and Hilda Gjika

Prostaglandins (PG's) are widely distributed in mammalian tissues and biologic fluids. The numerous and diverse biologic activities ascribed to PG's[4,5] have generated interest in the development of techniques for their quantitation. Until recently, measurement of these compounds had

[1] Publication No. 949 from the Graduate Department of Biochemistry, Brandeis University. Supported in part by Research Grant No. 1C-10L from the American Cancer Society, Inc.
[2] Supported by Training Grant No. CA-05174 from the National Cancer Institute, N. I. H.
[3] American Cancer Society Professor of Biochemistry (Award No. PRP-21).
[4] J. W. Hinman, *Annu. Rev. Biochem.* **41**, 161 (1972).
[5] J. R. Weeks, *Annu. Rev. Pharmacol.* **12**, 317 (1972).

been difficult because of their low concentration in tissues (<1 $\mu g/g$) and in most biologic fluids (<1 $\mu g/ml$). In the past the quantitation of PG's has been carried out primarily using gas chromatography and mass spectrometry. However, the development of serologic techniques, which can be used for measuring PG's[6-10] or PG metabolites,[11-13] has provided investigators with a relatively simple and useful analytical tool whose applications are not limited to clinical investigations but can be extended to the study of the biosynthesis and breakdown of the pharmacologically active compounds.

Preparation of Antigen

Since PG's are low molecular weight compounds, they must be linked covalently to a carrier protein or polypeptide to render them immunogenic. Conjugation of the PG (in this instance, $PGF_{2\alpha}$) to poly(L-lysine) is achieved by the use of a water-soluble carbodiimide. The dehydration reaction results in the formation of an amide bond between the carboxyl groups of the PG's and ϵ-amino groups of poly(L-lysine). However, interaction between the hydroxyl and keto groups of the PG's cannot be ruled out.

Details of the coupling procedure are as follows. Ten milligrams of PG[14] are dissolved in 0.5 ml of dimethylformamide and are added dropwise while stirring to a solution of poly(L-lysine) hydrobromide (MW = 92,000, NEN, Boston, Mass.) which is at a concentration of 10 mg in 0.5 ml of water. If necessary, the pH of the poly(L-lysine) solution is adjusted to approximately 7 with 0.1 M NaOH prior to the addition of PG. Ten milligrams of 1-ethyl-3-(3-dimethylaminopropyl)carbodiimide (Ott Chemical Co., Muskegon, Mich.) are added to this mixture. If the reaction mixture becomes cloudy, one or more drops of dimethylformamide are added until the cloudiness disappears. Following overnight incubation at room temperature under nitrogen, the mixture is dialyzed

[6] L. Levine and H. Van Vunakis, *Biochem. Biophys. Res. Commun.* **41**, 1171 (1970).
[7] B. M. Jaffe, J. W. Smith, W. T. Newton, and C. W. Parker, *Science* **171**, 494 (1971).
[8] B. V. Caldwell, S. Burstein, W. A. Brock, and L. Speroff, *J. Clin. Endocrinol. Metab.* **33**, 171 (1971).
[9] K. T. Kirton, J. C. Cornette, and K. L. Barr, *Biochem. Biophys. Res. Commun.* **47**, 903 (1972).
[10] F. Dray, E. Maron, S. A. Tillson, and M. Sela, *Anal. Biochem.* **50**, 399 (1972).
[11] L. Levine and R. M. Gutierrez-Cernosek, *Prostaglandins* **2**, 281 (1972).
[12] E. Granström and B. Samuelsson, *FEBS Lett.* **26**, 211 (1972).
[13] L. Levine and R. M. Gutierrez-Cernosek, *Prostaglandins* **3**, 785 (1973).
[14] We thank Dr. J. E. Pike of the Upjohn Company, Kalamazoo, Mich., who supplied us with the PG's.

against 2 liters of 0.15 M NaCl, 0.005 M phosphate buffer, pH 7, for 3–5 days in the cold room with a minimum of five changes. When possible, trace amounts of the corresponding [³H]PG should be included in the coupling reactions to determine the extent of coupling of the PG to the polymer. In the case of PGE$_1$, 1 molecule of PGE$_1$ was incorporated per 10–20 lysine residues of the polymer.

Since poly(L-lysine) is a positively charged macromolecule, it must be complexed to a negatively charged macromolecule (i.e., succinylated hemocyanin) for successful production of antibodies. The hemocyanin is succinylated according to the method of Gounaris and Perlmann.[15] Four hundred milligrams of hemocyanin (from Giant Keyhole Limpet, Schwarz/Mann, Orangeburg, N.Y.) are dissolved in 10 ml of cold 0.1 M NaCl. The pH is adjusted to 8 with 2 M NaOH while stirring in a water-ice bath. Slowly add 600 mg of succinic anhydride and insure that a pH of 8–9 is maintained. Dialyze against 0.15 M NaCl, 0.005 M phosphate buffer, pH 7. Lyophilize, dissolve in distilled water, and redialyze against distilled water. Following lyophilization, the material is dissolved in water at a concentration of 5 mg/ml.

The poly(L-lysine) conjugate is complexed with the succinylated hemocyanin by adding the latter dropwise to a solution of the conjugate (approximately 200–500 μg of succinylated hemocyanin are added per 1 mg of conjugate). Prior to immunization, the complex is emulsified with an equal volume of complete Freund's adjuvant (Difco Laboratories, Detroit, Mich.). One milligram of the prepared antigen is sufficient for injection of two rabbits or monkeys.

Immunization

Antibodies against certain PG's and PG metabolites have been induced in rabbits,[6–8] goats,[9,10] monkeys,[11] and guinea pigs.[16] The following route and schedule has been used for injection of rabbits. The antigenic conjugate (about 1 mg) is injected into the hind toe pads and muscles. The dose is repeated a week later. One week after the second injection and each succeeding week for 2 weeks the rabbits are bled by the ear. Every 4 weeks, the animals receive a booster injection intramuscularly. Monkeys are injected subcutaneously or intramuscularly and blood is withdrawn from the femoral vein 7–10 days after immunization. A booster injection is administered intramuscularly 3–4 weeks after the initial injection and the animals are bled 7, 9, and 11 days afterward. The monkeys can be boosted again after 1 month.

[15] A. D. Gounaris and G. E. Perlmann, *J. Biol. Chem.* **242**, 2739 (1967).
[16] L. Levine, *Pharmacol. Rev.* **25**, 293 (1973).

Radioimmunoassay

The commercial availability of [³H]PG's has made it possible to employ the radioimmunoassays for the analysis of the antisera. However, other methods which have been used are microcomplement fixation[6] and "chemically modified bacteriophage" techniques.[10] In general, radioimmunoassays are set up as follows. Diluted rabbit antiserum (0.1 ml) is incubated with 0.1 ml of [³H]PG (New England Nuclear, Boston, Mass.)[17] containing 10,000–12,000 cpm in the presence or absence of increasing concentrations of unlabeled PG at 37° for 1 hour. The total volume of the reaction mixture is 0.3 or 0.4 ml. The Tris-gelatin buffer used for adjusting the volume and as a diluent consists of 0.01 M Tris, 0.14 M NaCl, 5×10^{-4} M MgSO$_4$, 1.5×10^{-4} M CaCl$_2$, pH 7.5, and 0.1% gelatin. To determine the extent of nonspecific binding of the [³H]PG to serum proteins a set of tubes which contain 0.1 ml of diluted serum obtained from the rabbit prior to immunization in place of the antiserum are included in the assay. After the 1-hour incubation period 0.1 ml of carrier rabbit γ-globulin (serum obtained from the rabbit prior to immunization at a 1:100 dilution) is added to tubes containing <1 μl of antiserum followed by addition of 0.1 ml of undiluted goat anti-rabbit γ-globulin (previously calibrated to be in antibody excess with respect to the rabbit γ-globulin in the reaction mixture).[18] The second antigen–antibody system is used to separate the antibody-bound hapten from the free hapten. After incubation overnight at 4°, the antibody-bound [³H]PG is collected as a precipitate by centrifugation at 1000 g for 30 minutes. The precipitate is dissolved in 0.2 ml of 0.1 N NaOH and counted in a modified Bray's solution.

Other procedures that have been used for separating the free PG from the antibody-bound PG are (1) nitrocellulose membrane filtration,[19] (2) dextran-coated charcoal adsorption,[8] and (3) ammonium sulfate precipitation.[7]

Specificities of Antibodies

The use of a radioimmunoassay for quantitating PG's depends primarily on the ability of the antibodies to differentiate among the various types (A, E, F, and B, based on the kind of functional groups present

[17] [³H]PGB₁ is prepared by heating a solution of [³H]PGE₁ whose pH has been adjusted to 12 with 1 N NaOH for 5 minutes at 100°. After cooling in an ice bath 1 N HCl is added to adjust the pH to 7.5.

[18] If monkey antiserum is used, carrier monkey γ-globulin and rabbit anti-monkey γ-globulin are added.

[19] H. Gershman, E. Powers, L. Levine, and H. Van Vunakis, *Prostaglandins* 1, 407 (1972).

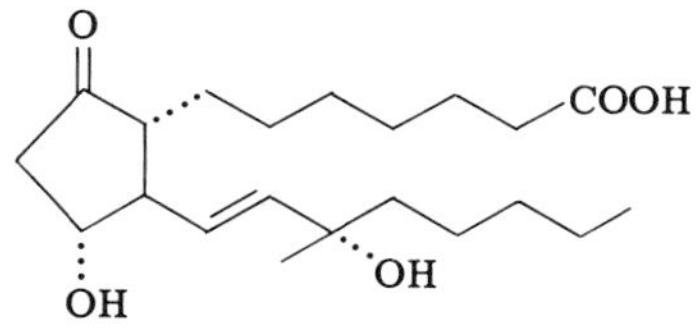

PGE_1

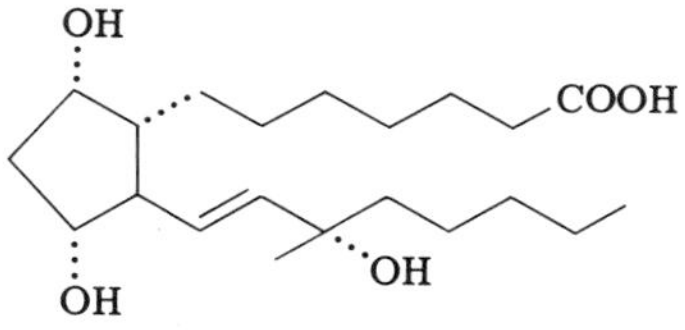

$PGF_{1\alpha}$

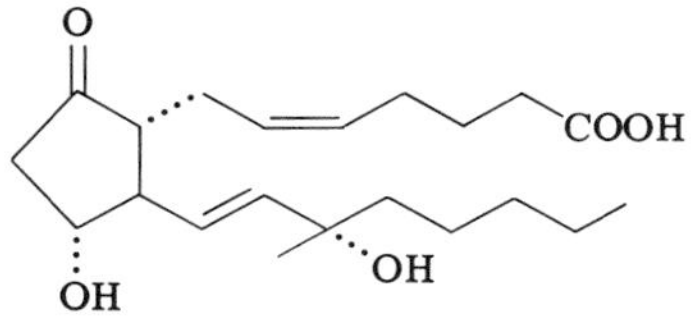

PGE_2

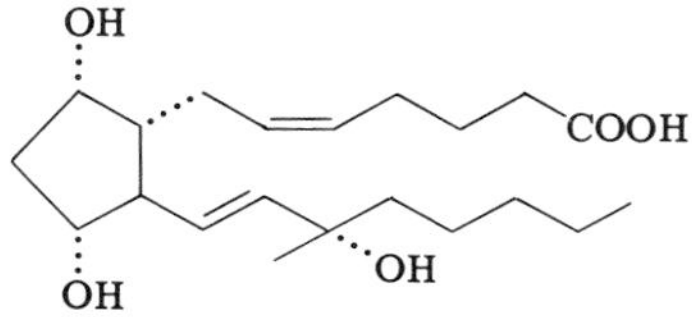

$PGF_{2\alpha}$

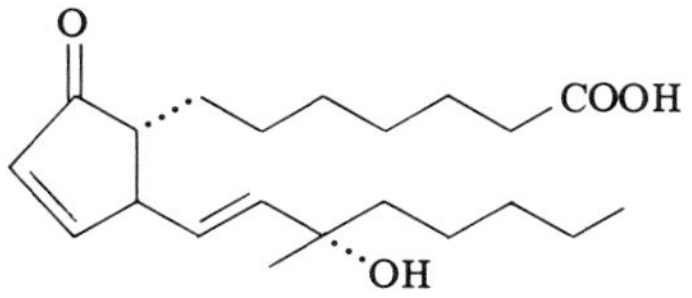

PGA_1

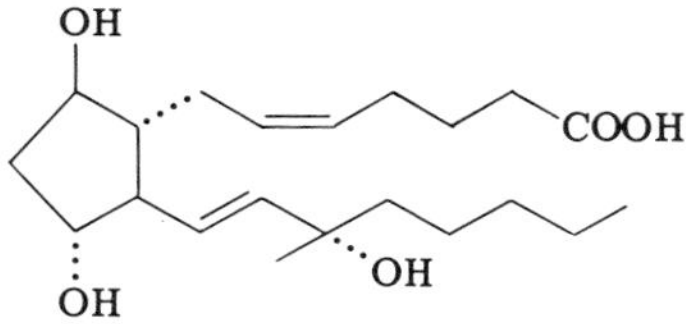

$PGF_{2\beta}$

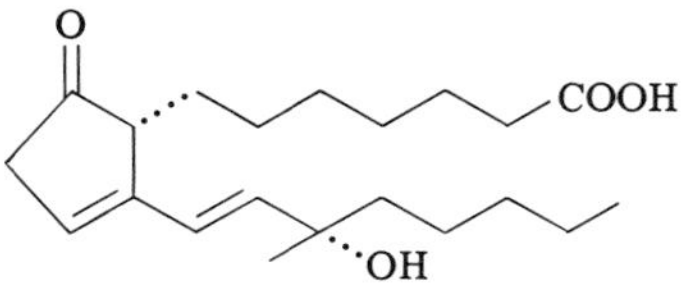

PGC_1

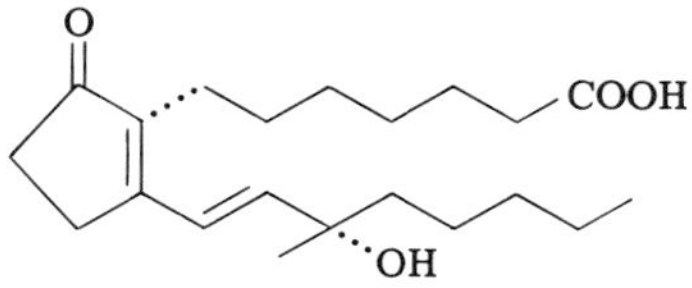

PGB_1

FIG. 1. Structural formulas of some PG's.

on the cyclopentane ring) and classes ($PGF_{1\alpha}$, $PGF_{2\alpha}$, etc., based on the degree of unsaturation of the aliphatic side chains) of PG's. The chemical structures of some of the major PG's are shown in Fig. 1.

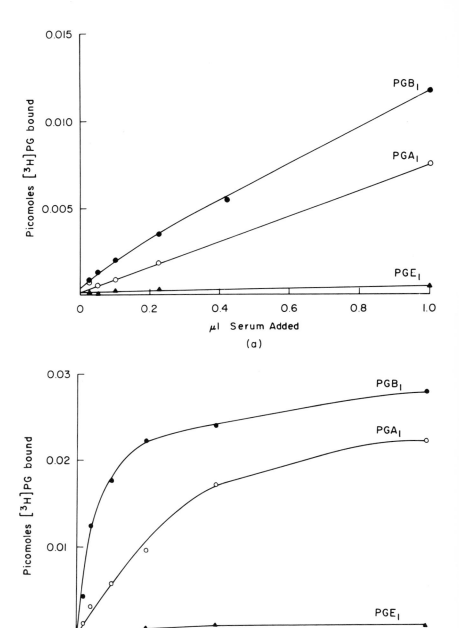

292

Figure 2 illustrates the binding capacity of antisera from a rabbit obtained during the early course of immunization with PGE_1 conjugate and after it had received several booster injections. Prior to immunization, there was very little binding of [³H]PG to diluted serum (1% or less). However, after two injections, the binding of the serum increased significantly (Fig. 2a); the binding increased approximately 60-fold over the control when 1 μl of antiserum was added to the assay system. Even at this early stage of immunization, the antiserum exhibited preferential binding to [³H]PGB_1. After the booster injections, the specificity and binding capacity of the antiserum was further enhanced (Fig. 2b); 0.25 μl of the antiserum yielded more than a 100-fold increase in binding. The most effective binding was obtained with [³H]PGB_1, followed by [³H]PGA_1. Binding to [³H]PGE_1 was very low. The increse in antibody titer and in specificity as immunization progressed indicates that the proteins binding the PG's are antibodies. Since PG's are hydrophobic and can bind nonspecifically to proteins, it is essential to demonstrate that the binding proteins are indeed antibodies. Fractionation of rabbit anti-PGE_1 on Sephadex G-200 indicated that the anti-PG activity is present in the 7 S immunoglobulin fraction of the serum proteins.[20]

Antibodies produced as a result of immunization of rabbits, guinea pigs, and monkeys with PGE or PGA conjugate are all directed toward PGB. The immunogens are probably converted enzymatically to PGB conjugates by dehydrases and isomerases present in tissues and/or serum.[21-23] Figure 3 and Table I illustrate the serologic specificity of an antiserum obtained from a rabbit immunized with a PGE_1 conjugate. Although PGA_1 (9-keto-prosta-10,13-dienoic acid) differs from PGB_1 (9-keto-prosta-8(12), 13-dienoic acid) only by the location of the double bond in the cyclopentane ring, anti-PGB_1 can readily distinguish between these two PG's. Modification of the cyclopentane ring as in PGE_1 (which has a hydroxyl group at C-11) or $PGF_{1\alpha}$ (which has hydroxyl groups

[20] L. Levine, R. M. Gutierrez-Cernosek, H. Polet, and H. Gershman, in "Third Conference on Prostaglandins in Fertility Control" (S. Bergström, K. Green, and B. Samuelsson, eds.), p. 38. Karolinska Institutet, Stockholm, 1972.

[21] E. Horton, R. Jones, C. Thompson, and N. Poyser, Ann. N.Y. Acad. Sci. **180**, 351 (1971).

[22] H. Polet and L. Levine, Biochem. Biophys. Res. Commun. **45**, 1169 (1971).

[23] R. L. Jones and S. Cammock, Advan. Biosci. **9**, 61 (1973).

FIG. 2. Binding of [³H]PG's by the serum of a rabbit after (a) a short course of immunization with PGE_1 conjugate and (b) six subsequent injections of the same rabbit. Each reaction mixture contained either [³H]PGE_1 (0.16 pmole), [³H]PGA_1 (0.153 pmole), or [³H]PGB_1 (0.16 pmole), the indicated volume of anti-PGE_1, and sufficient Tris-gelatin buffer to adjust the total volume to 0.3 ml.

TABLE I

DECREASE IN FREE ENERGIES OF BINDING OF PG'S TO ANTI-PGB₁

Prostaglandin	Nanograms required for 50% inhibition	$\Delta(\Delta F^\circ)$[a] (kcal/mole)
B_1	0.080	—
B_2	0.260	−0.43
A_1	2.3	−1.77
A_2	5	−2.25
E_1	860	−5.42
E_2	1500	−5.76
$F_{1\alpha}$	>2000	> −5.94
$F_{2\alpha}$	>2000	> −5.94

[a] $\Delta(\Delta F^\circ) = -RT \ln$ 50% inhibition (heterologous)/50% inhibition (homologous).

at C-9 and C-11) dramatically decreases the level of inhibition. PGB₁ is also a better inhibitor than PGB₂ which contains a second double bond at the C-5 to C-6 position. There is a threefold difference between the concentration of homologous hapten required for 50% inhibition and that of PGB₂. Thus, this antiserum can distinguish the different types and classes of PG's. Furthermore, anti-PGB₁ is not only highly specific but also exhibits great sensitivity. It can detect as little as 20 pg of PGB₁.

Another example of the specific antibodies elicited in rabbits by injection with a PG conjugate is the antiserum to PGF₂α. Anti-PGF₂α binds best with the homologous hapten and poorly with [³H]PGF₁α (Fig. 4). The serologic specificity of PGF₂α anti-PGF₂α reaction is shown in Fig. 5 and Table II. The homologous hapten is the most effective inhibitor.

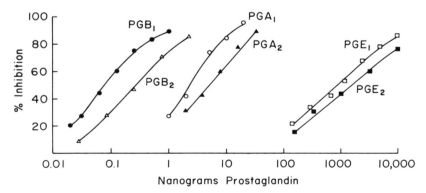

Fig. 3. Inhibition of [³H]PGB₁ anti-PGE₁ binding by prostaglandins. Each reaction mixture contained 0.18 pmole of [³H]PGB₁, 0.1 ml of a 1:10,000 dilution of antiserum, unlabeled PG's, Tris-gelatin buffer, in a final volume of 0.3 ml.

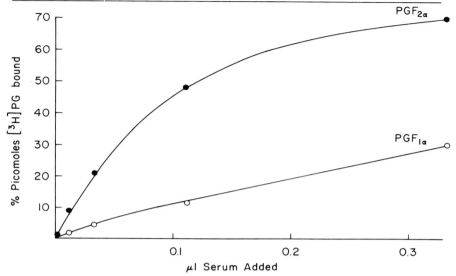

Fig. 4. Binding of [³H]PGF$_{2\alpha}$ and [³H]PGF$_{1\alpha}$ by rabbit anti-PGF$_{2\alpha}$. Each reaction mixture contained either 0.8 pmole of [³H]PGF$_{2\alpha}$ or 0.1 pmole of [³H]PGF$_{1\alpha}$, the indicated volume of antiserum, Tris-gelatin buffer, in a final volume of 0.3 ml.

Whereas 2.5 ng of PGF$_{2\alpha}$ produce 50% inhibition, 55 ng of PGF$_{1\alpha}$ are required to yield the same amount of inhibition. Other PG's or PG metabolites cross react very poorly. Their concentration must exceed

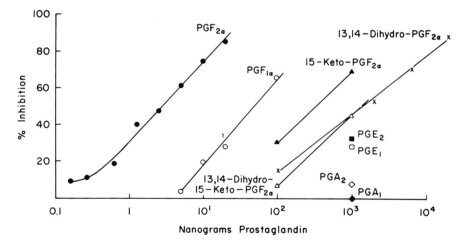

Fig. 5. Inhibition of PGF$_{2\alpha}$ anti-PGF$_{2\alpha}$ binding. Each reaction mixture contained 0.8 pmole of [³H]PGF$_{2\alpha}$, 0.1 ml of a 1:1000 dilution of antiserum, unlabeled PG's, Tris-gelatin buffer, in a final volume of 0.3 ml.

TABLE II

DECREASE IN FREE ENERGIES OF BINDING OF PG'S TO ANTI-PGF$_{2\alpha}$

Prostaglandin	Nanograms required for 50% inhibition	$\Delta(\Delta F^\circ)$[a] (kcal/mole)
F$_{2\alpha}$	2.5	—
F$_{1\alpha}$	55	−1.8
15-Keto-F$_{2\alpha}$	330	−2.8
F$_{2\beta}$	600	−3.2
(+)-7-Oxa-F$_{1\alpha}$	1 × 10^3	−3.5
(−)-7-Oxa-F$_{1\alpha}$	>1 × 10^{3b}	
13,14-Dihydro-15-keto-F$_{2\alpha}$	1.3 × 10^3	−3.6
13,14-Dihydro-F$_{2\alpha}$	1.4 × 10^3	−3.7
E$_2$	3 × 10^3	−4.1
E$_1$	5 × 10^3	−4.4
A$_2$	10 × 10^3	−4.8
A$_1$	>10 × 10^3	>−4.8

[a] $\Delta(\Delta F^\circ) = -RT \ln$ 50% inhibition (heterologous)/50% inhibition (homologous).
[b] 11% with 1 μg.

100-fold over that of PGF$_{2\alpha}$ in order to yield 50% inhibition. Thus, antiserum to PGF$_{2\alpha}$ is both sensitive and highly specific since it can measure pg concentrations of PGF$_{2\alpha}$ and distinguishes the different types and classes of PG's and initial PGF$_{2\alpha}$ metabolites. On the basis of the cross reactions of the various PG's with anti-PGF$_{2\alpha}$ one can conclude that the antibodies recognize (1) modifications in the cyclopentane ring, i.e., substitution of the C-9 hydroxyl group by a keto group (PGE) or a change in the spatial orientation of the C-9 hydroxyl group (PGF$_{2\beta}$); (2) the C-5 to C-6 double bond (absent in PGF$_{1\alpha}$); (3) the C-15 hydroxyl group (oxidation of PGF$_{2\alpha}$ to 15-keto-PGF$_{2\alpha}$ dramatically decreases its inhibitory capacity); and (4) the C-13 to C-14 double bond (absent in 13,14-dihydro-PGF$_{2\alpha}$ or 13,14-dihydro-15-keto-PGF$_{2\alpha}$).

Antibodies to PGF$_{1\alpha}$ exhibit comparable sensitivity but are less specific. As shown in Fig. 6, PGF$_{2\alpha}$ cross reacts appreciably with the [^{3}H]PGF$_{1\alpha}$ anti-PGF$_{1\alpha}$ binding reaction. Therefore, recognition of the C-5 to C-6 double bond is not as striking as that by antiserum to PGF$_{2\alpha}$. However, anti-PGF$_{1\alpha}$ can distinguish other types of PG's and initial PGF$_{2\alpha}$ metabolites.

Recently, several investigators obtained antibodies to certain PG metabolites.[11−13,16] Since the pharmacologically active PG's are readily inactivated by metabolic transformations,[24,25] antibodies to the metabo-

[24] E. Granström, Eur. J. Biochem. 27, 462 (1972).
[25] B. Samuelsson, Fed. Proc., Fed. Amer. Soc. Exp. Biol. 31, 1442 (1972).

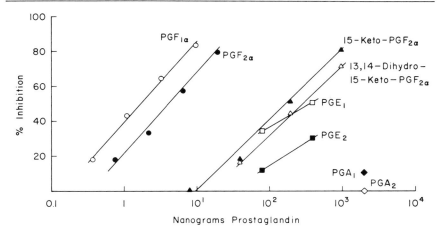

FIG. 6. Inhibition of $PGF_{1\alpha}$ anti-$PGF_{1\alpha}$ binding. Each reaction mixture contained 0.1 pmole of [³H]$PGF_{1\alpha}$, 0.1 ml of a 1:100 dilution of antiserum, unlabeled PG's, Tris-gelatin buffer, in a final volume of 0.3 ml.

lites are proving to be extremely useful; for example, measurement of urinary metabolites of the respective PG provides a good reflection of the production of that PG *in vivo*. With these antibodies one can also study activities of metabolic enzymes in different tissues or biologic fluids more easily.

Prostaglandins, unlike other fatty acids, have been shown to be good immunogens when coupled to a carrier. Thus, it is possible to obtain antisera which are specific and sensitive. In instances where significant cross reactions can occur, unknown samples can be assayed with several antisera. Measurement of $PGF_{2\alpha}$ presents no problem, since anti-$PGF_{2\alpha}$ does not cross react strongly with $PGF_{1\alpha}$ and very poorly with other PG types and metabolites tested. In measuring $PGF_{2\alpha}$ in serum, possible interference by $PGF_{1\alpha}$ has been ruled out on the basis of the results obtained when samples were assayed with anti-$PGF_{2\alpha}$ and anti-$PGF_{1\alpha}$.[26]

Determination of PGE, PGA, or PGB with our antiserum is more complex; for example, if a sample reacts with the anti-PGE_1 system, it can be treated with NaOH or methanolic KOH to convert PGE and/or PGA to PGB and assayed before and after alkaline treatment. If the degree of competitive inhibition does not change following alkaline treatment, then only PGB_1 and/or PGB_2 is present. A change in the amount of competitive inhibition would indicate a mixture, perhaps of all three types of PG's. The presence of PGE can be established by the order of

[26] R. M. Gutierrez-Cernosek, L. M. Morrill, and L. Levine, *Prostaglandins* 1, 71 (1972).

magnitude of change in competitive inhibition occurring after conversion to PGB (Fig. 3). Since PGE's are poor inhibitors of this antibody system, the sample is assayed at a dilution at which PGE's do not cross react. To determine if PGE_1 and/or PGE_2 are present, the sample can be treated with $NaBH_4$ which quantitatively converts PGE_2 to $PGF_{2\alpha}$ and $PGF_{2\beta}$ (approximately at a 50:50 ratio) or PGE_1 to $PGF_{1\alpha}$ and $PGF_{1\beta}$. Then the sample can be analyzed with anti-$PGF_{2\alpha}$ and anti-$PGF_{1\alpha}$ before and after $NaBH_4$ treatment to establish if PGE_1 and/or PGE_2 are present. The presence of $PGF_{2\beta}$ and $PGF_{1\beta}$ does not present a problem since they cross-react poorly with the respective antisera.

If the concentration of PGE's derived from the assay of a $NaBH_4$-treated sample is known, this amount is subtracted from the concentration of PGB obtained after alkaline treatment with anti-PGE_1 system. The remaining value of PGB represents that amount present in the sample plus PGB converted from PGA. With this particular antiserum it is necessary to separate the individual PGA's and PGB's by reverse phase chromatography[27] before assaying in order to accurately quantitate these compounds. The availability of a specific antiserum to PGA would eliminate the need for such a separation. The successful production of highly specific antibodies to PGA_1 has been reported.[28]

The number of laboratories employing immunologic techniques to determine PG concentrations in biologic systems has been increasing rapidly. The radioimmunoassay has allowed investigators who do not have access to a gas chromatograph mass spectrometer to measure PG's by a relatively simple technique.

[27] M. Hamberg and B. Samuelsson, *J. Biol. Chem.* **241**, 257 (1966).
[28] W. A. Stylos and B. Rivetz, *Prostaglandins* **2**, 103 (1972).

[36] Enzymatic Determination of Microquantities of Acetate[1]

By MARVIN SCHULMAN and HARLAND G. WOOD

A variety of methods for determination of acetate in biologic fluids have been previously described,[2-4] but none of these provides a rapid, specific means of determining microquantities of acetate. The method[5]

[1] This work was supported in part by USPHS NIH Grant 11839.
[2] I. A. Rose, M. Grunberg-Manogo, S. R. Carey, and S. Ochoa, *J. Biol. Chem.* **211**, 737 (1954).
[3] H. Wierzlicka, A. B. Legacki, and J. Pawekiewicz, *Acta Polon.* **14** (1967).
[4] H. U. Bergmeyer and H. Moellering, *Biochem. Z.* **344**, 167 (1966).
[5] M. Schulman and H. G. Wood, *Anal. Biochem.* **39**, 505 (1971).

to be described is rapid, quite specific, and about twenty times more sensitive than those previously described.

Principle. The spectrophotometric determination of acetate involves the conversion of acetate to citrate via the following reactions.

$$\text{Succinyl-CoA} + \text{acetate} \xrightarrow{\text{CoA-transferase}} \text{acetyl-CoA} + \text{succinate} \quad (1)$$

$$\text{Malate} + \text{NAD}^+ \xrightarrow{\text{malate dehydrogenase}} \text{oxalacetate} + \text{NADH} + \text{H}^+ \quad (2)$$

$$\text{Oxalacetate} + \text{acetyl-CoA} \xrightarrow{\text{citrate synthase}} \text{citrate} + \text{CoA} \quad (3)$$

$$\text{Succinyl-CoA} + \text{acetate} + \text{malate} + \text{NAD}^+ \rightleftarrows \text{citrate}$$
$$+ \text{ succinate} + \text{CoA} + \text{NADH} + \text{H}^+ \quad (4)$$

Acetate is converted to acetyl-CoA by CoA-transferase from *Propionibacterium shermanii* [reaction (1)]. The enzyme has an apparent K_m of 7×10^{-3} M for acetate.[6] When coupled with malate dehydrogenase [reaction (2)] and citrate synthase [reaction (3)], the conversion of acetate to acetyl-CoA and then to citrate is virtually complete. The amount of acetate is equal to the NADH formed and is measured spectrophotometrically at 340 nm.

Reagents

Tris-Cl (Trizma base) (Sigma) 1.0 M pH 8.0
Sodium DL-malate 0.3 M
β-DPN (NAD) (Sigma) 0.02 M
DPNH (NADH) 0.005 M
Malate dehydrogenase (5 mg of protein per milliliter, 1100 IU/mg of protein, Boehringer)
Citrate synthase (2 mg of protein per milliliter, 70 IU/mg of protein, Boehringer)
Succinyl-CoA prepared from recrystallized succinic anhydride (J. L. Baker) and CoA (P. L. Biochemicals) by the method of Simon and Shemin[7] as modified by Swick and Wood.[8] An approximate determination of succinyl-CoA is made by the hydroxamate method of Lipman and Tuttle[9] at the time of preparation and a final quantitative determination by spectrophotometric assay with CoA-transferase as described by Schulman and Wood.[10]

[6] S. H. G. Allen, R. W. Kellermeyer, R. L. Stjernholm, and H. G. Wood, *J. Bacteriol.* **87**, 171 (1964).
[7] E. J. Simon and D. Shemin, *J. Amer. Chem. Soc.* **75**, 2520 (1953).
[8] R. W. Swick and H. G. Wood, *Proc. Nat. Acad. Sci. U.S.* **42**, 28 (1960).
[9] F. Lipmann and L. C. Tuttle, *J. Biol. Chem.* **159**, 21 (1945).
[10] M. Schulman and H. G. Wood, [28].

Procedure

The determination is conducted in 0.5 ml cuvettes with a 10-mm light path and 2 mm width at 25°.

Reactants

(1) 0.1 ml of Mixture 1 which contains in micromoles per milliliter: Tris-Cl pH 8.0, 25; sodium malate, 7.5; NAD, 20; and NADH, 0.25.

(2) 0.01 ml of Mixture 2 which contains in IU per milliliter: malate dehydrogenase 220 and citrate synthase 35 in 0.1 M phosphate buffer, pH 6.8.

(3) 0.01 ml of ~0.014 M succinyl-CoA

(4) Acetate to be determined (4–55 nmoles)

(5) Sufficient water to bring the volume to 0.24 ml

(6) 0.01 ml containing 0.7 unit of CoA-transferase in 0.1 M phosphate buffer, pH 6.8 containing 0.1 mM EDTA.

All but the CoA-transferase is added and the initial absorbance is recorded. A blank cuvette containing all the reactants except the acetate sample is prepared for each assay. The initial optical density (OD) of the cuvettes is approximately 0.6 as a result of the NADH in the mixtures.[11] The spectrophotometer is adjusted so that the recorder output of the initial absorbance of the cuvettes reads near zero so the full scale of the recorder is available for measurement of absorbance changes. The CoA-transferase is added and the changes in OD recorded with time. The time course of the reaction is shown in Fig. 1. The time required for complete reaction is usually 15–20 minutes.

Calculations. The amount of acetate in the sample is calculated as follows using the data of Fig. 1:

$$\Delta E_B = E_B - E_{B0} = 0.255\text{--}0.065 = 0.190$$
$$\Delta E_R = E_R - E_{R0} = 0.920\text{--}0.040 = 0.880$$
$$\Delta E_{acetate} = \Delta E_R - \Delta E_B = 0.880\text{--}0.190 = 0.690$$
$$\mu\text{moles acetate} = \Delta E_{acetate}/E = \frac{0.690}{24.9} = 0.0277$$

[11] D. J. Pearson [*Biochem. J.* **95,** 23c (1965)] pointed out that this type of determination is subject to substantial error because of differences in the concentration of NADH at the beginning and end of the assay which result from a change in the relative concentrations of oxalacetate and malate at the beginning and end of the reaction. This error has been reduced by adjusting the concentrations of malate and NAD to appropriate levels and by adding NADH to the initial reaction mixture.[5]

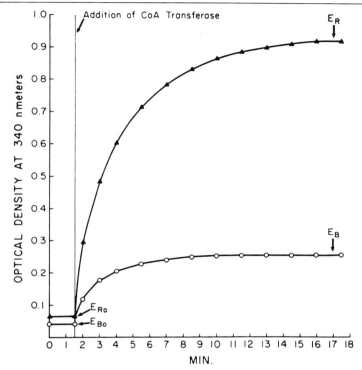

Fig. 1. Time course of the spectrophotometric determination of acetate. ($\triangle$) Assay cuvette and ($\bigcirc$) blank cuvette which does not contain acetate. E_{Bo} and E_{Ro} denote the absorbance of the blank and assay mixture before initiation of reaction by addition of CoA-transferase. E_B and E_R denote absorbances after completion of the reaction. Thirty nanomoles of sodium acetate were determined with 0.46 IU CoA-transferase as described in the text. Reproduced from Schulman and Wood,[5] with permission of the publisher.

E is the absorbance of 1 μmole NADH at 340 nm in 0.25 ml and a 1-cm light path.

Accuracy and Sensitivity. The accuracy and sensitivity of the spectrophotometric determination has been determined with standard acetate solutions by Schulman and Wood.[5] The method is applicable over a wide range of acetate concentrations (5–55 nmoles) and the accuracy generally exceeds 90%.

Specificity. The method is quite specific for acetate. Formate, propionate, lactate, pyruvate, butyrate, α-ketobutyrate, succinate, glucose, and methyltetrahydrofolate are without effect. Acetate has been assayed in a complex bacterial growth medium containing yeast extract, tryptone, and tomato juice.[5]

[37] Enzymatic Determination of Acetate

By ROBERT W. GUYNN and RICHARD L. VEECH

Several enzymatic acetate assays have been previously described.[1-8] However, those methods which have been used to determine acetate in tissue extracts have required the distillation or diffusion of the acetate from the extract first before measurement. Besides being time consuming the distillation or diffusion procedures carry with them the very real risk of breakdown of tissue acetyl compounds to acetate during the relatively prolonged, harsh conditions used. Since significant quantities of labile acetyl compounds exist in animal tissues, the breakdown of such compounds is a potential source of error in acetate measurement. In contrast, the following spectrophotometric and fluorimetric procedures for acetate measurement avoid the diffusion step and can be performed directly in the tissue extract.

Assay Methods

Principle. The method is based upon the enzyme sequence:

$$\text{Acetate}^- + \text{ATP}^{4-} + \text{CoA} = \text{AMP}^{2-} + \text{PP}_i^{4-} + \text{acetyl-CoA} + \text{H}^+$$
$$\text{Acetyl-CoA} + \text{oxalacetate}^{2-} = \text{citrate}^{3-} + \text{CoA} + \text{H}^+$$
$$\text{Malate}^{2-} + \text{NAD}^+ = \text{oxalacetate}^{2-} + \text{NADH} + \text{H}^+$$

$$\text{Net: acetate}^- + \text{ATP}^{4-} + \text{malate}^{2-} + \text{NAD}^+ = \text{AMP}^{2-} + \text{PP}_i^{4-} + \text{citrate}^{3-} + 3\text{H}^+ + \text{NADH}$$

NADH production is followed spectrophotometrically or fluorimetrically.

The assay depends upon the conversion of acetate to acetyl-CoA and the measurement of the acetyl-CoA by a coupled system of malate dehydrogenase and citrate synthase. The assay is specific since citrate synthase is specific for acetyl-CoA. A potential lack of stoichiometry in

[1] R. W. von Korff, *J. Biol. Chem.* **210,** 539 (1954).
[2] I. A. Rose, M. Grunberg-Manago, S. R. Korey, and S. Ochoa, *J. Biol. Chem.* **211,** 737 (1954).
[3] I. A. Rose, Vol. 1 [97].
[4] M. Soodak, Vol. 3 [48].
[5] F. Lundquist, U. Fugmann, and H. Rasmussen, *Biochem. J.* **80,** 393 (1961).
[6] H. U. Bergmeyer and H. Moellering, *Biochem. Z.* **344,** 167 (1966).
[7] M. Schulman and H. G. Wood, *Anal. Biochem.* **39,** 505 (1971).
[8] F. J. Ballard, O. H. Filsell, and I. G. Jarrette, *Biochem. J.* **126,** 193 (1972).

this system accounts for the lack of stoichiometry in the previously reported use of this system as an acetate assay.[1] This lack of stoichiometry has been discussed a number of times.[6,9-11] From a practical point of view, however, the inclusion of NADH in the reagent cocktail for the spectrophotometric method produces a stoichiometric assay as judged by the excellent agreement (within 2%) between the assay of standard potassium acetate by the procedure outlined below with both the value expected from the dry weight of the salt and with the value obtained by the acetate assay of Ballard et al.[8]

Enzymes and Reagents

1. Enzymes. Purified beef heart mitochondria acetyl-CoA synthetase (EC 6.2.1.1) is prepared by the method of Webster[12] (specific activity 35 units[13]/mg of protein at 38°C). Malate dehydrogenase (EC 1.1.1.37) from beef heart (1100 units/mg of protein) and citrate synthase (EC 4.1.3.7) from pig heart (110 units/mg protein) are commercially available. The enzymes used for the assay should have no detectable activity of acetyl-CoA deacylase, NADH oxidase, oxalacetate decarboxylase, carnitine acetyltransferase (EC 2.3.1.7), or ATP citrate lyase (EC 4.1.3.8).

2. Reagents. Stable for at least a month unless otherwise noted.

(a) Stored at 0–4°

Tris-HCl, 0.2 M, pH 8.0 containing KCl, 0.1 M

Magnesium chloride, 0.1 M

NADH, 5.0 mM in 0.3% (w/v) $KHCO_3$, stable for 5–7 days

Perchloric acid, 1.2 M

(b) Stored below —15°C

Adenosine 5'-triphosphate, 27 mM, Tris, potassium or sodium salt

NAD^+, free acid, 24 mM

Malate, potassium salt, 0.1 M

Coenzyme A, free acid, 1.2 mM containing mercaptoethanol, 3.0 mM, prepare fresh weekly

Bovine serum albumin, 100 mg/ml

Tissue Preparation. Tissue samples obtained with rapid freezing tech-

---

[9] W. Buckel and H. Eggerer, *Biochem. Z.* **343**, 29 (1965).

[10] D. J. Pearson, *Biochem. J.* **95**, 23c (1965).

[11] R. W. Guynn, H. J. Gelberg, and R. L. Veech, *J. Biol. Chem.* **248**, 6957 (1973).

[12] L. T. Webster, Jr., *J. Biol. Chem.* **240**, 4158 (1965).

[13] One unit of enzyme is the amount which will utilize 1 μmole of a substrate per minute under optimal conditions.

niques are used.[14,15] The frozen tissue is powdered under liquid nitrogen, homogenized in 4 volumes of ice-cold perchloric acid (1.2 M) and centrifuged at 35,000 g for 15 minutes at 0°. The resulting supernatant is carefully neutralized to pH 6.5 with KOH taking care not to allow the extracts to become basic. After standing at 0° for half an hour, the potassium perchlorate precipitate is packed by centrifugation and the resulting supernatant is used for the assay.

Instruments. The spectrophotometric assay is designed for either a dual or single beam instrument holding cuvettes of 1 cm light path. The fluorimetric procedure is designed for an instrument capable of handling 1.0 ml volumes. For the fluorimetric assay quinine sulfate in sulfuric acid is used as the reference standard[16]; however, for each assay a standard curve of potassium acetate is run simultaneously.

Spectrophotometric Procedure

Perchloric acid extracts of tissue containing 0.008–0.1 μmole acetate per milliliter can be satisfactorily measured with the spectrophotometric procedure. This range corresponds to tissue concentrations of approximately 0.04–0.5 μmole/g wet weight tissue. Considerably higher concentrations can be measured although the time required for completion will be prolonged making it more convenient in such a situation to dilute the extract.

A reagent cocktail is prepared containing: 90 mM Tris pH 8.0, 2.7 mM ATP, 0.075 mM NADH, 0.6 mM NAD$^+$, 0.12 mM CoA, 0.6 mM potassium malate, 4.5 mM MgCl$_2$, 0.3 mM mercaptoethanol, 45 mM KCl, 50 mg/ml dialyzed bovine serum albumin, 12 units/ml malate dehydrogenase, and 0.5 unit/ml citrate synthase. In the complete reagent cocktail acetate contamination from the reagents is expected to be less than 3 nmoles/ml.

Two parts cocktail are combined with one part tissue extract in a cuvette of 1 cm light path. Usually a final volume of 2.1 ml has been found convenient; however, the lower limit for the final volume is restricted only by the size of the cuvette (of 1 cm light path) which can be used with a particular spectrophotometer. The blank cuvette contains cocktail plus water. After an initial reading at 340 nm, 0.075 unit of acetyl-CoA synthase is added per milliliter of the complete reaction mix-

[14] A. Wollenberger, O. Ristau, and G. Schoffa, *Pflüegers Arch. Gesamte Physiol. Menschen Tiere* **270**, 399 (1960).

[15] R. L. Veech, R. L. Harris, D. Veloso, and E. H. Veech, *J. Neurochem.* **20**, 183 (1973).

[16] O. H. Lowry and J. V. Passonneau, "A Flexible System of Enzymatic Analysis," pp. 7–8. Academic Press, New York, 1972.

ture, and the reaction is followed to completion (15–20 minutes typically). The reaction is stoichiometric; therefore, the concentration acetate in the tissue extract can be calculated directly from the ΔOD and the extinction coefficient of NADH ($\epsilon = 6.22 \times 10^6$ cm²/mole).[17]

Concentration of acetate in tissue extract (in μmoles/ml)

$$= \frac{(\Delta OD)}{6.22 \text{ (fraction of final volume due to extract)}}$$

When measuring acetate in tissue extracts the ΔOD is relatively low, often in the range of 0.030–0.120. However, in spite of the relatively low ΔOD, high precision and accuracy are possible. On triplicate samplings of the same frozen tissue powder the standard error of the mean of the three determinations with the spectrophotometric assay is less than 1%. Assays of standard acetate solutions show similar precision. The precision of the method seems not to be as much a function of the assay as it is a function of the care with which the tissue is prepared. Using standard acetate solutions the linearity of the assay has been confirmed between optical density changes of 0.030 and 0.400. The accuracy of the method is supported by the finding that the values obtained for the standard acetate solutions agree within 2% of the values expected from the dry weight of the standard potassium acetate and the values obtained by the method of Ballard et al.[8]

Fluorimetric Procedure

For low concentrations of acetate in a tissue extract (0.003–0.012 μmole/ml, corresponding to tissue concentrations or approximately 0.015–0.05 μmole/g wet weight) a fluorimetric procedure is preferred. The reagent cocktail is modified to contain 66 mM Tris-HCl pH 8.0, 33 mM KCl, 0.050 mM potassium malate, 0.025 mM NAD⁺, 3.0 mM MgCl₂, 0.2 mM mercaptoethanol, 0.04 mM CoA, 1.0 mM ATP, 0.1 mg/ml bovine serum albumin, 3.5 units/ml malate dehydrogenase and 0.15 unit/ml citrate synthase. One milliliter cocktail is combined with 100 μl extract (equivalent to approximately 15 mg wet weight tissue). After a base line reading 0.03 unit of acetyl-CoA synthetase is added and the reaction is followed to completion (40–60 minutes). A standard curve of potassium acetate is run simultaneously and used to calculate the acetate concentrations in the extracts. Often brain extracts assayed fluorimetrically can be used directly after neutralization. However, treatment of the neutralized extracts with Florisil[18] (50 mg/ml extract) followed by centrifuga-

[17] B. L. Horecker and A. Kornberg, J. Biol. Chem. **175**, 385 (1948).
[18] 100–200 mesh, Fisher Scientific Co., Fairlawn, N.J.

tion at 0° and 15,000 g for 20 minutes reduces the background fluorescence of an extract without a detectable loss of acetate. If tissues containing a large amount of flavins are to be analyzed fluorimetrically, the preliminary treatment of the extract will be necessary.

The fluorimetric procedure is usually less precise than the spectrophotometric assay. The standard error of the mean of triplicate tissue samples is 3–5%. The higher standard error of the mean in the fluorimetric procedure reflects the relatively greater significance of the acetate contaminations of the reagent cocktail at these very low concentrations of acetate.

Variations of the Procedures

The actual concentrations of the activating ions Tris and K^+ do not seem to be critical although the total is best kept above 80 mM.[19,20] On the other hand, Na^+ concentrations above 5 mM should be avoided even in the presence of optimal K^+ or Tris. High Na^+ concentrations will slow the assay making the end point less distinct and producing the potential hazard of an underestimation of the amount of acetate present. Such a problem can be readily detected by using internal standards. The assay procedures as described have been developed for use in such tissues as liver and brain; however, the procedures should have general applicability to the perchloric acid extract of any tissue. Difficulty might be anticipated, however, with tissues with a high sodium content (e.g., blood). If Na^+ inhibition is considered important in such cases, the concentration of acetyl-CoA synthetase can be increased as needed. The extracts of most solid tissues, however, contain sufficiently low concentrations of Na^+ that even $Na_2ATP \cdot 2\frac{1}{2}H_2O$ can be used in the reagent cocktail without serious interference.

The concentrations of malate and NAD^+ should not be significantly increased in the spectrophotometric assay since the stoichiometry of the assay may be compromised. The usual concentrations of malate and NAD^+ in tissue extracts do not interfere.

Recoveries

Known amounts of potassium acetate equivalent to tissue concentrations are added to frozen tissue powder. The tissue is then extracted with perchloric acid and assayed by the standard procedure. The endogenous

[19] R. W. von Korff, *J. Biol. Chem.* **203**, 265 (1953).
[20] F. Campagnari and L. T. Webster, Jr., *J. Biol. Chem.* **238**, 1628 (1963).

concentration of acetate is determined separately on another sample of the same tissue powder. The recovery of the small quantities of acetate added to the frozen tissue is 95–98%. If acetyl-CoA, acetylcarnitine, or N-acetyl-aspartate is added to the frozen tissue in amounts equal to those normally found in tissue, they do not interfere with the determination of the acetate. There is no evidence for breakdown of these compounds to acetate during the extraction procedure or the assay.

Discussion

The relatively low K_m of acetyl-CoA synthetase for acetate makes the direct measurement of acetate in tissue extracts practical (K_m = 0.8 mM).[12] Normal acetate values vary considerably among tissues and dietary states; for example, the following values have been found in rat liver: starved 48 hours, 75–130 nmoles/g wet weight; fed *ad libitum*, 125–200 nmoles/g wet weight; meal-fed (3 hours daily), 175–275 nmoles/g wet weight; starved and then refed a no-fat high sucrose diet, 20–30 nmoles/g wet weight. In rat brain lower values have been found and there is less obvious variation with diet: starved 48 hours or fed *ad libitum*, 20–30 nmoles/g wet weight.[21]

[21] For further details of normal tissue values, recoveries, and the procedure see R. W. Guynn and R. L. Veech *Anal. Biochem.* (1974) (in press).

[38] Determination of Acetate by Gas–Liquid Chromatography

By M. WADKE and J. M. LOWENSTEIN

Analysis of aqueous solutions of short-chain free fatty acids (C_2 to C_6) by gas–liquid chromatography has been reported by numerous workers.[1-6] The quantitative determination of acetate in biologic fluids is subject to a number of errors. Potentially serious sources of error arise from the tailing of peaks and the formation of ghost peaks. Tailing of peaks is easily recognized and has been discussed elsewhere.[7,8] The ghost-

[1] R. G. Ackman, *J. Chromatogr. Sci.* **10**, 560 (1972).

[2] D. M. Ottenstein and D. A. Bartley, *J. Chromatogr. Sci.* **9**, 673 (1971).

[3] D. A. M. Geddes and M. N. Gilmour, *J. Chromatogr. Sci.* **8**, 394 (1970).

[4] V. Mahadevan and L. Stenroos, *Anal. Chem.* **39**, 1652 (1967).

[5] R. G. Ackman and J. C. Sipos, *J. Chromatogr.* **13**, 337 (1964).

[6] R. Turner and M. N. Gilmour, *Anal. Biochem.* **13**, 552 (1965).

[7] O. E. Schupp, *in* "Gas Chromatography" *Tech. Org. Chem.* (E. S. Perry and A. Weissberger, eds.), p. 13. Wiley (Interscience), New York (1968).

[8] D. M. Ottenstein, *Advan. Chromatogr.* **3**, 141 (1968).

ing phenomenon is most simply visualized by the appearance of a peak upon the injection of pure solvent, in this case water, after running an unknown mixture or a standard through the column. The ghost peak often has a retention time similar to that of the unknown or standard solute used in the previous injection. The size of the ghost peak decreases with successive injections of solvent. Numerous treatments have been recommended to suppress ghosting and tailing. Among these the addition of formic acid vapor to the carrier gas or to the sample itself appears to be simple and effective,[1,3,5] although it does not eliminate ghosting entirely.

Method

The following procedure is used for the quantitative determination of acetate in biologic fluids. Propionic acid is used as internal standard. This is the internal standard of choice after it has been determined that the unknown samples do not themselves contain propionate. If propionate is present in the samples to be assayed, some other acid, such as iso-butyric acid, can be used as internal standard. Trichloroacetic acid is used to deproteinize the biologic fluid.

Trichloroacetic acid itself produces several major peaks on the chromatogram, but these emerge well before acetic acid and do not interfere with the acetate analysis. In addition, trichloroacetic acid produces a very small peak at a position on the chromatogram which coincides exactly with the position of acetic acid. When we first observed this peak we thought that it may be caused by a very small contamination of the reagent grade trichloroacetic acid used by us with acetic acid. However, the relative size of this unknown peak remained unchanged after the trichloroacetic acid had been recrystallized twice from heptane. The unknown peak is proportional to the concentration of trichloroacetic acid in the solution injected into the column. If this impurity were acetic acid it would represent 0.001% of the trichloroacetic acid in the solution that is injected onto the column. This amount of contamination is so small that it can be neglected when assaying concentrations of acetate greater than 0.7 mM. However, it must be taken into account when acetate concentrations of the order of 0.1 to 0.5 mM are being measured. Since the area of the impurity peak is constant, the area of the impurity peak observed in blanks is simply substracted from the area of the peak in the test sample. This impurity is of such a minor nature that we have not attempted to establish conditions under which it could be separated from acetic acid.

The use of perchloric acid to deproteinize the biologic fluid was also investigated. This reagent itself also yields a number of peaks, including a relatively large one that runs in the region of acetate and another large one that runs in the region of propionate. We have avoided the use of perchloric acid to deproteinize our biologic samples, because of the occurrence of the two unknown peaks.

Apparatus

A Perkin-Elmer gas chromatography apparatus, model 900, equipped with flame ionization detector is used by us, but many other suitable instruments are commercially available. The glass columns [1.8 m (6 ft) $\times$ 2 mm i.d.] are packed with 10% SP-1200 plus 1% H_3PO_4 on Chromosorb W (80–100 mesh).[2] This material is available from Supelco Inc., Bellefonte, Pennsylvania 16823. The columns are plugged at each end with silanized glass wool. The injection port is glass-lined. Nitrogen is used as carrier gas. The columns are conditioned overnight at 200° with carrier gas flowing at a rate of 40 ml/minute. Before use the columns are injected several times with 2–3 μl of water to clear out any extraneous material. The columns are operated isothermally at 105°. The injection port and detector are operated at a temperature of 200°.

Procedure

The following procedure is for a sample volume of 0.5 ml. It can easily be adapted to other volumes.

1. 0.5 ml of biologic fluid or standard (e.g., 10 mM sodium acetate in 4% serum albumin).
2. Add 0.1 ml of internal standard (25 mM sodium propionate or sodium isobutyrate).
3. Add 0.1 ml of 20% trichloracetic acid.
4. Mix well for 1–2 minutes.
5. Centrifuge at 2000 rpm for 10 minutes.
6. Inject 0.4 μl of supernatant into the injection port of the gas chromatography apparatus.
7. After the last peak has emerged (usually the internal standard), inject 2 μl of water. If ghosting is observed, repeat injection of water. If ghost peaks of acetate or standard are observed, their area is added to those of the original acetate and standard peaks, respectively.

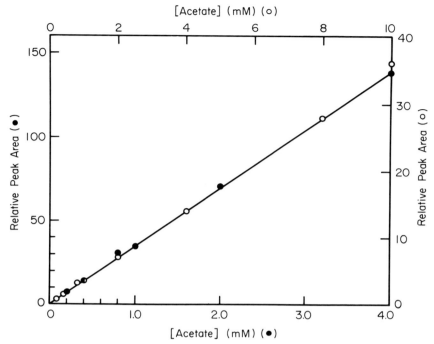

FIG. 1. Calibration curve for acetate in the range 0.2–10 mM. Various known amounts of acetate were added to samples of a perfusate which had circulated through a liver for 2 hours, and which contained no measurable amounts of acetate. The area of the small peak in the acetate region which occurs in blanks containing no perfusate and no acetate, and which results from a breakdown product of or an impurity in trichloracetatic acid, has been subtracted. It is equivalent to 0.1 mM acetate.

Calculations and Results

Areas under each peak are determined gravimetrically or by some other suitable method. A ratio of unknown to internal standard is obtained, and the concentration of acetate in the sample is calculated from the ratio. A calibration curve is shown in Fig. 1. Chromatograms of serial samples of a rat liver perfusion are shown in Fig. 2.

Note

The procedure calls for the injection of 0.4 μl of aqueous sample. The response of the flame ionization detector to acetic acid may vary with the amount of water injected.[9] Although this need not be of consequence

[9] R. B. H. Wills, J. Chromatogr. Sci. **10**, 582 (1972).

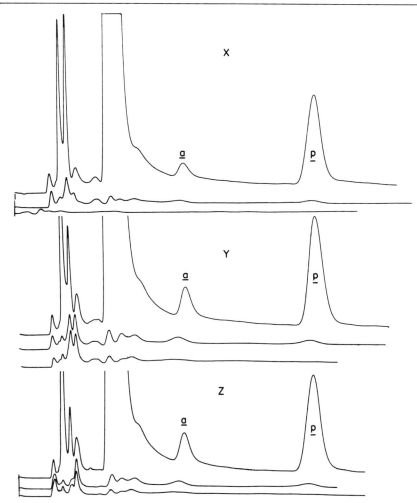

Fɪɢ. 2. Chromatograms showing the appearance of acetate during a rat liver perfusion. Acetate began to appear in response to the addition of 15 mM ethanol 60 minutes after the start of the perfusion. Samples X, Y, and Z were taken 15, 45, and 60 minutes, respectively, after the addition of ethanol. Each acetate peak is indicated by a. Propionate was added as internal standard; each propionate peak is indicated by p.

when internal standards are used with each sample, we prefer to use a constant volume for a given set of unknowns, because for a constant volume of water injected the peak area is a linear function of the acetic acid concentration.

[39] Measurement of Malonyl Coenzyme A

By Robert W. Guynn and Richard L. Veech

Malonyl-CoA is a key intermediate in fatty acid synthesis. It is generated in the cytoplasm by acetyl-CoA carboxylase (EC 6.4.1.2)[1] and utilized both in fatty acid synthesis *de novo* (by fatty acid synthase multienzyme complex)[2-4] and in chain lengthening processes (by the microsomal chain lengthening system).[5] Therefore, since the steady state concentration of malonyl-CoA will be a function of the rate of production by acetyl-CoA carboxylase and the rate of utilization by the fat synthesizing systems, measurement of this metabolite is a useful tool in the study of the control of fatty acid synthesis *in vivo*. Measurement of malonyl-CoA has proved to be especially useful when combined with an estimate of the rate of fatty acid synthesis *in vivo*.[6]

Principle

Malonyl-CoA is measured enzymatically in tissue extracts with purified rat liver fatty acid synthase which, using NADPH and acetyl-CoA or another suitable primer, converts malonyl-CoA to fatty acids. The oxidation of NADPH is followed spectrophotometrically at 340 nm, 2 molecules of NADPH being oxidized for each molecule of malonyl-CoA consumed.

$$\text{Acetyl-CoA} + n \text{ malonyl-CoA}^- + 2n \text{ NADPH} + 3n \text{ H}^+ \rightarrow$$
$$\text{CH}_3(\text{CH}_2\text{CH}_2)_{n-1}\text{CH}_2\text{CH}_2\text{COOH} + 2n \text{ NADP}^+$$
$$+ n \text{ CO}_2 + n \text{ H}_2\text{O} + (n + 1) \text{ CoA}$$

Tissue Preparation

The liver is freeze-clamped and powdered under liquid nitrogen. A 1:5 extract of the powdered liver is prepared in 0.85 M HClO$_4$ and centrifuged at 35,000 g for 15 minutes at 0°. The resulting supernatant is carefully neutralized with KOH, taking care not to exceed pH 7.0, and the

[1] S. J. Wakil, E. B. Titchener, and D. M. Gibson, *Biochim. Biophys. Acta* **29**, 225 (1958).
[2] S. J. Wakil and J. Ganguly, *J. Amer. Chem. Soc.* **81**, 2597 (1959).
[3] R. O. Brady, R. M. Bradley, and E. G. Trams, *J. Biol. Chem.* **235**, 3093 (1960).
[4] F. Lynen, *Biochem. J.* **102**, 381 (1967).
[5] S. Abraham, I. L. Chaikoff, and W. M. Bortz, *Nature (London)* **192**, 1287 (1961).
[6] R. W. Guynn, D. Veloso, and R. L. Veech, *J. Biol. Chem.* **247**, 7325 (1972).

samples are allowed to stand in ice for half an hour. The KClO₄ precipitate is packed by centrifugation and the supernatant is used for the assay. Perchloric acid extracts of tissue, although very inhibitory of fatty acid synthase, are convenient to use and permit extraction of malonyl-CoA without significant hydrolysis. The recovery of malonyl-CoA added to tissue was found to be 96%.

Assay Reagents

   All prepared in deionized water
   1 M potassium phosphate buffer, pH 7.0
   0.125 M EDTA, pH 7.0
   2 mM acetyl-CoA (free acid or trilithium salt)
   Activating solution: 1 M potassium phosphate buffer containing 20 mM dithiothreitol and 2 mM EDTA, pH 7.0

Enzyme

Rat liver fatty acid synthase is prepared by the method of Nepokroeff *et al.*[7] The dialyzed enzyme is usually satisfactory for the assay after the second ammonium sulfate fractionation. The preparation must be checked for NADP⁺ or NADPH-using enzymes. The enzyme was assayed spectrophotometrically at 25°.[7,8] Occasionally batches of the enzyme will contain significant glucose-6-phosphate dehydrogenase activity which interferes with the malonyl-CoA assay since it reduces NADP⁺ formed during the fatty acid synthase reaction. In spite of the different molecular weights of glucose-6-phosphate dehydrogenase and fatty acid synthase they are not separated by Sephadex chromatography. Glucose-6-phosphate dehydrogenase contamination may be minimized by following the procedure of Nepokroeff *et al.* exactly. Special attention should be given to the feeding schedule and size of rat.[7]

Before use the enzyme is activated by incubating for 30 minutes at 38° with an equal volume of activating solution.

Assay Procedure

For convenience a reagent cocktail is prepared. An amount sufficient for five assays consists of 5 mg of NADPH dissolved in 2.0 ml 1 M potassium phosphate buffer, pH 7.0; 2.0 ml 0.125 M EDTA, pH 7.0; and 0.2 ml 2 mM acetyl-CoA. Although useful in activating the enzyme,

---

[7] C. M. Nepokroeff, M. R. Lakshmann, and J. W. Porter, this volume [6].
[8] A unit of enzyme is defined as that amount of protein necessary to catalyze the utilization of 1 μmole of malonyl-CoA per minute.

dithiothreitol is not added to the cuvette since it has been found to accelerate a nonenzymatic oxidation of the NADPH in the presence of the tissue extract. High concentrations of NADPH seem to be necessary to insure a stoichiometric assay. Lowering the NADPH concentration by more than half results in a less reliable or nonlinear assay.

Duplicate cuvettes of 1 cm light path are prepared, each containing 2.0 ml of neutralized tissue extract and 400 μl of the reagent cocktail. The cuvettes are placed in a dual beam spectrophotometer and recorder of sufficient stability at high optical densities to measure relatively small changes (0.010–0.080 OD unit). With full-scale expansion to 0.1 OD units a base line recording is taken. Since the cuvettes should be identical this base line will be flat. A significant slope to the base line which cannot be related to drift in the recording instrument implies that the cuvettes are not identical. The only satisfactory course in such a situation is to reprepare the cuvettes. If the base line is suitable, $2–5 \times 10^{-2}$ unit[8] of fatty acid synthase is added to one cuvette and an equal volume of diluted (1:2) activating solution to the other cuvette. The optical density change is followed by continuous recording until completion (approximately 30 minutes). At the end of the reaction, second additions of enzyme and diluted activating solution are made to determine the correction for absorption by the enzyme. Standard malonyl-CoA can then be added to both cuvettes to check the stoichiometry of the assay. In all cases once the enzyme has been added mixing of the cuvettes should be done gently since the enzyme is readily inactivated at interfaces.

Calculations

The reaction is stoichiometric; 2 NADPH molecules are oxidized for each malonyl-CoA molecule incorporated. Therefore, the concentration of malonyl-CoA in the neutralized extract can be calculated directly from the change in optical density, the final volume in the cuvette, the extract sample volume, and the extinction coefficient of NADPH ($\epsilon = 6.22 \times 10^6$ cm²/mole)[9]:

μmoles malonyl-CoA/ml neutral extract

$$= \left(\frac{\Delta OD}{2}\right) \frac{(2.4 \text{ ml final volume})}{(2.0 \text{ ml sample volume}) \, 6.22}$$

Sensitivity

Under the conditions of the assay the reaction is linear with 2–40 nmoles of added malonyl-CoA per cuvette.

[9] B. L. Horecker and A. Kornberg, *J. Biol. Chem.* **175**, 385 (1948).

Standardization

The standard malonyl-CoA is assayed in pure solution by the same procedure. In the absence of $KClO_4$ the reaction is much more rapid than in tissue extracts. The value obtained has been found to agree well with the amount of CoA released by complete hydrolysis of the standard under mildly aklaline conditions in the presence of dithiothreitol.

Normal Values

The concentration of malonyl-CoA in liver tissue varies considerably under different conditions. The following values have been found in rat liver: Rats starved 24 to 48 hours or fed a high fat diet have 0.004–0.006 μmole/g wet weight, rats fed *ad libitum* on a diet containing 4% fat have 0.013–0.015 μmole/g wet weight, and rats meal-fed 3 hours a day on the same diet have 0.020–0.038 μmole/g wet weight immediately after eating.

Discussion

The assay so described has proved very useful in liver, but coenzyme A derivatives in other tissues are present in lower concentrations; for example, the concentration of malonyl-CoA in brain is below the sensitivity of the assay as described.

[40] Determination of the Positional Distribution of Fatty Acids in Glycerolipids

By H. Brockerhoff

Introduction

Glycerol has one secondary and two primary hydroxyl groups. The two primary groups are stereochemically distinguishable in spite of the fact that there is a plane of symmetry in the glycerol molecule, because esterification of one or the other primary group, e.g., with phosphoric acid, leads to optically active derivatives, and so does the esterification of each hydroxyl with different fatty acids. In the recommended and generally used nomenclature, glycerol is viewed in a Fischer projection with its secondary hydroxyl to the left (in L configuration), and the three carbons are numbered from top to bottom. Glycerol so numbered is called *sn*-glycerol, for stereospecifically numbered.

The analytical problem is the distinction between the three groups, in our case, the determination of the fatty acid composition in each of the positions of a glyceride or phosphoglyceride.

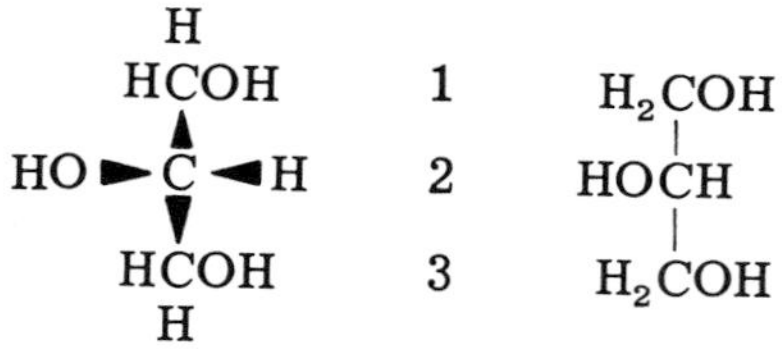

$$
\begin{array}{ccc}
\text{H} & & \\
\text{HCOH} & 1 & \text{H}_2\text{COH} \\
\text{HO}\!\blacktriangleright\!\text{C}\!\blacktriangleleft\!\text{H} & 2 & \text{HOCH} \\
\text{HCOH} & 3 & \text{H}_2\text{COH} \\
\text{H} & &
\end{array}
$$

Fatty acid distribution analyses are most often performed on natural glyceride mixtures consisting of numerous "species" of glycerides, i.e., glycerides that have the same fatty acid composition, and each of these species can be a mixture of "positional" isomers, two in the case of phospholipids and six in the case of a "triacid" triglyceride. (In the abbreviated formulas, A, B, and C are fatty acids, P is the phosphate group, and X a substituent.)

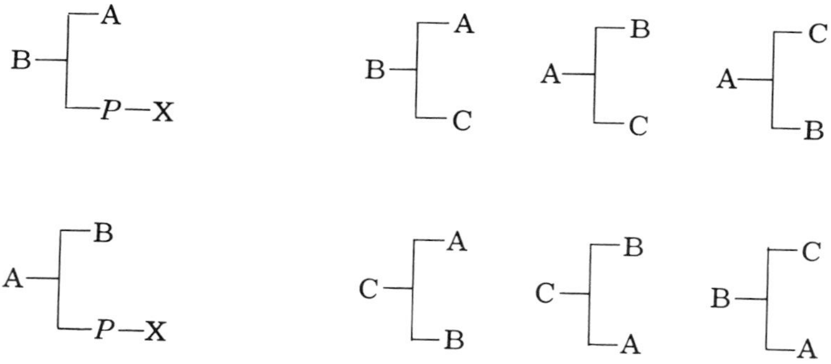

Analyses on mixtures of species yield information only on the statistical positional distribution of fatty acids; this is sufficient for the purposes of many investigations. For a quantitation of the individual glycerides, the species have to be isolated by multiple chromatographic procedures. Even then, the isomers of a triacid triglyceride cannot be quantitated by the methods later described; but the problem is not intractable, and a complete analysis of the isomers of a triglyceride mixture has been published[1] as well as a general scheme for the exhaustive analysis of even complex natural fats.[2]

This article describes the analysis of fatty acid distribution as applied to naturally occurring phospholipid and triglyceride mixtures. The methods can be applied equally well to the analysis of lipid "species";

[1] J. Sampugna and R. G. Jensen, *Lipids* **3**, 519 (1968).
[2] E. G. Hammond, *Lipids* **4**, 246 (1969).

for the methods of isolation of such species the review literature should be consulted.[3,4] Information on the triacid triglyceride problem and the "total" analysis of fats may be obtained from the original papers.[1,2,5-7]

Phosphoglycerides

Since all natural phosphoglycerides have the *sn*-glycerol-3-phosphate structure, the analytical problem is narrowed down to the determination of the fatty acid distribution over positions 1 and 2. Two methods, both enzymatical, are known: first, the selective hydrolysis of the ester bond 1 by pancreatic lipase[8]; second, the hydrolysis of ester 2 by a phospholipase A_2. (1, 2, and 3 stand for the fatty acids in these positions. Free OH groups of glycerol are not written out.)

$$
2\underset{P-X}{\overset{+\,1}{\bigg|}} \xleftarrow[\text{lipase}]{\text{pancreatic}} 2\underset{P-X}{\overset{\underset{\Big|}{1}}{\bigg[}} \xrightarrow{\text{phospholipase } A_2} \underset{P-X}{\overset{\underset{\Big|}{1}}{\bigg[}} +\,2
$$

$$\text{(I)} \qquad\qquad\qquad\qquad \text{(II)}$$

The generally suitable method is the second one; unfractionated snake venom is usually used as the source of the phospholipase. It is a peculiarity of the reaction that it can be carried out in organic solvents, especially ether, although it will also proceed in aqueous micellar solutions of phospholipids. The position-specific hydrolysis is easily carried to completion, and both fatty acid 2 and 1 (in the remaining lysophospholipid) can be recovered and analyzed.

Procedures

Hydrolysis of Phosphatidyl Choline in Ether

The procedure follows that described by Wells and Hanahan.[9] The substrate, phosphatidylcholine (lecithin), 5–25 mg, is dissolved in 2 ml

---

[3] L. J. Morris, *J. Lipid Res.* **7**, 717 (1966).

[4] G. Jurriens, *in* "Analysis and Characterization of Oils, Fats and Fat Products" (H. A. Boekenoogen, ed.), Vol. 2, p. 217. Wiley (Interscience), New York, 1968.

[5] H. Brockerhoff, *Lipids* **1**, 162 (1966).

[6] W. E. M. Lands and P. M. Slakey, *Lipids* **1**, 295 (1966).

[7] H. Brockerhoff, *Lipids* **6**, 942 (1971).

[8] A. J. Slotboom, G. H. de Haas, P. P. M. Bonsen, G. J. Burbach-Westerhuis, and L. L. M. van Deenen, *Chem. Phys. Lipids* **4**, 15 (1970).

[9] M. A. Wells and D. J. Hanahan, *Biochemistry* **8**, 414 (1969).

of 95% diethyl ether (peroxide-free) and 5% methanol. The enzyme solution is prepared by dissolving 1 mg of lyophilized snake venom (*Crotalus adamanteus, Ophiophagus hannah. Agkistrodon piscivorus*) in 0.1 ml of 0.22 M NaCl, 0.02 M CaCl$_2$, 0.001 M EDTA, and 0.05 M MOPS,[10] pH 7.2. Of this solution, 25 μl are added to the substrate; the solutions are mixed by vigorous manual shaking (30 seconds), and then left standing at room temperature. In most cases, lysophosphatidylcholine will precipitate after a few minutes. The incubation mixture is left standing for 2 hours with occasional shaking; hydrolysis is then usually complete, but the course of the reaction can be monitored (see below). The solvent is removed in a stream of nitrogen and the reaction products separated as described later.

The amount of water (i.e., enzyme solution) in the ether–methanol solvent, 25 μl per 2 ml, is quite critical; if more enzyme is required, a more concentrated enzyme solution must be prepared. Acidic phospholipids, or lipid fractions recovered from silicic acid chromatography, are often only hydrolyzed if the pH is properly adjusted. This can be done according to Long and Penny[11] by adding one drop of 0.1% bromocresol green in ether to the etheral phospholipid solution and then adding etheral ammonia or triethylamine (50 μmoles/ml) until the tint matches that of the same indicator in an aqueous buffer of pH 5.2 (the pH in ether is then 7.2).

The advantages of the reaction in ether are the visible indication of the proceeding hydrolysis and the ease with which the reaction products can be recovered. However, some phospholipids are only slowly, or not at all, attacked, usually because they are not soluble in ether. The aqueous micellar system is more generally applicable.

Hydrolysis in Micellar Solution

The phosphoglyceride, 3–20 mg, is dissolved in 0.5 ml of ether, and 5 ml of aqueous 0.1 M MOPS, 0.001 M CaCl$_2$, pH 7.2, containing 0.3–2.0 mg of snake venom, are added. The mixture is gently shaken for 6–12 hours. When the hydrolysis is complete, the water is removed either by freeze-drying or by evaporation (after addition of 3 volumes of isobutanol to prevent foaming) at the rotary evaporator. Alternatively, the reaction mixture may be acidified with 0.1 ml of acetic acid and the lipids extracted within 10 ml of chloroform–methanol 1:1 followed twice by 5 ml of chloroform. These solvents are then evaporated.

[10] Here, MOPS stands for morpholinopropane sulfuric acid.
[11] C. Long and J. F. Penny, *Biochem. J.* **65**, 382 (1952).

Monitoring the Reaction

Microscopic slides are coated by dipping into a suspension of silica gel G in chloroform and drying in air. One drop of the reaction mixtures (the ether system must be stirred up to obtain a fair sample of the precipitated lipids) is removed, clarified with a drop or two of methanol containing 1% acetic acid, and run on the slide in chloroform–methanol–water 62:34:4. Since free fatty acids run near the front, the plates vapor; the disappearance of the phosphatidylcholine (or other diacyl phosphoglyceride) signals the completion of the reaction.

Separation of Products

The lipids are separated on a silica gel G plate with chloroform–methanol–water 62:34:4. Since free fatty acids run near the front, the plates should be prewashed with solvent, or the fatty acids should be rechromatographed with hexane–ether 1:1. Alternatively, the plate may be run with ether–hexane–acetic acid 90:9:1; the fatty acid is then localized by spraying the upper third of the plate with dichlorofluorescein. Then the plate is run with chloroform–methanol–water–acetic acid 54:40:5:1, and the development is interrupted before the solvent front reaches the fatty acid spot. Spraying the whole plate with the indicator now locates the lysophospholipid. All lysocompounds have R_f values lower than the starting phosphoglyceride. To prevent the autoxidation of polyenoic lipids, 0.02% α-tocopherol can be added to all solvents. (Tocopherol does not overlap with any fatty acid methyl esters in subsequent gas–liquid chromatography.)

Preparation for Gas–Liquid Chromatography

The lipids are localized with dichlorofluorescein (0.1% in methanol). The spots are scraped into small columns and the lipids eluted with methanol into stoppered test tubes. To the eluate of phospholipids (1–3 ml) 1 ml of M methanolic KOH is added, followed after 2 minutes by 3 ml of 0.2 M aqueous H_2SO_4. The methyl esters are extracted, in the test tubes, by 3 times 2 ml of hexane or petrol ether. The extracts are collected with a Pasteur pipette and, if necessary, concentrated in a stream of nitrogen.

Eluates of free fatty acids are freed of the solvent and boiled with 2 ml of 10% borontrifluoride in methanol for 2 minutes. Cool, dilute with 4 ml of water, and extract the methyl esters with hexane.

Triglycerides: α,β-Fatty Acid Distribution

The distribution analysis most frequently performed on triglycerides determines the fatty acid composition in the primary (1 and 3 or α) and secondary (2 or β) groups, without distinguishing between the fatty acids in positions 1 and 3. Two methods are known: the random deacylation of a triglyceride with the help of a Grignard reagent[12] (III); and the selective hydrolysis of esters 1 and 3 by pancreatic lipase, followed by analysis of the remaining 2-monoglyceride (IV). (The various fatty acids bound

in 1 and 3 are not always released with the same speed by lipase and cannot be used for the analysis.) In both methods, the breakdown products are isolated by thin-layer chromatography. Only the method using pancreatic lipase will be described here; it is, at present, more accurate and universally used. The Grignard deacylation, on the other hand, is more useful in the stereospecific analysis described later.

Procedure

Lipolysis

The fat, 0.1 g, is dissolved in 0.2 ml of hexane which serves to liquify the fat and randomize the interfacial structure; 1.5 ml M tris, pH 8,

[12] M. Yurkowski and H. Brockerhoff, *Biochim. Biophys. Acta* **125**, 55 (1966).

0.05 ml 40% aqueous $CaCl_2$, 2 drops of bromophenol blue indicator solution, and 50 mg of porcine pancreatic powder (Steapsin) suspended in 0.3 ml buffer are added. The mixture is homogenized for a few seconds with a vibrating laboratory mixer and then gently shaken in a water bath of 37°. The reaction is monitored, as for phospholipids, on microscopic slides coated with silica gel G, with hexane–ether–acetic acid 80:19:1 as the developing solvent and iodine vapor as detector. The desired maximal yield of monoglyceride is usually obtained at the time when the triglyceride spot (fastest running) has almost disappeared. This end point can also be roughly estimated by noting the time when the calcium soaps first start to flocculate and give the mixture a grainy appearance; the incubation should then be continued for an equal length of time. The mixture is then acidified with 1 ml M H_2SO_4, and the lipids are extracted with 3 times 3 ml of ether. Separation on silica gel G with hexane–ether–acetic acid 80:19:1 leaves the monoglycerides near the origin, from where they are extracted with ether and transesterified as described for the phospholipids. Analysis by gas–liquid chromatography (GLC) gives the fatty acid composition in position 2. The composition in positions 1 and 3 is obtained from the triglyceride by subtracting position 2, i.e., $(3 \times TG - MG)/2$, where TG is the fatty acid composition of the triglyceride and MG is the composition of the 2-monoglyceride in mole percent.

The procedure can, of course, be scaled down.

Equally efficient procedures are described by Mattson and Volpenhein[13] and by Luddy et al.,[14] who use a dentist's amalgamator to achieve efficient homogenization of the fat.

Triglycerides: Stereospecific Analysis

In order to distinguish between fatty acids 1 and 3 of triglycerides use is made of stereospecific enzymes. In one method (V), triglycerides are degraded to a mixture of 1,2- and 2,3-diglycerides by pancreatic lipase. The enzyme ATP:1,2-sn-diglyceride 3-phosphotransferase, EC 2.7.1.- (diglyceride kinase), of E. coli phosphorylates only the 1,2 isomer, in position 3, and the reaction products are isolated and analyzed.[15] Position 2 is analyzed from the 2-monoglyceride; position 1, from the phosphatidic acid by subtraction of 2; and position 3 from the triglyceride minus 1 and 2.

[13] F. H. Mattson and R. A. Volpenhein, J. Lipid Res. 2, 58 (1961).

[14] F. E. Luddy, R. A. Barford, S. F. Herb, P. Magidman, and R. W. Riemenschneider, J. Amer. Oil Chem. Soc. 41, 693 (1964).

[15] W. E. M. Lands, R. A. Pieringer, P. M. Slakey, and A. Zschocke, Lipids 1, 444 (1966).

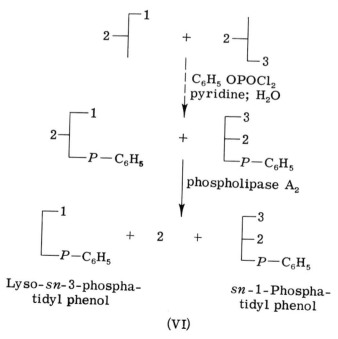

(V)

In the second method (VI), the racemic diglycerides are prepared by the action of pancreatic lipase, or isolated from the reaction products of a Grignard deacylation, and then converted into a racemic phospholipid, sn-1 and sn-3 phosphatidyl phenol.[16,17] Only the sn-3 phospholipid is de-

Lyso-sn-3-phospha-
tidyl phenol

sn-1-Phospha-
tidyl phenol

(VI)

[16] H. Brockerhoff, *J. Lipid Res.* **6,** 10 (1965).
[17] H. Brockerhoff, R. J. Hoyle, and P. C. Hwang, *Can. J. Biochem.* **44,** 1519 (1966).

graded by snake venom phospholipase A_2. Fatty acid 1 is analyzed in the lysophosphatidyl phenol; fatty acid 2 in the 2-monoglyceride obtained by hydrolysis of the triglyceride with lipase, and also as the fatty acid hydrolyzed from position 2 of the sn-3 phospholipid; fatty acid 3 is calculated from the unattacked sn-1 phospholipid by subtracting 2, and from the triglyceride by subtracting 1 and 2.

A third method[18] (VII) starts with the 1,3-diglyceride that can be isolated from the reaction products of the Grignard deacylation. This is converted to an sn-2-phosphatidyl phenol (a "β-phosphoglyceride") from which phospholipase A_2 will remove fatty acid 1 only.

$$
\begin{array}{c} -1 \\ -3 \end{array} \xrightarrow[\text{pyridine; H}_2\text{O}]{\text{C}_6\text{H}_5\,\text{OPOCl}_2} \quad \text{C}_6\text{H}_5\!-\!P\!-\!\begin{array}{c}-1 \\ -3\end{array} \xrightarrow[\text{A}_2]{\text{phospholipase}} \quad \text{C}_6\text{H}_5\!-\!P\!-\!\begin{array}{c}\\-3\end{array} \; +1
$$

(VII)

Method VII, it can be seen, allows the direct analysis of fatty acid 3 also, thus eliminating the danger of additive errors in the indirect determinations. It would be the method of choice were it not for the fact that the 1,3-diglyceride from the Grignard deacylation is usually contaminated with 10% isomerized 1,2(2,3)-diglyceride.[12,19] In method V, both positions 1 and 3 are determined indirectly, and there is no provision to check the accuracy of the result. The most reliable and most often used method is VI. Although fatty acid 3 is not analyzed directly, it can be determined in two independent ways, while fatty acid 2 can be directly analyzed in two different products. The agreement of these determinations is a measure for the accuracy of the analysis. If a direct analysis of fatty acid 3 should still be required (for instance, to ascertain the presence of fatty acids in trace amounts) the sn-1 phosphatidyl phenol might be degraded with pancreatic lipase.[8] Unpurified enzyme (Steapsin) could be used since the pancreatic phospholipase would not attack the

$$
\begin{array}{c} -3 \\ -2 \\ -P\!-\!\text{C}_6\text{H}_5 \end{array} \xrightarrow[\text{lipase}]{\text{pancreatic}} \quad \begin{array}{c} -2 \; + \; 3 \\ -P\!-\!\text{C}_6\text{H}_5 \end{array}
$$

sn-1 phospholipid. Method VI will be described here in detail. For details of the other methods the original literature should be consulted.

[18] H. Brockerhoff, J. Lipid Res. **8**, 167 (1967).
[19] W. W. Christie, and J. H. Moore, Biochim. Biophys. Acta **176**, 445 (1969).

Procedures

Analysis of Triglyceride and 2-Monoglyceride

These are performed as described in the previous section for the determination of the fatty acid distribution in positions 2 and $1 + 3$.

1,2-Diglycerides Obtained with Pancreatic Lipase[20]

The lipolysis of triglyceride is carried out as described above but interrupted at the time when the amount of diglyceride is near its maximum, as judged by the appearance of grainy calcium soaps in the incubation mixture or by visual inspection of the thin-layer chromatographic plate that monitors a pre-run. At this point there is still one-third to one quarter of the triglyceride left. The reaction is stopped by adding ethanol (one-half of the incubation volume), the mixture is adjusted to pH 4 (bromothymol blue) with M H_2SO_4 and the lipids are extracted 3 times with ether. They are applied to silica gel G plates which have been prepared with a 3% aqueous boric acid solution and activated for 1 hour at 125°, and developed with hexane–ether 1:1. The diglyceride band is located with a spray of dichlorofluorescein (0.1% in methanol), collected, and eluted with ether.

1,2-Diglycerides Obtained by Grignard Degradation

Three milliliters of (commercial) 3 M ethylmagnesium bromide in ether are added, with stirring, to 1 g of triglyceride in 50 ml of ether. Acetic acid (1 ml) is added after 30 seconds, water (10 ml) after further 30 seconds. The ether layer is separated and washed successively with 10 ml of water, 10 ml of 2% $NaHCO_3$ and 10 ml of water, and dried over Na_2SO_4. The lipids are developed in silica G–boric acid plates, first with hexane–ether 92:8, then (on the same plate) with hexane–ether 60:40. Three bands appear in the middle of the plate: tertiary alcohols, 1,3-diglycerides, and 1,2-diglycerides (the slowest band). The fatty acid composition should be analyzed to ascertain that the degradation has been random; the theoretical percentage for each acid is

$$1,2\text{-DG} = (3 \times \text{TG} + \text{MG})/4.$$

The procedure, and all following ones, can be easily scaled down.[19]

Synthesis of the Racemic Phospholipids

The diglyceride (60–100 mg) is dissolved in 1 ml of ether and added to a mixture of 1 ml of pyridine (refluxed over BaO), 1 ml of ether (dry),

[20] H. Brockerhoff, *Arch. Biochem. Biophys.* **110**, 586 (1965).

and 0.5 ml of phenyldichlorophosphate (vacuum distilled). After 1 hour at room temperature, 2 ml of pyridine, 2 ml of ether, and several drops of water are added, with cooling. The contents of the flask are then added to a mixture of 30 ml of methanol, 25 ml of water, 30 ml of chloroform, and 1 ml of triethylamine. After shaking, the lower layer of chloroform is recovered and the solvent evaporated below 40°. The completion of the reaction can be checked, and the phospholipid purified, on silica gel H plates with ether–methanol–concentrated ammonia 80:18:2; however, small amounts of unreacted diglyceride do not interfere with the following reaction.

Reaction with Phospholipase

The phospholipid is dissolved in 1.5 ml of ether, and 15 ml 0.1 M MOPS[10] containing 1 drop of 0.1 M CaCl$_2$ and 10 mg of snake venom are added, and the mixture is shaken gently overnight. The water is then evaporated below 40°, together with 20 ml of isobutanol to prevent foaming; or, 0.2 ml of acetic acid is added and the lipids are extracted with 12 ml of methanol and 15 ml of chloroform, followed by 15 ml of chloroform. The lipids are applied to a silica gel G–boric acid plate and developed with hexane–ether 40:60. The fatty acids, in the upper third of the plate, can be localized with dichlorofluorescein. The plate is then kept in a jar over concentrated ammonia for 10 minutes and developed in ether–methanol–ammonia 80:18:2; the solvent front is allowed to sweep over the fatty acid spot. Approximately final R_f values are: fatty acid, 0.8; unhydrolyzed sn-1 phosphatidyl phenol, 0.5; lysophosphatidyl phenol, 0.1; 1,2-diglycerides, if present, 0.95. The lipids are eluted with chloroform–methanol 1:1 and analyzed by GLC. To ascertain that the enzymatic reaction has gone to completion, the phosphorus content in the two phospholipids can be determined; there should be equal quantities.

The fatty acid composition of the free fatty acid (position 2) should be identical with that of the monoglyceride from lipolysis, but the monoglyceride analysis must be considered more reliable. The lysolipid yields position 1. The sn-1 phospholipid minus the monoglyceride gives position 3. Position 3 is also calculated by $3 \times$ TG minus MG minus lysophospholipid (i.e., $3 \times (1 + 2 + 3) - 2 - 1$). The analyses should closely agree. Analyses with discrepancies, for the 3 position, of not larger than 2% (absolute) for minor acids ($<10\%$) or 3% for major acids, i.e., $\pm1\%$ and $\pm1.5\%$, have been considered acceptable.[20]

[41] Microdetermination of Stereoisomers of 2-Hydroxy and 3-Hydroxy Fatty Acids

By Sven Hammarström

2-Hydroxy acids and 3-hydroxy acids are intermediates in animal and plant degradation of fatty acids.[1] 3-Hydroxy acids are also intermediates in the biosynthesis of fatty acids[1] and 2-hydroxy acids are constituents of brain cerebrosides and of certain waxes.[2] It is possible to determine the enantiomeric composition of microgram amounts of these acids by gas–liquid chromatography of diastereoisomeric derivatives.[3,4]

Derivatives of 2-Hydroxy Acids

2-Hydroxy acid methyl esters are analyzed as $O(-)$-menthyloxycarbonyl derivatives. About 100 μg of hydroxy acid is methylated with diazomethane[5] and subsequently treated with $(-)$-menthyl chloroformate.[3]

$(-)$-*Menthyl Chloroformate.*[6] $(-)$-Menthol (15.63 g, 0.1 mole; Kebo AB Stockholm, Sweden) and 11.8 ml (0.1 mole) of freshly distilled quinoline (b.p., 118–120°/20 mm) are added to 100 ml of 12.5% (w/v) phosgene in benzene (Matheson, Coleman & Bell) at 0° in a well ventilating fume hood.[6a] The mixture is stirred overnight (0°) with the outlet of the reaction vessel connected, via an empty wash bottle, to a wash bottle containing 20% w/v NaOH in water. The quinoline hydrochloride is filtered off in the fume hood, and the excess phosgene is removed by bubbling nitrogen through the filtrate for about 5 hours (the outlet of the reaction vessel should be connected to the trap with 20% NaOH). The reagent solution is transferred to a 100-ml flask with ground glass stopper which contains calcium carbonate. It contains about 1 μmole $(-)$-menthyl chloroformate per milliliter and can be kept without noticeable decomposition at +4° for a year.

$(-)$-*Menthyloxycarbonyl Derivatives.*[3] To approximately 100 μg of

[1] S. J. Wakil, *in* "Lipid Metabolism" (S. J. Wakil, ed.), pp. 1–48. Academic Press, New York, 1970.

[2] D. T. Downing, *Rev. Pure Appl. Chem.* **11**, 196–211 (1961).

[3] S. Hammarström, *FEBS Lett.* **5**, 192–195 (1969).

[4] S. Hammarström and M. Hamberg, *Anal. Biochem.* **52**, 169–179 (1973).

[5] H. Schlenk and J. L. Gellerman, *Anal. Chem.* **32**, 1412–1414 (1960).

[6] J. W. Westley and B. Halpern, *J. Org. Chem.* **33**, 3978–3980 (1968).

[6a] Caution: Phosgene is very toxic. Symptoms may be delayed up to 24 hours.

2-hydroxy acid methyl ester in 40 μl of dry benzene is added 60 μl of (—)-menthyl chloroformate solution and 12 μl of dry pyridine. The mixture is left for 30 minutes at room temperature. During this time, the color of the solution gradually changes from pale yellow to deep purple. Two milliliters of benzene is added and the solution is washed with 3×2 ml of water in the reaction tube. The benzene phase is evaporated to dryness after addition of absolute ethanol, and the residue, dissolved in CS_2, is quantitatively applied as a band $(30 \times 4$ mm$)$ to a thin-layer chromatography plate. The plate should be coated with 0.25 mm of silica gel G and activated at 120° for 45 minutes before use. Benzene–dioxane, 97:3 (v/v), is used as the developing solvent. The compounds are made visible in ultraviolet light by spraying the chromatogram with 0.2% (w/v) 2′,7′-dichlorofluorescein in ethanol. The diastereoisomeric 2-hydroxy acid derivatives do not separate during the chromatography but move as a single zone $(R_f$ 0.85) closely behind a large zone of excess reagent $(R_f$ 0.95–1.0). The 2-hydroxy acid derivatives are recovered by transferring the zone of silica gel to a glass column and eluting it with 3×10 ml of diethyl ether. The solvent is removed and the purified derivative is dissolved in 100 μl of CS_2 for gas–liquid chromatography.

Derivatives of 3-Hydroxy Acids

3-Hydroxy acid methyl esters are converted to 2-D-phenylpropionate derivatives. About 100 μg of hydroxy acid is methylated with diazomethane[5] and then treated with 2-D-phenylpropionyl chloride.[4]

2-D-Phenylpropionyl Chloride.[4] 2-DL-Phenylpropionic acid (3 g, 20 mmole; K & K Laboratories Inc., N.Y.) in 60 ml of acetone is added to a solution of (+)-1-phenylethylamine in 20 ml of acetone. The mixture is kept at 70° for 5 minutes and then cooled to —20°. This yields a crystalline salt (ca. 2.8 g) which is recrystallized 3 times from acetone at +4°. The yield of the (+)-1-phenylethylammonium salt of 2-D-phenylpropionic acid is about 0.3 g. The salt is dissolved in water, treated with 2 N hydrochloric acid and extracted 3 times with diethyl ether to give 2-D-phenylpropionic acid as a pale yellowish syrup in 11% of the theoretical yield ($[\alpha]_D^{25} = +88.0°$; C 1, benzene; previously reported $[\alpha]_D^{20} = +92.5°$ benzene[6]). Two-hundred milligrams of 2-D-phenylpropionic acid and 0.24 ml of thionyl chloride (newly distilled from beeswax) are mixed at 0° and kept at 70° for 30 minutes. Dry benzene is added and the mixture is evaporated to dryness. This step is repeated once to remove the last traces of thionyl chloride. The residue is dissolved in 2.40 ml of dry benzene and stored at +4° in a flask with ground glass stopper.

2-D-Phenylpropionate Derivatives.[4] To about 100 μg of 3-hydroxy acid methyl ester is added 90 μl of 2-D-phenylpropionyl chloride solution and 20 μl of dry pyridine. The mixture is left for 2 hours at room temperature in a dessicator. After evaporation of the solvents the residue is dissolved in chloroform and applied to a thin-layer plate (see above). The chromatogram is developed with chloroform (E. Merck AG., *pro analysi,* containing 0.6–1.0% ethanol as stabilizer), and the compounds are detected with 2′,7′-dichlorofluorescein. The diastereoisomeric 3-hydroxy acid derivatives move as a single zone (R_f 0.80). They are eluted as described above but using ethyl acetate instead of diethyl ether.

Steric Analyses

The derivatives described above can be resolved on packed columns containing OV-210 (Pierce Chemical Co.) or QF-1 on 100–120 mesh Gas Chrom Q (Applied Science). We use U-shaped glass columns (i.d. 2 or 3 mm, length 180 cm) in an F & M model 400 Biomedical Gas chromatograph equipped with a hydrogen flame ionization detector or circular glass columns (i.d. 3 mm, length 180 cm) in an LKB model 9000 gas chromatograph–mass spectrometer. Our column packings contain about 1.5% OV-210 or 5% QF-1 and are generally prepared with the Hi-Eff fluidizer (Applied Science) as described before.[7]

A gas chromatogram of (−)-menthyloxycarbonyl derivatives of a mixture of racemic 2-hydroxy acid methyl esters (C_{14}–C_{26}) is shown in Fig. 1a. The derivatives of optical isomers give rise to pairs of partially separated peaks. Figure 1b shows a gas chromatogram of the same mixture after addition of (−)-menthyloxycarbonyl derivatives of 2-L-hydroxypalmitic acid methyl ester, 2-L-hydroxyarachidic acid methyl ester, and 2-L-hydroxylignoceric acid methyl ester. It is clear that *the first peaks* of the pairs *represent the 2-L-hydroxy acids* and the second peaks the D-enantiomers. Figure 2 shows a gas–liquid chromatogram of 2-D-phenylpropionate derivatives of five racemic 3-hydroxy acid methyl esters (C_{10}–C_{18}). Addition of the same derivative of 3-D-hydroxycapric acid methyl ester showed that *the first peaks* of the pairs *represent the 3-D-hydroxy acids* and the second peaks of the 3-L-hydroxy acids.

(−)-Menthyloxycarbonyl derivatives are not suitable for steric analyses of 3-hydroxystearic, 9-hydroxystearic, 12-hydroxystearic, or 17-hydroxystearic acid methyl ester on the columns described above. Neither are 2-D-phenylpropionate derivatives of 2-hydroxy acid methyl esters suitable for steric analyses on these columns. The latter derivatives,

[7] E. C. Horning, *in* "Gas Phase Chromatography of Steroids" (K. B. Eik-Nes and E. C. Horning, eds.), pp. 1–71. Springer-Verlag, Berlin and New York, 1968.

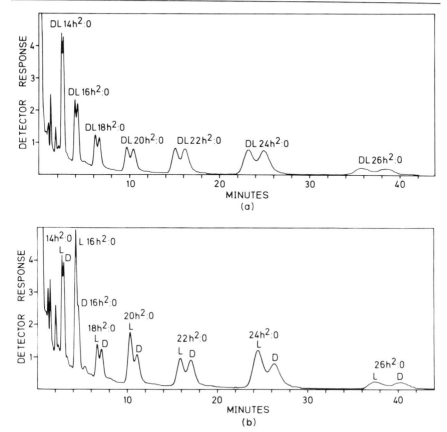

FIG. 1. (a) Gas–liquid chromatogram of (−)-menthyloxycarbonyl derivatives of 2-DL-hydroxy 14:0 Me, 2-DL-hydroxy 16:0 Me, 2-DL-hydroxy 18:0 Me, 2-DL-hydroxy 20:0 Me, 2-DL-hydroxy 22:0 Me, 2-DL-hydroxy 24:0 Me, and 2-DL-hydroxy 26:0 Me. Conditions for gas–liquid chromatography (GLC): Instrument, F&M model 400; stationary phase, 1.4% OV-210 on 100–120 mesh Gas Chrom Q; column temperature, 250°; carrier gas, helium with a flow rate of 30 ml/minute. (b) The same chromatogram as in (a) after addition of appropriate amounts of (−)-menthyloxycarbonyl derivatives of 2-L-hydroxy 16:0 Me, 2-L-hydroxy 20:0 Me, and 2-L-hydroxy 24:0 Me. (Reproduced from Ref. 3.)

however, are useful also for analysis of 15-hydroxystearic, 16-hydroxystearic, and 17-hydroxystearic acid as well as for 2-alkanols and 3-alkanols.[4]

The retention times of the (−)-menthyloxycarbonyl derivatives of the 2-hydroxy acid methyl esters in Fig. 1 are expressed as triglyceride carbon units in Table I. Triglyceride carbon units are obtained by plot-

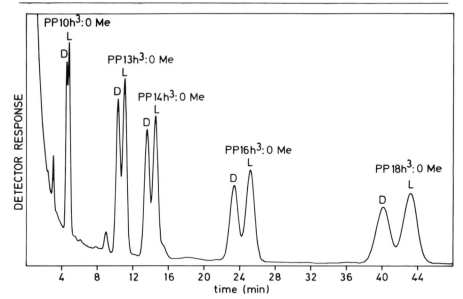

FIG. 2. Gas–liquid chromatogram of 2-D-phenylpropionate derivatives of 3-DL-hydroxy 10:0 Me, 3-DL-hydroxy 13:0 Me, 3-DL-hydroxy 14:0 Me, 3-DL-hydroxy 16:0 Me, and 3-DL-hydroxy 18:0 Me. Conditions for GLC: Instrument, F&M model 400; stationary phase, 5% QF-1 on 100–120 mesh Gas Chrom Q; column temperature, 200°; carrier gas, helium with a flow rate of 30 ml/minute. (Reproduced from Ref. 4.)

TABLE I

RETENTION DATA FOR (−)-MENTHYLOXYCARBONYL DERIVATIVES OF RACEMIC 2-HYDROXY ACID METHYL ESTERS EXPRESSED AS TRIGLYCERIDE CARBON UNITS[a]

Hydroxy acid	Triglyceride carbon units	
	1.4% OV-210	5% QF-1
2-DL-Hydroxy 14:0	23.20 and 23.50	22.80 and 23.10
2-DL-Hydroxy 16:0	25.40 and 25.70	24.90 and 25.20
2-DL-Hydroxy 18:0	27.50 and 27.80	27.05 and 27.40
2-DL-Hydroxy 20:0	29.70 and 30.00	29.05 and 29.40
2-DL-Hydroxy 22:0	31.80 and 32.15	31.20 and 31.50
2-DL-Hydroxy 24:0	33.85 and 34.20	33.25 and 33.60
2-DL-Hydroxy 26:0	35.85 and 36.20	35.25 and 35.60

[a] The stationary phase was either 1.4% OV-210 or 5% QF-1 on 100–120 mesh Gas Chrom Q, the carrier gas was helium (30 ml/minute), and the column temperature was 250°.

TABLE II

C Values for 2d-Phenylpropionate Derivatives of Racemic
3-Hydroxy Acid Methyl Esters[a]

Hydroxy acid	C values	Separation factor (ratio of retention times of diastereoisomers)
3-DL-Hydroxy 10:0	20.70 and 20.90	1.06
3-DL-Hydroxy 13:0	23.65 and 23.90	1.08
3-DL-Hydroxy 14:0	24.60 and 24.85	1.08
3-DL-Hydroxy 16:0	26.55 and 26.85	1.08
3-DL-Hydroxy 18:0	28.50 and 28.80	1.08

[a] The stationary phase was 5% QF-1 on 100–120 mesh Gas Chrom Q, the carrier gas was helium (30 ml/minute), and the column temperature was 200°. (Reproduced from Ref. 4.)

ting the logarithm of the retention time of synthetic triglycerides (e.g., tricaprylin, tricaprin, and trilaurin) against their *total* numbers of carbon atoms (27, 33, and 39) and interpolating the logarithm of the retention time of the compound in question. C values are obtained in the same way but using normal fatty acid methyl esters instead and plotting the logarithm of the retention time against the number of carbon atoms in the fatty acids. Table II gives the retention times, expressed as C values, of the 2-D-phenylpropionate derivatives shown in Fig. 2. This table also gives the separation factors for the isomeric compounds.

Figures 3 and 4 show partial mass spectra of the compounds in Figs. 1 and 2, respectively. In all cases, the mass spectra of diastereoisomeric compounds are mutually indistinguishable. Figure 5 shows structural formulas of the (—)-menthyloxycarbonyl derivative of 2-D-hydroxystearic acid methyl ester and of the 2-D-phenylpropionate derivative of 3-D-hydroxystearic acid methyl ester. It also shows some of the fragments which are formed on electron impact. The ions M-137, M-169, M-182, M-199, M-231, M-241, and M-286 in the mass spectra of (—)-menthyloxycarbonyl derivatives (Fig. 3) and the ions M, M-76, M-118, M-149, M-181, M-199, and M-223 in the mass spectra of 2-D-phenylpropionate derivatives (Fig. 4) retain the fatty acid part of the molecule whereas the ions at m/e 214, 138, and 123 (Fig. 3) and those at m/e 297, 167, 150, 137, 132, 129, 125, 121, 111, and 105 (Fig. 4) are formed by elimination of this part of the molecule.

The methods described can be used for steric analyses of radioactive hydroxy acids in tracer amounts.[8]

[8] A. J. Markovetz, P. K. Stumpf, and S. Hammarström, *Lipids* 7, 159–164 (1972).

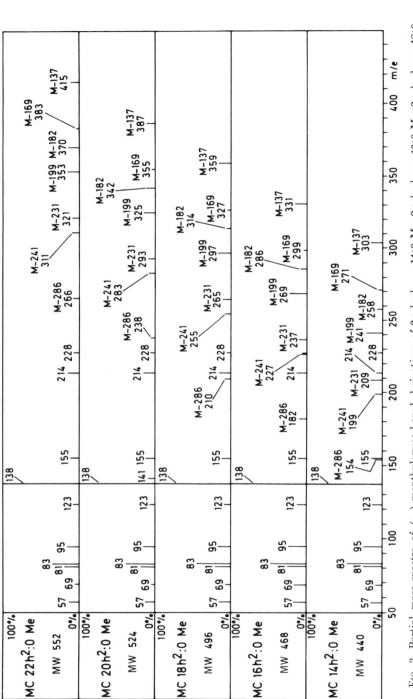

Fig. 3. Partial mass spectra of (−)-menthyloxycarbonyl derivatives of 2-ᴅ-hydroxy 14:0 Me, 2-ᴅ-hydroxy 16:0 Me, 2-ᴅ-hydroxy 18:0 Me, 2-ᴅ-hydroxy 20:0 Me, and 2-ᴅ-hydroxy 22:0 Me. Conditions for mass spectrometry: Instrument, LKB model 9000 gas chromato-graph–mass spectrometer; electron energy 22.5 eV; trap current, 60 μA; accelerating voltage 3.5 kV.

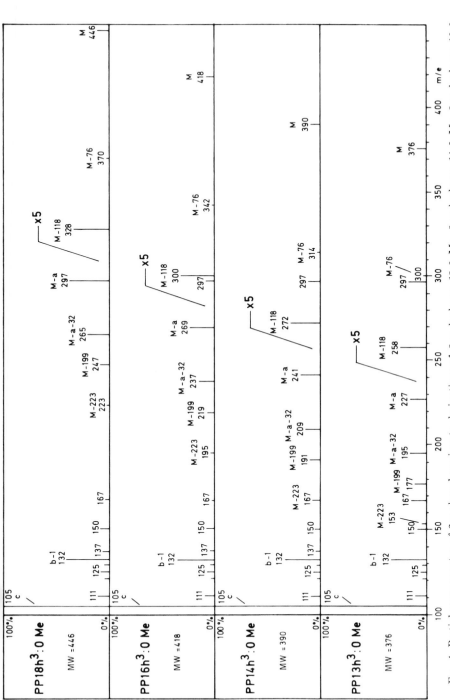

Fig. 4. Partial mass spectra of 2-D-phenylpropionate derivatives of 3-DL-hydroxy 13:0 Me, 3-DL-hydroxy 14:0 Me, 3-DL-hydroxy 16:0 Me, and 3-DL-hydroxy 18:0 Me (all unresolved diastereoisomers). Conditions for mass spectrometry: See legend for Fig. 3. (Reproduced from Ref. 4.)

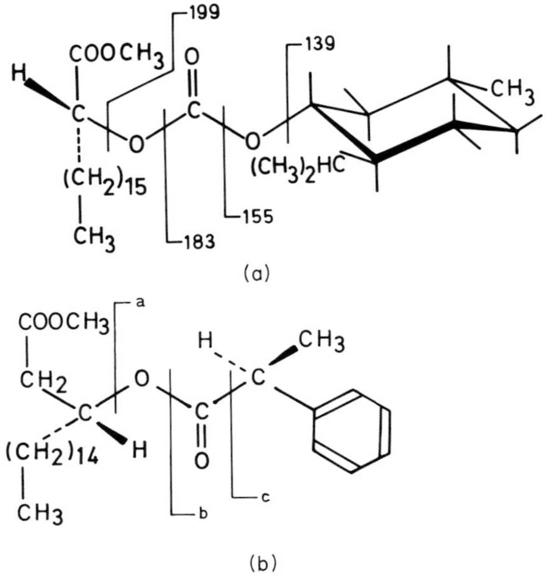

Fig. 5. Structural formulas of (a) the (−)-menthyloxycarbonyl derivative of 2-D-hydroxystearic acid methyl ester and (b) the 2-D-phenylpropionate derivative of 3-D-hydroxystearic acid methyl ester.

[42] Measurement of Protein in Lipid Extracts

By Lewis C. Mokrasch

In the course of studies of organic-solvent-soluble proteins, a variety of classic protein assay methods have been found to be inapplicable or involving excessive effort to reduce interference as a result of the presence of lipids in the mixture. The simplest resource, extracting the lipids away from the protein by the use of organic solvents, fails because of the solubility of the protein in organic solvents. Attempts to extract the protein from the lipid mixture by the use of aqueous systems are also prone to failure because of unfavorable partition characteristics of the protein, to emulsion formation, turbidities and other problems which preclude the possibility of simple procedures.

Historically the protein contents of lipid extracts have been estimated by the use of manometric procedures,[1,2] ultraviolet absorbance,[3] the biuret procedure,[4] and the Lowry modification[5] of the Folin procedure.[6] This latter method has been justifiably popular, even though it is susceptible to the interferences which affect both the biuret procedure and the Folin procedure. To the limitations and interferences described by Lowry et al., the recent literature has added many others, among which is a demonstration that the most successful adaptation[7] of the Lowry-Folin procedure is still susceptible to the chromogenicity of the lipids themselves.[8]

A modification of the method of Satake et al.[9] has proved valuable in the estimation of protein in the presence of amines and amino acids.[10] A further modification has been developed[11] which permits a reliable estimation of protein in protein-containing lipid extracts[12] from a variety of tissues.

The procedure presented here is a further evolvement of the earlier procedures with simplifications and improvements.

Materials

2,4,6-Trinitrobenzene sulfonic acid (TNBS) of a suitable quality is obtained commercially as a pale yellow crystalline material. It may be recrystallized from 1 M HCl[13] or from 2-butanone-petroleum ether mixtures. Since the crystalline material may have a variable degree of hydration, the stock solution of 0.03 M TNBS in butanol (tertiary:secondary, 9:1 v/v) is conveniently diluted from a stronger solution after an alkalimetric check on its concentration. Organic solvents and potassium tetra-

[1] D. D. Van Slyke, J. Biol. Chem. 83, 425 (1929).
[2] J. P. Teters and D. D. Van Slyke, in "Quantitative Clinical Chemistry," Vol. 2, p. 385. Williams & Wilkins, Baltimore, Maryland, 1932.
[3] J. Folch-Pi, in "Brain Lipids and Lipoproteins" (J. Folch-Pi and H. J. Bauer, eds.), p. 18. Amer. Elsevier, New York, 1963.
[4] A. G. Gornall, C. J. Bardawill, and M. M. David, J. Biol. Chem. 177, 751 (1949).
[5] O. H. Lowry, N. J. Rosebrough, A. L. Farr, and R. J. Randall, J. Biol. Chem. 193, 265 (1951).
[6] O. Folin and V. Ciocalteau, J. Biol. Chem. 177, 751 (1949).
[7] H. H. Hess and E. Lewin, J. Neurochem. 12, 205 (1965).
[8] J. Eichberg and L. C. Mokrasch, Anal. Biochem. 30, 386 (1969).
[9] K. Satake, T. Okuyama, M. Ohashi, and T. Shinoda, J. Biochem. (Tokyo) 47, 654 (1960).
[10] L. C. Mokrasch, Anal. Biochem. 18, 64 (1967).
[11] L. C. Mokrasch, Anal. Biochem. 36, 273 (1970).
[12] L. C. Mokrasch, Prep. Biochem. 2, 1 (1972).
[13] A. R. Goldfarb, Biochemistry 5, 2570 (1966).

borate are the analytical reagent grade or equivalent. The 6 M HCl used by the author is the easily prepared 6.1 M azeotrope; a dilution of 12 M HCl would be equally satisfactory.

The solutions used in this procedure are stable. The TNBS solution is stable indefinitely when stored at 0–2°. No extraordinary precautions need be taken with respect to light or the presence of heavy metals in the reagents. Avoidance of heat, evaporation, and ammonia are sufficient precautionary measures. The choice of solvents for TNBS is arbitrary; tertiary butanol is relatively inert and secondary butanol is present as an antifreeze for storage at 0°.

Brij-35 (Calbiochem) is a dodecylpoly(ethylene-oxy)ethanol. Any similar nonionic and nonaromatic detergent would probably be similarly suitable. It can be used as a solution in chloroform–methanol mixtures or in whatever solvent the lipid sample is dissolved.

Principles

TNBS, like picric acid, has been used for many years to characterize organic amino compounds. In aqueous solution, TNBS reacts with primary and secondary amines (and ammonia) to form an arylated amine, or trinitroaniline derivative. The reaction of TNBS with amines and its hydrolysis to picric acid are pH-dependent reactions[13]; the pH of the borate buffer[10] has proved to provide a satisfactory control over the competing reactions. The absorption spectrum of the trinitrophenylamines has an absorbance maximum near 345 nm. Once formed, the trinitrophenylated amine derivative or trinitrophenylated protein is comparatively stable.

Methods

Guidelines for the selection of sample size and solvents will be detailed below in the section entitled "Variations and Limitations." The procedure outlined here has proved to be reliable for brain and liver extracts without introducing any significant departures in methodology.

Replicate samples containing less than 10 mg of nonvolatile solutes are added to tubes containing 0.05 ml of Brij-35 (50 g liter^{-1} in 1:1 (v/v) chloroform–methanol). Tubes containing protein standards (0.1 or 0.2 mg, either aqueous or nonaqueous solution) and reagent blanks containing the detergent, but no sample, should also be prepared.

The organic solvents are driven off by placing the tubes in an oven near the boiling point of the solvent; 70° gives essentially complete removal of chloroform–methanol from samples in 1 hour.

To the dried, cooled samples, add 1 ml 0.1 M $K_2B_4O_7$. Agitate the tubes violently with a mechanical "buzzer" or similar device. To each tube add 0.2 ml 0.03 M TNBS in butanol. Mix again vigorously and repeat the mixing at 10 minute intervals; alternatively, the rack of tubes may be mounted on a mechanical shaker during the color development.

At the end of 30 minutes of color development at 23–26° add 3.0 ml n-butanol (water saturated) to each tube and mix vigorously. Centrifuge the tubes for a minute or two at 1000 g to separate the phases cleanly. Aspirate off and discard the organic phase.

To each tube add 0.1 ml 6 M HCl and 3 ml n-butanol. Again mix vigorously, centrifuge, and discard the organic phase. The protein will probably be observed as yellow flakes at the interface.

To the remaining aqueous phases containing the solution or suspension of trinitrophenylated protein add 2.0 ml 98% formic acid (caution: concentrated formic acid is a powerful vesicant). Cap or stopper the tubes and heat at 100° for 30 minutes. Cool and read against the reagent blanks at 345 nm.

Variations and Limitations

Lipid solvents which have been used successfully in this procedure are the common solvents with a boiling point less than 100°: chloroform, methanol (and mixtures of the two), carbon tetrachloride, ethyl acetate, diethyl ether, acetone, 2-butanone, ethanol, 1- and 2-propanol, and petroleum ether. Petroleum ether is a poor solvent for Brij but when the detergent is added as a chloroform–methanol solution, no problems of solubility arise.

Detergents other than Brij probably can be used. Brij is clearly superior to sodium dodecyl sulfate in promoting the dispersal and suspension of the lipid residue in the arylation medium.

The drying procedure permits several options. Removing lipid solvents with a stream of nitrogen, carbon dioxide, or air has been used. It is a cumbersome procedure compared to the low temperature evaporation in the oven. With any of the evaporation procedures, care must be exercised with dangerously flammable and toxic vapors.

The TNBS and $K_2B_4O_7$ may be combined and added as a single solution provided they are mixed immediately before use.

The identity of the extractant solvent in this procedure is critical. Many solvents (carbon tetrachloride, toluene, dibutyl ether, and diisobutyl ketone among others) set up an intractable emulsion during the first post-arylation extraction. Ethyl acetate, n-pentanol, ethylene trichloride, and water-saturated n-butanol have been used without difficulty in the presence of Brij.

The heating period for the solution of the trinitrophenyl (TNP) protein in formic acid depends upon the source of protein being analyzed. Typical protein standards such as ovalbumin, casein, γ-globulin, and serum albumin are easily soluble in formic acid after trinitrophenylation. The typical membrane protein, such as rat brain proteolipid protein, may require 30 minutes of heating, and others such as that from rat kidney may require 90 minutes of heating to effect solution. The color of TNP protein decreases at the rate of about 10% per hour in 66% formic acid at 100°. If the protein proves to have properties like ovalbumin, 88% formic acid may be added to the last aqueous phase instead of 98% and the heating step eliminated.

All ninhydrin-reactive lipids are also reactive with TNBS. Their TNP derivatives are removed in the extraction steps. Similarly, amines and amino acids deriving from any sources as breakdown products are also converted into TNP derivatives which are removed in the extraction steps. In any case, small molecules such as biogenic amines and amino acids are ordinarily not present in concentrations near millimolar in lipid extracts and are easily removed by partition against aqueous systems.

Similarly, small molecules with a significant absorption at 345 nm, such as nucleotides and carotenoids, are not soluble in *both* a total lipid extract (especially if washed by partitioning against an aqueous phase) *and* the last aqueous phase of the TNBS system.

Turbidities may be observed following the extraction of the TNP reactive substances other than protein. Most often, if the turbidity is not discharged by the addition of formic acid to the aqueous phase, a change in the organic solvent used to extract the TNP derivatives is indicated.

The allowable quantities of lipid and protein in each tube cannot be specified for all probable applications of this procedure. In the past using brain preparations with lipid quantities between 3 and 10 mg, 0.050 to 0.5 mg protein has been estimated with a precision of 2–5%. Quantities of lipids greater than 10 mg make the emulsification of the dried lipid sample in the arylation medium difficult without modification of the volumes used.

As with other colorimetric methods of any sort, the accuracy of the analyses becomes increasingly poor with smaller quantities of protein, below 50 μg. A serious problem in any analysis of a minority component in a mixture is the falsely high values obtained when the minority component is in very low concentration; this destroys the applicability of a method to test for traces of one component in the presence of another. In this laboratory, when lipid extracts (rat liver and lung) known to be free of polypeptides have been analyzed by the TNBS system, the results demonstrated that in 10–15 mg of lipids, the color values for the protein determinations were not significantly different from the reagent blanks.

Discussion

The value of this procedure for protein estimation, apart from its wide applicability, is that it permits the estimation of protein in the presence of lipids from which the protein is not easily separable. Obviously, if a lipid extract were prepared and partitioned against an aqueous phase with transfer of all the protein to the aqueous phase, the likelihood is great that any protein analysis method could be used for its estimation in the aqueous phase. However, if simple physical separation procedures do not effect the separation of proteins from lipid extracts, then the method outlined here should be tested.

Concerning the chromogenicity of various proteins, the TNBS method depends principally on the free amino contents of the protein. A typical mammalian water-soluble protein contains about 8.4 mole-% lysine; a typical mammalian proteolipid protein contains 3.9 mole-% of lysine.[12] The popular protein standard, bovine serum albumin, contains 12.8 mole-% lysine. Obviously, problems in chromogenicity are handled by a judicious selection of protein standards, the ideal being a pure sample of the protein being measured. Nevertheless, the lysine content of different proteins tends to be less variable than their aromatic amino acid content, which favors the TNBS protein method over those depending upon the aromatic moieties of the proteins in a mixture.

Unlike the original modification of the TNBS reaction for coestimation of amines, amino acids, and proteins,[10] the method presented here is not designed for the coestimation of proteins and amino lipids. Nevertheless, it has been used in this laboratory to obtain a reasonable appraisal of the content of phosphatidylserine and phosphatidylethanolamine in chromatographic fractions of total lipid extracts during the purification of proteolipid proteins. Since the two amino lipids are not quantitatively separated in the extraction procedure after trinitrophenylation, the concurrent use of thin-layer chromatography is required.

[43] Gas Chromatography–Mass Spectrometry of Esters of Perdeuterated Fatty Acids

By James A. McCloskey

Mass spectrometry is potentially well-suited for studies of lipid metabolism or biosynthesis in which fully deuterated fatty acids can be employed as substrates. These compounds can be obtained by chemical synthesis using catalytic exchange reactions[1,2] or by isolation from the alga *Scenedesmus obliquus*[3-5] which has been cultured in D_2O.[4,6] Examples of applications have been in the biosynthesis of 4-hydroxysphinganine[7] and studies of chain elongation in bacterial fatty acids.[8,9]

This chapter discusses the gas chromatographic and mass spectrometric properties of the esters of perdeuterated fatty acids with emphasis on the uses of these techniques for structural characterization.

Gas Chromatography

Gas chromatography is normally a highly useful tool for work with fatty acid esters, but care must be exercised when working with perdeuterated analogues because of anomalous changes in relative retention time. Deuterated components elute earlier than their protium forms[3-5,10] in accordance with similar behavior of other classes of compounds (e.g., Falconer and Cretanović[11,12]). This characteristic is shown by the retention time data from various protium–deuterium pairs of esters indicated in Fig. 1. Labeled and unlabeled analogues can generally be separated using packed columns. If a sufficient number of points are available the

[1] N. Dinh-Nguyen and E. Stenhagen, *Acta Chem. Scand.* **20**, 423 (1966).
[2] N. Dinh-Nguyen and E. A. Stenhagen, French Patent 1,466,088; *Chem. Abstr.* **67**, 63814 (1967).
[3] J. A. McCloskey, A. M. Lawson, and F. A. J. M. Leemans, *Chem. Commun.* p. 285 (1967).
[4] G. Graff, P. Szczepanik, P. D. Klein, J. R. Chipault, and R. T. Holman, *Lipids* **5**, 786 (1970).
[5] G. Wendt and J. A. McCloskey, *Biochemistry* **9**, 4854 (1970).
[6] A. J. Williams, A. T. Morse, and R. S. Stuart, *Can. J. Microbiol.* **12**, 1167 (1966); see also references therein.
[7] A. J. Polito and C. C. Sweeley. *J. Biol. Chem.* **246**, 4178 (1971).
[8] E. Oldfield, D. Chapman, and W. Derbyshire, *Chem. Phys. Lipids* **9**, 69 (1972).
[9] E. Oldfield, *Chem. Commun.* p. 719 (1972).
[10] W. K. Rohwedder, C. R. Scholfeld, H. Rakoff, J. Nowakowska, and H. J. Dutton, *Anal. Chem.* **39**, 820 (1967).
[11] W. E. Falconer and R. J. Cretanović, *Anal. Chem.* **34**, 1064 (1962).
[12] R. Bentley, N. C. Saha, and C. C. Sweeley, *Anal. Chem.* **37**, 1118 (1965).

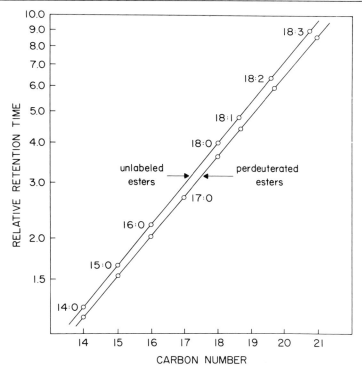

FIG. 1. Comparison of gas chromatographic behavior of unlabeled and perdeuterated fatty acid esters.[5] Conditions: 6 feet × 4 mm, 15% EGS; 170°; 40 cc N_2/minute. Carbon numbers are equal to the number of carbon atoms in the parent acid. [Reprinted with permission from G. Wendt and J. A. McCloskey, *Biochemistry* **9**, 4854 (1970). Copyright by the American Chemical Society.]

usual linear semilogarithmic plot can be obtained for the perdeuterated esters and used for the preliminary identification of homologues. The nature and polarity of the liquid phase which is employed is important in determining the degree of separation which is observed, but prediction of the extent of the effect may not be possible. Parallel plots as in Fig. 1 have been reported for the polar liquid phase EGS,[4,5] while decreasing separation with increasing chain length, and convergence near C_{18}, was indicated for the medium-polarity phase OV-17.[4] However, the low-polarity phase OV-1 was reported in one instance to produce parallel plots[4] and in another to result in decreasing separation with decreasing chain length with convergence in the vicinity of C_{10}.[5]

The most effective use of gas chromatography is in direct combination with mass spectrometry.[13] The former technique can be used for prelimi-

[13] J. A. McCloskey, Vol. 14 [50].

nary identification of members of a homologous series, and is particularly valuable for establishing the occurence of chain branching by comparison of retention time with that of the corresponding normal chain isomer, after mass spectrometric establishment of molecular weight.

Mass Spectrometry

Measurement of Deuterium Content. By measurement of ion abundance ratios, the mass spectrum provides (a) the total deuterium content of the molecule, often expressed as a percentage of the maximum possible content; (b) the molar percentage of each labeled species; and (c) the distribution or location of residual protium. A list of parameters required and procedure for the calculation has been published.[5] For small numbers of measurements the calculations are easily done manually; for routine measurements a simple computer program has been written.[5]

If the sample is introduced into the mass spectrometer by gas chromatograph, partial fractionation of the various isotopic species will result. However, model studies have shown that in the case of perdeuterated esters the errors introduced in the deuterium content calculations are minor, regardless of the exact position on the chromatographic peak at which the spectrum is recorded.[5]

Interpretation of Mass Spectra. Peaks in the mass spectrum of perdeuterated esters appear in the mass spectrum at higher mass values than their protium analogues, the exact shifts reflecting the number of deuterium atoms present in the ion. Structure correlations are generally straightforward and are facilitated by the considerable body of published data pertaining to unlabeled fatty acid esters.[13-15] In addition, partial or complete mass spectra of esters from a number of perdeuterated fatty acids have been reported: hexadecanoic acid-d_{31},[3,7] nonadecanoic acid-d_{37},[13] docosanoic acid-d_{43},[1] and a total of twenty normal chain, branched chain, and unsaturated acids isolated from *Scenedesmus obliquus* grown in D_2O.[4,5] Closely related spectra include lauric acid-d_{23},[8] and esters of several perdeuterated acids labeled with protium in the alkyl chain.[8,16]

Perdeuterated acids can be esterified to produce either methyl or methyl-d_3 esters. For structural work the fully deuterated derivative is preferred because recognition of peaks arising from artifacts or ions which contain residual protium is greatly facilitated. This is because the masses

[14] J. A. McCloskey, *Top. Lipid Chem.* **1**, 372 (1970).
[15] G. Odham and E. Stenhagen, *in* "Biochemical Applications of Mass Spectrometry" (G. R. Waller, ed.), p. 211. Wiley (Interscience), New York, 1972.
[16] N. Dinh-Nguyen, A. Raal, and E. Stenhagen, *Chem. Scripta* **1**, 117 (1971).

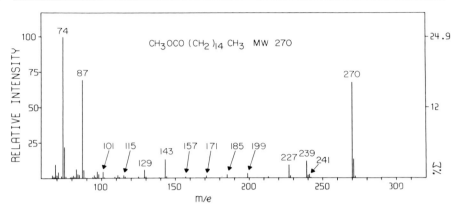

Fig. 2. Mass spectrum of methyl palmitate (CEC-21-110B, 70 eV).

of all perdeuterated ions are even numbers, with the exception of ^{13}C and other minor isotope peaks.

Comparison of methyl palmitate and perdeuterated methyl palmitate can be made from Figs. 2 and 3. The fully labeled molecular ion (D_{34}) appears at m/e 304 in Fig. 2, and corresponds to an incorporation level of 98.1%, with a hydrogen isotope distribution of 52% D_{34}, 33% $D_{33}H$, 12% $D_{32}H_2$, and 2.5% $D_{31}H_3$. Although the lower protium-containing isotope peaks at m/e 303 and 302 give the visual impression of large amounts of protium, the deuterium content corresponds to 98.6% of the maximum possible value. In some instances the presence of lower isotope peaks may be misleading if the peaks contain contributions from fragment ions. A test for isotopic homolgues vs. fragment-related ions can be made using the multiple ion monitoring technique.[17] Peaks resulting from different isotopic species will be associated with gas chromatographic retention times which differ by several seconds, while peaks related by fragmentation processes must exhibit identical retention times.[5]

Molecular ion enhancement (m/e 270 in Fig. 2 vs. sum of m/e 302, 303, and 304 in Fig. 3) is a characteristic consequence of perdeuteration,[3] otherwise the patterns in Figs. 1 and 2 are very similar. Structures from the principal ion species[18,19] are represented in Scheme 1. Peaks resulting from loss of OCH_3 and C_2H_5 occur at separate mass values in Fig. 2 but fall at the same nominal mass (m/e 270) in Fig. 3. Otherwise peaks

[17] C. C. Sweeley, W. H. Elliott, I. Fries, and R. Ryhage, *Anal. Chem.* **38,** 1549 (1966).
[18] R. Ryhage and E. Stenhagen, *Ark. Kemi* **13,** 523 (1959).
[19] N. Dinh-Nguyen, R. Ryhage, S. Ställberg-Stenhagen, and E. Stenhagen, *Ark. Kemi* **18,** 393 (1961).

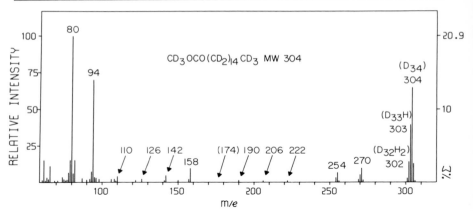

FIG. 3. Mass spectrum of methyl palmitated-d_{34} (CEC 21-110B, 70 eV; taken in part from McCloskey *et al.*[3]).

are simply shifted in accordance with deuterium content, and it is evident that identification of normal chain perdeuterated esters can be made by direct correlation with unlabeled esters.

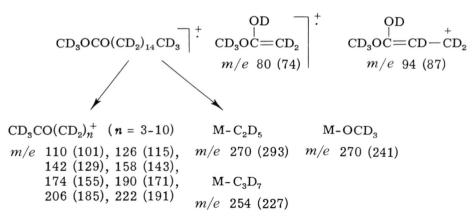

SCHEME 1. Principal ions from the mass spectrum of methyl palmitated-d_{34} (Fig. 3). Values in parentheses refer to analogous ions in Fig. 2.

Determination of sites of branching in the alkyl chain is made in the same manner as for unlabeled fatty acid esters: (1) by enhancement of members of the $CD_3OCO(CH_2)_n{}^+$ series as a result of higher stability of the branched ion, or (2) by characteristic elimination of methanol from the branched ion but not from isomeric normal chain ions of the $CD_3OCO(CD_2)_n{}^+$ series[20]; for example, in the spectrum of perdeuterated

[20] R. Ryhage and E. Stenhagen, *Ark. Kemi* **15**, 291 (1960).

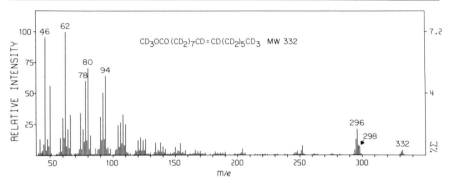

FIG. 4. Mass spectrum of methyl oleate-d_{36} (LKB 9000, 70 eV).[5] [Reprinted with permission from G. Wendt and J. A. McCloskey, *Biochemistry* **9**, 4854 (1970). Copyright by the American Chemical Society.]

methyl 5,9-dimethylpentadecanoate, m/e 142 is more abundant than in normal chain esters and further decomposes by loss of CD_3OD to m/e 106.[5] The simple cleavage product m/e 222 did not show unusual enhancement, but expulsion of CD_3OD to form m/e 186 provided clear evidence for the presence of a second methyl branch at C-9.

$$
\begin{array}{c}
106 \xleftarrow{\;-\,CD_3OD\;} 142 \\
CD_3OCO(CD_2)_3CD \;-\!|\; CD_2CD_2CD_2CD \;-\!|\; (CD_2)_5CD_3 \\
\qquad\qquad |\; \\
\qquad\quad CD_3 \qquad\qquad\qquad CD_3 \\
186 \xleftarrow{\;-\,CD_3OD\;} 222
\end{array}
$$

Unsaturation is readily detected by the downward shift of the molecular ion four mass units (2 D) per double bond. As shown by the mass spectrum of methyl oleate-d_{36} (Fig. 4),[5] the molecular ion value m/e 332 corresponds to an 18-carbon acid containing one double bond. In analogy to the spectrum of methyl oleate,[21] the upper mass range shows peaks resulting from loss of the CD_3O radical (m/e 298) and CD_3OD (m/e 296). The lower portion of the spectrum shows the familiar ions m/e 80 and 94 (see Scheme 1) and intense peaks at m/e 46, 62, 78, 94 . . . , arising from the unsaturated hydrocarbon series C_nD_{2n-1}.

Assignment of double bond position can be made by conventional ozonolysis and analysis of the products by gas chromatography.[4] Alter-

[21] B. Hallgren, R. Ryhage, and E. Stenhagen, *Acta Chem. Scand.* **13**, 845 (1959).

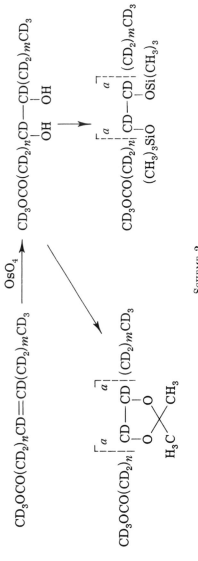

Scheme 2.

TABLE I

MASS DOUBLETS FROM MASS SPECTRA OF PERDEUTERATED FATTY ACID ESTERS

Doublet[a]	Mass difference (millimass units)
H_2–D	1.5
CD–^{13}CH	2.9
C_3–D_2O_2	18.0
CD_2–O	33.3
OD_4–C_2	51.8
D_6–C	84.6

[a] Higher mass member listed first.

natively, effective mass spectrometric methods consist of stereospecific oxidation using osmium tetroxide followed either by trimethylsilylation[5,22] or condensation with acetone.[5,23] Original double bond positions are marked by peaks in the mass spectrum representing favored cleavages adjacent to oxygen (see Scheme 2, ions *a* above). The silylation method has also been applied to perdeuterated dienoic esters. Double bond stereochemistry in perdeuterio-monoenes can be established from the gas chromatographic and mass spectrometric behavior of the respective *erythro*- and *threo*-*O*-isopropylidene derivatives of *cis*- and *trans*-monoenoic esters.[5]

High Resolution Mass Spectrometry. Measurement of exact mass can be used effectively to establish the elemental composition of a perdeuterated ester if that information is not clearly implied from examination of the low resolution spectrum. Table I shows the most commonly encountered mass doublets,[5] which are different from those normally encountered with unlabeled fatty acid derivatives. Because of the small mass differences differentiation between ^{13}CH and CD or D and H_2 (if residual protium is present) will not usually be possible. The most important doublet which is encountered is CD_2 vs. O, which can be resolved at $R < 10,000$. Most computer programs which are used for high resolution mass spectrometry can be used for perdeuterated molecules but require two changes: (1) the exact mass of hydrogen (1.00782) must be replaced by that of deuterium (2.01410), and (2) tests for identification of doubly charged ions must be deleted or modified. The latter requirement is because of the high fractional masses of ions which contain large numbers of deuterium atoms that would therefore fall in the region where doubly charged ions are normally found; for example, the molecular mass

[22] P. Capella and C. M. Zorzut, *Anal. Chem.* **40**, 1458 (1968).
[23] J. A. McCloskey and M. J. McClelland, *J. Amer. Chem. Soc.* **87**, 5090 (1965).

of methyl palmitate-d_{34} is 304.4692, the unusually high fractional value of which is easily recognized even at low resolving power, especially in the presence of fluorocarbon mass standards which exhibit small negative mass defects.

[44] Mass Spectrometry of Triglycerides

By RONALD A. HITES

Mass spectrometry has long been a valued tool for the lipid chemist.[1] In particular, the elucidation of fatty acid structures has greatly benefited by the application of this technique. Less well known but no less valuable is the application of mass spectrometry to the structural elucidation and quantitative analysis of triglycerides (both pure and in mixtures). It is the purpose of this review, first, to explain the nature of triglyceride mass spectra and, second, to review the past applications of mass spectrometry to this compound class.

Nature of Triglyceride Mass Spectra

Mass spectra of triglycerides can be obtained by inserting the material directly into the ion source by way of a direct introduction inlet (the

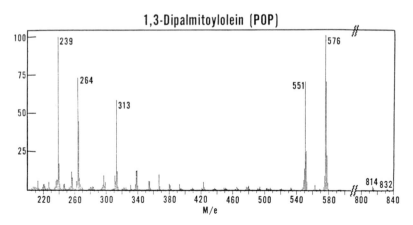

FIG. 1. Mass spectrum of 1,3-dipalmitoylolein. For experimental conditions, see Hites.[9]

[1] J. A. McCloskey, Vol. 14 [50].

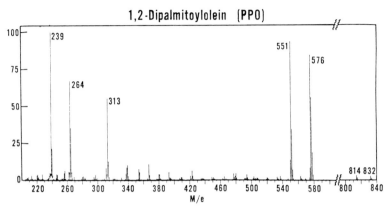

FIG. 2. Mass spectrum of 1,2-dipalmitoylolien. For experimental conditions, see Hites.[9]

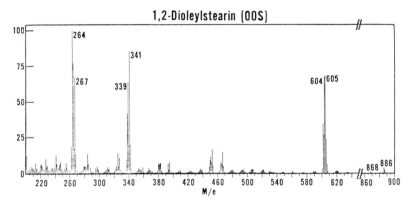

FIG. 3. Mass spectra of 1,2-dioleylstearin. For experimental conditions, see Hites.[9]

"probe" inlet).[2] Vaporization at 170 to 230° usually produces spectra without evidence of thermal decomposition. Figures 1 to 3 show three spectra of triglycerides obtained in this manner. Since these spectra show the characteristic features of the spectra of all triglycerides, they will serve as a basis for the following discussion.

Molecular ions (M+) are shown in Figs. 1 to 3 at m/e 832 and 886 and are generally observed in the mass spectra of all triglycerides except for those of the very short chain fatty acids.[3] For longer chain fatty acids,

[2] M. Barber, T. O. Merren, and W. Kelly, *Tetrahedron Lett.* p. 1063 (1964).
[3] W. M. Lauer, A. J. Aasen, G. Graff, and R. T. Holman, *Lipids* 5, 861 (1970).

molecular ions are easily detectable, although they are usually of low abundance.[2-4]

An unusual feature of the mass spectra of triglycerides is an ion corresponding to the loss of water, the $(M-18)^+$ peak.[2-4] This fragmentation is unique among the mass spectra of esters. The intensity of the $(M-18)^+$ ion is about equal to the intensity of the molecular ion for triglycerides with saturated fatty acids, but its intensity decreases to almost zero as the degree of unsaturation increases (compare Figs. 2 and 3). The mechanism of this electron impact-induced loss of water has been studied by deuterium labeling techniques.[4] Enolization followed by 1,4 elimina-

tion of water is consistent with the data,[4] where R is the glyceryl diacyl moiety. This mechanism is similar to that proposed for the loss of water from cyclohexanone.[5]

Among the more intense ions are several which are characteristic of the individual fatty acids. These ions fall into two general classes: (1) those ions containing two complete fatty acid chains and (2) those containing only one complete chain. There are two important class 1 ions: one corresponding to the ion remaining after the loss of one acyloxy group from the molecular ion[2-4] [abbreviated as $(M-RCOO)^+$], and the other corresponding to the $(M-RCOO)^+$ ion minus one additional hydrogen atom,[2-4] $(M-RCOOH)^+$. If the three fatty acids are different, three different $(M-RCOO)^+$ and $(M-RCOOH)^+$ ions are observed corresponding to the loss of each different acyloxy group.

In general, the $(M-RCOOH)^+$ ion is less than half as intense as the $(M-RCOO)^+$ ion if the remaining two fatty acid moieties are saturated; if either (or both) of the remaining fatty acid moieties are unsaturated, the intensity of the $(M-RCOOH)^+$ ion is equal to or, usually, greater than the intensity of the $(M-RCOO)^+$ ion.

Tentative structures for these ions have been suggested on the basis of deuterium labeling experiments.[3,4] If the 1 or 3 position acyloxy group is lost, an ion such as

[4] A. J. Aasen, W. M. Lauer, and R. T. Holman, *Lipids* **5**, 869 (1970).
[5] D. H. Williams, H. Budzikiewicz, Z. Pelah, and C. Djerassi, *Monatsh. Chem.* **95**, 166 (1964).

could result. The $(M-RCOOH)^+$ ion could have the structure[3,4]:

$$RCO-\begin{array}{c}O\\||\end{array}\overset{\cdot+}{\underset{O}{\overset{O}{\bigcirc}}}=CHCH_2R' \quad (M-RCOOH)^+$$

If the 2-position acyloxy group is lost, the suggested structures are[3,4]:

$$\underset{\underset{O}{||}}{RCOH_2C}\overset{\overset{+}{O}}{\bigcirc}-CH_2CH_2R' \quad (M-RCOO)^+$$

$$\underset{\underset{O}{||}}{RCOH_2C}\overset{\overset{\cdot+}{O}}{\bigcirc}=CHCH_2R' \quad (M-RCOOH)^+$$

By way of clarification and example, the identities of the $(M-RCOO)^+$ and $(M-RCOOH)^+$ ions in the mass spectra of the triglycerides shown in Figs. 1 to 3 are detailed. Both 1,3- and 1,2-dipalmitoylolein have a molecular weight of 832 amu, the mass of the palmitoyl acyloxy group ($C_{15}H_{31}COO$) is 255 amu, and the oleyl acyloxy group ($C_{17}H_{33}COO$) is 281 amu. Thus, the loss of a palmitoyl acyloxy group gives an ion containing a palmitoyl and oleyl group at m/e 577 (832 minus 255). Since this ion contains some unsaturation, a somewhat more intense $(M-RCOOH)^+$ ion at m/e 576 is also observed. On the other hand, the loss of an oleyl group gives an ion at m/e 551 (832 minus 281) containing two palmitoyl groups which are, of course, fully saturated. Thus, the corresponding $(M-RCOOH)^+$ ion at m/e 550 is less intense than m/e 551. 1,2-Dioleylstearin (molecular weight 886), in addition to the oleyl acyloxy groups (m/e 281), contains a stearyl acyloxy group ($C_{17}H_{35}COO$) of 283 amu. Thus, the $(M-RCOO)^+$ ions appear at m/e 603 (886 minus 283) and m/e 605 (886 minus 281). Since both of these $(M-RCOO)^+$ ions contain at least one unsaturated fatty acid group, m/e 602 and 604 are of equal or greater intensity than m/e 603 and 605, respectively.

The other class of intense ions characteristic of the individual fatty acid chains is that group of ions containing only one of the fatty acid moieties. The simplest of these class 2 ions are the acyl ions, $RC\equiv O^+$ (abbreviated RCO^+). Since each fatty acid chain produces an RCO^+ ion, if two or three of the chains are different, two or three different RCO^+

ions will result. As was the case for the $(M\text{-}RCOO)^+$ ion, if the RCO^+ ion is unsaturated, an additional hydrogen atom is lost producing a ketene-like ion,[3] which is usually more intense than the corresponding RCO^+ ion. Figures 1 to 3 show RCO^+ ions corresponding to the palmitoyl, oleyl, and stearyl groups at m/e **239**, **265**, and **267**, respectively. In the case of the oleyl group, the $(RCO\text{-}1)^+$ ion at m/e **264** is more intense than m/e **265**. It is interesting to note that the spectrum of glyceryl triocta-deca-9,12-dienoate exhibits intense RCO^+, $(RCO\text{-}1)^+$, $(RCO\text{-}2)^+$, and $(RCO\text{-}3)^+$ ions.[3]

Another class 2 ion of relatively high intensity is that corresponding to the acyl ion plus 74 amu.[2-4] High resolution mass measurements indicate that $C_3H_6O_2$ is the elemental composition of the 74 amu, and deuterium labeling experiments suggest that this three-carbon moiety corresponds to the glyceryl "backbone" of the triglyceride.[3] A possible route to this $(RCO + 74)^+$ ion is the loss of a ketene from the $(M\text{-}RCOO)^+$ ion, giving an ion of the following hypothetical structure[4]:

$$RCO-\overset{\underset{\|}{O}}{}\diamondsuit\overset{+}{O}H \qquad (RCO+74)^+$$

Unlike the RCO^+ and $(M\text{-}RCOO)^+$ ions, the loss of an additional hydrogen, if the alkyl group is unsaturated, usually does *not* predominate over the $(RCO + 74)^+$ ion. Thus, referring to Figs. 1 to 3, the $(RCO + 74)^+$ ions for the palmitoyl, oleyl, and stearyl groups are observed at m/e **313**, **339**, and **341**, respectively.

A class 2 ion which is particularly prominent (50–100% relative intensity) only in the mass spectra of fully saturated triglycerides occurs at a mass 128 amu above the RCO^+ ion. Based on high-resolution mass spectrometry and deuterium labeling, the following structure of this ion has been suggested[3,4]:

$$\overset{\overset{O}{\|}}{RCO}-\diamondsuit\hspace{-0.3em}\begin{array}{c}O\\\\O\end{array}\hspace{-0.3em}=CHCH_2^+ \qquad (RCO+128)^+$$

The 128 mass units thus have the elemental composition of $C_6H_8O_3$, which include the glyceryl backbone and the first three carbons from one other alkyl chain. Very weak ions are also observed at masses corresponding to $(RCO + 128 + 14n)^+$, where $n = 1, 2, 3 \ldots$.[3,4] It has been demonstrated that this series of ions can serve to locate double bond positions by way of deuteration in unmixed triglycerides.[3] The utility of this procedure has yet to be fully realized.

Another ion of medium intensity (10–20% relative intensity) which also appears only in the spectra of fully saturated triglycerides corresponds to $(RCO + 115)^+$. Its suggested structure is[2,3]

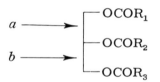

By way of partial summary, it should now be apparent that there are at least three types of ions which will give a self-consistent indication of the masses of the individual fatty acids. These ions are the $(M-RCOO)^+$, RCO^+, and $(RCO + 74)^+$ ions. If any of the fatty acids are unsaturated, the $(M-RCOOH)^+$ and $(RCO-1)^+$ ions will also be intense. If all of the fatty acids are saturated, the $(RCO + 115)^+$ and $(RCO + 128)^+$ ions will have a relative intensity of 10–20% and 50–90%, respectively.

Although the above set of ions will nicely specify the individual fatty acids, they will not specify the order of these acids on the glyceryl backbone. This information can be obtained, however, from a weak ion which corresponds to the electron impact-induced cleavage of the glyceryl moiety with charge retention on the larger fragment. Consider a fully mixed triglyceride with fatty acids R_1, R_2, and R_3:

$$a \longrightarrow \begin{array}{l} \text{—OCOR}_1 \\[4pt] \text{—OCOR}_2 \\[4pt] \text{—OCOR}_3 \end{array} \quad b \longrightarrow$$

Cleavage of bond a or b will produce a $(M-R_1COOCH_2)^+$ or $(M-R_3COOCH_2)^+$ ion, respectively. There is no cleavage that will produce an $(M-R_2COOCH_2)^+$ ion. Examination of Figs. 1 and 2 will demonstrate the utility of these ions. The mass of the $(M-RCOOCH_2)^+$ ion will be m/e 563 if R is palmitoyl or m/e 537 if R is oleyl. For 1,3-dipalmitoylolein, the only possible $(M-RCOOCH_2)^+$ ion would result from the loss of a palmitoyl group and would occur at m/e 563. Note that m/e 563 is much more intense than m/e 537. Conversely, 1,2-dipalmitoylolein exhibits ions at both m/e 537 and 563 since both $(M-RCOOCH_2)^+$ ions can be formed.

Applications of Mass Spectrometry to the Structural Elucidation of Triglycerides

Armed with the general analysis of the mass spectra of triglycerides presented above, it is instructive to review the applications of this technique. This is not a difficult task since there have been only a few papers reporting the use of this method.

In one case, an unknown optically active triglyceride was isolated from the seed oil of the Chinese tallow tree, *Sapium sebiferum*.[6] Together with infrared spectrophotometry, mass spectrometry of both the natural product and its fully saturated derivative played an important role in the structural elucidation of this compound. The mass spectrum of the hydrogenated unknown triglyceride showed a molecular ion at m/e 920 and two different ion series containing the RCO^+, $(RCO + 74)^+$, $(RCO + 128)^+$, and $(M-RCOO)^+$ ions. These series corresponded to acyl groups of 267 and 297 amu, which were interpreted to have elemental compositions of $C_{18}H_{35}O$ and $C_{18}H_{33}O_3$, respectively. The three oxygen atoms in the second ion series indicated an extra ester functionality in one fatty acid group.[6] An intense acyl ion at m/e 155, but without the corresponding $(RCO + 74)^+$ or $(RCO + 128)^+$ ions, suggested a ten-carbon atom acyl group in the ester-containing moiety. An ion of m/e 593 formed by the loss of $C_{17}H_{33}O_2COOCH_2$ from the molecular ion indicated that the ester-containing group was bound to a primary hydroxyl group of glycerol. The following structure results from these observations[6]:

$$
\begin{array}{l}
\quad\quad\quad\quad\overset{\displaystyle O}{\overset{\displaystyle \|}{}} \\
CH_2-O\overset{O}{\overset{\|}{C}}(CH_2)_{16}CH_3 \\
| \quad\quad\quad\quad O \\
\quad\quad\quad\quad \| \\
CH-O\overset{}{C}(CH_2)_{16}CH_3 \\
| \quad\quad\quad\quad O \quad\quad\quad\quad O \\
\quad\quad\quad\quad \| \quad\quad\quad\quad \| \\
CH_2-O\overset{}{C}(CH_2)_7 o\overset{}{C}(CH_2)_8CH_3
\end{array}
$$

The mass spectrum of the original natural product, although not obtained on a pure sample, was of considerable use in deducing the degree and position of unsaturation.[6] The natural product exhibited a molecular ion at m/e 902, thus indicating that it contained nine double bonds. Ions at m/e 261 and 263, corresponding to linolenic and linoleic acyl ions, respectively, along with the appropriate $(RCO + 74)^+$ ions were observed. The only $(M-RCOO)^+$ ion observed was at m/e 597, and it corresponded to the loss of the ester-containing fatty acid moiety from the molecular ion. The ten-carbon acyl ion at m/e 155 in the spectrum of the saturated molecule was shifted to m/e 151 in the spectrum of the natural product, thus indicating two double bonds in this section of the molecule.

---

[6] H. W. Sprecher, R. Maier, M. Barber, and R. T. Holman, *Biochemistry* **4**, 1856 (1965).

The above information, when taken together with other data on the intact and degraded molecule, led to the following structure:

$$CH_2—O\overset{\overset{\textstyle O}{\|}}{C}(CH_2)_7CH=CHCH_2CH=CHCH_2CH=CHCH_2CH_3$$

$$CH—O\overset{\overset{\textstyle O}{\|}}{C}(CH_2)_7CH=CHCH_2CH=CH(CH_2)_4CH_3$$

$$CH_2—O\overset{\overset{\textstyle O}{\|}}{C}(CH_2)_3—CH=C=CHCH_2—O\overset{\overset{\textstyle O}{\|}}{C}CH=CHCH=CH(CH_2)_4CH_3$$

As part of an investigation into the heat-induced flavors in processed milk, the triglyceride 1-(5-hydroxy)-dodecanoyl-2,3-dihexadecanoyl glycerol was synthesized and its mass spectrum determined.[7] The expected RCO^+, $(RCO + 74)^+$, $(RCO + 128)^+$, and $(M-RCOO)^+$ ions were all observed (although some were of low intensity). In addition, an intense ion at m/e 256 was observed which corresponds to the loss of the 5-hydroxy group from the $(RCO + 74)^+$ ion containing the hydroxy dodecanoyl moiety (at m/e 273).

It is interesting to note that neither the hydroxyl group in this compound nor the ester group in the previous compound changed the nature of the mass spectra very much; that is, ions corresponding to RCO^+, $(RCO + 74)^+$, $(RCO + 128)^+$, and $(M-RCOO)^+$ were observed, even though the R group was not a hydrocarbon.

In another application, the eye of the sand trout *Cynoscion arenarius* was shown to contain a reflecting layer of lipid spherules consisting almost exclusively of glyceryl tridocosahexenoate.[8] Although the mass spectral data were not given in detail, the authors reported the observation of the $(RCO + 74)^+$, $(RCO + 115)^+$, $(RCO + 128)^+$, and $(M-RCOO)^+$ ions in the spectra of both the natural product and its saturated derivative.[8] In this case mass spectrometry offered confirmation rather than insight.

Analysis of Triglyceride Mixtures by Mass Spectrometry

Although the above three cases dealt with relatively pure triglycerides, mixtures and, in fact, quite complex mixtures are the more usual situation encountered in nature; for example, a natural fat containing only three fatty acids may consist of a mixture of eighteen different tri-

[7] A. L. Dolendo, J. C. Means, J. Tobias, and E. G. Perkins, *J. Dairy Sci.* 52, 21 (1968).

[8] J. A. C. Nicol, H. J. Arnott, G. R. Mizuno. E. C. Ellison, and J. R. Chipault, *Lipids* 7, 171 (1972).

TABLE I

Carbon No.	Double bonds	MW	Kokum butter 0.41		Cocoa butter 0.46		Olive oil 0.99		Peanut oil 1.11	
			OB[b]	TH[b]	OB	TH	OB	TH	OB	TH
48	2	802					0.7	0.1		
48	1	804					0.8	0.3		
48	0	806								
50	4	826					0.4	0.1		
50	3	828			0.8	0.4	1.2	1.2	0.3	0.1
50	2	830			2.3	1.2	2.3	4.2	0.6	0.9
50	1	832			13.4	13.2	2.5	4.5	0.4	1.6
50	0	834								
52	6	850								
52	5	852					1.3	0.4	0.6	0.4
52	4	854			1.1	0.5	8.5	4.6	5.4	3.4
52	3	856			3.0	2.4	14.7	16.0	7.5	9.7
52	2	858	1.4	0.5	10.1	9.8	14.3	18.8	4.3	9.1
52	1	860	5.4	3.6	30.7	34.6	0.7	1.7	0.3	1.3
52	0	862								
54	9	872								
54	8	874								
54	7	876							0.3	0.4
54	6	878					2.0	0.6	4.1	3.5
54	5	880			0.7	0.1	9.7	5.8	21.1	12.2
54	4	882			1.5	0.9	18.6	18.0	22.4	20.3
54	3	884	5.1	1.2	4.9	3.2	19.5	20.0	14.2	15.1
54	2	886	18.9	19.2	11.2	10.7	2.2	3.4	1.1	4.2
54	1	888	69.2	75.4	20.4	22.8	0.4	0.2	0.3	0.7
54	0	890								
Correlation coefficient			0.99+		0.99+		0.96		0.92	

Average number header above: 0.41, 0.46, 0.99, 1.11

[a] Reprinted from Hites.[9] Copyright 1970 by the American Chemical Society.
[b] OB, observed; TH, theoretical.

glycerides. The identification and quantitation of the individual components in these mixtures has been a difficult task usually requiring large amounts of sample and time. The application of mass spectrometry in this area has resulted in a quantitative method that is both rapid (i.e., 30 minutes/sample) and sensitive (less than 100 μg of fat is needed) and involves no pretreatment of the sample.[9]

As can be seen from Figs. 1 to 3, there are no intense ions within at least 200 amu of the molecular ion (if the fatty acid groups contain

[9] R. A. Hites, *Anal. Chem.* **42**, 1736 (1970).

OBSERVED AND THEORETICAL TRIGLYCERIDE COMPOSITIONS
FOR VARIOUS VEGETABLE OILS[a]

Cotton-seed oil		Corn oil		Soybean oil		Sunflower oil		Safflower oil		Linseed oil	
double bonds											
1.30		1.46		1.58		1.62		1.65		2.13	
OB	TH	OB	TH	OB	TH	OB	TH	OB	TH	OB	TH
0.8	0.8										
0.3	0.4										
1.4	1.1										
0.8	1.5	0.4	0.1	0.6	0.2					0.6	0.5
1.9	8.0	1.3	1.9	0.7	1.5	0.3	0.8	0.2	1.0	0.3	0.1
0.5	2.7	0.3	0.9	0.2	0.6	0.1	0.2	0.1	0.2	0.3	0.2
				0.7	0.2					8.5	5.4
1.7	0.8	1.4	0.8	4.6	3.0					5.3	3.2
31.5	22.0	17.7	12.0	13.5	11.2	12.9	11.2	13.5	13.1	4.8	4.6
8.0	14.9	6.0	11.1	4.2	8.4	2.2	4.8	2.0	5.0	1.1	1.9
1.3	4.5	1.1	3.7	1.0	2.8	0.4	1.6	0.2	1.3	0.7	1.0
0.4	0.7	0.5	0.5	0.5	0.4	0.6	0.2	0.6	0.2	0.5	0.2
				0.0	0.1					14.2	15.2
		0.0	0.1	1.4	1.1					11.9	13.6
		1.5	1.8	9.0	7.9					20.0	21.5
26.9	15.3	31.5	19.7	30.1	22.5	46.4	39.0	56.8	42.5	14.9	14.4
16.4	15.8	26.4	26.2	21.1	22.6	25.9	25.0	19.6	24.3	10.6	10.4
6.1	8.3	9.5	15.5	9.2	12.3	9.4	12.8	5.7	9.8	3.7	5.1
1.3	2.6	2.1	5.0	2.5	4.1	1.0	3.6	0.9	2.3	1.6	1.9
0.7	0.5	0.4	0.9	0.6	0.8	0.4	0.7	0.3	0.3	0.6	0.6
		0.2	0.1	0.2	0.1	0.2	0.1			0.4	0.1
0.90		0.92		0.96		0.99		0.97		0.98	

Reprinted by permission of the copyright owner.

more than about twelve carbon atoms each). This lack of interferences makes it possible to measure the molecular weight distribution of a triglyceride mixture from its mass spectrum. Because the molecular weight gives only the total number of carbon atoms and double bonds, it is impossible to distinguish between positional isomers or between isologs. In spite of this limitation, the method seems quite useful.

The method is briefly summarized here; details can be found elsewhere.[9] From 100 to 500 μg of the fat of interest is placed directly into the ion source of the mass spectrometer ("probe" inlet) and heated to a known and constant temperature (in this case, $289 \pm 4°$) until thermal equilibrium is established and the total ion current is constant. At least three mass spectra are then recorded over the mass range of about m/e 700 to 900. The resulting intensities are corrected for contributions of ^{2}H, ^{13}C, ^{17}O, and ^{18}O; the M^+ and corresponding $(M\text{-}18)^+$ intensities are summed, and the replicates, if any, are averaged. The fractionation correction factors (see below) are applied, and the resulting intensities are normalized to a total of 100%.

Because a direct introduction inlet system is used, a correction must be made for the consequent fractionation caused by molecular distillation. It can be shown[9] that the logarithm of this correction factor, F_i, is a linear function of the number of carbon atoms in the triglyceride, C_i.

$$\log F_i = -kC_i + k' \tag{1}$$

By performing the above measurements on a mixture consisting of several triglycerides of known relative concentration, the constants in Eq. (1) can be determined.

To determine the accuracy of this method, the oils listed in Table I were analyzed. The observed results were then compared to those values predicted by the 1,3-saturates–2-unsaturates theory of triglyceride distribution.[10,11] In simplified terms this theory assumes that unsaturated fatty acids are preferentially esterified at the 2 position of glycerol. The theoretical amounts of the various triglycerides which are calculated from the known fatty acid composition in a straightforward manner[9] are given in Table I under the columns headed "TH." These values should be compared to the observed values in the columns headed "OB." The agreement between the observed and theoretical values throughout Table I is gratifying, considering that the theoretical assumptions are not exact, that no molar sensitivity corrections were used, and that the measurements cover a wide dynamic range.

Applications of this technique have been demonstrated in various areas of triglyceride research. These include following an interesterification reaction (randomization), comparing an artificial cocoa butter to the natural product, and determining the level of adulteration in a commercial oil.[9] In addition, Table I and data like it can serve as a means of qualitatively identifying an unknown oil.

A more recent method of quantitating triglyceride mixtures makes use of combined gas chromatography–mass spectrometry.[12] The gas chro-

[10] M. H. Coleman, *Advan. Lipid Res.* **1**, 1 (1963).
[11] H. J. Dutton, *Progr. Chem. Fats Other Lipids* **6**, 313 (1963).
[12] T. Murata and S. Takahashi *Anal. Chem.* **45**, 1816 (1973).

matographic inlet separates the triglycerides into fractions on the basis of the total number of carbon atoms, and the mass spectrum taken during these gas chromatographic (GC) fractions identifies the individual fatty acids comprising that fraction. Further, the intensities of the RCO$^+$ and (RCO + 128)$^+$ ions can be used to semiquantitate the abundance of each fatty acid within a GC fraction. By combining knowledge of the total carbon number of the fatty acids (obtained from GC retention time and molecular weight) and of the identities and amounts of the individual fatty acids, one can deduce the identities and abundances of the triglyceride types present in each GC fraction. (A triglyceride type is defined by its three constitutive fatty acids.) Because of overlapping peaks this method[12] cannot be applied with any confidence to those triglycerides containing unsaturated fatty acids. However, because of the prior GC separation, this technique can cover a wider range of carbon numbers than the direct introduction method outlined above.[9] Analyses of the triglycerides found in butter fat, tallow, rat blood, rat liver, head oil, and coconut oil have been carried out using this gas chromatographic–mass spectrometric method.[12]

Acknowledgment

The author is grateful to the National Research Council for a Postdoctoral Research Associateship during 1968–69 held at the Northern Regional Research Laboratory, Peoria, Illinois. The data in Figs. 1 to 3 were obtained during this associateship.

[45] Mass Spectrometry of Prostaglandins

By P. F. CRAIN, DOMINIC M. DESIDERIO, JR.,
and JAMES A. McCLOSKEY

The structure elucidation of prostaglandins[1,2] and their metabolites has been dominated by the use of gas chromatography–mass spectrometry, and in fact represents one of the most successful applications of mass spectrometry to biologic structural problems.[3,4] In particular, the

[1] T. O. Oesterling, W. Morozowich, and T. J. Roseman, *J. Pharm. Sci.* **61**, 1861 (1972).

[2] J. W. Hinman, *Annu. Rev. Biochem.* **41**, 161 (1972).

[3] L. L. Engel and J. C. Orr, *in* "Biochemical Applications of Mass Spectrometry" (G. R. Waller, ed.), pp. 538–545. Wiley (Interscience) New York, 1972.

[4] M. F. Grostic, *in* "Biochemical Applications of Mass Spectrometry" (G. R. Waller, ed.), Chapter 20, p. 587. Wiley (Interscience), New York, 1972.

TABLE I

MASS SPECTRA OF PROSTAGLANDINS

Ring type	Structure code[a] and references[b]
A	20-4-2,[5-8] 20-4-3[9]
B	16-4-2,[10,11] 16-5-2,[11] 16-6-2,[11] 18-4-2,[10] 20-3-1,[12] 20-4-2,[6,7,10,13,14] 20-4-3,[14] 20-5-2[3,15]
E	16-5-0,[11,16] 16-5-1,[11] 16-6-0,[11] 16-7-0,[11,17-19*] 18-5-0,[16] 18-5-1,[16] 20-5-0,[16,20] 20-5-1,[3,16,21-27] 20-5-2,[3,16,18,22,27,28*,29*,30*,31,32] 20-5-3[3,22]
F	13-5-0,[33] 13-6-0,[33] 14-4-1,[34] 14-6-0,[34,35] 14-6-1,[34,36] 14-7-0,[34,35] 15-5-0,[37] 16-4-0,[38] 16-5-0,[3,3*,11,38,39,39*,40-44] 16-5-1,[43] 16-6-0,[42,42*,43,45] 16-6-1,[43,45] 16-7-0,[41,42,43,45,46] 16-7-1,[38,45] 18-4-1,[34] 18-5-0,[47] 18-5-1,[10,43,45,46] 18-6-1,[45] 18-7-1,[45,46] 20-5-0,[37] 20-5-1,[21,23,24,29*,30*,37,48-52] 20-5-2,[23,28*,29*,30*,32,50,52,53*,54*]
Other	20-4-0,[55] 20-4-3,[6] 20-5-0,[55,56] 20-5-1,[16,55] 20-5-2[56,57]

[a] See text for details.
[b] Asterisk denotes presence of a stable isotope in the molecule.

sensitivity of the technique, the objectivity of the information derived, and the ability to work without rigorous sample purification have contributed greatly to development of its routine use, particularly by Samuelsson and his collaborators in Sweden.

The extensive role of mass spectrometry in this field can be judged by the number of entries in Table I,[5-57] which provides a comprehensive guide to published full mass spectra[58] of prostaglandins and their chemical and biologic derivatives. Entries are grouped according to structure

[5] D. M. Desiderio and K. Haegele, *Chem. Commun.* p. 1074 (1971).

[6] R. L. Jones, *J. Lipid Res.* **13**, 511 (1972).

[7] B. J. Sweetman, J. C. Frölich, and J. T. Watson, *Prostaglandins* **3**, 75 (1973).

[8] B. S. Middleditch and D. M. Desiderio, *Prostaglandins* **4**, 31 (1973).

[9] J. B. Lee, *Prostaglandins, Proc. Nobel Symp., 2nd, 1966* p. 197 (1967).

[10] M. Hamberg, *Eur. J. Biochem.* **6**, 135 (1968).

[11] K. Gréen, *Biochemistry* **10**, 1072 (1971).

[12] B. Samuelsson and G. Ställberg, *Acta Chem. Scand.* **17**, 810 (1963).

[13] C. B. Struijk, R. K. Beerthius, H. J. J. Pabon, and D. A. van Dorp, *Rec. Trav. Chim. Pays-Bas* **85**, 1233 (1966).

[14] B. S. Middleditch and D. M. Desiderio, *Lipids* **8**, 267 (1973).

[15] U. Israelsson, M. Hamberg, and B. Samuelsson, *Eur. J. Biochem.* **11**, 390 (1969).

[16] K. Gréen, *Chem. Phys. Lipids* **3**, 254 (1969).

[17] M. Hamberg and B. Samuelsson, *J. Amer. Chem. Soc.* **91**, 2177 (1969).

[18] M. Hamberg and B. Samuelsson, *J. Biol. Chem.* **246**, 6713 (1971).

[19] M. Hamberg, *Biochem. Biophys. Res. Commun.* **49**, 720 (1972).

[20] E. Änggård and B. Samuelsson, *J. Biol. Chem.* **239**, 4097 (1964).

[21] S. Bergström, L. Krabisch, B. Samuelsson, and J. Sjövall, *Acta Chem. Scand.* **16**, 969 (1962).

[22] S. Bergström, F. Dressler, R. Ryhage, B. Samuelsson, and J. Sjövall, *Ark. Kemi* **19**, 563 (1962).

[23] F. Vane and M. G. Horning, *Anal. Lett.* **2**, 357 (1969).

[24] M. Hamberg and U. Israelsson, *J. Biol. Chem.* **245**, 5107 (1970).

[25] B. Samuelsson, E. Granström, K. Gréen, and M. Hamberg, *Ann. N. Y. Acad. Sci.* **180**, 138 (1971).

[26] M. A. Marrazzi, J. E. Shaw, F. T. Tao, and F. M. Matschinsky, *Prostaglandins* **1**, 389 (1972).

[27] B. S. Middleditch and D. M. Desiderio, *J. Org. Chem.* **38**, 2204 (1973).

[28] U. Axen, K. Gréen, D. Hörlin, and B. Samuelsson, *Biochem. Biophys. Res. Commun.* **45**, 519 (1971).

[29] K. Gréen, *Advan. Biosci.* **9**, 91 (1973).

[30] K. Gréen, E. Granström, and B. Sameulsson, *in* "Third Conference on Prostaglandins in Fertility Control" (S. Bergström, K. Gréen, and B. Samuelsson, eds.), p. 92. Karolinska Institutet, Stockholm, 1972.

[31] R. J. Light and B. Samuelsson, *Eur. J. Biochem.* **28**, 232 (1972).

[32] L. Baczynskyj, D. J. Duchamp, J. F. Zieserl, Jr., and U. Axen, *Anal. Chem.* **45**, 479 (1973).

[33] M. Hamberg and B. Samuelsson, *J. Biol. Chem.* **242**, 5336 (1967).

[34] E. Granström, *Advan. Biosci.* **9**, 49 (1973).

[35] E. Granström, *Eur. J. Biochem.* **25**, 581 (1972).

[36] E. Granström and B. Samuelsson, *J. Amer. Chem. Soc.* **94**, 4380 (1972).

[37] E. Granström, *Eur. J. Biochem.* **20**, 451 (1971).

[38] K. Svanborg and M. Bygdeman, *Eur. J. Biochem.* **28**, 127 (1972).

[39] E. Granström and B. Samuelsson, *Eur. J. Biochem.* **10**, 411 (1969).

[40] E. Granström and B. Samuelsson, *J. Amer. Chem. Soc.* **91**, 3398 (1969).

[41] M. Hamberg and B. Samuelsson, *Biochem. Biophys. Res. Commun.* **34**, 22 (1969).

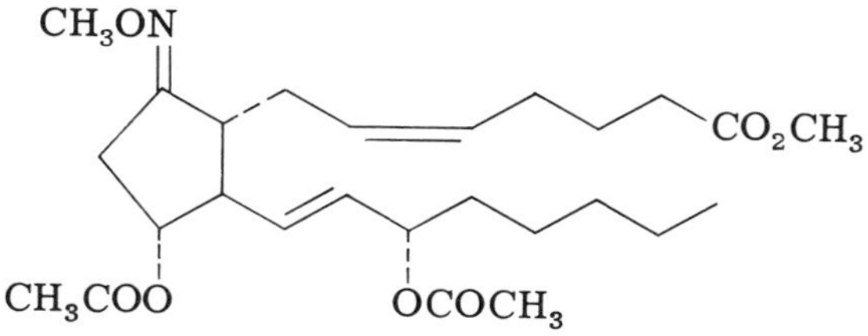

(I)

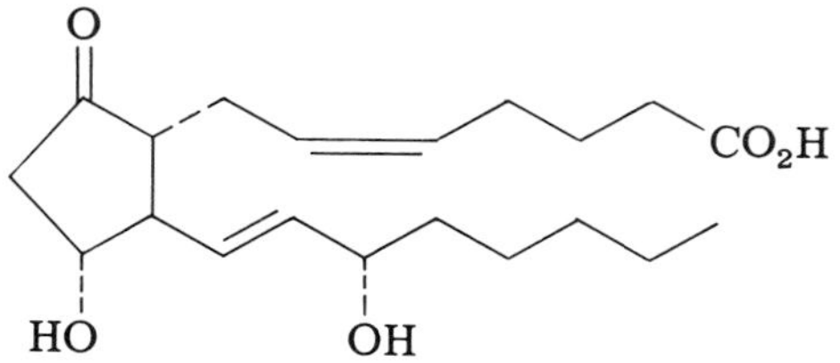

(II)

of the five-membered ring without regard to the stereochemistry of substituents. The code preceding literature citations gives (a) the carbon
atom chain length, (b) the number of oxygen atoms in the molecule, and

[42] E. Granström and B. Samuelsson, *J. Biol. Chem.* **246**, 5254 (1971).

[43] K. Gréen, *Biochim. Biophys. Acta* **231**, 419 (1971).

[44] M. Hamberg and B. Samuelsson, *J. Biol. Chem.* **247**, 3495 (1972).

[45] E. Granström and B. Samuelsson, *J. Biol. Chem.* **246**, 7470 (1971).

[46] E. Granström, *in* "Third Conference on Prostaglandins in Fertility Control"
(S. Bergström, K. Gréen, and B. Samuelsson, eds.), p. 18. Karolinska Institutet,
Stockholm, 1972.

[47] H. Kindahl and E. Granström, *Biochim. Biophys. Acta* **280**, 466 (1972).

[48] E. Granström, W. E. M. Lands, and B. Samuelsson, *J. Biol. Chem.* **243**, 4104
(1968).

[49] E. Granström, *Biochim. Biophys. Acta* **239**, 120 (1971).

[50] C. Pace-Asciak and L. S. Wolfe, *J. Chromatogr. Sci.* **56**, 129 (1971).

[51] E. Granström, *Eur. J. Biochem.* **27**, 462 (1972).

[52] B. S. Middleditch and D. M. Desiderio, *Anal. Biochem.* **55**, 509 (1973).

[53] E. Änggård and B. Samuelsson, *Biochemistry* **4**, 1864 (1965).

[54] E. Änggård and B. Samuelsson, *Mem. Soc. Endocrinol.* **14**, 107 (1966).

[55] P. S. Foss, C. J. Sih, C. Takeguchi, and H. Schnoes, *Biochemistry* **11**, 2271
(1972).

[56] C. Pace-Asciak and L. S. Wolfe, *Biochemistry* **10**, 3657 (1971).

[57] C. Pace-Asciak, *Biochemistry* **10**, 3664 (1971).

[58] Full mass spectra are defined as those which include detailed ion abundance
data from m/e 100 or lower through the molecular ion peak, and which are
represented in plotted form.

(c) the number of double bonds in the molecule, excluding carbonyl groups. These values refer to the parent compound and are independent of protective groups added for mass spectrometry. An asterisk denotes the presence of a stable isotope in the molecule; for example, the code for the methyl ester, O-methyl oxime diacetyl derivative I of PGE$_2$[59] (II) is 20-5-2, listed under ring designation E. The data in Table I may not be complete because of the wide distribution of prostaglandin mass spectra in the literature.

The remainder of this chapter deals with the two principal uses of mass spectrometry: the interpretation of spectra in terms of prostaglandin structure, and the quantitative estimation of prostaglandins by the deuterium carrier technique.

Acquisition and Interpretation of Mass Spectra

Following initial isolation and separation procedures, prostaglandins can be submitted directly to mass spectrometry without the intervention of gas chromatography, but this approach is currently seldom used in view of the inherent advantages of combination gas chromatography–mass spectrometry.[60,61] Because of the polarity of prostaglandins, conversion to volatile derivatives prior to gas chromatography is mandatory. Examples of derivatives which have been proposed for this purpose are given in Table II.[16,21,23,32,33,37,50,51,62,63]

The most frequently used and generally successful derivatives are the ME-MO-TMS and ME-MO-Ac,[59] and will therefore be used as examples. TMS esters can also be employed,[23] but the more stable methyl esters are usually prepared prior to chromatographic separations during isolation. The methoxime derivative is particularly advantageous when working with PGE to prevent degradation of the thermally sensitive β-ketol system during gas chromatography. The use of two derivatives (e.g., TMS vs. Ac), when permitted by sample quantity, provides cross correla-

[59] Abbreviations used are: PGA, 15α-hydroxy-9-oxo-10,13-*trans*-prostadienoic acid; PGB, 15-α-hydroxy-9-oxo-8(12),13-*trans*-prostadienoic acid; PGE$_1$, 11α,15α-dihydroxy-9-oxo-13-*trans*-prostenoic acid; PGE$_2$, 11α,15α-dihydroxy-9-oxo-5-*cis*, 13-*trans*-prostadienoic acid; PGF$_{2\alpha}$, 9α11α,15α-trihydroxy-5-*cis*,13-*trans*-prostadienoic acid; ME, methyl ester; MO, O-methyl oxime (methoxime); TMS, trimethylsilyl; and Ac, acetyl.

[60] J. A. McCloskey, Vol. 14 [50].

[61] J. T. Watson, *in* "Ancillary Techniques in Gas Chromatography" (L. S. Ettre and W. H. McFadden, eds.), Chapter 5, p. 145. Wiley (Interscience), New York, 1969.

[62] C. J. Thompson, M. Los, and E. W. Horton, *Life Sci., Part I* 9, 983 (1970).

[63] P. Foss, C. Takeguchi, H. Tai, and C. Sih, *Ann. N.Y. Acad. Sci.* 180, 126 (1971).

TABLE II
CHEMICAL DERIVATIVES OF PROSTAGLANDINS FOR GAS
CHROMATOGRAPHY–MASS SPECTROMETRY

Carboxyl group
　　Methyl ester[16]
　　Trimethylsilyl ester[23]
Carbonyl group
　　O-Methyl oxime[16,23,37]
　　O-Benzyl oxime[32,51]
Hydroxyl group
　　Trimethylsilyl ether[16,33]
　　Acetyl[16,37]
　　Trifluoroacetyl[62]
　　Phenylboronyl[50]
　　Methyl ether[21,63]

tions of structural features derived by interpretation of their mass spectra. In addition, applicability of the four above protecting groups is enhanced by the availability of deuterium-labeled reagents. Spectra of the labeled derivatives thus produced are highly useful in corroborating data obtained from the spectra of unlabeled derivatives and in further understanding the details of fragmentation reactions which occur in the mass spectrometer.

Preparation of Derivatives for Gas Chromatography–Mass Spectrometry[16,17]

a. Methyl Ester (ME). 100–200 μg of prostaglandin is dissolved in CH$_3$OH (0.3 ml) and treated with diazomethane in ether.

b. O-Methyl Oxime (MO). ME derivatives are dissolved in pyridine (0.2 ml) containing methoxylamine hydrochloride (5 mg) and allowed to stand in a closed vessel 15–20 hours at room temperature. If methoxime derivatives are to be used they are prepared prior to acetylation or trimethylsilylation.

c. Trimethylsilyl Ether (TMS). Trimethylchlorosilane (20 μl) and hexamethyldisilazane (40 μl) are added to the final solution from step *b* above, then allowed to stand 30–60 minutes in a desiccator. The sample is evaporated to dryness under vacuum and immediately dissolved in carbon disulfide, and injected into the gas chromatograph.

d. Acetate (Ac). To prepare ME-Ac derivatives, the product from step *a* above is taken to dryness, dissolved in pyridine (0.3 ml) and acetic anhydride (0.3 ml) is added. If the ME-MO-Ac derivative is prepared, the residue from step *b* is extracted 3 times with ether (0.5 ml) and the combined extracts are evaporated to dryness prior to addition of pyridine and acetic anhydride. After standing overnight, the reaction mixture is diluted with ice water (1 ml), acidified, then extracted with ether 3 times. The combined extracts are washed with small volumes of 0.2 N hydrochloric acid, 10% sodium bicarbonate, and water, until neutral. The solution is evaporated to dryness, and redissolved in carbon disulfide prior to injection into the gas chromatograph.

e. Derivatives Labeled with Deuterium in the Functional Groups. Methyl-d_3 esters are prepared[16] by transmethylation of the ME derivative by treatment with 0.5 N NaOCD$_3$ (0.25 ml) at room temperature for 10 minutes. The solution is acidified with 0.3 N HCl (1 ml) and extracted with ether 3 times; the ether phase is washed with water until neutral and then evaporated to dryness.

O-Methyl-d_3 oxime, trimethylsilyl-d_9, and acetyl-d_3 derivatives are prepared by the normal procedure, using deuterated reagents. Labeled reagents for silylation and acetylation are available commercially; O-methyl-d_3-oxime can be prepared by reaction between potassium hydroxylamine disulfonate and methyl-d_3 iodide.[18]

Gas Chromatography and Mass Spectrometry. Both ME-MO-Ac and ME-MO-TMS derivatives are thermally stable and exhibit reasonably good gas chromatographic properties. Using nonpolar phases such as SE-30 these derivatives elute at temperatures corresponding to carbon numbers[64] in the approximate range 20–31 (e.g., Hamberg,[10] Gréen,[43] and Hamberg et al.[65]). If 10% EGSS-X is employed it has been noted that methoxime derivatives will separate according to the number of derivatized carbonyl functions.[24] By comparison of acetyl and trimethylsilyl derivatives, differences in carbon number can be used to estimate the number and types of functional groups in the molecule.[10,43]

Virtually all applications of the combination technique to date have involved the jet-orifice carrier gas superator of Ryhage.[66] Although other types of separators may also be satisfactory for prostaglandin work, the low levels of adsorption encountered with the jet separator are important when working with small quantities of relatively polar prostaglandins.

A minor disadvantage of methoxime derivatives is the tendency to form both syn and anti isomers, which generally leads to two gas chromatographic peaks, or four if two methoxime functions are present.[16,23,65] As indicated by comparison of the two peaks associated with ME-MO-

---

[64] Expressed relative to the retention behavior of normal chain fatty acid esters; e.g., the carbon number of methyl docosanoate = 22.0.

[65] M. Hamberg, U. Israelsson, and B. Samuelsson, Ann. N.Y. Acad. Sci. 180, 164 (1971).

[66] R. Ryhage, Ark. Kemi 26, 305 (1967).

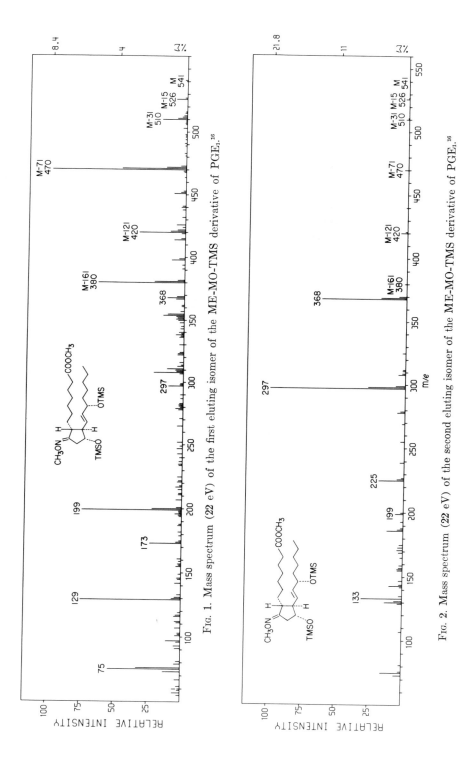

FIG. 1. Mass spectrum (22 eV) of the first eluting isomer of the ME-MO-TMS derivative of PGE₁.[16]

FIG. 2. Mass spectrum (22 eV) of the second eluting isomer of the ME-MO-TMS derivative of PGE₁.[16]

TMS derivatives of PGE_1 (III), their mass spectra are typically similar (Figs. 1 and 2) but show differences in relative intensities of most ions.[16] It is not presently known which chromatographic peak (or spectrum) corresponds to the *syn* or *anti* isomer, and whether a predictable retention time characteristic for the two isomers will hold for all prostaglandin MO derivatives.

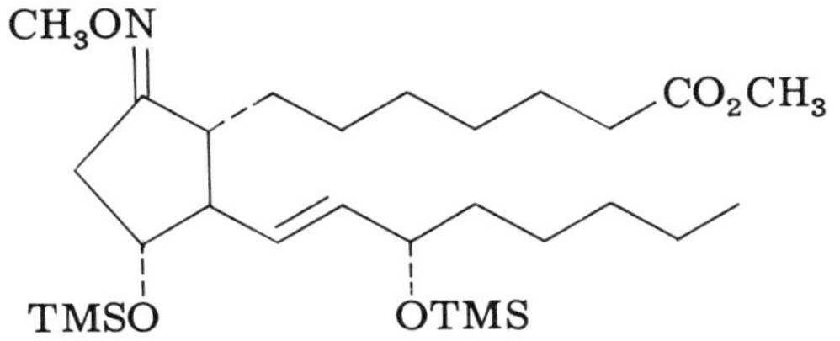

M.W. 541

(III)

The quantity of derivatized prostaglandin required for a conventional mass spectrum will normally be 0.1–1 μg, depending upon experimental conditions, although in favorable instances samples as small as $\sim$12 ng have been utilized.[62]

Basic structural information provided by mass spectra include molecular weight, chain length—in particular the hydrocarbon side chain, number of double bonds, and number of carbonyl, hydroxyl, and carboxyl groups. These data are derived from comparisons using different derivatives and deuterium-labeled functional groups (e.g., $-Si(CH_3)_3$ vs. $Si(CD_3)_3$[67]) and are readily deduced because of constraints dictated by molecular weight and certain basic structural assumptions such as the presence of a five-membered ring. Stereochemical features are reflected in the relative abundances of some ions but cannot normally be determined directly from the mass spectrum. These characteristics are shown by the minor (1:3), earlier eluting isomer of III represented in Fig. 1. As shown in Scheme 1, the molecular weight and C_5H_{11} side chain are represented by a series of upper mass ions derived from fragmentation reactions of the functional groups. Fragmentation is extensive and the molecular ion (M) peak is small. For this reason mass spectra are usually recorded in the relatively low electron energy region around 22 eV. The alternative use of O-benzyl oxime derivatives apparently produces more abundant molecular ions[51] as a result of charge localization in the aro-

[67] J. A. McCloskey, R. N. Stillwell, and A. M. Lawson, *Anal. Chem.* **40**, 233 (1968).

matic ring. The chemical ionization technique is another approach which is useful in enhancing peaks in the molecular ion region of prostaglandin spectra.[4]

The M-15 peak in Fig. 1 is produced by loss of CH_3 from a TMS group, while OCH_3 is lost from both the ester and methoxime groups to form M-31.[16] The spectrum of the second isomer (Fig. 2) is dominated by peaks at m/e 368 and 297. The former peak consists of two ion species, one proposed to arise by cleavage of the cyclopentyl ring and the other by hydrogen rearrangement accompanied by loss of OCH_3 from the methoxime group.[16] The m/e 297 ion is most likely generated by loss of the C_5H_{11} side chain from one of the m/e 368 species,[27] although a mechanism involving loss of the chain from C-15 and expulsion of

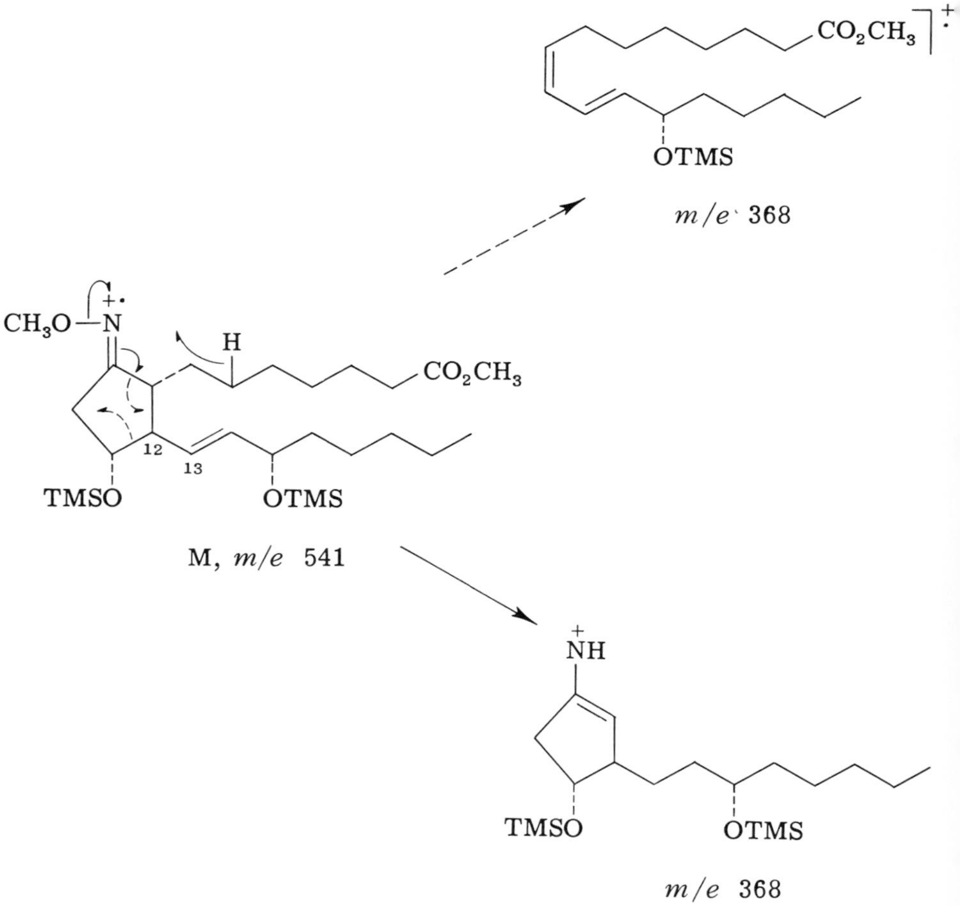

m/e 368

m/e 368

M, m/e 541

m/e 368

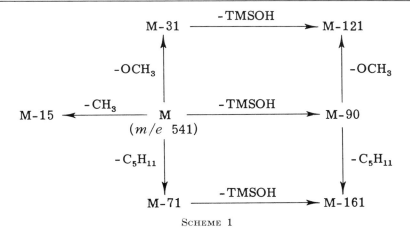

M-31 ——————$-$TMSOH——————▶ M-121

▲   ▲
$-$OCH$_3$ $-$OCH$_3$

M-15 ◀——$-$CH$_3$—— M ——$-$TMSOH——▶ M-90
         (m/e 541)

$-$C$_5$H$_{11}$ $-$C$_5$H$_{11}$

M-71 ——————$-$TMSOH——————▶ M-161

SCHEME 1

CH$_3$NO has also been proposed.[16] The prominent peak at m/e 199 in Fig. 1 is produced by cleavage of C-12,13[16] in a cyclization step, followed by expulsion of C$_2$H$_2$ to produce m/e 173.[27]

(III) —————▶ [structure with O, + TMS] —————▶ [structure with TMSO, +]

m/e 199 m/e 173

These pathways may be considerably modified in the mass spectra of prostaglandins having other functional groups or additional double bonds. However many of the basic ions often remain, suitably shifted in mass; for example, the ions m/e 368 and 297 shown in Figs. 1 and 2 appear 2 mass units lower in the spectrum of the ME-MO-TMS derivative of PGE$_2$ (IV).[16]

[structure IV with CH$_3$ON, CO$_2$CH$_3$, TMSO, OTMS]

M. W. 539

(IV)

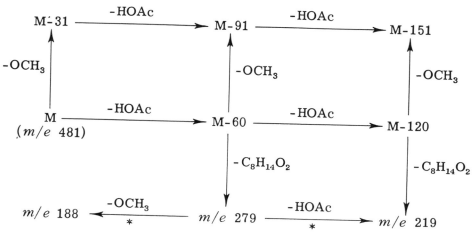

M. W. 481

(V)

Preparation of the O-acetyl analogue (e.g., V) provides an independent means of correlating gas chromatographic retention behavior and major features derived from the mass spectrum of III. The spectrum of the earlier eluting isomer of V, shown in Fig. 3, exhibits a molecular ion peak of very low abundance, but the molecular mass is easily verified from the characteristic network of ions represented in Scheme 2. The principal pathways arise from losses of OCH_3 and acetic acid, but unfortunately, unlike Scheme 1, do not include losses of the side chain moiety. The actual paths followed may be less redundant than that shown (i.e., M-31-60 vs. M-60-31), but few relevant metastable ion data (denoted by an asterisk in Scheme 2) have been reported.

Additional major ions relating to loss of the ester-containing side chain with hydrogen rearrangement, presumably by the γ-hydrogen migration proposed[16] for III above, are m/e 279, 219, and 188. The spectrum of the later eluting isomer (not shown) exhibits most of the same ions as in Fig. 3 but with altered relative intensities.

SCHEME 2

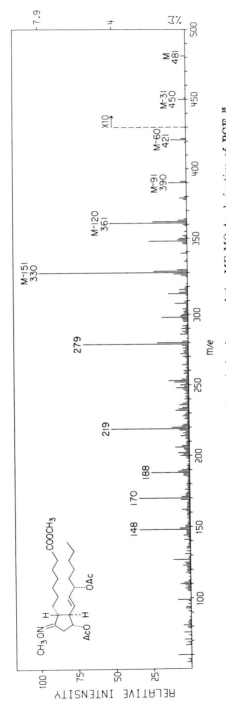

Fig. 3. Mass spectrum (22 eV) of the first eluting isomer of the ME-MO-Ac derivative of **PGE**.[16]

Ancillary Derivatization Reactions. Detailed structural features can often be determined by employing selective chemical reactions followed by mass spectrometry of the products. Examples of reactions which have been used for this purpose are listed below.

1. Determination of double bond position by oxidative ozonolysis[45,68]:

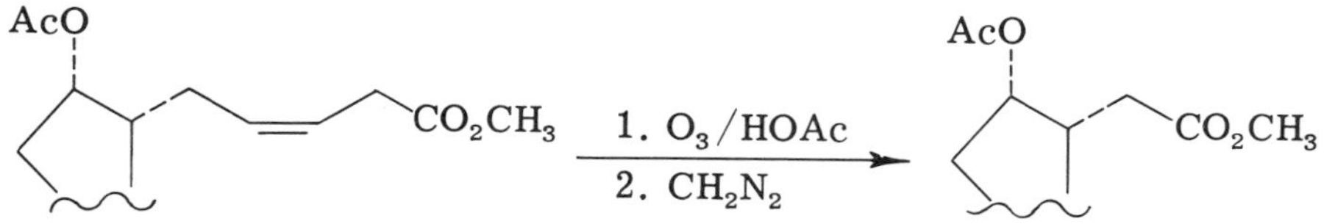

2. Determination of presence and number of double bonds by catalytic hydrogenation[20,34]:

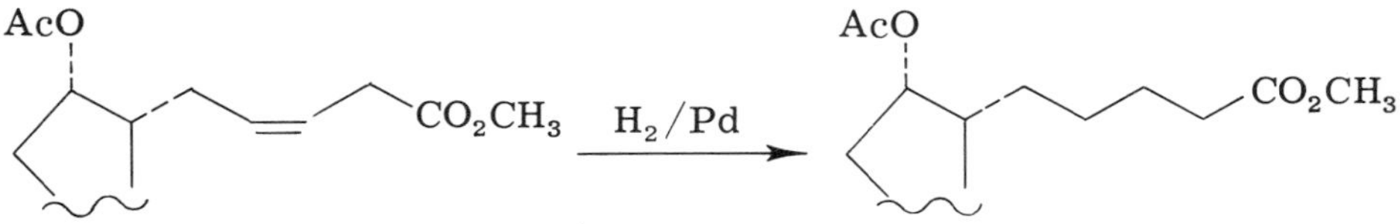

3. Ring transformations[69,70]:

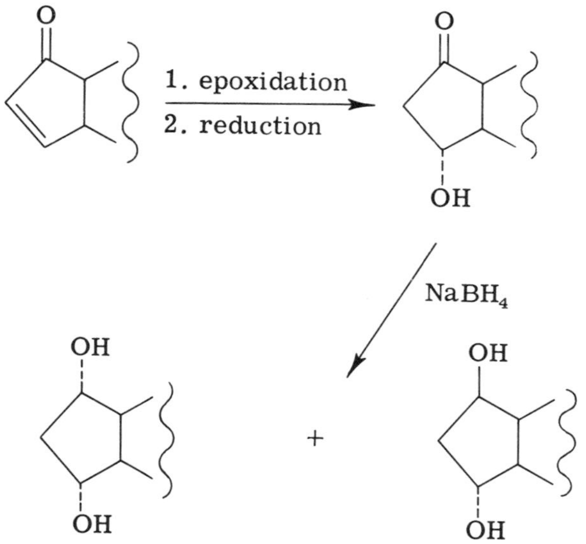

[68] M. Hamberg and B. Samuelsson, *J. Biol. Chem.* **241,** 257 (1966).
[69] G. Bundy, F. Lincoln, N. Nelson, J. Pike, and W. Schneider, *Ann. N.Y. Acad. Sci.* **180,** 76 (1971).
[70] G. L. Bundy, W. P. Schneider, F. H. Lincoln, and J. E. Pike, *J. Amer. Chem. Soc.* **94,** 2123 (1972).

4. Determination of the number of keto groups[20,48]:

$$\xrightarrow[\text{(R = H, D)}]{\text{NaBR}_4}$$

5. Determination of the number of carboxyl or ester groups[45]:

$$\xrightarrow[\text{2. silylate}]{\begin{array}{c}\text{1. LiAlR}_4\\ \text{(R = H, D)}\end{array}}$$

6. Establishment of keto-carboxyl or hydroxyl-carboxyl group proximity by γ-lactone formation:

$$\xrightarrow[\text{2. CH}_2\text{N}_2]{\begin{array}{c}\text{1. NaBR}_4\\ \text{(R = H, D)}\end{array}}$$

(Granström and Samuelsson[45])

$$\xrightarrow[\text{2. CH}_2\text{N}_2]{\text{1. OH}^-}$$

(Granström and Samuelsson[36])

7. Verification of lactone structure in metabolites by ring opening[45]:

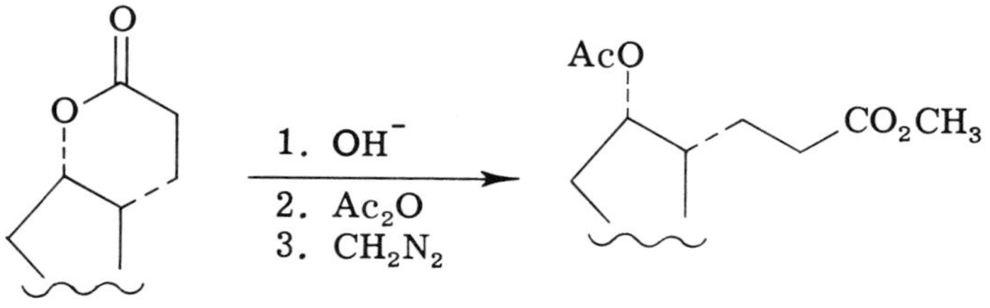

Incorporation of deuterium through use of labeled reagents such as sodium borodeuteride provides an additional means for mass spectrometric verification of the products which are obtained. In a similar fashion, structure correlations for prostaglandin metabolites can be made by employing specifically deuterated precursors, e.g., in the conversion of $PGF_{2\alpha}\text{-}d_7$ (VI) in the female human to the hexadeuterio metabolite VII[34]:

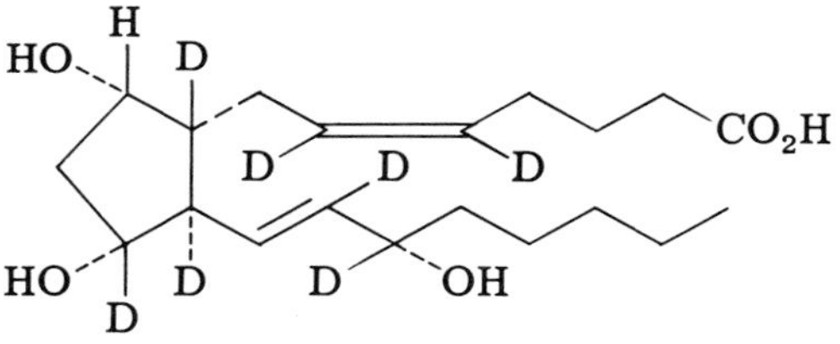

(VI)

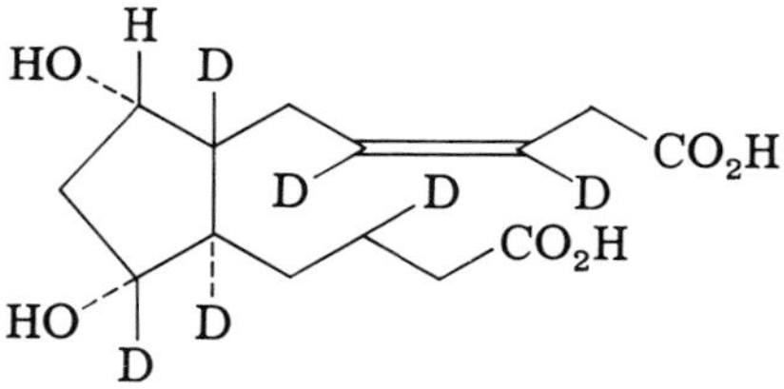

(VII)

The location and extent of deuteration can then generally be determined by careful analysis of the mass spectrum.

Quantitative Determination of Prostaglandins

Gas chromatography–mass spectrometry provides a unique and highly effective means for quantitative measurement of prostaglandins and their metabolites. The technique[28,71,72] as it is most commonly used involves the addition of a large excess (1:250–1:1000) of a deuterium-labeled form of the sample being examined, preferably before isolation.[28] If a sample containing a known quantity of the carrier is subjected to mass spectrometry, measurement of the relative mass spectrometric response from labeled and nonlabeled components of the mixture provides a quantitative measure of the unlabeled component. Since both forms of the molecule are subject to the same pretreatment, the method compensates for errors which are usually associated with working in the nanogram range: losses during isolation and on the gas chromatographic column, and variations of yield in the derivatization reactions. One inherent disadvantage of the method is the requirement for synthesis of the deuterated carrier prostaglandin or metabolite under investigation. However, the method is highly selective and suitable for routine use at the picomole level; for example, quantitative measurement of 0.7 pmole of the ME-MO-Ac derivative of PGE_2 has been reported, with standard deviations ($n = 10$) of 5.9% for 2.8 pmoles and 1.4% for 11.2 pmoles.[28]

The method of measurement relies on the multiple ion detection technique,[73] in which the mass spectrometer serves as a highly selective gas chromatographic detector by rapid, sequential monitoring of ions unique to the prostaglandin and its labeled carrier during elution from the gas chromatograph. For this purpose ions are chosen which are represented by intense peaks in the mass spectrum and have no contributions at that mass from other components, gas chromatographic column bleed or other sources of background. Ions from the high mass region of the spectrum are therefore most suitable for measurement.

An example of the technique is given by the work of Gréen,[29] in which the concentration of $PGF_{2\alpha}$ as the ME-Ac derivative VIII was de-

(VIII) R = H
(IX) R = D

[71] B. Samuelsson, M. Hamberg, and C. C. Sweeley, *Anal. Biochem.* **38**, 301 (1970).
[72] K. Gréen, E. Granström, B. Samuelsson, and U. Axen, *Anal. Biochem.* **54**, 434 (1973).
[73] C. C. Sweeley, W. H. Elliott, I. Fries, and R. Ryhage, *Anal. Chem.* **38**, 1549 (1966).

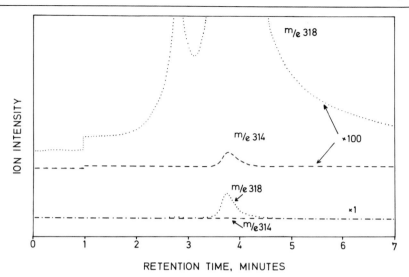

FIG. 4. Elution profiles of the ME-Ac derivatives of PGF$_{2\alpha}$ and its 3,3,4,4-d_4 carrier (1:1,000), represented by fragment ions m/e 314 and 318.

termined relative to the derivative of PGF$_{2\alpha}$-3,3,4,4-d_4 (IX) as carrier. The ions m/e 314 (from VIII) and 318 (from IX) were chosen for monitoring. These ions correspond to elimination of three molecules of acetic acid from the molecular ion and are the most intense peaks in their respective spectra. The elution profiles are shown in Fig. 4. Compounds VIII and IX were mixed in the ratio 1:1,000, and 100 ng of the mixture was injected into the combination instrument. Systematic variation of the ratio of unlabeled to labeled prostaglandin provides data points for the calibration curve shown in Fig. 5. Assuming equal mass spectrometer response from labeled and unlabeled molecules the theoretical slope of the calibration curve is 45°. If the carrier material contains a small amount of unlabeled compound there will be a "blank" and the experimental calibration curve will intercept the ordinate at a value above zero. For this reason systematic errors may arise if carrier dilutions greater than 1000:1 are employed.

Quantitative measurement of the E prostaglandins can also be carried out by methods involving their chemical conversion in high yield to PGA or PGB.[16,74,75] Volatile derivatives of the latter compounds exhibit more

[74] J. C. Frölich, B. J. Sweetman, K. Carr, J. Splawinski, J. T. Watson, E. Änggård, and J. A. Oates, *Advan. Biosci.* **9**, 321 (1973).
[75] G. H. Jouvenaz, D. H. Nugteren, R. K. Beerthius, and D. A. van Dorp, *Biochim. Biophys. Acta* **202**, 231 (1970).

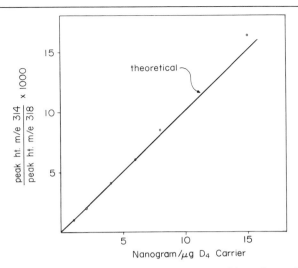

FIG. 5. Calibration curve derived from 100 ng quantities of standard mixtures of derivatized PFG$_{2\alpha}$ and its 3,3,4,4-d_4 carier, after corection from carrier blank.[28]

favorable gas chromatographic characteristics than the more polar PGE$_1$ or PGE$_2$, resulting in improved sensitivity in the sub-nanogram range.

The success of the deuterium carrier method has led to the application of computer techniques to various aspects of prostaglandin quantification, including acquisition and manipulation of data from gas chromatography–mass spectrometry, and calculation of prostaglandin concentration relative to the internal standard.[29,32,74,76–78] The speed and simplicity of use of these computerized systems assures their future use and further development for the routine quantitative measurements of prostaglandins and their metabolites.

Acknowledgments

The authors are grateful to Dr. L. Baczynskyj, Dr. K. Gréen, Dr. B. Samuelsson, Dr. J. T. Watson, and Dr. M. H. Wilson for their assistance with literature citations and in providing copies of unpublished manuscripts.

[76] J. F. Holland, C. C. Sweeley, R. E. Thrush, R. E. Teets, and M. A. Bieber, *Anal. Chem.* **45,** 308 (1973).

[77] J. T. Watson, D. R. Pelster, B. J. Sweetman, J. C. Frölich, and J. A. Oates, *Anal. Chem.* **45,** 2071 (1973).

[78] B. J. Sweetman, J. T. Watson, K. Carr, J. A. Oates, and J. C. Frölich, *Prostaglandins* **3,** 385 (1973).

[46] Chromatography on Lipophilic Sephadex

By ERNST NYSTRÖM and JAN SJÖVALL

I. Introduction

Column chromatography with lipophilic Sephadex derivatives is a valuable tool for separation of compounds soluble in organic solvents.[1,2] The technique has several advantages: mild conditions, high recoveries of trace amounts of labile compounds, simplicity in use, and possibility to prepare permanent column systems for use over long periods of time. This chapter describes methods for preparation of different lipophilic Sephadex derivatives which permit the creation of straight- or reversed-phase partition and adsorption systems and of systems based on specific group interactions. Examples of the use of some derivatives in preparative and analytical chromatography are given.

II. Preparation of Sephadex Derivatives

This section gives synthetic procedures for the preparation of (A) nonselective derivatives, i.e., materials of different polarity capable of swelling in organic solvents; and (B) selective derivatives, i.e., materials containing specific functional groups and capable of swelling in organic solvents. A schematic summary of different synthetic routes is given in Scheme 1.

All reactions should be carried out with moderate stirring to avoid mechanical damage of the beads and formation of fines. A magnetic stirrer is not suitable, and it is often sufficient to shake the reaction vessel intermittently. If a column material with a narrow and defined particle size is required, the starting material (e.g., Sephadex G-25 or Sephadex LH-20) should be sieved in the dry state. Further particle fractionation may be carried out by flotation in water.

[1] J. Sjövall, E. Nyström, and E. Haahti, *Advan. Chromatogr.* **6**, 119 (1968).
[2] M. Joustra, B. Söderqvist, and L. Fischer, *J. Chromatogr.* **28**, 21 (1967).

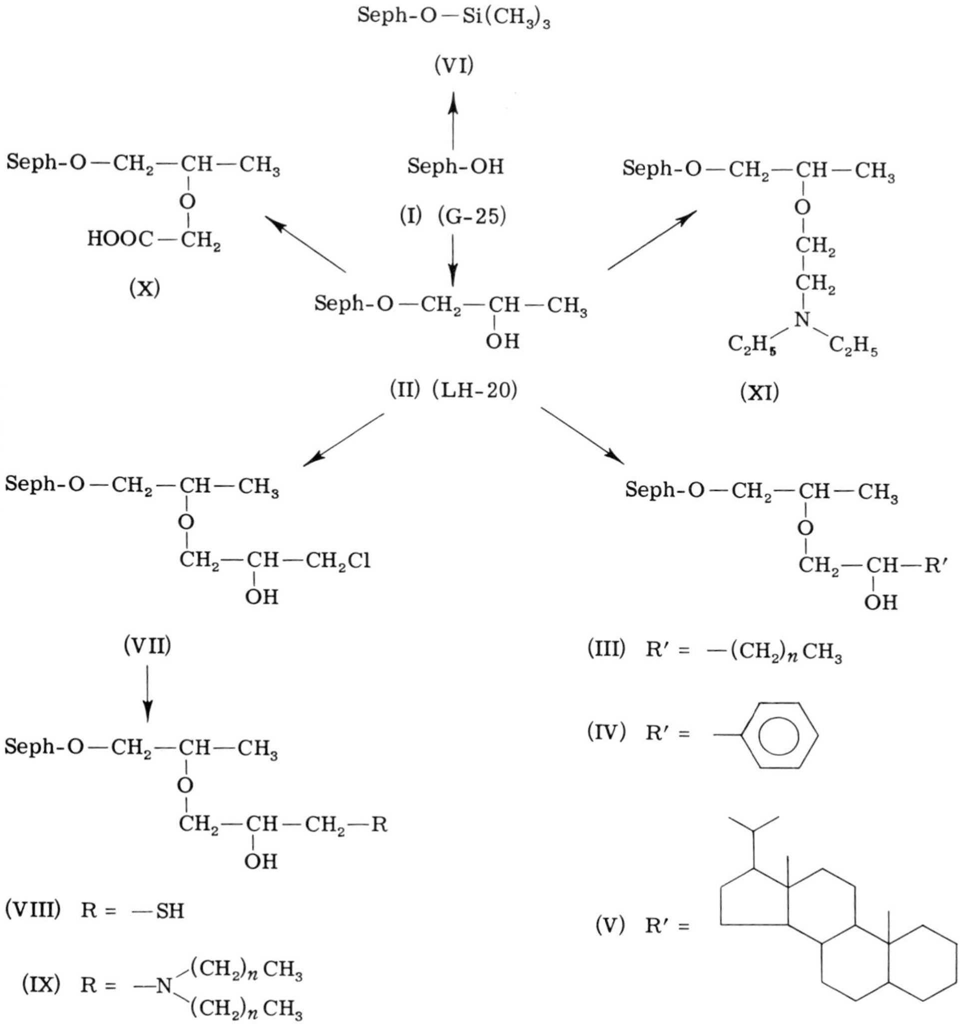

$$Seph-O-Si(CH_3)_3$$

(VI)

$$Seph-O-CH_2-CH-CH_3$$
with O and $HOOC-CH_2$ substituents

(X)

Seph-OH

(I) (G-25)

$$Seph-O-CH_2-CH-CH_3$$
with O–CH_2–CH_2–N(C_2H_5)(C_2H_5) substituent

(XI)

$$Seph-O-CH_2-CH-CH_3$$
with OH substituent

(II) (LH-20)

$$Seph-O-CH_2-CH-CH_3$$
with $O-CH_2-CH-CH_2Cl$ and OH substituents

(VII)

$$Seph-O-CH_2-CH-CH_3$$
with $O-CH_2-CH-CH_2-R$ and OH substituents

(VIII) R = —SH

(IX) R = $-N \underset{(CH_2)_n CH_3}{\overset{(CH_2)_n CH_3}{}}$

$$Seph-O-CH_2-CH-CH_3$$
with $O-CH_2-CH-R'$ and OH substituents

(III) R' = $-(CH_2)_n CH_3$

(IV) R' = phenyl

(V) R' = steroid structure

SCHEME 1. Summary of synthetic routes for preparation of lipophilic Sephadex derivatives.

A. Nonselective Derivatives

Methyl Sephadex. Methyl Sephadex was used in the early work with lipophilic Sephadex.[3,4] It has been replaced by the somewhat more polar

[3] E. Nyström and J. Sjövall, *Biochem. J.* **92**, 10P (1964).
[4] E. Nyström and J. Sjövall, *Anal. Biochem.* **12**, 235 (1965).

β-hydroxypropyl Sephadex, which is simpler to prepare. A detailed description of the synthesis of methyl Sephadex is given in Nyström.[5]

Hydroxypropyl Sephadex. Hydroxypropyl Sephadex (II, Scheme 1) may be prepared from Sephadex G-25 or G-50 by reaction with propylene oxide in aqueous sodium hydroxide (see Ellingboe et al.[6]). Two types of β-hydroxypropyl Sephadex are commercially available (Pharmacia Fine Chemicals, Uppsala, Sweden): Sephadex LH-20[2] and LH-40 which have been prepared from Sephadex G-25 and G-50, respectively. The hydroxypropyl group content corresponds to full substitution of the original hydroxyl groups of the glucose units. These preparations contain varying amounts of impurities and may be purified by refluxing for 4 hours in chloroform–methanol, 1:1 (v/v), prior to use.

Hydroxyalkoxypropyl Sephadex (III–V).[6,7] Sephadex LH-20 is used as starting material in the synthesis of hydrophobic alkyl derivatives. Ten grams are soaked in 100 ml of methylene chloride. Boron trifluoride ethyletherate (48% BF$_3$) is added, and thorough mixing is achieved by shaking. After 15 minutes, 15 ml portions of a 20% (v/v) solution of the alkyl epoxide Nedox 1114 (chain length 11–14, Ashland Chemical Co., Columbus, Ohio) in methylene chloride are added every 10 minutes with shaking and cooling of the reaction flask in ice water. When 150 ml of the epoxide solution have been added, the reaction is allowed to proceed for another 20 minutes. The product is then filtered on a Buchner funnel and is washed with successive portions of chloroform, ethanol, and chloroform–methanol, 1:1 (v/v). The product is then refluxed with stirring in chloroform–methanol, 1:1 (v/v), for 4 hours, washed with chloroform–methanol, 1:1, and benzene, and refluxed in benzene for 4 hours. After a final washing in benzene, the product (22.1 g) is dried *in vacuo* at room temperature. The weight increase corresponds to a hydroxyalkyl content of 55% (w/w). The degree of hydroxyalkylation can be selected by varying the amount of alkyl olefin oxide in the reaction mixture (Fig. 1). Different epoxides have been used: Nedox 1114, Nedox 1518, phenyl ethylene oxide,[6] 23,24-oxido-5β-cholane,[8] and cyclohexene oxide.[9] Other epoxides are also available. Two hydroxyalkoxypropyl derivatives, Lipidex 5000 (50% hydroxyalkyl groups) and Lipidex 1000 (10% hydroxyalkyl groups), are commercially available from Packard-Becker B.V. (Chemical Division, Groningen, Holland). Sephadex LH-40 may be used as starting material with minor modifications of the above procedure.

[5] E. Nyström *Ark. Kemi* **29**, 99 (1968).
[6] J. Ellingboe, E. Nyström, and J. Sjövall, *J. Lipid Res.* **11**, 266 (1970).
[7] J. Ellingboe, E. Nyström, and J. Sjövall, *Biochim. Biophys. Acta* **152**, 803 (1968).
[8] R. A. Anderson, C. J. W. Brooks, and B. A. Knights, *J. Chromatogr.* **75**, 247 (1973).
[9] R. A. Anderson, B. A. Knights, and C. J. W. Brooks, *J. Chromatogr.* **82**, 337 (1973).

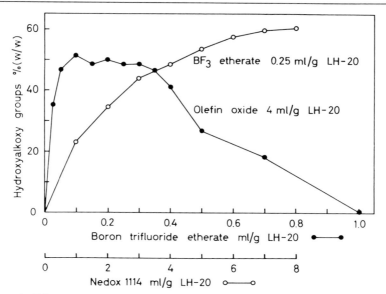

FIG. 1. Effect of amount of boron trifluoride etherate and Nedox 1114 on the degree of substitution of Sephadex LH-20 (from Ellingboe et al.[6] by permission).

It should be noted that the amount of boron trifluoride ethyletherate added is critical and should be about 0.25 ml/g hydroxypropyl Sephadex (Fig. 1).

Trimethylsilyl Sephadex (VI).[10] Sephadex G-25 is washed thoroughly with methanol and dried overnight at 50° *in vacuo.* Forty grams are placed in a three-necked 1-liter flask containing 600 ml of dried pyridine–toluene, 1:1 (v/v). Trimethylchlorosilane, 125 g, is added dropwise with mild stirring and the temperature is slowly raised to 80°. After 2 hours at this temperature the mixture is cooled and 100 ml of methanol is added. The reaction product is filtered off and washed with 200 ml of 2% sodium acetate–methanol and 300 ml of methanol. The product is dried *in vacuo,* and the yield is 78 g corresponding to 17.8% silicone and substitution of about 2.5 hydroxyl groups per glucose unit.

The material is stable in 50% aqueous methanol for over 72 hours at room temperature. It is readily hydrolyzed in acidic media.

B. Selective Derivatives

Chlorohydroxypropyl Sephadex (VIII).[11] Sephadex LH-20, 76 g, is soaked in 200 ml of methylene–chloride. Boron trifluoride ethyletherate,

[10] H. Tanaka and K. Konishi, *J. Chromatogr.* **64,** 61 (1972).
[11] J. Ellingboe, B. Almé, and J. Sjövall, *Acta Chem. Scand.* **24,** 463 (1970).

19 ml, is added, and the mixture is shaken. After 15 minutes, 50 ml of a solution of epichlorohydrin in methylene chloride (35%, v/v) are added slowly (1–2 ml/minute) and the mixture is left for 40 minutes with occasional shaking. The product is filtered, washed with chloroform and ethanol, and dried at 40° to constant weight [98.2 g, 8.57% (w/w) chlorine]. The degree of substitution can be controlled by the amount of epichlorohydrin used in the reaction. The degree of lipophilicity can be controlled by adding an alkyl olefin oxide to the reaction mixture before, during, or after the addition of epichlorohydrin.

The chlorohydroxypropyl Sephadex is reactive, and there are several possibilities for further substitution reactions. Under basic conditions, the vicinal halogen and hydroxyl groups undergo reactions typical of epoxides, whereas otherwise only the hydroxyl or the halogen may react. Thus, it is possible to synthetize derivatives with covalently bound substituents which, when used in gel chromatography, show greater selectivity toward solutes than hydroxypropyl Sephadex and hydroxyalkoxypropyl Sephadex. This is exemplified by the preparation of derivatives containing sulfhydryl or amino groups. The amino derivatives offer possibilities for ion-exchange chromatography in organic solvents; derivatives containing primary amino groups might also be useful as matrices to which enzymes, antigens, and antibodies could be bound.

Mercaptohydroxypropoxypropyl Sephadex (VIII).[11] Thirty grams of chlorohydroxypropylated (10.5% w/w) Sephadex LH-20 are treated with a solution of 10 g of sodium hydrosulfide in 100 ml of absolute ethanol and 200 ml of ethylene glycol for 2 hours at 100°. The product is washed with water and ethanol and dried at 40°. The product has a sulfur content of 14.3 μg/mg, and has a strong affinity for phenyl mercuric hydroxide and mercuric acetate. The potential usefulness of this derivative in separations of unsaturated lipids has not been explored.

Dibutylaminohydroxypropoxypropyl Sephadex (IX).[11,12] Chlorohydroxypropyl Sephadex LH-20, 13.6 g (8.57% Cl), is suspended for 30 minutes in a 15-fold molar excess of dibutylamine (0.51 mole, 68 ml). A solution of 2.87 g of potassium hydroxide in 118 ml of methanol is added to make the final concentration of potassium hydroxide 0.22 M and the final volume (in milliliters) 15 times the weight (in grams) of the starting material. After heating to 55° for 3 hours with occasional shaking, the product is collected on a Büchner funnel and washed with ethanol, 0.1 M potassium hydroxide in ethanol–water, 9:1 (v/v), and ethanol until the eluate is neutral. Primary and secondary amines can be prepared in a similar way substituting dibutylamine with ammonia and butyl amine, respectively.

[12] B. Almé and E. Nyström, *J. Chromatogr.* **59**, 45 (1971).

Carboxymethoxypropyl Sephadex (*X*).[5] Sephadex LH-20, 50 g, is suspended in 750 ml of isopropanol. Sodium hydroxide, 60 g, is then added; the suspension is stirred for 15 minutes, and sodium chloroacetate, 60 g, is then added in small portions. After 12 hours of continuous stirring, the Sephadex derivative is collected on a filter and washed consecutively with water, ethanol, chloroform, ethanol, water, 0.1 N hydrochloric acid, 0.001 N hydrochloric acid, ethanol and diethyl ether. The degree of substitution is estimated by electrometric titration of about 1 g of the Sephadex derivative, in 10 ml of 0.2 M KCl, using 0.2 M aqueous KOH.

Diethylaminoethoxypropyl Sephadex (*VI*).[5] This derivative is prepared by an adaptation of the procedure described by Peterson and Sober.[13] Sephadex LH-20, 50 g, is carefully mixed with aqueous sodium hydroxide (33.2 g in 142 ml of water) in an ice-cooled flask. After 30 minutes, 2-chlorotriethylamine hydrochloride (29.2 g in 37.5 ml of water) is added slowly while stirring, and the temperature is then raised to 80–85° for 45 minutes. The mixture is then cooled, and the product is filtered off and washed with water, 10% ammonium hydroxide, ethanol, 0.1 N hydrochloric acid, ethanol and diethyl ether. The degree of substitution, as determined by electrometric titration is approximately 0.8 mEq/g.

Properties of Lipophilic Sephadex Derivatives. Solvent regain values for some representative derivatives are given in Table I. The derivatives have been chosen to give chromatography systems as inert and chemically stable as possible. Thus, lipophilic esters of Sephadex, although simple to prepare, have been avoided. The hydroxypropyl and alkoxyhydroxypropyl derivatives are sufficiently stable for continuous use in columns over periods of several months to a year without change in properties. Deterioration of column performance is usually the result of contamination or formation of peroxides. It has been shown that the addition of 0.1% dimethyldisulfide to the solvent makes it possible to keep columns of hydroxyalkoxypropyl Sephadex for continuous use in triglyceride separations at 40° for at least one year.[14] To avoid formation of peroxides the derivatives may be stored at +4° in ethanol containing 0.1% BHT (2,6-di-*tert*-butyl-4-methyl-phenol). If a preparation shows a positive peroxide test with potassium iodide, the peroxide groups may be reduced with sodium borohydride in ethanol. If the derivatives are purified and stored as described above, they give very little column bleed, and the column eluate will not interfere in gas chromatography–mass spectrometry (GC-MS) analyses or protein binding reactions. When large volumes of effluent from hydroxyalkoxypropyl (C_{11}–C_{14}) Sephadex

[13] E. A. Peterson and H. A. Sober, *J. Amer. Chem. Soc.* **78**, 751 (1956).
[14] B. Lindqvist, J. Sjögren, and R. Nordin, *J. Lipid Res.* **15**, 65 (1974).

TABLE I

SOLVENT REGAIN OF SEPHADEX DERIVATIVES IN DIFFERENT SOLVENTS[a]

Derivative	Water	Ethanol	Acetone	Methylene chloride	Benzene	Heptane
Sephadex LH-20	2.3	1.6	0.7	2.0	0.4	0.2
Methyl Sephadex G-25 (–OCH$_3$,[b] 41.5%)	2.5	1.6	1.4	4.1	2.2	0.3
Hydroxyalkoxypropyl Sephadex G-25 (C$_{11}$–C$_{14}$,[c] 63%)	0.3	(0.3)[d]	—	(1.7)[e]	1.5	0.8
Hydroxyalkoxypropyl Sephadex G-50 (C$_{11}$–C$_{14}$,[c] 52%)	0.5	1.1	0.9	6.6	4.4	2.5
Trimethylsilyl Sephadex G-25 (Si,[f] 18%)	—	0.5	0.7	1.5	—	1.2
Carboxymethoxypropyl Sephadex G-25						
0.6 mEq/g, H$^+$ form	2.1	1.9	—	1.5	0.7	0.2
0.6 mEq/g, K$^+$ form	4.6	2.0	—	1.3	0.4	0.1

[a] Amount of solvent imbibed in the gel beads after removal of interstitial solvent by centrifugation (grams solvent/gram dry Sephadex derivative).

[b] Methoxyl group content, percent of weight of Sephadex derivative.

[c] Hydroxyalkyl group content, percent of weight of Sephadex derivative.

[d] Solvent regain in methanol.

[e] Solvent regain in ethylene chloride.

[f] Silicone content, percent of weight of Sephadex derivative.

columns are concentrated, trimethylsilylated, and analyzed by GC-MS, a "chemical" background may be seen at about 200–220°. The spectrum of this background shows ions of mass 229, 243, 257, and 271 which correspond to the structure $(CH_3)_3Si\overset{+}{O}\!\!=\!\!CH\!-\!(CH_2)_nCH_3$ where n is 8–11. The background appears to result from extraction of polymeric compounds which decompose on GC and bleed slowly through the column. Only very small amounts of monomeric compounds are seen as separate GC peaks.

III. Column Chromatography

Lipophilic Sephadex derivatives can be used to support the stationary phase in straight- and reversed-phase chromatography with two-phase solvent systems.[15,16] They can be adapted to thin-layer chromatography using a descending technique in a closed chamber where the sample is injected onto the equilibrated layer.[17] At present, however, lipophilic Sephadex derivatives have been found to be most useful in column chromatography employing miscible solvent systems. The following description will therefore be limited to such column systems which are well suited for work in the picogram to gram range.

Glass Columns. Ordinary all-glass jacketed columns (4–60 × 100–1000 mm) can be used as well as more sophisticated commercially available solvent-resistant equipment with adjustable end pieces. The columns are packed using a slurry of the Sephadex derivative in the solvent to be used. The volume of the gel bed will change with temperature or solvent composition (especially water content). Consequently, if a closed column system is used, careful equilibration is essential before inserting the top end piece. Gas bubbles, which may form in work with low-boiling solvents, can be avoided in several ways: (a) by degassing solvent components, (b) by keeping the column at a temperature 10° lower than that of the solvent entering the column, and (c) by keeping the end of the column outlet tubing at a level above that of the column top. The columns should preferably be thermostated (usually 20 or 25° ± 1°).

Recycling Chromatography. This technique is valuable for preparative separations when separation factors are small.[18] The solvent is pumped with a peristaltic pump (using Viton tubing unless ketonic solvents are used) through a jacketed glass column with adjustable end

[15] E. Nyström and J. Sjövall, *J. Chromatogr.* **17,** 574 (1965).
[16] B. H. Shapiro and F. G. Péron, *J. Chromatogr.* **65,** 568 (1972).
[17] E. Nyström and J. Sjövall, *Acta Chem. Scand.* **21,** 1974 (1967).
[18] E. Nyström and J. Sjövall, *Ark. Kemi* **29,** 107 (1968).

pieces. The effluent may then be passed through a suitable detector before entering a three-way selector valve. This valve permits the effluent to be directed into the column again or to a fraction collector. Formation of gas bubbles is avoided as indicated above.

Teflon Spaghetti Column.[1,7,18,19] This type of column is useful in analytical work or for isolation of small amounts of material prior to mass spectrometric analysis. High efficiency columns (HETP < 0.2 mm) may be obtained using a small and uniform particle size (e.g., 20 ± 2 μm in the dry state). Teflon tubing (0.5–2 m $\times$ 1.5 mm i.d., 2.5 mm o.d.) is supported in a vertical position. A Teflon net with a fine mesh size is folded over the end of a piece of stainless steel tubing ($\frac{1}{16}$ inch o.d.) which is pushed into the lower end of the Teflon tubing and serves as the column bed support. The tubing is filled with the solvent to be used, and the upper end is connected to a reservoir containing a slurry of the gel in the solvent. A column bed is formed slowly when pressure (1–8 atm) is applied to the reservoir and the column is vibrated. An injection port[19] is attached to the top of column bed and connected to a solvent reservoir to which pressure is applied to give a solvent flow rate of 0.2–0.6 ml/hour. Samples are injected through a silicone membrane in the injection port in 1–10 μl of a suitable solvent.

Mechanisms of Chromatographic Separations. Although detailed studies of the mechanisms for chromatographic separations on lipophilic Sephadex have not been carried out, the following factors may be assumed to be of importance; for a more detailed discussion, see Sjövall et al.[1]

1. Difference in solvent composition between mobile phase and gel phase. When mixtures of solvents are used, preferential incorporation of certain solvent components into the gel phase may occur depending on gel–solvent interactions. Thus, the average composition of the solvent mixture in the stationary inner gel phase may be different from that of the mobile phase. This is best exemplified by the chromatographic behavior of lipids in reversed-phase systems with hydroxyalkoxypropyl Sephadex which closely resembles that observed in conventional liquid–liquid partition chromatography with two immiscible solvent phases.[6,20]

2. Gel matrix–solute interactions. Specific interactions may be obtained between solutes and specific functional groups introduced into the matrix (e.g., ion exchange). Nonspecific interaction (e.g., hydrogen bonding, "aromatic adsorption") may occur with hydroxyl and ether groups in the substituted polysaccharide network. This is most important in nonpolar solvent systems, and it is difficult to distinguish adsorption from

[19] E. Nyström and J. Sjövall, *J. Chromatogr.* **24**, 212 (1966).
[20] K. Beijer and E. Nyström, *Anal. Biochem.* **48**, 1 (1972).

liquid–liquid partition in these systems. An additional effect is the result of the presence of negatively charged fixed ions in the Sephadex (about 10 μEq/g). When charged solutes are chromatographed on these derivatives in certain solvent systems, there may be a relative exclusion of acids and retardation of bases. When an electrolyte is incorporated into non-aqueous solvents, the elution volume of charged molecules will depend on the nature of the counterion.[21]

3. Fraction of the solvent in the gel phase available to the solute. Solutes with a larger effective molecular size have access to less stationary phase and may thus be less retarded than solutes with a smaller molecular size (gel filtration, gel permeation, molecular sieving).

Solvent Systems. In choosing suitable solvent mixtures the above factors should be taken into consideration. A prerequisite for a good solvent system is that it swells the Sephadex derivative; otherwise, broad and tailing peaks are obtained indicating adsorption and lack of penetration of the solute into the inner gel phase. The solvent regain properties of different derivatives are given in Table I. In most applications, polar single solvent systems should be avoided, and the aim should be to use solvent systems which create a large difference in polarity between the stationary gel phase and the mobile phase. With the relatively polar hydroxypropyl Sephadex (LH-20), efficient systems are obtained with methylene chloride,[22] methanol–chloroform–heptane,[18] methanol–ethylene chloride,[23] methanol–benzene,[24] chloroform–Skellysolve B,[25] and similar solvent mixtures. In these cases the more polar solvent component seems to be enriched in the gel phase, giving a straight-phase system where polar solutes are retarded. With the hydrophobic hydroxyalkoxypropyl derivatives, both straight- and reversed-phase systems can be designed. Thus, mixtures of aliphatic and chlorinated hydrocarbons create straight-phase systems. Solvents having a polar (e.g., methanol–water) and a non-polar component (e.g., a hydrocarbon or a chlorinated hydrocarbon), give a reversed-phase system. With such systems the total volume of the less polar components in the mixture (i.e., aliphatic, aromatic, and haloge-nated hydrocarbons, medium- and long-chain alcohols) should be 5–20% of the volume of the more polar components (e.g., methanol–water), which have a lower affinity to the gel.[6,20]

Chromatographic Conditions. Retention volumes are influenced by the

[21] J. Sjövall and R. Vihko, *Acta Chem. Scand.* **20**, 1419 (1966).

[22] P. Eneroth and E. Nyström, *Biochim. Biophys. Acta* **144**, 149 (1967).

[23] B. E. P. Murphy, *Nature (London)* **232**, 21 (1971).

[24] G. Mikhail, C. H. Wu, M. Ferin, and R. L. Van de Wiele, *Steroids* **15**, 333 (1970).

[25] M. F. Holick and H. H. DeLuca, *J. Lipid Res.* **12**, 460 (1971).

temperature, particularly in solvent systems where liquid–liquid partition effects are important.[20] However, if the temperature is constant, the elution volumes are highly reproducible. The best resolution is obtained with columns packed with small particles with a narrow size distribution. Commercial Sephadex LH-20 may be sieved in the dry state prior to use. The fractions 110–140 mesh and 140–180 mesh give good flow rates without need for pump equipment. The same fractions are used as starting materials for preparation of hydroxyalkoxypropyl derivatives. The flow rate should be 0.1–0.5 ml/minute and cm^2 column cross section area; this will give HETP values of about 0.2–0.6 mm (using particle fraction 140–170 mesh). The samples should preferably be applied to the column in the mobile phase. If another solvent has to be used, the effect of this should be investigated since the elution volumes might be influenced and the peaks become broad and tailing. It should be noticed that the sample can often be applied in a large volume of the eluting solvent[20] (up to 15% of the bed volume) without serious effect on the column efficiency, provided the retention volume is sufficiently large (e.g., >5 total column volumes).

IV. Applications

During the last few years a large number of papers have appeared reporting the use of lipophilic Sephadex derivatives. Therefore, the aim of the following section is to give a brief survey of the field of applications with references of general interest.

Group Separations. Free fatty acids and triglycerides can be separated on Sephadex LH-20 using chloroform with 0.2% (v/v) glacial acetic acid.[26] This provides a complementary method to silicic acid chromatography, where separation is difficult if the fatty acids are present in large excess. Chromatography on methyl Sephadex or Sephadex LH-20 may be used as a simple and mild method to obtain a phospholipid fraction. When serum lipids are chromatographed on methyl Sephadex G-25 (36%, w/w, of methoxyl groups), phospholipids are eluted ahead of triglycerides, which are followed by a mixture of cholesteryl esters, cholesterol, and fatty acids.[1] Similarly, galactolipid extracts can be purified on Sephadex LH-20 prior to galactose determination.[27] Shimojo *et al.*[28] have pointed out that the cations bound to the phospholipids affect the formation of micelles in organic solvents, thus determining the elution behavior on lipophilic Sephadex columns. Using a dibutylaminohydroxypropyl

[26] R. F. Addison and R. G. Ackman, *Anal. Biochem.* **28**, 515 (1969).

[27] M. A. B. Maxwell and J. P. Williams, *J. Chromatogr.* **35**, 223 (1968).

[28] T. Shimojo, H. Kanoh, and K. Ohno, *J. Biochem.* (*Tokyo*) **69**, 255 (1971).

derivative of Sephadex LH-20, Almé and Nyström separated an egg phospholipid preparation into lysolecithin, lysophosphatidylethanolamine, lecithin, and phosphatidylethanolamine with the solvent chloroform–methanol–water (20:65:35, v/v).[12] In contrast to the cellulose ion-exchange columns used in the separation of phospholipids,[29] the lipophilic Sephadex ion exchangers show excellent chromatographic properties in relatively nonpolar solvent mixtures.[12,30] Mokrash[31] has used Sephadex LH-20 with chloroform–methanol–concentrated ammonia, 50:50:1 (by volume), for rapid isolation of proteolipids, and this group has been further separated in chloroform–methanol mixtures.[32,33]

Steroid sulfates in lipid extracts from biologic materials can be isolated as one class by chromatography on Sephadex LH-20.[21] Using the solvent chloroform–methanol, 1:1 (v/v), containing 0.01 mole/liter of sodium chloride, steroid monosulfates and disulfates have elution volumes of about 2.9 and 12, respectively, relative to cholesterol and free steroids. Minor separations may be observed within the groups, the more polar compounds being eluted later than the less polar ones. The importance of having the steroid sulfates in a single salt form is stressed since separations are probably determined by the partition coefficients of the undissociated steroid sulfate salts. The ion exchanging properties of Sephadex therefore make it necessary to add an electrolyte to the solvent when small amounts of material are chromatographed.

In another type of group separation it is possible to remove a high-boiling solvent or a contaminating polymer from a sample by filtration through a column of lipophilic Sephadex. A polymer is excluded from the gel and is thus eluted with the void volume. A high-boiling solvent may be separated from the solutes, which are thus transferred into the low-boiling solvents used in chromatography on lipophilic Sephadex.

Neutral Lipids. Using Sephadex LH-20 with chloroform as eluting solvent, Joustra et al.[2] separated tristearin, tricaproin, and triacetin. The separation was probably mainly the result of a gel filtration effect, and similar chromatographies have been reported by Calderon and Bauman, who separated glycerol and ethanediol esters and ethers using chloroform[34] and ethanol[35] as solvents. Joustra et al.[2] also reported a separation

[29] H. S. Hendrickson and C. E. Ballou, *J. Biol. Chem.* **239**, 1369 (1964).

[30] J. C. Dittmer, *J. Chromatogr.* **43**, 512 (1969).

[31] L. C. Mokrasch, *Life Sci.* **6**, 1905 (1967).

[32] E. F. Soto, J. M. Pasquini, R. Plácido, and J. L. LaTorre, *J. Chromatogr.* **41**, 400 (1969).

[33] K. J. Cattell, C. R. Lindop, J. G. Knight, and R. B. Beechey, *Biochem. J.* **125**, 169 (1971).

[34] M. Calderon and W. J. Baumann, *Biochim. Biophys. Acta* **210**, 7 (1970).

[35] M. Calderon and W. J. Baumann, *J. Lipid Res.* **11**, 167 (1970).

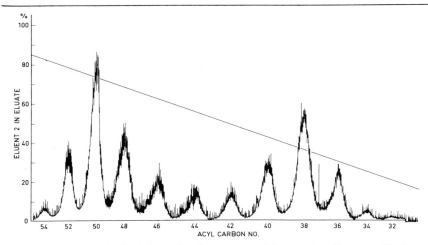

FIG. 2. Separation of triglycerides from 0.5 g of butter fat. Column: Hydroxy-(C$_{15}$–C$_{18}$) alkyl (55% (w/v)) Sephadex LH-20, 850–1000 × 25 mm, gradient elution. Eluent 1: isopropanol–chloroform–heptane–water, 115:15:2:35 (by volume); eluent 2: heptane–acetone–water, 4:15:1 (by volume). Temperature 40°, flow rate 75 ml/hour (from Lindqvist *et al.*[34] by permission).

of 1,2- and 1,3-dipalmitin on LH-20 in chloroform, which was based on the structure rather than on the molecular weight of the substances.

Since Sephadex LH-20 cannot be used to obtain nonpolar reversed-phase systems it is not well suited for separation of neutral lipids. However, hydroxyalkoxypropyl Sephadex permits the use of a variety of reversed-phase systems which give remarkably good separations of substances differing by number of methylene groups. Monoglycerides are separated by reversed-phase chromatography in isopropanol–water–chloroform–heptane, 70:70:9:1 (by volume).[6] Diglycerides are best separated as trimethylsilyl ethers since hydroxyalkoxypropyl Sephadex in aqueous solvents may catalyze isomerization of 1,2- into 1,3-diglycerides.[36] Use of hydroxyalkoxypropyl Sephadex [55% (w/w) hydroxyalkyl groups] in acetone–water–heptane–pyridine, 87:13:10:1 (by volume), gives nearly quantitative yields of the separated 1,2-diglyceride trimethylsilyl ethers if the fractions are collected in heptane so that water is separated from the silyl ethers.[36] Separation of 1,2- from the corresponding 1,3-diglycerides is best achieved by straight-phase chromatography of the unsilylated compounds using the same gel in heptane–chloroform, 100:5 (v/v). Triglycerides with long-chain fatty acids may be separated in sys-

[36] T. Curstedt and J. Sjövall, *Biochim. Biophys. Acta* (1974) (in press).

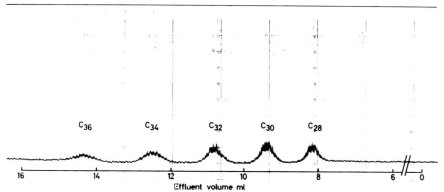

FIG. 3. Separation of straight-chain waxes synthesized from octadecanol and the appropriate acyl chlorides. Column: 285×4.4 mm, hydroxyalkoxypropyl Sephadex (made from Sephadex G-25, bead size 27.5 ± 5 μ and containing 34% (w/w) hydroxyalkyl (C_{11}–C_{14}) groups). Solvent: heptane–acetone–methanol, 1:2:4. Flow rate 0.65 ml/hour. The effluent was monitored with a moving-chain flame ionization detector[1] (from Ellingboe et al.[6] by permission).

tems similar to that used for diglyceride trimethylsilyl ethers.[6] However, such systems are unsuitable for separation of short-chain triglycerides and by modifying and combining the solvent systems for monoglycerides and triglycerides[6] into a gradient system, Lindqvist et al.[14] obtained excellent separations at 40° of triglycerides with acyl carbon numbers between 28 and 56 (Fig. 2). The column load in these systems was as high as 0.5 g/cm² column cross section area. In all these reversed-phase systems the mobility of a compound increases to about the same extent by introduction of a double bond as by removal of two CH_2 groups.

Examples of other nonpolar neutral lipids which are difficult to separate by liquid chromatography are the long-chain waxes and steryl esters. Figure 3 shows the separation of waxes in methanol–acetone–heptane,

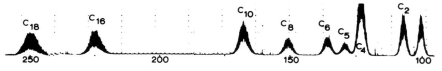

FIG. 4. Separation of cholesterol and its esters with C_2–C_{18} fatty acids. Numbers on the horizontal axis refer to percent of total column volume eluted. Column: 3000×1.5 mm, hydroxyalkoxypropyl Sephadex (same as in Fig. 3). Solvent: heptane–acetone–water, 4:15:1 (by volume). Flow rate 2.9 μl/minute. The effluent was monitored with a moving-chain flame ionization detector[1] (from Ellingboe et al.[7] by permission).

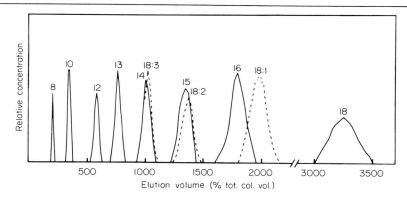

FIG. 5. Separation of methyl esters of C_8–C_{18} fatty acids. Column: 443 × 13 mm, hydroxyalkyl Sephadex LH-20 [140–170 mesh; hydroxyalkyl group (C_{11}–C_{14}) content 55%, w/w] used with a peristaltic pump and a sample loop injection. Solvent: methanol–water–ethylene chloride, 80:20:10 (by volume). Flow rate: 30 ml/hour and cm² (from Beijer and Nyström,[20] by permission).

4:2:1 (by volume),[6] and Fig. 4 illustrates a separation of cholesterol and its C_2–C_{18} esters in acetone–water–heptane, 15:1:4 (by volume).[6]

Phospholipids. Two different approaches for separation of individual species of phospholipids on lipophilic Sephadex have been described. Conversion of phosphatidylcholines into mercuric acetate adducts and chromatography on Sephadex LH-20 in benzene–chloroform–methanol, 3:3:4 (by volume), with increasing concentration of acetic acid (0.001–0.03%) permitted a separation into anenoic, monoenoic, dienoic, and polyenoic species.[37] However, some overlap occurred, probably resulting from incomplete adduct formation. In another attempt to separate individual phosphatidylcholines Almé *et al.*[38] used the less polar dibutylaminohydroxypropyl Sephadex LH-20 in a reversed-phase system, methanol–water–chloroform, 65:35:15 (by volume), and obtained a separation based on acyl carbon number and degree of unsaturation. In addition to the partition chromatographic separation, this ion exchanger gives a group separation of neutral basic and acidic compounds in the lipid extract. It should be very useful in a number of applications where separation both into groups and into individual species within the groups is desired.

Fatty Acids, Alcohols, and Prostaglandins. Reversed-phase systems for fatty acids and their methyl esters are based on hydroxyalkyl Sephadex LH-20 and methanol–water–ethylene chloride (or chloroform) mixtures, e.g., 90:10:10–70:30:10 (by volume) (Fig. 5).[6,20] These systems

---

[37] R. J. King and J. A. Clements, *J. Lipid Res.* 11, 381 (1970).
[38] B. Almé, J. Sjövall, and P. P. M. Bonsen, *Anal. Lett.* 4, 695 (1971).

are also useful for separation of labile terpenoid compounds.[39,40] Prostaglandins are too polar for these systems. They can be separated if the polarity of the solvent is increased by addition of n-butanol and the polarity of the gel is increased by reduction of the number of alkyl chains. Thus, reversed-phase separations of prostaglandins by number of double bonds and number and nature of substituents may be achieved with hydroxyalkyl (12%, w/w) Sephadex LH-20 in methanol–water–chloroform–butanol, 40:60:3:7 (by volume).[41] When solvent systems containing these large amounts of water are used it may be advantageous to swell the gel in the mixture of organic solvents first and then add the appropriate volume of water before the column is packed.

Straight-phase systems are less suited for free acids but may be used for methyl esters. By modifying solvent systems for steroids, Änggård and Bergkvist separated prostaglandins by number of hydroxyl and keto groups using Sephadex LH-20 in chloroform–heptane–ethanol, 10:10:1 (by volume).[42]

Steroids and Bile Acids. The use of methylated Sephadex in separations of steroids was reported in 1964 and 1965,[4,5] and since then numerous papers dealing with separations on lipophilic Sephadex of sterols, steroid hormones, and bile acids have appeared. Three main types of systems have been employed: (1) straight-phase systems consisting of Sephadex LH-20 in chlorinated hydrocarbon–hydrocarbon–alcohol mixtures,[4,18,22] (2) straight-phase systems consisting of hydroxyalkoxypropyl Sephadex in hydrocarbon–chlorinated hydrocarbon or alcohol mixtures,[6] and (3) reversed-phase systems consisting of hydroxyalkoxypropyl Sephadex in alcohol–water–chlorinated hydrocarbon–hydrocarbon mixtures.[6] Several modifications of the original solvent systems have been described for specific separations. Because of the commercial availability of Sephadex LH-20 most publications deal with this gel. However, hydroxyalkoxypropyl Sephadex is much more versatile and at present the particular usefulness of Sephadex LH-20 seems to be limited to the group separation of steroid sulfates mentioned above[21,43] and to the separation of polar steroids.[18] The systems for steroid sulfates have been successfully used in a modified form for separation of groups of estrogen conjugates[44] and for isolation of bile acid sulfates.[45] Several straight-phase systems with

[39] L. Ahlquist, B. Olsson, A.-B. Ståhl, and S. Ställberg-Stenhagen, *Chem. Scripta* **1**, 237 (1971).

[40] D. S. Stevenson and C. J. W. Brooks, *J. Chromatogr.* **75**, 308 (1973).

[41] E. Nyström and J. Sjövall, *Anal. Lett.* **6**, 155 (1973).

[42] E. Änggård and H. Bergkvist, *J. Chromatgr.* **48**, 542 (1970).

[43] O. Jänne, R. Vihko, J. Sjövall, and K. Sjövall, *Clin. Chim. Acta* **23**, 405 (1969).

[44] M. J. Tikkanen and H. Adlercreutz, *Acta Chem. Scand.* **24**, 3755 (1970).

[45] T. Cronholm, I. Makino, and J. Sjövall, *Eur. J. Biochem.* **26**, 251 (1972).

Sephadex LH-20 have been described for purification of free steroids prior to analysis by protein binding and radioimmunoassay for estrogens [methylene chloride–methanol, 95:5 (v/v),[23] benzene–methanol, 85:15 (v/v)[24]] androgens,[23,46,47] progestins,[23,47] and corticoids.[23,48-52] Suitable systems for corticoids are hexane–chloroform–ethanol, 5:5:1 (by volume) saturated with water,[18] methylene chloride–methanol, 98:2 (v/v),[22,23] and cyclohexane–methanol, 8:2 (v/v).[52] However, if androgens and progestins are to be purified, a much better resolution of these less polar steroids is obtained on hydroxyalkoxypropyl [50% (w/w)] Sephadex in hexane–chloroform, 9:1 (v/v), or hexane–tert-butanol, 95:5 (v/v).[6,53]

Nonpolar sterols are best separated by reversed-phase chromatography on hydroxyalkoxypropyl Sephadex. With this method preparative separations of 24α-methylcholesterol (campesterol) and 24α-ethylcholesterol (β-sitosterol) which frequently occur in mixtures can be achieved.[6,20,54] Methanol–water–ethylene chloride, 90:10:20 (by volume), is a suitable solvent.[20] The same solvents in proportions 95:5:25 (by volume) have been used for oxygenated cholesterol metabolites.[55] Generally, solvent systems of the type methanol–water–chloroform, 50:50:10–80:20:10 (by volume), may be used with hydroxyalkoxypropyl Sephadex [hydroxyalkyl group content 10–50% (w/w)] to obtain reversed-phase systems for steroids with 2–4 oxygen substituents. Alternatively, straight-phase systems may be employed consisting of mixtures of hexane and chloroform, e.g., in proportions 9:1–7:3, depending on the polarity of the steroids. Autoxidation products of cholesterol have also been separated on Sephadex LH-20 in methylene chloride,[56] and in this system the specific exclusion of ketonic steroids from the gel phase may be advantageous.[1,22] If partition-adsorption effects are undesirable, these may be diminished by the use of benzene or benzene–isopropanol, 75:25 (v/v), with hydroxyalkoxypropyl Sephadex [50% (w/w) hydroxyalkyl

[46] J. Ahmed, *Clin. Chim. Acta* **43**, 371 (1973).
[47] B. R. Carr, G. Mikhail, and G. L. Flickinger, *J. Clin. Endocrinol. Metab.* **33**, 358 (1971).
[48] T. Seki and T. Sugase, *J. Chromatogr.* **42**, 503 (1969).
[49] W. Waldhäusl, H. Haydl, and H. Frischauf, *Steroids* **20**, 727 (1972).
[50] T. Ito, J. Woo, R. Haning, and R. Horton, *J. Clin. Endocrinol. Metab.* **34**, 106 (1972).
[51] H. H. Newsome, Jr., A. S. Clements, and E. H. Borum, *J. Clin. Endocrinol. Metab.* **34**, 473 (1972).
[52] K. D. R. Setchell and C. H. L. Shackleton, *Clin. Chim. Acta* **47**, 381 (1973).
[53] T. Holmdahl and J. Sjövall, *Steroids* **18**, 69 (1971).
[54] P. M. Hyde and W. H. Elliott, *J. Chromatogr.* **67**, 170 (1972).
[55] L. Aringer and P. Eneroth, *J. Lipid Res.* **14**, 563 (1973).
[56] J. E. Van Lier and L. L. Smith, *J. Chromatogr.* **41**, 37 (1969).

groups]. A detailed study of gel filtration of terpenoid compounds in these systems has been made by Brooks and Keates.[57]

Although bile acids and their methyl esters were shown to separate on Sephadex LH-20 in chloroform,[1] free bile acids are best separated by reversed-phase chromatography on hydroxyalkoxypropyl Sephadex. Methanol–water–ethylene chloride (or chloroform) 60:40:10 (upper phase) and 70:30:10 (v/v) are used with a gel containing 50% (w/w) hydroxyalkyl groups for separation of di- and monooxygenated bile acids.[6] These systems are too nonpolar for trihydroxycholanoic acids which can be separated on a gel containing 10–15% hydroxyalkyl groups using the solvent methanol–water–chloroform–n-butanol, 50:50:5:5 (by volume). Mixtures of bile acid methyl esters prepared from a biologic extract may be conveniently fractionated by stepwise elution of a gel containing 50% (w/w) hydroxyalkyl groups with hexane–chloroform 8:2, 7:3, and 6:4, each solvent being used in a volume corresponding to five total column volumes.

Lipid-Soluble Vitamins. The separation of a homologous isoprene series of vitamins $K_2(40\text{-}10)$ can be accomplished on a Teflon spaghetti column of methyl Sephadex in heptane–chloroform–methanol, 2:1:1 (by volume),[19] and Sephadex LH-20 can be used in similar systems.[58] Vitamin D_3, 25-hydroxy vitamin D_3, and the more polar hydroxylated metabolites can be separated on Sephadex LH-20 in Skellysolve B–chloroform mixtures [35:65–70:30 (v/v)].[25]

A method for quantitative analysis of tocopherols in extracts of tissues has been described where the tocopherols are separated on hydroxyalkoxypropyl Sephadex [50% (w/w) C_{11}–C_{14} hydroxylalkyl groups] by elution with hexane–benzene, 70:30, followed by hexane–benzene–ethyl acetate, 67:30:3 (by volume). The effluent can be monitored with a spectrophotofluorimeter.[59] The use of hydrophobic Sephadex derivatives for separation of lipid-soluble vitamins has also been reported by Hayashi *et al.*[60]

[57] C. J. W. Brooks and R. A. B. Keates, *J. Chromatogr.* **44**, 509 (1969).
[58] J. M. Cambell and R. Bentley, *Biochemistry* **7**, 3323 (1968).
[59] J. N. Thompson, P. Erdody, and W. B. Maxwell, *Anal. Biochem.* **50**, 267 (1972).
[60] Y. Hayashi, K. Kondo, N. Nakajima, and I. Kusumi, *Bitamin* **42**, 39 (1970).

[47] Thin-Layer Chromatography of Neutral Glycosphingolipids

By VLADIMIR P. SKIPSKI

I. Introduction

In this section will be presented the methods for separation by thin-layer chromatography of glycosphingolipids which do not contain sialic acid and are called neutral glycosphingolipids. These glycosphingolipids include not only genuine neutral glycosphingolipids [e.g., monoglycosyl-ceramides (phrenosine, kerasin, etc.), diglycosylceramides (cytolipin H), and different oligoglycosylceramides] but also sulfatides, psychosines, and ceramides. Although some of these lipids are not neutral or are not glycolipids, they are closely related chemically and biochemically to neutral glycosphingolipids. To some extent information concerning the separation of monogalactosyl diglycerides and digalactosyl diglycerides will be included in spite of the fact that their molecules do not contain sphingosine. Common features of this heterogeneous group of compounds are that they are lipids (contain in their molecule long-chain hydrocarbon derivatives and are soluble in mixture of chloroform–methanol, 2:1 v/v); and when their chloroform–methanol solution is partitioned with water or proper salt solution according to Folch *et al.*[1] and other investiga-

---

[1] J. Folch, M. Lees, and G. H. Sloane-Stanley, *J. Biol. Chem.* **226**, 497 (1957).

tors[2-4b] they will be in the lower phase (chloroform layer), whereas more polar glycosphingolipids which contain siliac acid(s) in their molecules (e.g., gangliosides and hematosides) will be in the upper phase (water–methanol layer).

The lower phase of the Folch partition of animal tissue extracts will also contain, besides neutral glycosphingolipids, almost all phospholipids and "neutral lipids" i.e., simple lipids (e.g., triglycerides and lower glycerides, free fatty acids, cholesterol, cholesteryl esters, and hydrocarbons), and may contain some of the simplest hematosides. It is easy to separate most neutral (simple) lipids from glycosphingolipids and phospholipids by thin-layer chromatography or column chromatography. In conventional thin-layer chromatographic systems developed with appropriate polar solvents all neutral lipids move with the solvent front or near the solvent front.[4c] Phospholipids and neutral glycosphingolipids form discrete spots between the solvent front and the origin of chromatogram (point of the application of sample). However, as a result of the similarity in the polarity of some neutral glycosphingolipids and some phospholipids, their spots will overlap.

There are two practical approaches for separation of neutral glycosphingolipids from phospholipids. The first approach is the selection of specific adsorbents with different affinity strengths toward glycosphingolipids (weak) and phospholipids (strong), combined with developing solvents which move ("elute") only neutral glycosphingolipids leaving phospholipids at the origin of the chromatogram. This is *direct thin-layer chromatography of neutral glycosphingolipids*. The second approach is to remove phospholipids (completely or at least partially) from lipid extracts prior to thin-layer chromatography of neutral glycosphingolipids or, in other words, to enrich specifically the sample with neutral glycosphingolipids. This method of separation of neutral glycosphingolipids is called *thin-layer chromatography of neutral glycosphingolipids in enriched samples*. The second approach is more complicated and time consuming, but under certain conditions it is the only acceptable method.

Choice of the procedure—direct vs. enriched samples—is determined by the type of tissue studied and the goal of the investigations. Most animal tissues contain rather small quantities of neutral glycosphingolipids,[5] only a fraction of percent of total lipids. Therefore, in order to

---

[2] S. Gatt, *J. Neurochem.* **12**, 311 (1965).

[3] M. W. Spence and L. S. Wolfe, *J. Neurochem.* **14**, 585 (1967).

[4] K. Suzuki, *J. Neurochem.* **12**, 629 (1965).

[4a] C. Entenman, Vol. 3 [**55**].

[4b] N. S. Radin, Vol. 14 [**44**].

[4c] V. P. Skipski and M. Barclay, Vol. 14 [**54**].

[5] E. Mårtensson, *Progr. Chem. Fats Other Lipids* **10**, 365 (1969).

avoid overloading the thin-layer chromatogram with total lipids extract, in some cases the method utilizing samples enriched with neutral glyco-sphingolipids is the only practical approach.

II. Direct Thin-Layer Chromatography of Neutral Glycosphingolipids

A. General Description of the Procedure

Only a few procedures have been developed permitting direct thin-layer chromatography of neutral glycosphingolipids on total lipid extracts from animal tissues.

Skipski et al.[6] took advantage of the fact that magnesium silicate adsorbs phospholipids more strongly than neutral lipids and neutral glycosphingolipids, and that pyridine and acetone do not appreciably move phospholipids from the application point on magnesium silicate or silica gel plates whereas they do considerably move both neutral lipids and neutral glycosphingolipids. This combination of properties of adsorbents and developing solvents permitted the design of a thin-layer chromatographic system for separation of neutral glycosphingolipids. In this system all neutral lipids move with the developing solvent front (or near it), phospholipids stay at the origin of the chromatogram (or near it), and spots of neutral glycosphingolipids remain between.

B. Technique of Separation of Neutral Glycosphingolipids by Thin-Layer Chromatography

1. Preparation of Chromatoplates

General techniques for thin-layer chromatography are described by Skipski and Barclay.[4c]

A slurry prepared from 10 g of magnesium silicate and 30 g of silica gel in 100–110 ml of 0.05 M aqueous Na_2CO_3 was spread evenly over 200×200 mm glass plates (preferably of Pyrex glass). The magnesium silicate used in this procedure was purchased from M. Woelm, Eschwege, W. Germany, grade for "thin-layer chromatography" (obtained through Alupharm Chemicals, New Orleans, La.) and silica gel, fine (D-O type) without $CaSO_4$ binder was purchased from Camag Inc. (parent company: 4132 Muttenz, B. L., Homburger Str. 24, Switzerland; U.S. Firm: New

[6] V. P. Skipski, A. F. Smolowe, and M. Barclay, J. Lipid Res. **8,** 295 (1967).

Berlin, Wisc. 53151). The plates should dry in the air for about 45 minutes and then be activated in an oven at 150–160° for at least 4 hours before use. Activation of the chromatoplates for several days usually results in a better separation of glycosphingolipids. (Plates may be left in the oven up to 10 days at 150–160° without deleterious effects on the efficiency of the glycolipid separation.) The activated chromatoplates are cooled in a desiccator before application of samples.

2. Application of Samples and Development of Chromatograms

To control the content of water in the adsorbent the application of samples is performed under nitrogen. The authors used the Desaga-Brinkmann Application Box, Model DB (Brinkmann Instruments Inc., Westbury, N.Y. 11590). It is advisable to apply samples in the form of small bars, a series of small spots that form a band, especially when content of glycosphingolipids is low in samples to be tested. Unknown samples and reference compounds to be tested are dissolved in chloroform–methanol, 2:1 or 1:1 (v/v), or pure methanol. The amounts of reference compounds applied to chromatoplates may range from 5 to 25 μg, whereas total lipids extracted from animal tissues are applied as a single spot in 500–1500 μg quantities. When samples to be tested are applied in a bar, considerably higher quantities can be used.

The developing tank is lined on three sides with filter paper wetted with the developing solvent. For development of thin-layer chromatograms three different mixtures of solvents are used sequentially. The first solvent is acetone–pyridine–chloroform–water, 40:60:5:4 (v/v), which is allowed to run to about 8 cm above the application points of the samples. The plates are removed from the developing tank and dried in a vacuum oven for 30 minutes at room temperature. Then the chromatoplates are developed in the second solvent mixture, diethyl ether–pyridine–ethanol–2 M NH$_4$OH, 65:30:8:2 (v/v). This solvent is run to approximately 1 cm below the top of the plates. This two-step development system assures separation of glycolipids from each other, but free fatty acids, if they are present in the original lipid mixture, will overlap with tetraglycosyl-ceramides. To remove fatty acids it is necessary to develop the chromatograms in a third solvent mixture. The chromatograms are dried for 30 minutes in a vacuum oven at room temperature and then are developed in the third solvent mixture, diethyl ether–acetic acid, 100:3 (v/v). This third developing solvent is run to the top of the plate. It carries the free fatty acids to the upper edge of the chromatoplates without affecting the position of the glycosphingolipids. This three-solvent development procedure usually takes over 8 hours; therefore, it is practical to develop the chromatograms in the third solvent overnight (16 hours). This does not interfere with the separation of glycolipids.

3. Results Obtained

Figure 1 illustrates the separation of reference neutral glycosphingolipids applied individually and as a mixture. The following compounds were separated, from top to bottom of the chromatogram: ceramides (two spots), monoglycosylceramides (two spots), sulfatides (two spots), diglycosylceramides, psychosines, triglycosylceramides, and tetraglycosylceramides (globosides). Cardiolipin, which may interfere with the separation of oligoglycosylceramides, moved up very slightly. Free fatty acid (oleic acid) and monoglyceride (1-monoolein) were also applied to this chromatoplate as controls, since these lipids often interfere with the separation of glycolipids in many solvent systems tried by Skipski et al.[6] They moved with the solvent front to the very top of the chromatogram and therefore are not visible (Fig. 1, lane 8). All other simple lipids also moved with the solvent front, whereas all phospholipids tested, e.g., phosphatidic acid, phosphatidylglycerol, phosphatidylethanolamine, phosphatidylserine, phosphatidylinositol, phosphatidylcholine, lysophosphatidylcholine, and sphingomyelin remained at the point of application of samples. A mixture of gangliosides also stayed at the point of origin.

In some chromatograms the spots of sulfatides and diglycosylceramides are very close to each other. Their separation can be improved, if necessary, by increasing the 2 M NH$_4$OH content in the second developing solvent from 2 parts to 3 parts. This change in composition of the second developing solvent may interfere slightly with the separation of other glycosphingolipids.

Direct thin-layer chromatography has been applied successfully in our laboratory for analysis of neutral glycosphingolipids in lipid extracts of several animal organs such as rat brain, spleen, and kidney.

4. Evaluation of Procedures

Direct thin-layer chromatography of neutral glycosphingolipids performed on total lipids extracted from animal tissues permits separation of monoglycosylceramides, oligoglycosylceramides up to tetraglycosylceramides, sulfatides (sulfate esters of monoglycosylceramides), ceramides, and psychosines. However, it does not permit separation of oligoglycosylceramides with more than four sugars (e.g. pentaglycosylceramides and hexaglycosylceramides). Another limitation of direct thin-layer chromatography of neutral glycosphingolipids is inherent in the lipids of the tissues. Thus, tissues which contain extremely small amounts

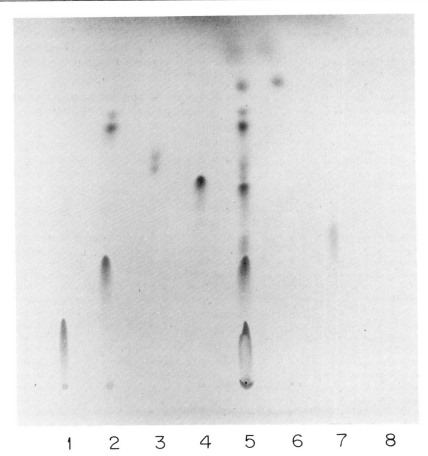

1   2   3   4   5   6   7   8

Fig. 1. Separation of neutral glycosphingolipids by thin-layer chromatography. Samples applied individually or as a mixture. The following approximate amounts of lipids were applied: lane 1, tetraglycosylceramides (globoside), 8 µg; lane 2, monoglycosylceramides (two upper spots), 8 µg; and triglycosylceramides (lower spot), 11 µg; lane 3, sulfatides (two fused spots), 5 µg; lane 4, diglycosylceramide (cytolipin H), 8 µg; lane 5, mixture of 1–4, 6, 7, and cardiolipin, 5 µg (at the origin of chromatogram); lane 6, ceramides (two spots), 8 µg; lane 7, psychosine, 7 µg; and lane 8, oleic acid, 10 µg and 1-monoolein, 7 µg (both of these compounds moved to the front of chromatogram and are not visible). Detection of spots: 40% H_2SO_4 spray. From Skipski et al.[6]

of neutral glycosphingolipids cannot be analyzed by this procedure but require enrichment of samples with glycolipids by column chromatography or other suitable procedures prior to analysis.

C. Review of Other Direct Thin-Layer Chromatographic Systems for Separation of Neutral Glycosphingolipids

Neskovic *et al.*[7] prepared the chromatoplates with silica gel H (Merck, Darmstadt, West Germany) containing 10% magnesium silicate (Florisil, 60–100 mesh). Reproducible results were obtained when magnesium silicate was milled in a small electric blender for 5 minutes prior to addition to silica gel. The thickness of adsorbent on the plate was 0.5–1 mm; activation time was 30 minutes at 110° done immediately before use of chromatoplate. The authors used a two-step development procedure. The first developing solvent was chloroform–acetone–pyridine–20% aqueous ammonia–water, 20:30:60:2:2 (v/v). This solvent was run 7.5 cm from the application of the samples (samples were applied 1.5 cm from the bottom of chromatoplates). The chromatograms were dried in a stream of air for 20 minutes. After drying, the lower part of the adsorbent (a 2.5-cm band starting at the bottom of the plates) was scraped off and the chromatograms were developed in the second solvent. This second developing solvent, a mixture of chloroform–methanol–water, 65:25:4 (v/v), was allowed to run to the top of the plate.

Neskovic *et al.*[7] obtained good separation of monoglycosylceramide, diglycosylceramide, triglycosylceramide, and tetraglycosylceramide applied as a mixture of reference compounds as well as in total lipids extracted (10–15 mg) from human serum for which they used adsorbent at a 1.0-mm thickness. This system has some advantages for the separation of glycosylceramides over the system described in Section II,B: The spots showed less tailing and the whole procedure requires only two developing solvents.

However, this system apparently does not separate such lipids as ceramides, psychosines, and sulfatides, which also are usually present in lipid extracts of animal tissues, from other glycosphingolipids. For separation of other glycolipids, Neskovic *et al.*[7,8] suggested a different second development solvent than described above. If the chromatograms were developed with the first solvent until it was 8–10 cm from the bottom of the chromatoplate, dried and then developed in a second solvent, chloroform–acetone–methanol–acetic acid–water, 65:35:11:4:1.5 (v/v), to the top of the plate, monogalactosyl diglyceride, monoglycosylceramides (two spots which according to the authors are kerasin and phrenosine), and sulfatides are separated from top to bottom of chromatograms. The authors applied this procedure as a quantitative method for analysis of neutral glycolipids in brains from humans and rats.[8]

[7] N. M. Neskovic, J. L. Nussbaum, and P. Mandel, *J. Chromatogr.* **49**, 255 (1970).
[8] N. Neskovic, L. Sarlieve, J.-L. Nussbaum, D. Kostic, and P. Mandel, *Clin. Chim. Acta* **38**, 147 (1972).

Eberlein and Gercken[9] used direct thin-layer chromatography to separate monoglycosylceramides, diglycosylceramides, and triglycosylceramides. They used an adsorbent consisting of a mixture of silica gel HR–magnesium silicate, 4:1 (w/w), and as a development solvent, a mixture of tetrahydrofuran–water, 5:1 (v/v). Chromatoplates were activated overnight at 150°. The resulting spots are clear but located very close to each other.

Each method for direct thin-layer chromatography described above has its advantages and disadvantages. The nature of the problem studied will determine what system is best. It is necessary to remember that the separation of some neutral glycosphingolipids may also be obtained with one- or two-dimensional thin-layer chromatography routinely used for separation of phospholipids. Thus, the earliest one-dimensional thin-layer chromatographic system for phospholipids published by Wagner et al.[10] also separates monoglycosylceramides. The thin-layer chromatographic system devised for separation of acidic phospholipids by Skipski et al.[11] (see also Skipski and Barclay[4c]) separates monoglycosylceramides and diglycosylceramides together with sulfatides. Two-dimensional thin-layer chromatographic systems for separation of polar lipids usually separate several glycolipids. Thus, two-dimensional systems described by Rouser et al.[12-14a] (see also Skipski and Barclay[4c]) separate monoglycosylceramides and sulfatides. Both monoglycosylceramides and sulfatides formed two spots on the chromatograms—one for compounds with normal fatty acids, another for compounds with hydroxy fatty acids. The system developed by Lepage[15,16] (see Skipski and Barclay[4c]) for separation of plant phospholipids separates monogalactosylglycerides, digalactosylglycerides, plant sulfolipids, and esterified and nonesterified sterol glycosides.

D. Detection Procedures for Glycosphingolipids and Other Lipids on Thin-Layer Chromatograms

Two types of tests are available for the detection of glycosphingolipid spots on a thin-layer chromatogram: (a) *General tests*, which reveal the

[9] K. Eberlein and G. Gercken, *J. Chromatogr.* **61**, 285 (1971).

[10] H. Wagner, L. Hörhammer, and P. Wolff, *Biochem. Z.* **334**, 175 (1961).

[11] V. P. Skipski, M. Barclay, E. S. Reichman, and J. J. Good, *Biochim. Biophys. Acta* **137**, 80 (1967).

[12] G. Rouser, S. Fleischer, and A. Yamamoto, *Lipids* **5**, 494 (1970).

[13] G. Rouser, C. Galli, and G. Kritchevsky, *J. Amer. Oil Chem. Soc.* **42**, 404 (1965).

[14] G. Rouser, G. Kritchevsky, and A. Yamamoto, *Lipid Chromatogr. Anal.* **1**, 99 (1967).

[14a] G. Rouser and S. Fleischer, Vol. 10 [69].

[15] M. Lepage, *J. Chromatogr.* **13**, 99 (1964).

[16] M. Lepage, *Lipids* **2**, 244 (1967).

presence of all lipids. Some of these tests are not destructive and, therefore, glycosphingolipids can be recovered by elution and analyzed further. (b) *Specific tests.* All of these tests are destructive, and hence lipids cannot be recovered. In most cases these tests are specific for particular classes of lipids.

More detailed and general descriptions of the different detection tests are given by Skipski and Barclay[4c]. Described here are tests which are useful for detection and identification of neutral glycosphingolipids or bear a direct relationship to this problem.

a. General Detection Tests

1. Sulfuric Acid

Chromatograms are sprayed with a 40 or 50% (v/v) aqueous solution of sulfuric acid followed by heating either on a hot plate (temperature up to 360°) for a few minutes or in an oven at 180° for 0.5–1.0 hour. All lipids, including neutral glycosphingolipids, as well as other organic compounds nonvolatile at these temperatures, will form dark brown or black spots on the white background. Glycosphingolipids usually produce comparative lighter spots than an equal weight of most neutral lipids or phospholipids.

2. Potassium Dichromate in Sulfuric Acid[17,18]

Chromatograms are sprayed with a saturated solution of $K_2Cr_2O_7$ in a 70% (v/v) aqueous solution of sulfuric acid followed by heating the chromatograms in an oven for 25–30 minutes at 180°. This spray is very useful for quantitation of glycosphingolipids by densitometry (see below).

3. Water Spray[19]

After development dried chromatograms are sprayed intensively with distilled water. The lipids, including neutral glycosphingolipids, become visible as white spots against a semitransparent background. The sensitivity of this test is very low. It is usually applied in preparative chromatography.

[17] O. S. Privett and M. L. Blank, *J. Amer. Oil Chem. Soc.* **39,** 520 (1962).
[18] M. L. Blank, J. A. Schmit, and O. S. Privett, *J. Amer. Oil Chem. Soc.* **41,** 371 (1964).
[19] E. Svennerholm and L. Svennerholm, *Biochim. Biophys. Acta* **70,** 432 (1963).

4. Primuline[20],[21]

Primuline is a spray reagent used originally by Wright[20] primarily for the detection of steroids and other neutral lipids on thin-layer chromatograms. However, our experiments[21] demonstrated that this reagent is the best of all known nondestructive, nonspecific sprays for the detection of glycosphingolipids on silica gel thin-layer chromatograms. The primuline spray is prepared by mixing 1 ml stock solution of primuline (0.1 g primuline in 100 ml water) with 100 ml of acetone–water (4:1, v/v) mixture. Primuline was obtained from Eastman Organic Chemicals, Eastman-Kodak Co., catalog No. T 1039, Rochester, N.Y. 14650, U.S.A., through Fisher Scientific Co., 711 Forbes Ave., Pittsburgh, Pa., 15219. The chromatograms, free of developing solvent, are sprayed very lightly with primuline solution. The spots are viewed under ultraviolet light at a wavelength of 365 nm. Lipids are revealed as light blue or yellowish spots on the dark blue–violet background. Primuline will detect all lipids, neutral lipids, phospholipids, and glycosphingolipids, without any negative discrimination toward glycosphingolipids. The sensitivity of the test is high (e.g., 0.05 μg of phrenosine and 0.05 μg of cytolipin H). Primuline spray is recommended especially when one intends to elute glycosphingolipids and characterize them further since this reagent does not interfere with determinations of monomeric carbohydrates, sphingosines, or fatty acids by gas–liquid chromatography or with the colorimetric determination of sphingosines by the procedure of Yamamoto and Rouser.[22]

5. Iodine Vapor[23-25]

Most lipids can be detected with iodine vapor on thin-layer chromatograms. However, glycosphingolipids should be present in relatively high quantities to be detected. The chromatogram is placed in a jar, or development tank, containing a trough filled with crystals of iodine. In several minutes most of the lipids will be revealed as brown spots on a pale yellow background. Glycosphingolipids usually give a slightly bluish brown color.

6. Oil Red O[6]

Chromatograms are sprayed with 0.1% oil red O in absolute ethanol. Most lipids including glycosphingolipids give dark red spots.

[20] R. S. Wright, J. Chromatogr. 59, 220 (1971).
[21] V. P. Skipski, A. Dnistrian, and M. Barclay, in preparation.
[22] A. Yamamoto and G. Rouser, Lipids 5, 442 (1970).
[23] H. K. Mangold and D. C. Malins, J. Amer. Oil Chem. Soc. 37, 383 (1960).
[24] H. K. Mangold, J. Amer. Oil Chem. Soc. 38, 708 (1961).
[25] R. P. A. Sims and J. A. G. Larose, J. Amer. Oil Chem. Soc. 39, 232 (1962).

7. Other Detection Tests

Some investigators used rhodamine B or rhodamine 6G, or 2′,7′-dichlorofluorescein or bromothymol blue for the detection of neutral glycosphingolipids. Details of preparation of these sprays are given by Skipski and Barclay.[4c]

b. Specific Detection Tests

8. Sphingolipids[6,26]

Sphingomyelin, neutral glycosphingolipids, and gangliosides give positive tests with Clorox–benzidine spray.

Reagents

REAGENT I. Fifty milliliters of benzene are mixed with 5 ml of Clorox bleach (tradename of commercial bleach; active reagent, sodium hypochlorite) and 5 ml of acetic acid (glacial).

REAGENT II. Two-hundred milligrams benzidine dihydrochloride (or benzidine) and a small crystal of potassium iodide are dissolved in 50 ml of 50% ethanol (filter, if necessary).

Procedure. Chromatograms are sprayed with freshly prepared reagent I (immediately after its preparation), left at room temperature for 30 minutes, dried in air at approximately 80° for 10 minutes, and then sprayed in hood with reagent II. (Extreme care should be taken to avoid any contact with this reagent; use a mask when spraying.) Reagent II should be used within 2 hours after its preparation, and it should be protected from direct light. Most sphingolipids appear almost immediately as blue spots on a white background.

9. Glycolipids–Orcinol Test[27]

Orcinol–sulfuric acid spray is used for the detection of lipid sugars. Two-hundred milligrams of orcinol are dissolved in 100 ml of H_2SO_4–H_2O, 3:1 (v/v), and stored in the dark under refrigeration. The spray is stable for about a week. Chromatograms are sprayed until the whole surface becomes obviously moist and then they are placed in an oven at 100° for approximately 15 minutes. Most glycolipids, including neutral glycosphingolipids, gangliosides, and sulfatides, will give pinkish violet spots on a white background.

[26] M. D. Bischel and J. H. Austin, *Biochim. Biophys. Acta* **70,** 598 (1963)
[27] L. Svennerholm, *J. Neurochem.* **1,** 42 (1956).

10. Glycolipids–Diphenylamine Test[10,28,29]

Diphenylamine spray detects neutral glycosphingolipids, gangliosides, and sulfatides. This reagent is prepared by mixing 20 ml of 10% diphenylamine solution in ethanol, 100 ml of concentrated hydrochloric acid, and 80 ml of acetic acid.

Chromatograms are sprayed extensively with diphenylamine reagent. A cardboard (or plastic) frame is put around the sprayed chromatogram and then covered with another clean glass plate. Both chromatogram and glass plate are clamped together and placed into an oven at 105–110° for 30–60 minutes. Glycolipids will give gray-bluish spots on a slightly grayish background.

11. Glycolipids–α-Naphthol Test[30]

A 0.5% solution of α-naphthol (purified by recrystallization from hexane–chloroform, 1:1, v/v) in methanol–water, 1:1 (v/v), is sprayed until the whole thin-layer chromatogram is damp. Chromatograms are allowed to air-dry and are then sprayed very lightly with a fine spray of 95% (by weight) sulfuric acid. The chromatograms are covered with clean glass plates (fastened with clamps) to prevent drying and heated in an oven at 120° until maximum color is developed.

All glycolipids (monoglycosylceramides, oligoglycosylceramides, sulfatides, gangliosides, and glycosyl diglycerides) give bluish purple spots and phosphatidylethanolamine gives a yellow spot. Other lipids usually do not produce any "spots" or may give light yellowish spots (cholesterol may produce a gray-red spot).

12. Gangliosides–Resorcinol Test[31,32]

Resorcinol spray is specific for sialic acid and, therefore, all gangliosides and hematosides produce positive tests with this reagent.

Reagents. The resorcinol stock solution is prepared by dissolving 2 g resorcinol in 100 ml water. Ten milliliters of the stock solution of resorcinol and 0.25 ml 0.1 M copper sulfate (2.5 g $CuSO_4 \cdot 5H_2O$ in 100 ml water) are added to 80 ml concentrated HCl. The total volume of the resorcinol spray is brought to 100 ml with water. This spray should be prepared at least 4 hours before use. It must be kept refrigerated, and it is stable for about 2 weeks.

[28] Z. Dische, *Microchemie* **7**, 33 (1929).
[29] H. Jatzkewitz and E. Mehl, *Hoppe-Seylor's Z. Physiol. Chem.* **320**, 251 (1960).
[30] A. N. Siakotos and G. Rouser, *J. Amer. Oil Chem. Soc.* **42**, 913 (1965).
[31] L. Svennerholm, *Biochim. Biophys. Acta* **24**, 604 (1957).
[32] L. Svennerholm, *Acta Chem. Scand.* **17**, 239 (1963).

Procedure. Chromatograms are sprayed with resorcinol reagent, covered with glass plates of the size of the chromatoplate (keep them together with clamps), and placed into an oven at 140° for about 25–35 minutes. Gangliosides and hematosides give purple-blue spots on a white background.

This spray is useful to detect whether or not chromatograms of neutral glycosphingolipids are contaminated with gangliosides or hematosides (gangliosides which lack hexosamines in their molecules).

13. Test for Phospholipids

There are several specific reagent sprays for the detection of phospholipids on thin-layer chromatograms. These reagents produce a color reaction with phospholipid phosphorus. Below is described one of these tests for phospholipids recently devised by Vaskovsky and Svetashev.[33]

Reagents

REAGENT I. Ten grams of sodium molybdate is dissolved in 100 ml of 3–4 N HCl.

REAGENT II. One gram of hydrazine hydrochloride is dissolved in 100 ml of water.

Mix both reagent I and reagent II and heat in a boiling water bath for 5 minutes. After cooling, the total volume is adjusted to 1 liter with water and the final reagent is ready for use. This reagent is stable for several months.

Procedure. Chromatograms are sprayed with the final reagent. Phospholipids produce blue spots on a white background immediately after the application of this spray. Later (1–2 hours) after spraying a blue background will develop. Spraying the chromatogram with 10% sulfuric acid in methanol preserves the white background. It is necessary to test chromatograms of neutral glycosphingolipids for the possible contamination by phospholipids.

14. Other Tests for Phospholipids

Among other specific tests for the detection of phospholipid it is necessary to mention the test developed by Dittmer and Lester[34] (see also Skipski and Barclay[4c]) and by Vaskovsky and Kostetsky.[35]

---

[33] V. E. Vaskovsky and V. I. Svetashev, *J. Chromatogr.* **65**, 451 (1972).
[34] J. C. Dittmer and R. L. Lester, *J. Lipid Res.* **5**, 126 (1964).
[35] V. E. Vaskovsky and E. Y. Kostetsky, *J. Lipid Res.* **9**, 396 (1968).

III. Thin-Layer Chromatography of Lipid Samples Enriched with Neutral Glycosphingolipids

A. General Description of the Methods for Enrichment of Samples

Most lipids extracted from animal tissues, except nervous tissue, contain relatively small amounts of glycolipids, especially neutral glycosphingolipids. Direct thin-layer chromatography, at best, allows the detection and analysis of only the major species of neutral glycosphingolipids. Therefore, to overcome this problem, lipids extracted from tissues are enriched with neutral glycosphingolipids. Two critical issues in this enrichment process are quantitative recovery and types of impurities which may still be present. There are several approaches to this enrichment process: (a) lipids extracted from tissues are chromatographed on columns packed with silicic acid, Florisil, or ion-exchange materials [diethylaminoethyl (DEAE)-cellulose and triethylaminoethyl (TEAE)-cellulose] to remove neutral (simple) lipids (e.g., triglycerides and lower glycerides, cholesterol, cholesteryl esters and free fatty acids) and all or most of phospholipids; or (b) removal of neutral (simple) lipids and phospholipids by chemical destruction (alkaline hydrolysis or methanolysis) of glycerolipids; or (c) reversible chemical alteration of neutral glycosphingolipids (e.g., their carbohydrate moieties) by chemical treatment (acetylation) which alters their chromatographic properties and insures chromatographic separation from unaltered phospholipid impurities.

Detailed descriptions will be given of two methods for enrichment of lipid samples. In the first procedure chromatographic separation is made without chemical treatment of samples. This procedure involves a method previously published for thin-layer chromatography of neutral glycosphingolipids (described in Sections II,A and B) converted to column chromatography.

Another procedure for enrichment of lipid samples with glycosphingolipids will be presented in detail. This is from Hakomori's group and involves acetylation of lipid samples.

B. Enrichment of Lipid Samples with Neutral Glycosphingolipids by Column Chromatography Followed by Thin-Layer Chromatography

1. Column Chromatography of Lipid Extracts[21]

After Folch partition, the lower phase, containing certain lipids extracted from animal tissues, is evaporated to a small volume or to dryness and then redissolved in chloroform–methanol, 2:1 (v/v). These "lower phase" lipids include neutral glycosphingolipids, neutral (or simple) lipids, phospholipids, and may contain some monosialohematosides. To enrich the samples with neutral glycosphingolipids all neutral lipids and most phospholipids (more than 95%) are removed by column chromatography. The lipid extracts are subjected to chromatography on a mixed adsorbent, silicic acid–Florisil (magnesium silicate), 1:1 (w/w).

The silicic acid is purchased from Malinckrodt Chemical Works (St. Louis, Mo. 63160) SilicAR, CC-7, special for column chromatography. Florisil mesh 100–200 is from Floridin Co. and distributed by Fisher Scientific Co. (711 Forbes Ave., Pittsburgh, Pa. 15219). Prior to use, both adsorbents are washed extensively by stirring with chloroform–ethanol mixture (1:1, v/v) twice, and then 5–6 times with ethanol, until the odor of chloroform disappears. They are then washed with doubly distilled water. The adsorbents are stirred during the last 2–3 washings with water, allowed to settle for 2 minutes, and the water phase containing unsettled particles is discarded. This removes very small particles of adsorbents which considerably slow down the flow rates of eluting solvents. The washed adsorbents, transferred to glass trays, are dried overnight in an oven at 105–110° to remove most of the water. The adsorbents are then transferred to the original jars and the drying is continued for another 48 hours at 175°. The washed, cooled adsorbents (room temperature) are mixed in the proportion 1:1 (w/w) and kept in a desiccator. Before using in chromatographic columns, the mixed adsorbents are again activated at 175° for at least 3 hours and again cooled to room temperature in a desiccator.

If lipid samples to be enriched contain up to 100 mg of original tissue lipid extract it is necessary to use 20 g of silicic acid–Florisil mixture for packing the chromatographic column. The ratio of height to diameter of the working adsorbent in the column should be 12 to 1. The 20 g of silicic acid–Florisil (weighed in ground glass stoppered Erlenmeyer flask) are suspended in 200 ml of chloroform, vigorously shaken, and transferred rapidly to the chromatographic column reservoir. The reservoir is supplied with a ground glass stoppered drying tube filled with a moisture absorbent [silica gel (Tell Tale) or Drierite (calcium sulfate anhydrous)]. The silicic acid–Florisil should pack slowly and evenly into the column; therefore, a vibrator is employed continuously. To prevent sticking of silicic acid–Florisil to the reservoir it is advisable to wash the walls with a small amount of chloroform. After the column is packed it is transferred to a cold room (5–6°). One or two hours later the column is conditioned by putting 250 ml of cold chloroform through it. Up to 100 mg of total lipid extract in 1.0–2.0 ml of chloroform–methanol (2:1, v/v) are applied to the column. The lipids are eluted from the column with 500 ml of chloroform. This eluate contains neutral (simple) lipids. The second elution solvent is 121 ml of acetone–pyridine–water 4:6:1 (v/v). This solvent mixture will elute all neutral glycosphingolipids and a small amounts of phospholipids (crude neutral glycosphingolipid fraction). The phospholipids, present in the eluate as contaminants that may interfere with the separation of glycosphingolipids in the thin-layer chromatographic systems used are trace amounts of phosphatidylethanolamine, and substantial amounts of phosphatidylserine and phosphatidylinositol. Over 95% of the total phospholipids stay on the chromatographic column. Unesterified fatty acids are also present in the crude neutral glycosphingolipid fraction. The eluate containing neutral glycosphingolipids is evaporated under vacuum while bubbling nitrogen through the eluate to approximately 1 ml. The temperature during evaporation should not exceed 40°. A small amount of chloroform–methanol, 2:1 (v/v), is added, and the concentrated eluate is transferred quantitatively to a volumetric flask. In cases where large samples were originally chromatographed (100 mg of original lipid extract), the final volume of crude glycolipid fraction is adjusted to 5 ml (20 mg equivalent to original extract per milliliter). The crude neutral glycosphingolipid fraction can be subjected either directly to analytical thin-layer chromatography for separation, identification, and quantitative analysis of individual glycolipids or purified further by preparative thin-layer chromatography.

2. Preparative Thin-Layer Chromatography

When it is desirable to have pure (or almost pure) fractions of neutral glycosphingolipids the following method of preparative thin-layer chromatography is used. A silica gel "slurry" is prepared by suspending 40 g of silica gel, D-O, in 90 ml of water, and this "slurry" is applied to glass plates with the applicator set for 0.5 mm thickness. The prepared plates are allowed to dry in air for about 1 hour and then activated in the oven at 110° for 1–2 hours. Chromatoplates are prewashed with chloroform–methanol–water, 25:75:10 (v/v). Prior to use, chromatoplates are again activated in the oven at 110° for at least 1½ hours. The plates are cooled in a desiccator. Before application the sample is concentrated to a smaller volume by partial evaporation of solvent under nitrogen in a water bath at approximately 40°. Methanol is added during the evaporation several times to remove as much as possible any pyridine if it is present in the sample solution. The sample and standard reference compounds are applied under nitrogen in a Desaga-Brinkmann Application box. Applied with the sample are available reference neutral glycosphingolipids and phospholipids (e.g., phosphatidylserine, phosphatidylinositol, and phosphatidylethanolamine likely to be present in the sample). The sample is usually applied as a band (series of small spots close to each other). The developing tank is lined with filter paper on three sides and paper is wetted with developing solvent. Chromatograms are developed by using consecutively two developing solvents. The first is a mixture of chloroform–methanol–acetone–water, 65:30:2:3 (v/v), which is allowed to run up to a level approximately 10 cm above the bottom of the chromatoplate (8 cm above the points of sample applications). Plates are removed from the developing tank and dried in a vacuum oven for 10 minutes at room temperature. Chromatoplates are then developed in the second solvent mixture: chloroform–methanol–acetone–water, 65:20:2:1 (v/v). This solvent is allowed to run to the top of the plates (approximately 19 cm). The developing system used in this preparative thin-layer chromatogram is devised to insure the proper separation of glycosphingolipids from phospholipids which are present in small quantities in the crude fraction of neutral glycosphingolipids.

Lipid spots on chromatograms are visualized under ultraviolet light after the primuline reagent is sprayed on the plates (see Section II,D,4). However, it is advisable, when a sufficient amount of material is available, to run additional preparative thin-layer chromatograms with small amounts of sample and test them as follows: for glycolipids by the orcinol reagent (Section II,D,9), for sialic acid containing glycolipids by resorcinol reagent (Section II,D,12), for phospholipid impurities (Section II,D,13), and for organic material in general (Section II,D,1 or II,D,2). (The test for organic material can be run on the same chromatoplate after orcinol, resorcinol, or primuline sprays.)

After positive identification of positions of neutral glycosphingolipid and phospholipid spots on thin-layer plates sprayed with primuline (the principal plate), the silica gel areas containing phospholipids are scraped off and discarded. The areas containing neutral glycosphingolipids may be scraped off individually for each lipid, for subsequent elution and chemical analysis, or as a group for analytical thin-layer chromatography. In both cases silica gel containing neutral glycosphingolipids is transferred to ground glass stoppered, heavy walled centrifuge tubes (15 ml or larger) and subjected to the elution. The following eluting solvents and their quantities are used consecutively: pyridine, 5–8 ml; pyridine–methanol, 1:1 (v/v), 2–5 ml; chloroform–methanol, 1:3 (v/v), 2–5 ml; and methanol 2–5 ml used twice. During each elution the silica gel suspension in solvent is shaken vigorously for 15 minutes at 37–40° in a mechanical shaker. We used Rotary Evapo-Mix (Buchler Instruments, Fort Lee, N.J. 07024). Then, the silica gel suspensions are centrifuged and the eluates removed with disposable capillary pipettes. The eluates are combined and filtered through ultrafine fritted disc filter funnels to remove any residual silica gel. The combined eluates are evaporated under vacuum (nitrogen is bubbled through the solution) to small suitable volumes. The eluting solvent quantities presented above are suitable for up to 10 mg equivalent of total lipids extract.

3. Analytical Thin-Layer Chromatography

The concentrated eluate obtained from preparative thin-layer chromatography may be studied further by analytical thin-layer chromatography. Chromatoplates are prepared and activated in the same way as for preparative thin-layer chromatography. The sample is usually applied as a small band, in quantities equivalent to 4 mg of original lipid extract, along with reference neutral glycosphingolipids. The developing solvents used are chloroform–methanol–water, 60:30:7 or 60:25:4 (v/v). The spots of neutral glycosphingolipids can be visualized by any of the above-described suitable sprays (Section II,D). If it is important to get a general picture of glycolipids present in lipid extracts, orcinol spray (Section II,D,9) or sulfuric acid spray (Section II,D,1) are the best.

For quantitative determination of neutral glycosphingolipids by densitometric procedures it is necessary to use potassium dichromate in sulfuric acid spray (Section II,D,2).[36–38a] However, in this case it is advisable to run several analytical thin-layer chromatograms with development solvents of different polarity (change the content of methanol and/or water in chloroform–methanol–water developing solvent) in order to increase distances between spots in selected areas of chromatograms. This approach gives more reproducible and reliable quantative data.

If there is an intention to elute neutral glycosphingolipids from thin-layer chromatograms for further chemical characterization, or chemical quantitation of neutral glycosphingolipids (e.g., quantitative determination of sphingosine as a method for quantitative determination of individual glycosphingolipids[22] or to perform gas–liquid chromatography on separated neutral glycosphingolipids, primuline spray should be used. This spray does not interfere with all the above-mentioned tests.

[36] O. S. Privett and M. L. Blank, J. Amer. Oil Chem. Soc. 39, 520 (1962).

[37] M. L. Blank, J. A. Schmit, and O. S. Privett, J. Amer. Oil Chem. Soc. 41, 371 (1964).

[38] O. S. Privett, M. L. Blank, D. W. Codding, and E. C. Nickell, J. Amer. Oil Chem. Soc. 42, 381 (1965).

[38a] G. Rouser, G. Kritchevsky, C. Galli, and D. Heller, J. Amer. Oil Chem. Soc. 42, 215 (1965).

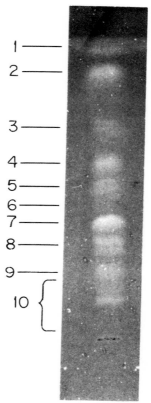

Fig. 2. Thin-layer chromatogram of the crude neutral glycosphingolipid fraction isolated from lipids extracted from Walker carcinosarcoma 256 by column chromatography (see text for details). The thin-layer chromatogram was developed with chloroform–methanol–water, 60:25:4 (v/v). Spots were detected by primuline spray and the chromatogram was photographed under long waves of ultraviolet light. Tentative identification of spots is as follows: 1. ceramides, 2. fatty acids, 3. monoglycosylceramides, 4. diglycosylceramides and traces of phosphatidylethanolamine, 5. sulfatides, 6. triglycosylceramides, 7. Phosphatidylserine and phosphatidylinositol, 8. tetraglycosylceramides, 9. pentaglycosylceramides, and 10. oligoglycosylceramides containing more than five sugars and neoproteolipid-W. The amount of the crude neutral glucosphingolipid fraction chromatographed corresponds to the amount from 4 mg of total lipid extracted from Walker carcinosarcoma 256.

To illustrate the above-described procedure, Fig. 2 shows a thin-layer chromatogram of the crude neutral glycosphingolipid fraction obtained from lipid extracted from Walker carcinosarcoma 256. This corresponds to a preparative thin-layer chromatogram, but in this case instead of

a two-step development system, a single development system was used
to make the data on Fig. 2 more comparable with those on Fig. 3. Figure
2 was taken after primuline spray (according to the procedure described)
and photographed at long waves of ultraviolet. On this figure one can

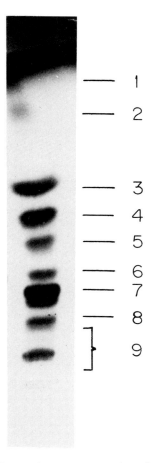

Fig. 3. Analytical thin-layer chromatogram of crude neutral glycosphingolipid
fraction isolated from lipids extracted from Walker carcinosarcoma 256 by column
chromatography and purified by preparative thin-layer chromatography. Tentative
identification of spots is as follows: 1. ceramides, 2. fatty acids, 3. monoglycosylceramides, 4. diglycosylceramides, 5. sulfatides, 6. triglycosylceramides, 7. tetraglycosylceramides, 8. pentaglycosylceramides, and 9. oligoglycosylceramides with more
than five sugars and neoproteolipid-W. The amount of neutral glycosphingolipids
chromatographed corresponds to that present in 4 mg of total lipid extracted
from Walker carcinosarcoma 256. Developing solvent: chloroform–metanol–water
60:30:7 (v/v). Detection: 40% sulfuric acid spray.

see some impurities (phospholipids) on thin-layer chromatogram. Figure 3 shows the final analytical thin-layer chromatogram after phospholipids were removed (see description of the procedure). The chromatogram in Fig. 3 was sprayed with sulfuric acid. In both Fig. 2 and Fig. 3 identification of spots is tentative.

The procedure described for thin-layer chromatography of neutral glycosphingolipids on enriched samples gave a quantitative recovery of these lipids. The mixture of standards, consisting of monoglycosylceramides (phrenosine isolated from bovine spinal cord[39]), diglycosylceramide (synthetic cytolipid H, Miles-Yeda, Miles Laboratories, Inc., P.O. Box 272, Kankakee, Ill., 60901), sulfatides (Applied Science P.O. Box 440, State College, Pa. 16801), and tetraglycosylceramides (isolated in our laboratory from the transplanted rat tumor, Walker carcinosarcoma 256), was subjected consecutively to column chromatography, preparative thin-layer chromatography, and analytical thin-layer chromatography as described above. Quantitative analysis of glycosphingolipids performed by a densitometric procedure on the final analytical thin-layer chromatogram against an original mixture of standards gave the following recoveries: monoglycosylceramides, 98.2%; diglycosylceramide, 103.0%; sulfatides, 98.0%; and tetraglycosylceramides, 96.3%. These data were obtained by averaging results from four completely independent experiments. Higher oligoglycosylceramides (pentoglycosylceramides) were tested and also gave quantitative recoveries.

4. Evaluation of the System for Enrichment and Separation of Neutral Glycosphingolipid Samples

The system described above[21] for enrichment of samples before thin-layer chromatography facilitates the separation of neutral glycosphingolipids on silica gel chromatoplates using chloroform–methanol–water as development solvents in different proportions.[10] An advantage of this method is that it is based exclusively on adsorption chromatography without any kind of chemical treatment for either neutral glycosphingolipids or phospholipid impurities. This practically excludes losses even of polyglycosylceramides (pentaglycosylceramides were checked). Success of the method results mainly from usage of silicic acid–magnesium silicate adsorbent with elution solvents selected to elute the neutral glycosphingolipid fractions which are only slightly contaminated with phospholipids and simple lipids. Successful removal of phospholipid and free

---

[39] V. P. Skipski, S. M. Arfin, and M. M. Rapport, *Arch. Biochem. Biophys.* **82**, 487 (1959).

fatty acids impurities from neutral glycosphingolipids by preparative thin-layer chromatography is facilitated greatly by application of primuline spray for detection of spots. Primuline spray, for detection of glycolipids and phospholipids, has not been previously used. This spray is not destructive, is very sensitive (detects 0.05 μg glycolipids), and does not interfere with further chemical analysis of neutral glycosphingolipids. However, even with the advantages cited for this method, this procedure is not necessarily the best for all circumstances. Several methods, which allow effective enrichment of samples with neutral glycosphingolipids, have been worked out in other laboratories.

5. Other Methods to Enrich Lipid Samples with Neutral Glycosphingolipids

Adsorption Chromatography. Rouser *et al.*[14,40] used silicic acid columns for selective elution of glycolipids from lipid extracts of human brain. The authors used sequentially: chloroform (elutes neutral lipids), acetone (elutes glycolipids with some phospholipids), and methanol (elutes phospholipids). According to the authors, the acetone fraction contains monoglycosylceramides (cerebrosides), sulfatides, oligoglycosylceramides, and some cardiolipin. However, the authors did not give either the identities of oligoglycosylceramides or their percent recovery. The recoveries of cerebrosides and sulfatides were quantitative. Acetone eluates also contained mono- and diglycosyl diglycerides, and sulfolipids when plant lipid extracts were applied to this system of chromatography.

Gray[41] applied lipids extracted from rat and rabbit lungs to silica gel H columns. Neutral lipids were eluted with chloroform–methanol, 49:1 (v/v), and tetrahydrofuran–benzene, 1:4 (v/v); and total neutral glycosphingolipids were eluted with tetrahydrofuran–methylal(dimethoxymethane)–methanol–water, 10:6:4:1 (v/v). The glycolipid fraction contained monoglycosylceramides, diglycosylceramides, triglycosylceramides, tetraglycosylceramides, and sulfatides. If gangliosides were not removed from the original extract, they also would be present in this fraction. However, this fraction contained about 12% total phospholipid phosphorus in the form of acidic phospholipids. The phospholipids were eliminated from the glycolipid fraction with mild alkaline methanolysis, using 0.2 M methanolic NaOH (final concentration in methanolysis mixture) for 2 hours at room temperature. Gray states[41] that glycolipids were completely unaffected by this treatment and were recovered quantitatively. Glycolipids were separated further from methanolysis by-products

[40] G. Rouser, G. Kritchevsky, G. Simon, and G. J. Nelson, *Lipids* **2**, 37 (1967).
[41] R. M. Gray, *Biochim. Biophys. Acta* **144**, 511 (1967).

with silicic acid column chromatography. The total pure fraction of glycosphingolipids was separated into its family species using the Wagner[10] system for thin-layer chromatography. The recovery of each of the glycosphingolipids was not less than 95%. However, mild alkaline methanolysis may not be effective for removal of all phospholipid impurities if original lipid extracts contain considerable amounts of phospholipids with ether bonds since these bonds are resistant to alkaline treatment.

The procedures of Rouser et al.[14,40] and especially of Gray[41] in which acetone or tetrahydrofuran, respectively, are used for selective elution of glycosphingolipids are similar to our approach in which pyridine is used for the same purpose (Section III,B,1). Tetrahydrofuran and pyridine are good solvents for glycolipids and most phospholipids are not soluble in them.

Vance and Sweeley[42] obtained fractions selectively enriched with glycosphingolipids from lipid extracts of human blood plasma and erythrocytes. They used silicic acid columns and eluted neutral lipids with chloroform, after which the crude total glycosylceramide fraction was eluted with acetone–methanol, 9:1 (v/v), a similar procedure to that of Rouser et al.,[14,40] (see above). Several minor phospholipid contaminants were removed by mild alkaline methanolysis, neutralized with HCl, and partitioned by addition of water and chloroform. The glycosphingolipids were recovered from the lower chloroform phase and subjected to preparative thin-layer chromatography. Vance and Sweeley used a method for thin-layer chromatography of Wagner et al.,[10] previously modified by Svennerholm and Svennerholm.[43] Plates coated with silica gel were heated for at least 2 hours at 90° just prior to applying the solution of mixed glycosphingolipids. The developing tank was equilibrated with developing solvent from 2 to 4 hours before the insertion of the chromatoplate into the solvent. The composition of the developing solvent was chloroform–methanol–water, 100:42:6 (v/v). Recovery of neutral glycosphingolipids from blood plasma varied with the complexity of the oligosaccharide moiety and ranged from 94% for monoglycosylceramides to 71% for tetraglycosylceramides.

These examples of procedures for enrichment of lipid extracts with neutral glycosphingolipids are based mainly on adsorption chromatography. Most of the authors cited subjected the fraction enriched with glycolipids to mild alkaline methanolysis to remove phospholipid impurities. Such purified fractions were subsequently chromatographed on thin-layer

[42] D. E. Vance and C. C. Sweeley, J. Lipid Res. 8, 621 (1967).
[43] E. Svennerholm and L. Svennerholm, Biochim. Biophys. Acta 70, 432 (1963).

silica gel chromatoplates using chloroform–methanol–water in different proportions.

Ion-Exchange Chromatography. Another approach to the problem of enrichment of lipid fractions with glycosphingolipids is the application of ion-exchange chromatography. Separation of lipids by this procedure is based primarily on exchange of the different ionic groups. However, differences in polarities provided by nonionic groups (e.g., hydroxyl groups) also influence the behavior of the molecules on ion-exchange columns. All lipids can be divided provisionally into three groups: nonionic, zwitterionic, and acidic.[44] They are separated readily on ion exchange into these three major groups. Within each group further separation may be achieved on the basis of differences in polarity and, when applicable, degree of ionization. Two ion-exchange celluloses are used generally for the chromatographic separation of lipids: DEAE (diethylaminoethyl)-cellulose and TEAE (triethylaminoethyl)-cellulose. A detailed description of the application of these ion-exchange celluloses to the separation of lipids is given by Rouser *et al.*[44a] (see also Rouser and co-workers).[14,38a]

The general properties and handling of ion-exchange cellulose are described by Peterson.[45]

Ion-exchange cellulose column chromatography produces quite a different order of elution for different lipids compared with that observed when silicic acid or Florisil columns are used. Therefore, in neutral glycosphingolipid fractions, the types of impurities will be different from those in adsorption chromatography. In certain instances this may facilitate the effective separation of neutral glycosphingolipids by thin-layer chromatography.

C. Acetylation of Neutral Glycosphingolipids

Recently, a new approach was developed for the separation of neutral glycosphingolipids: alteration by acetylation[46] of glycolipids prior to any kind of separation. This method results in changes in the polarities of glycolipids and secures their separation from phospholipids when column

[44] M. Kates, *in* "Laboratory Techniques in Biochemistry and Molecular Biology" (T. S. Work and E. Work, eds.), Vol. 3, p. .267. Amer. Elsevier, New York, 1972.

[44a] G. Rouser, G. Kritchevsky, A. Yamamoto, G. Simon, C. Galli, and A. J. Bauman, Vol. 14 [47].

[45] E. A. Peterson, *in* "Laboratory Techniques in Biochemistry and Molecular Biology" (T. S. Work and E. Work, eds.), Vol. 2, p. 225. Amer. Elsevier, New York, 1970.

[46] T. Saito and S. Hakomori, *J. Lipid Res.* **12**, 257 (1971).

chromatography on silicic acid is used. This approach is especially fruitful when it is necessary to study neutral glycosphingolipids with more than four sugars (from pentaglycosylceramides to octaglycosylceramides). Actually, this method had been used for some time for the isolation of blood-group glycolipids,[47-50] but at present is used also for other glycolipids.[51-53] After successful separation of acetylated glycosphingolipids by column chromatography they are deacetylated and then can be used in native form for thin-layer chromatography.

Saito and Hakomori[46] described two different procedures for separation of acetylated lipids. Procedure 1 is suitable for analysis of neutral glycosphingolipids, sulfatides, hematosides, and monosialogangliosides; but higher gangliosides containing the sialosylsialosil group do not give quantitative recoveries. A detailed description of this procedure is given below.

Procedure 1. A piece of tissue (less than 1 g) is homogenized with 10 ml of chloroform–methanol, 2:1 (v/v) and centrifuged. The tissue residue is again homogenized with chloroform–methanol, 2:1, and centrifuged. Both extracts are combined and evaporated to dryness under nitrogen. The crude lipid extract obtained is dried in a vacuum over phosphorus pentoxide. To the dry crude lipid extract is added 0.3 ml of dried pyridine and 0.2 ml of acetic anhydride. This mixture is allowed to stand for 18 hours at room temperature. Approximately 50 ml of dry toluene is added to the reaction mixture and the solvent is completely evaporated in a rotary evaporator. The dry residue (acetylated lipids) is dissolved in a mixture of 1,2-dichloroethane–hexane, 4:1 (v/v), and applied to the chromatographic column. The column (0.7 × 10 cm) had been packed with 5 g of Florisil, 50–100 mesh, which was suspended in 1,2-dichloroethane (ethylene chloride)–hexane, 4:1 (v/v). The chromatographic column is eluted successively with 15 ml each of the following solvents: 1,2-dichloroethane–hexane, 4:1 (v/v); 1,2-dichloroethane; 1,2-dichloroethane–acetone, 1:1 (v/v); and 1,2-dichloroethane–methanol–water, 2:8:1 (v/v). The first two solvents elute all neutral lipids. All the acetylated glycolipids are eluted by 1,2-dichloroethane–acetone, 1:1 (v/v) mixture, leaving the phospholipids on the column. Phospholipids can be eluted by 1,2-dichloroethane–methanol–water, 2:8:1 (v/v).

[47] S. Handa, *Jap. J. Exp. Med.* **33**, 347 (1963).
[48] S. Hakomori and G. D. Strycharz, *Biochemistry* **7**, 1279 (1968).
[49] S. Hakomori and H. D. Andrews, *Biochim. Biophys. Acta* **202**, 225 (1970).
[50] S. Hakomori, *Chem. Phys. Lipids* **5**, 96 (1970).
[51] S. Hakomori, *Proc. Nat. Acad. Sci. U.S.* **67**, 1741 (1970).
[52] W. J. Esselman, J. R. Ackermann, and C. C. Sweeley, *J. Biol. Chem.* **248**, 7310 (1973).
[53] C. G. Gahmberg and S. Hakomori, *J. Biol. Chem.* **248**, 4311 (1973).

The following acetylated glycolipids were eluted (experimental mixture of standards was used): ceramides, monoglycosylceramides, diglycosylceramides (lactosylceramide), triglycosylceramides, tetraglycosylceramides (globoside), pentaglycosylceramides, octaglycosylceramides, hematosides, and monosialogangliosides. Recovery of all these glycolipids ranged from 95 to 100%. The fraction of acetylated glycolipids also contained disialosylhematoside and trisialogangliosides, but recoveries were in the 80–85% range. Sulfatides, if present in the original lipid mixture, would be eluted together with other acetylated glycolipids with 95–100% recovery.

The eluates containing acetylated glycolipids are evaporated to dryness under vacuum, lipids dissolved in small volumes of chloroform–methanol, 2:1 (v/v), and transferred quantitatively to small test tubes, after which the solvents are evaporated to dryness under nitrogen. To the residues are added 100–200 μl chloroform–methanol, 2:1 (in order to redissolve the lipids), and 25–50 μl 5% sodium methoxide in methanol. The reaction mixtures are left for 30 minutes at room temperature and then neutralized with ethyl acetate. The neutralized solutions are evaporated to dryness, emulsified in water with slight warming, and dialyzed against ice water overnight. The dialyzed solutions are lyophilized or evaporated under nitrogen and the residues dissolved in suitable quantities of chloroform–methanol, 2:1. The solutions of glycolipids obtained are now ready for analysis by thin-layer chromatography. However, the presence of gangliosides together with neutral glycosphingolipids may complicate thin-layer chromatographic separations of the latter.

Procedure 2. Procedure 2 of Saito and Hakomori[46] permits the separation of neutral glycosphingolipids and gangliosides into two separate frac-
They partitioned the total lipid extracts from tissue with 0.2 volume of water according to Folch et al.[1] overnight at 4° (see also Radin[4b]). Neu-
Saito and Hakomori,[46] securing also the separate treatment of neutral glycosphingolipids and gangliosides. This modification is described here. Esselman et al.[52] subjected only neutral glycosphingolipids to acetylation. They partitioned the total lipid extracts from tissue with 0.2 volume of water according to Folch et al.[1] overnight at 4° (see also Radin[4b]). Neutral glycosphingolipids will be in the lower (chloroform) phase whereas sialic acid–containing glycolipids will be in the upper (methanol–water) phase. Only the lower phase was evaporated to dryness *in vacuo* and subjected to acetylation according to Procedure 1 described above. The upper phase from the Folch partition was evaporated to dryness *in vacuo*, emulsified in water, and dialyzed against several changes of distilled water at 4° for 24 hours. The resultant purified fraction of gangliosides was evaporated to dryness, redissolved in chloroform–methanol, 1:1

(v/v), and used for thin-layer chromatography. Esselman et al.[52] used silica gel chromatoplates activated at 125° for at least 2 hours for separation of both neutral glycosphingolipids and gangliosides. For separation of neutral glycosphingolipids (after deacetylation) they used chloroform–methanol–water, 80:42:6 (v/v), as developing solvents. For separation of gangliosides the developing solvent mixture of chloroform–methanol–2.5 N NH₄OH, 60:40:9 (v/v) was used. Lipids from chromatoplates were eluted with 10–15 ml of chloroform–methanol–water, 10:10:3 (v/v).

This modification suggested by Esselman et al.[52] considerably simplifies the final separation on thin-layer chromatograms of neutral glycosphingolipids and gangliosides. It is applicable also to Procedure 2 of Saito and Hakomori.[46]

D. Enrichment of Lipid Samples with Neutral Glycosphingolipids by Partial Hydrolysis or Alcoholysis

It was mentioned before in this chapter that the separation of neutral glycosphingolipids from neutral (simple) lipids does not present any problems. Both silica gel thin-layer chromatography and silicic acid column chromatography efficiently separate these two large groups of lipids. The major problem is the separation of neutral glycosphingolipids from phospholipids, since these have similar chromatographic behaviors. A number of investigators[41,42,54–58] took advantage of the differences in chemical properties of neutral glycosphingolipids and the majority of phospholipids. Neutral glycosphingolipids contain relatively alkali-resistant amide and glycosyl bonds, whereas most phospholipids are glycerolipids which contain alkali-labile ester linkages. Mild alkali treatment (hydrolysis or preferentially methanolysis) does not affect neutral glycosphingolipids, but glycerophospholipids undergo deacylation (removal of fatty acids esterified to glycerol). As a result of deacylation (alkaline methanolysis), glycerophospholipids are converted into methyl esters of fatty acids and water-soluble glycerylphosphoryl derivatives.[59] However, sphingomyelin, which is always present in total lipid extracts, will not be affected by mild alkali treatment and will stay intact. Since in most cases fractions of neutral glycosphingolipids obtained after silicic acid

[54] R. J. Desnick, C. C. Sweeley, and W. Krivit, J. Lipid Res. 11, 31 (1970).
[55] H.-D. Klenk and P. W. Choppin, Proc. Nat. Acad. Sci. U.S. 66, 57 (1970).
[56] T. Eto, Y. Ichikawa, K. Nishimura, S. Ando, and T. Yamakawa, J. Biochem. (Tokyo) 64, 205 (1968).
[57] K. C. Kopaczyk and N. S. Radin, J. Lipid Res. 6, 140 (1965).
[58] K. Suzuki and G. C. Chen, J. Lipid Res. 8, 105 (1967).
[59] R. M. C. Dawson, Lipid Chromatogr. Anal. 1, 163 (1967).

or Florisil column chromatography are subjected to mild alkaline methanolysis rather than total lipid extracts of tissues, these fractions usually
do not contain sphingomyelin. On the other hand, fractions of neutral
glycosphingolipids obtained as a result of ion-exchange cellulose chromatographic separation (Section III,B,5) may contain sphingomyelin.[14,44a,60]

Although most hydrocarbon chains in glycerophospholipids are attached to glycerol by ester bonds, (fatty acid esters of glycerol or O-acyl
group), some hydrocarbon chains (in most cases only one) are attached
to glycerol by ether bonds (O-alkyl and O-alk-1-enyl groups) and known
as glyceryl ether phospholipids and plasmalogens. The relative quantities
of glycerol ether phosphatides and plasmalogens may be substantial in
some normal tissues and malignant tumors.[61] Mild alkali treatment will
remove (deacylate) fatty acids from glyceryl ether phospholipids and
plasmalogens, but it will not affect their ether bonds. As a result these
products of mild alkali methanolysis retain their solubilities in lipid solvents by virtue of the ether-linked long-chain hydrocarbon groupings still
attached to the molecules.

After mild alkali methanolysis, the reaction mixture is neutralized and
subjected to Folch partition (or dialysis) to remove water-soluble products which have formed. After Folch partition the chloroform phase will
contain neutral glycosphingolipids, methyl esters of fatty acids, some
deacylated glyceryl ether phospholipids and plasmalogen residues, and
may also contain sphingomyelin (if present in the original mixture subjected to alkali methanolysis). Neutral glycosphingolipids can be separated easily from methyl esters of fatty acids (and even sphingomyelin)
by silicic acid or Florisil column chromatography (or thin-layer chromatography). However, the neutral glycosphingolipid fraction will contain
ether residue lipids as impurities. This is the disadvantage of the method
of purification of neutral glycosphingolipids by mild alkali methanolysis.

In Section III,B,5 are given two examples of the application of mild
alkali treatment to remove phospholipids from eluates that contain neutral glycosphingolipids.[41,42] The technique[42] for this mild alkali alcoholysis follows (see also Esselman et al.[61a]). The eluates containing neutral
glycosphingolipids (from 1 to 10 mg) are evaporated to dryness in test
tubes. One milliliter of chloroform and 1 ml 0.6 N methanolic NaOH
are added to the dry samples. The mixture is allowed to react at room
temperature for 1 hour. Then 1.2 ml 0.5 N methanolic HCl, 1.7 ml of
water, and 3.4 ml of chloroform are added to the reaction mixture, and
the biphasic system which results is mixed well. This is centrifuged and

[60] G. L. Feldman, L. S. Feldman, and G. Rouser, *Lipids* 1, 21 (1966).
[61] F. Snyder, *Prog. Chem. Fats Other Lipids* 10, 287 (1969).
[61a] W. J. Esselman, R. A. Laine, and C. C. Sweeley, Vol. 28 [8].

the lower chloroform layer containing neutral glycosphingolipids is removed. This lower layer is washed 2–3 times with Folch "pure upper phase" [methanol–water, 1:1 (v/v)], and the solvent is removed by drying under nitrogen. The samples are redissolved in small measured volumes of chloroform–methanol, 1:1 (v/v). The fractions are now ready for thin-layer chromatography of neutral glycosphingolipids.

E. Conclusions

The successful separation of neutral glycosphingolipids performed on lipid samples enriched with glycosphingolipids is dependent upon the quality of enriched samples. If enriched samples represent pure, or almost pure, fractions of neutral glycosphingolipids, their separations can be achieved without difficulty by using silica gel chromatoplates and a chloroform–methanol–water mixture as the developing solvent,[10] usually used for separation of glycosphingolipids. The type of silica gel, the composition of enriched samples, and the nature of the research problem will determine the ratio of ingredients in the above-mentioned developing solvent. Several variations in the composition of these developing solvents have been presented in this section of the book. Another developing solvent system which may be of great importance for the separation of neutral glycosphingolipids with four or more monomeric carbohydrates is the system of Müldner et al.,[62] which consists of a mixture of chloroform–methanol–10% NH_4OH, 60:35:8 (v/v). There are several modifications of this system; e.g., Tao and Sweeley[63] used chloroform–methanol–7% NH_4OH, 55:40:10 (v/v), for separation of neutral glycosphingolipids and hematosides. Some hematosides usually partition into the chloroform phase together with neutral glycosphingolipids during the Folch wash. In addition to the above-mentioned thin-layer chromatography systems, Yung and Hakomori[64] also used mixtures of tetrahydrofuran–2-butanone–water, 8:2:1 (v/v), and propanol–water, 7:3 (v/v), as developing solvents on silica gel H chromatoplates. Acetylated glycosphingolipids also can be separated on silica gel H with 1,2-dichloroethane–methanol–water, 90:10:0.1 (v/v).[64]

The ability of sugars to form borate complexes offers additional technical possibilities for separation of neutral glycosphingolipids from each other as well as from impurities.[65,66] Kean[66] prepared chromato-

[62] H. G. Müldner, J. R. Wherrett, and J. N. Cumings, J. Neurochem. 9, 607 (1962).
[63] R. V. P. Tao and C. C. Sweeley, Biochim. Biophys. Acta 218, 372 (1970).
[64] H. Yang and S. Hakomori, J. Biol. Chem. 246, 1192 (1971).
[65] O. M. Young and J. N. Kanfer, J. Chromatogr. 19, 611 (1965).
[66] E. L. Kean, J. Lipid Res. 7, 449 (1966).

plates from a slurry of 30 g of silica gel G in 65 ml 1% $Na_2B_4O_7 \cdot 10H_2O$ (pH 9.2) in water. Plates were activated for 1 hour at 125°. As developing solvent the author used a mixture of chloroform–methanol–water–concentrated NH_4OH (15 M NH_4OH), 280:70:6:1. Kean, as well as Young and Kanfer,[65] achieved separation of glucocerebrosides from galatocerebrosides, using borate-impregnated silica gel chromatoplates. Again, the selection of a particular developing solvent for thin-layer chromatography depends upon the goal of the problem and purity of enriched sample.

It is desirable that any reader who is going to work with thin-layer chromatography of neutral glycosphingolipids should consult the excellent chapter by Esselman et al.[61a] concerning isolation and characterization of glycosphingolipids.

Acknowledgments

The greatest appreciation is due Dr. Marion Barclay for her scientific and editorial assistance. The work was supported by the Damon Runyon Memorial Fund for Cancer Research, grant from the Elsa U. Pardee Foundation and grant CA 08748 from the National Cancer Institute.

Section VII

Synthesis or Preparation of Substrates

[48] Chemical Synthesis of Phospholipids and Analogues of Phospholipids Containing Carbon–Phosphorus Bonds

By Arthur F. Rosenthal

I. Introduction

The last decade and a half has seen a very extensive expansion in the field of chemical synthesis of phospholipids of unequivocal structure and their analogous. Obviously a review of this work is well outside the scope of this chapter, but the interested reader is referred to the reviews of van Deenen and de Haas,[1] Baer,[2] and Slotboom and Bonsen.[3]

[1] L. L. M. van Deenen and G. H. de Haas, *Advan. Lipid Res.* **2**, 167 (1964).
[2] E. Baer, *Trans. Roy. Soc. Can., Sect. 1, 2, 3* [4] **4**, 181 (1966).
[3] A. J. Slotboom and P. P. M. Bonsen, *Chem. Phys. Lipids* **5**, 301 (1970).

The intent of this chapter is to provide laboratory directions for the synthesis of at least one compound from each of the major classes of phospholipids by the best available procedure. Obviously, space limitations dictate that to a considerable degree the factors involved in the selection of what constitutes a major phospholipid class, which compounds in each class should be extensively discussed, and what is the best available synthesis must be left to the judgment of the present author. Guiding criteria have been the potential utility of each compound in the widest variety of applications, the variety of synthetic methods and their generality, i.e., their probable usefulness in the preparation of new compounds, as well as in the maximum diversity of types, including the presence of special structural features such as unsaturation, ether linkages, and optical activity.

Usually no one synthetic method is best from all points of view; for example, a synthetic procedure which will allow the preparation of a phospholipid containing two similar saturated fatty acid moieties may be ill-adapted to the preparation of unsaturated compounds or those containing two different fatty acid groups. For particularly important classes, therefore, more than one synthetic procedure is included. At the end of many of the synthetic procedures notes containing references to the synthesis of related compounds will be found.

The synthetic directions presented are mainly those given in the authors' original work; but if modifications are suggested in the work of subsequent authors or in the present author's experience, these are noted whenever appropriate.

II. Synthesis of Phosphoric Acid-Containing Phospholipids

A. Synthesis of Phosphatidic Acids

Early syntheses of phosphatidic acid employed either α,β-diglycerides[4,5] or α-iodo-β,γ acyloxypropanes.[6,7] Because of the instability of the former intermediates, routes using the latter compounds have been preferred in later works. In these cases the phosphoric acid function is always introduced as a disubstituted silver phosphate containing selectively removable protecting groups. More recently, a practical procedure for

[4] E. Baer, *J. Biol. Chem.* **189**, 235 (1951).

[5] J. H. Uhlenbroek and P. E. Verkade, *Rec. Trav. Chim. Pays-Bas* **72**, 395 (1953).

[6] L. W. Hessel, I. D. Morton, A. R. Todd, and P. E. Verkade, *Rec. Trav. Chim. Pays-Bas* **73**, 150 (1954).

[7] N. Z. Stanacev and M. Kates, *Can. J. Biochem. Physiol.* **38**, 297 (1960).

the synthesis of phosphatidic acids directly from α-glycerophosphoric acid has been developed; complete syntheses of both types are given below.

1. Synthesis of sn-3-Phosphatidic Acid via α-Iodoglycerol

This procedure allows the synthesis of saturated or unsaturated "mixed-acid" phosphatidic acids, as illustrated below. The great synthetic utility of the "mannitol" procedure for introducing optical activity into synthetic phospholipids requires that it be presented in full (see also Fig. 1).

The synthesis given above provides phosphatidic acids of the same configuration as that found in the phosphatidyl moiety of natural glycerophosphatides (i.e., the 1,2-diacyl-*sn*-glycerol-3-phosphoryl or L-α-phosphatidyl group).

*Isopropylidene-*D*-glycerol* (D-*Acetone Glycerol, sn-Glycerol-1,2-isopropylidene Ketal*)

*1,2,5,6-Diacetone-*D*-mannitol.* This procedure is that given by Baer,[s] with modifications as noted in the text.

To 1350 ml of dry acetone in a 2-liter Erlenmeyer flask are added 270 g of zinc chloride in stick form. The stoppered flask is gently swirled until the zinc chloride has dissolved (considerable heat is evolved), leaving only a small amount of insoluble material. The zinc chloride–acetone solution is cooled to room temperature and set aside for a while to allow most of the insoluble material to settle out.

In a 3-liter flask, arranged so that the contents can be mechanically stirred under anhydrous conditions, are placed 170 g of finely powdered (200 mesh preferable) D-mannitol, and the still slightly turbid zinc chloride–acetone solution is added by decantation, care being taken that most of the insoluble material remains behind. The mixture is stirred vigorously until the greater part of the mannitol has dissolved. This operation usually requires about 2 hours at room temperature (19–20°). The solution is filtered to remove the unreacted mannitol (approximately 40 g), and the filtrate is processed immediately as follows.

In a 5-liter round-bottomed flask equipped with an efficient motor-driven stirrer, a solution of 340 g of anhydrous potassium carbonate in 340 ml of water is prepared. After the solution has

[s] E. Baer, *Biochem. Prep.* 2, 31 (1952).

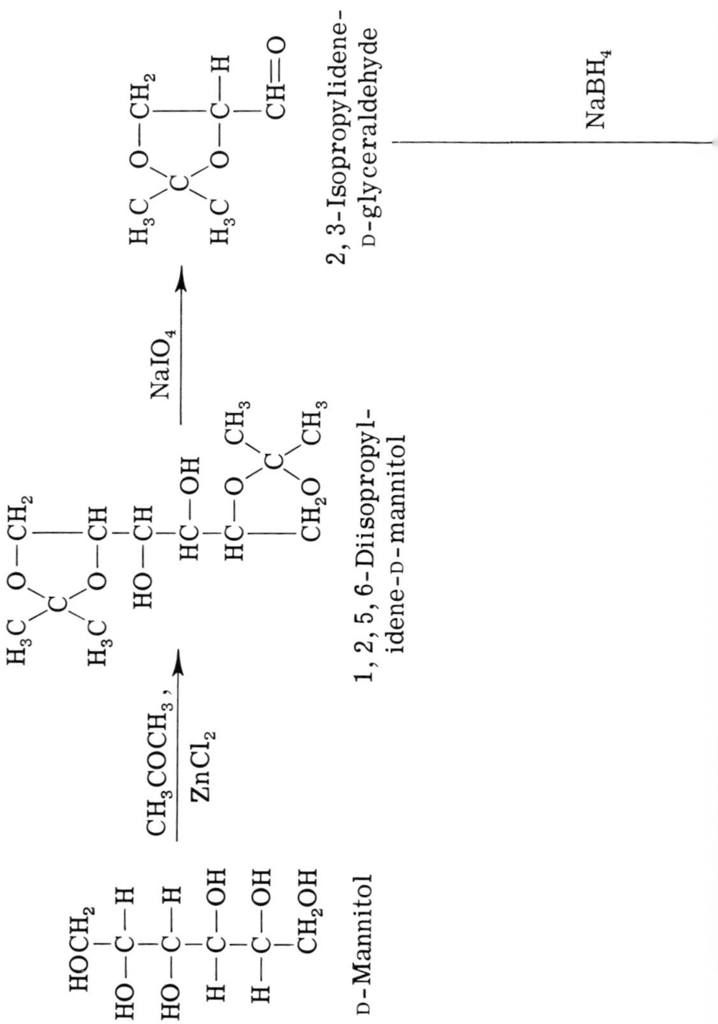

$$\begin{array}{c}
\text{HOCH}_2 \\
\text{HO--C--H} \\
\text{HO--C--H} \\
\text{H--C--OH} \\
\text{H--C--OH} \\
\text{CH}_2\text{OH}
\end{array}$$

D-Mannitol

$\xrightarrow[\text{ZnCl}_2]{\text{CH}_3\text{COCH}_3,}$

$$\begin{array}{c}
\text{H}_3\text{C} \quad \text{O--CH}_2 \\
\quad\quad \text{C} \\
\text{H}_3\text{C} \quad \text{O--CH} \\
\text{HO--CH} \\
\text{HC--OH} \\
\text{HC--O} \quad \text{CH}_3 \\
\quad\quad\quad\quad \text{C} \\
\quad\quad \text{CH}_2\text{O} \quad \text{CH}_3
\end{array}$$

1, 2, 5, 6-Diisopropyl-
idene-D-mannitol

$\xrightarrow{\text{NaIO}_4}$

$$\begin{array}{c}
\text{H}_3\text{C} \quad \text{O--CH}_2 \\
\quad\quad \text{C} \\
\text{H}_3\text{C} \quad \text{O--C--H} \\
\quad\quad\quad\quad \text{CH=O}
\end{array}$$

2, 3-Isopropylidene-
D-glyceraldehyde

$\xrightarrow{\text{NaBH}_4}$

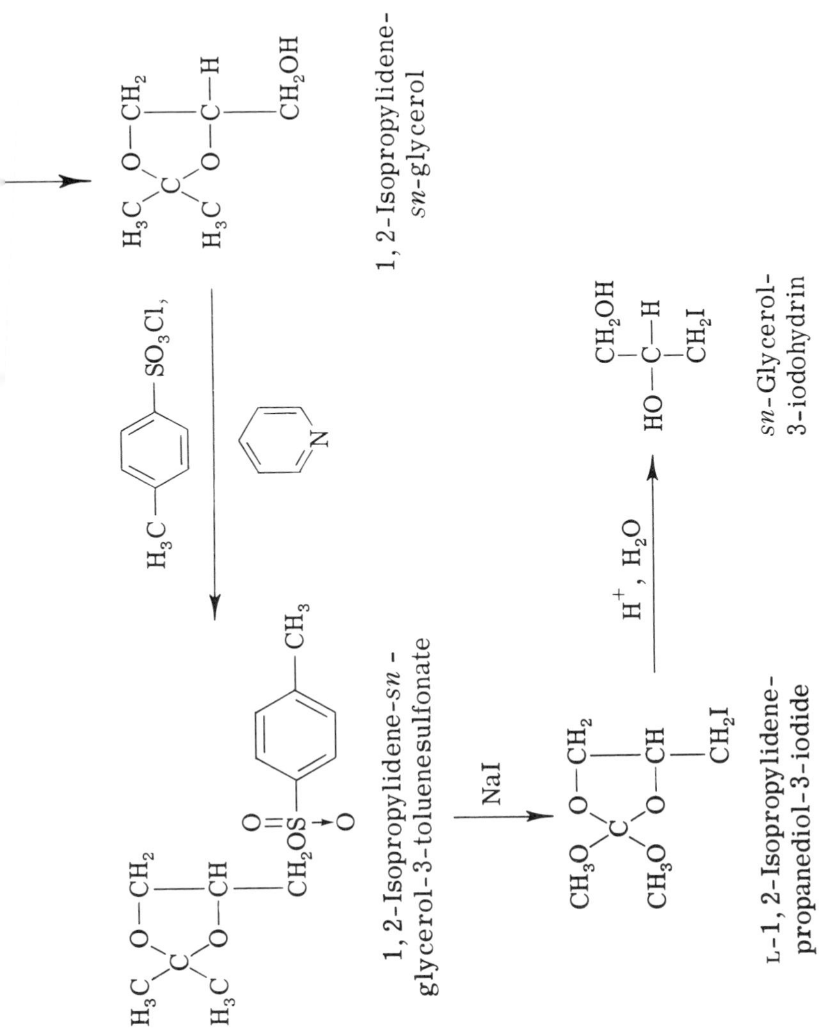

FIG. 1. Scheme for synthesis of *sn*-glycerol-α-iodohydrin (see text for details).

cooled to room temperature it is covered with 1350 ml of pure ether (free from ethanol). The mixture is stirred vigorously while the filtered acetone solution is added as rapidly as possible. The vigorous stirring is continued for a period of 30–40 minutes, after which the ether–acetone solution is decanted and the zinc carbonate pellets are washed with several portions (totaling 300–400 ml) of a 1:1 acetone–ether mixture. The combined filtrates and washings are evaporated under reduced pressure, and the residue is thoroughly dried in vaccum at 60–70° (water bath) for 2 hours.

Although Baer cautioned against running the ketalization at a temperature higher than 20°, the following modification is advocated by Bird and Chadha[9]:

D-Mannitol was stirred at 35–37° for 90 minutes in an acetone solution of anhydrous zinc chloride, filtered from unreacted mannitol (about 2% of starting material), and poured rapidly into cold potassium carbonate aqueous solution covered with ether (as in the Baer method). The combined ether–acetone extracts were concentrated to a low bulk and the resulting aqueous solution cooled to 0°, upon which the diisopropylidene mannitol crystallized out; after filtering and drying by suction, washing with hexane gave the desired product in 55% yield (purity shown by thin-layer chromatography (TLC) on silica gel G using chloroform–methanol–concentrated ammonia, 65:16:2 as eluent).

Baer[8] indicated that the purification of 1,2,5,6-diacetone (or diisopropylidene)-D-mannitol may also be accomplished by crystallization from hot dibutyl ether. In the experience of the present author a more convenient recrystallization solvent is a mixture of chloroform and hexane. The yields to be expected are in the range of 50–55%. The product melts over a 1–2.5° range beginning at 117 to 122°, depending on its degree of purity.

The compound is somewhat unstable in storage, particularly at room temperature. It should not be kept for more than a few weeks and then only < −10° in some form of desiccator. Stored material should always be recrystallized before use to remove re-formed mannitol and, presumably, monoisopropylidene-D-mannitol.

Isopropylidene-D-*glycerol*. The procedure below, developed by LeCocq and Ballou,[10] is much more convenient than the earlier[8] lead tetraacetate oxidation and catalytic hydrogenation.

To a cold solution of 40 g of sodium metaperiodate in 500 ml of water were added 34.2 g of 1,2,5,6-di-O-isopropylidene-D-man-

[9] P. R. Bird and J. S. Chadha, *Tetrahedron Lett.* p. 4541 (1966).
[10] J. LeCocq and C. E. Ballou, *Biochemistry* **3**, 976 (1964).

nitol. The oxidation of isopropylideneglyceraldehyde was complete in a few minutes. One liter of ethanol was added to precipitate the sodium iodate, which was removed by filtration. A solution of 10 g of sodium borohydride in water was added cautiously (foaming) to the cold filtrate. After 2 hours the solution was adjusted to pH 7 with acetic acid and the ethanol was removed by evaporation *in vacuo*. The resulting water solution was extracted 6 times with chloroform, and the combined chloroform extracts were dried over sodium sulfate and then concentrated to a syrup that weighed 28.5 g. The 1,2-*O*-isopropylidene-L-glycerol had $[\alpha]_D$ +14.5° (in substance) and was obtained in 85% yield.

The substance is distillable *in vacuo* using an oil bath, if necessary, according to the procedure of Baer,[8] who gave the following constants: b.p.$_{10}$ 77–78°; n_D^{20} 1.4347; d_4^{20} 1.0704; $[\alpha]_D$ + 13.98° (neat). In any case D-isopropylidene-D-glycerol is configurationally unstable on long standing and should be kept $< -10°$ and used within a few days.

L-*Glycerol-α-iodohydrin* (α-*Iodo*-L-*propylene Glycol*, L-*1,2-Dihydroxy-3-iodopropane*)

The procedure of Baer and Fisher,[11] given below, appears to have been used by virtually all subsequent authors to the present time.

α-(*p-Toluenesulfonyl*) D-*Acetone Glycerol*. To a gently agitated and cooled mixture of 38.0 g of *p*-toluenesulfonyl chloride and 20 ml of dry pyridine was added 26.4 g of freshly prepared D(+)-acetone glycerol ($[\alpha]_D$ 13.9°). The reaction mixture was kept in an ice bath until the reaction had subsided (approximately 1 hour). After standing 48 hours at room temperature a crystalline sludge had formed which was poured with stirring into 500 ml of ice water. The supernatant liquid was immediately decanted and the heavy oil, dissolved in ether, washed with sodium carbonate solution. The ether solution was dried with anhydrous sodium sulfate and concentrated *in vacuo* to a thick syrup. The residue (48.4 g) was dissolved in a mixture of 300 ml of dry ether and 600 ml of petroleum ether (b.p. 30–60°), the solution cooled to —70° and the thick sludge filtered with suction on a cooled Büchner funnel. The substance, which liquifies at room temperature, was freed *in vacuo* from adhering solvent. The yield of α-(*p*-toluenesulfonyl) D-acetone glycerol was 46.0 g (80.2%), n_D^{22} 1.5054; d^{24} 1.208, $[\alpha]_D^{24}$ —6.7° in substance; $[\alpha]_D^{24}$ —4.6° in dry ethanol (*c* 13).

α-*Iodoacetone* L-*Propylene Glycol*. A solution of 46.0 g of α-(*p*-toluenesulfonyl) D-acetone glycerol and 65 g of sodium iodide

[11] E. Baer and H. O. L. Fischer, *J. Amer. Chem. Soc.* **70**, 609 (1948).

in 300 ml of dry acetone was kept in a pressure bottle at 90° for 10 hours. The sodium p-toluenesulfonate was collected on a Büchner funnel and thoroughly washed with dry acetone. The combined filtrates were brought to dryness at reduced pressure and the reddish brown solid extracted with four 100-ml portions of ether. The ether extract was washed twice with 30 ml portions of a dilute sodium thiosulfate solution, dried with anhydrous sodium sulfate and concentrated under diminished pressure (bath to 35°). The fractional vacuum distillation of the residue yielded 30.0 g (77.4%) of pure α-iodoacetone L-propylene glycol, b.p. (6 mm) 68–69°; n_D^{25} 1.5022; $d^{20.5}$ 1.644; $[\alpha]_D^{23}$ +54.0° in substance; $[\alpha]_D^{23}$ +35.5° in dry ethanol (c 12.7).

α-Iodo-L-propylene Glycol. A solution of 30.0 g of α-iodoacetone L-propylene glycol in 75 ml of 85% ethanol was prepared and after the addition of 2.5 ml of 5 N sulfuric acid kept for 20 hours at room temperature. In order to ensure a complete hydrolysis the solution was diluted with 10 ml of water and refluxed for a period of 5 minutes. The cold solution was made neutral to litmus with aqueous baryta and concentrated *in vacuo* to a dry syrup at a bath temperature of 35–45°. The residue was extracted with four 75-ml portions of ether and the combined extracts, after drying with anhydrous sodium sulfate, brought to dryness under reduced pressure. The residue (20.2 g, m.p. 47–49°) was recrystallized from chloroform–petroleum ether (b.p. 30–60°) and yielded 19.3 g (77%) of analytically pure α-iodo-L-propylene glycol. Platelets, m.p. 48.5–49.5°, $[\alpha]_D$ —5.5° in dry ethanol (c 9.5).

1,2-Diacyl-L-glycerol-3-iodohydrin

The procedure of de Haas and van Deenen[12] for stepwise acylation of L-glycerol α-iodohydrin allows the synthesis of mixed-acid diacylglycerol iodohydrins containing virtually any desired acyl group in either 1 or 2 position. The synthetic procedure for one such compound is given below; essentially identical conditions are employed for diacyl iodohydrins containing other acyl groups, with simple alteration of the specific acid chloride used.

α-Stearoylglycerol-L-α-iodohydrin. To a solution of 28 g (0.138 mole) of glycerol-L-α-iodohydrin in 34 ml of chloroform (distilled immediately before use over phosphorus pentoxide) and 10.3 g (0.130 mole) of anhydrous pyridine, kept at 0°, was added dropwise over a period of 1 hour 32.5 g (0.107 mole) of freshly distilled stearoyl

[12] G. H. de Haas and L. L. M. van Deenen, *Rec. Trav. Chim. Pays-Bas* **80**, 951 (1961).

chloride dissolved in 44 ml of anhydrous chloroform. After storage in the dark for 18 hours at room temperature, the reaction mixture was diluted with 500 ml of ether, and subsequently poured into a solution of ice-cold 0.5 N sulfuric acid. The separated ether layer was washed several times with 0.5 N sulfuric acid, with water until neutral in reaction, subsequently with a solution of sodium thiosulfate, and finally with water. After drying over anhydrous sodium sulfate the solvent was distilled off under reduced pressure in a nitrogen atmosphere. The remaining white solid was dissolved in a minimal amount of dry ether, diluted with methanol until slightly turbid, and placed at −10° for crystallization. The first crop of crystals appeared to consist of a component which analyzed correctly for the unwanted distearoylglycerol-L-α-iodohydrin, m.p. 53–54°, $[\alpha]_D^{20}$ +2.5° in chloroform (c 10). Upon addition of methanol to the mother liquor the required component was precipitated. This white, crystalline material was collected with suction on a Büchner funnel, washed with a small amount of cold methanol, and recrystallized twice in the same way. After drying over phosphorus pentoxide in $vacuo$ in the dark we obtained 32.5 g of α-stearoylglycerol-L-α-iodohydrin; m.p. 65–66°; $[\alpha]_D^{20}$ +1.8° in chloroform (c 10).

The compound is highly light-sensitive. The yield of four batches of this compound prepared in this manner ranged from 45 to 55%. The melting curve of one sample showed a purity of at least 99%.

β-Oleoyl-α-stearoylglycerol-L-α-iodohydrin. The acylation of 29.3 g (0.063 mole) of α-stearoylglycerol-L-α-iodohydrin with 10.9 g (0.063 mole) of freshly distilled oleoyl chloride was carried out as described above for the monoacylation of glycerol-L-α-iodohydrin; the reaction mixture was, however, stored in the dark at room temperature for 48 hours and subsequently heated at 40° for 2 hours. The mixture on working up as described for the monoacyl iodohydrin gave 39.0 g (85%) of β-oleoyl-α-stearoylglycerol-L-α-iodohydrin; colorless crystals with m.p. 17°, $[\alpha]_D^{20}$ +3.1° in chloroform (c 10). The melting curves showed two metastable phases with transition at 15 and 30°. Yields of several batches varied from about 80 to 90%.

Silver Di-tert-butyl Phosphate

This substance, containing very acid-labile alkyl groups, allows the smooth introduction of the phosphate moiety at the glycerol 3 position in the diacyl iodohydrins independent of the nature of the specific acyl moieties.[13] Its synthesis is given as follows:

[13] P. P. M. Bonsen and G. H. de Haas, Chem. Phys. Lipids. 1, 100 (1967).

Di-tert-butyl Phosphite.[14] A solution of 82.4 g (0.6 mole) of phosphorus trichloride in 500 ml of petroleum ether (b.p. 30–60°) was cautiously dropped over a period of 1 hour into an ice–methanol chilled solution of 267 g (3.6 moles) of *tert*-butyl alcohol and 91.2 g (1.8 moles) of triethylamine in 3 liters of petroleum ether. Stirring was continued for 1 hour without external cooling, after which time the suspension was filtered, the cake being washed well with an additional 1 liter of petroleum ether. The solvent was removed on the steam bath at water aspirator pressure, the resultant yellow residue being rapidly distilled at 0.1 mm without fractionation. The distillate was distilled slowly (considerable frothing) giving 60 g (51%), b.p. 72–78° at 10–12 mm, n_D^{25} 1.4162. A final distillation gave a single fraction, b.p. 70–72° at 10 mm, n_D^{25} 1.4169, d^{25} 0.975. A portion of this product was redistilled at 4 mm, b.p. 62–62.5°. In various other preparations, samples of higher refractive index were obtained, but repeated slow distillations at 10–12 mm eventually produced a product of constant refractive index.

Barium Di-tert-butyl Phosphate. The oxidation of di-*tert*-butyl phosphite was performed according to the procedure described by Brown and Hammond[15] for other phosphites. In the case of the *tert*-butyl compound it is very important that contact of the product with acidic solutions be kept to the minimum possible time. The procedure of Brown and Hammond[15] was modified by de Haas, van Deenen and co-workers (van Deenen, private communication):

Di-*tert*-butyl phosphite (0.10 mole) and sodium bicarbonate (3.0 g) were suspended in 1:1 aqueous dioxan (100 ml) and potassium permanganate (10.5 g) in water (150 ml) added with stirring during 1 hour. The next morning the coagulated manganese dioxide was removed by filtration. Some unoxidized phosphite was removed by ether extraction and the aqueous layer at 0° was acidified with 0.5 N sulfuric acid. The acidic aqueous solution was extracted immediately 3 times with ice-cold chloroform and the chloroform extract was treated with methanol and $Ba(OH)_2$ solution for 1 hour and the solvents were removed *in vacuo*. The colorless solid residue was extracted with methanol and cleared by centrifugation. The methanol solution contained nearly pure barium di-*tert*-butyl phosphate which could be crystallized from acetone at −15° to remove a trace of contaminant.

Silver Salt of Di-tert-butyl Phosphate.[13] To a solution of the barium salt in water was added a solution of an equivalent amount of silver sulfate in water. The precipitated barium sulfate was centri-

[14] R. W. Young, *J. Amer. Chem. Soc.* **75**, 4620 (1953).
[15] D. M. Brown and F. R. Hammond, *J. Chem. Soc., London* p. **4229 (1960)**.

fuged off, and the clear supernatant evaporated *in vacuo*. Crystallization of the remaining material from methanol–acetone at −15° yielded the silver salt as white crystals in a yield of 75%; m.p. 185–187°.

1-Oleyl-2-myristoyl-sn-glycerol-3-phosphoric Acid

The synthesis of this mixed-acid phosphatidic acid illustrates the flexible method of Bonsen and de Haas[13] for the synthesis of phosphatidic acids of any fatty acid composition desired.

Glycerol-3-iodohydrin-1-oleate-2-myristate (β-myristoyl-α-oleyl-L-glycerol-α-iodohydrin) was synthesized by exactly the method given above for the synthesis of oleylstearoylglycerol-3-iodohydrin.

The condensation of the silver salt of di-*tert*-butyl phosphate and 1-oleoyl-2-myristoyl glycerol-3-iodohydrin was effected in dry chloroform at boiling temperature for 1 hour. The precipitated silver iodide was removed by centrifugation. The supernatant was evaporated *in vacuo*, dissolved in pentane, and subsequently washed with a solution of sodium bicarbonate and water. After drying over sodium sulfate, the pentane solution was evaporated and the oily residue dried *in vacuo* over KOH. This phosphotriester was not purified because of the lability of the *tert*-butyl esters. This material was dissolved in dry chloroform (freshly distilled over P_2O_5) through which, under cooling at 0°, a stream of anhydrous hydrogen chloride was passed for 1 hour. After evaporation of the chloroform, the phosphatidic acid was converted into the barium salt with barium acetate in methanol–water (2:1, v/v). The barium salt was chromatographed on silica with chloroform–methanol (95:5, v/v) as eluent. The phosphatidic acid was crystallized as the barium salt from chloroform–methanol at −15°, and obtained as a white solid in a yield of 70%. m. p. >360°, $[\alpha]_D^{20}$ +12.4° (*c* 8 in chloroform–methanol 95:5, v/v). Fatty acid analysis revealed a proportion of oleic acid to myristic acid of 1.01. This compound was partly converted into the disodium salt; m.p. 208–210°, $[\alpha]_D^{20}$ +5.70° (*c* 11 in chloroform). Another part was converted into the acid form with 0.5 N sulfuric acid and crystallized from acetone; m.p. 61–63° $[\alpha]_D^{18}$ +2.05° (*c* 6 in chloroform).

2. Synthesis of Phosphatidic Acids by Acylation of sn-Glycerol-3-phosphoric Acid

This more recent procedure[16] for the synthesis of phosphatidic acids has the great disadvantages of directness and simplicity, which arises

---

[16] Y. Lapidot, I. Barzilay, and J. Hajdu, *Chem. Phys. Lipids* **3**, 125 (1969).

from the fact that no protecting groups are used. The procedure can be used to give optically active or racemic phosphatidic acids containing unsaturated as well as saturated fatty acyl moieties. There seems to be no *a priori* reason why this type of synthesis could not be applied to the preparation of mixed-acid phosphatidic acids, but to date this has not been reported.

The function of the fatty acid salt in the procedure of Lapidot *et al.*,[16] given below, is apparently to suppress the formation of mixed phosphoric–carboxylic anhydrides with consequent phosphorylation of neighboring hydroxy groups, while at least one of the reasons for the superiority of the fatty anhydride over the corresponding acid chloride may be that the formation of glycerol chlorhydrin esters is thereby avoided.[17]

Dipalmitoyl-DL-α-glycerol Phosphate

Fatty Acid Anhydrides (General Procedure). A solution of dicyclohexylcarbodiimide (10 mmoles) in dry CCl_4 (50 ml) was added to a solution of the fatty acid (20 mmoles) in dry CCl_4 (150 ml). The reaction mixture was kept at room temperature for 5 hours. The dicyclohexylurea precipitate was removed by filtration and the solvent was removed from the filtrate by evaporation under reduced pressure. The residue can either be recrystallized or used as such.

Dipalmitoyl-DL-α-glycerol Phosphate. DL-α-Glycerol phosphate pyridinium salt (0.4 mmole) was added to a solution of tetraethylammonium palmitate (4 mmoles). The mixture was rendered anhydrous by repeated addition of dry pyridine and subsequent evaporation. In the last evaporation, suction was continued until a viscous gum remained. Palmitic anhydride (4 mmoles) was added, and the sealed reaction mixture was kept at 70–80° for 24 hours. Chloroform (20 ml) was added and the solution was shaken with Dowex 50 (pyridinium) ion exchange resin (15 g of dry resin) for 5 hours. The resin was removed by filtration, and the filtrate was passed through a column of the same resin (10 g), to ensure removal of the tetraethylammonium ions. The chloroform was removed by evaporation. The residue was dissolved in chloroform–methanol (2:1, 15 ml). Pyridine (13 ml) and water (3 ml) were added to hydrolyze the anhydrides. After 15 hours at room temperature the solvents were removed by evaporation in vacuum and the residue was dried in a vacuum desiccator over P_2O_5. The dry product was dissolved in chloroform (20 ml) and passed through a column (36 × 2 cm) of silicic acid (unisil). The column was eluted with chloroform to remove the

[17] R. Aneja and J. S. Chadha, *Biochim. Biophys. Acta* **239**, 84 (1971).

palmitic acid, and then with 10% methanol in chloroform (300 ml). The second fraction contained dipalmitoyl-DL-α-glycerol phosphate which could be isolated by solvent evaporation.

B. Notes

Related Compounds. Various ether-containing phosphatidic acids have been prepared.[18,19] Glycol analogues of phosphatidic acids were synthesized two decades ago.[20] A partial synthesis of a plasmalogenic phosphatidic acid has recently been reported.[21]

C. Synthesis of Lecithin (Phosphatidylcholine)

1. Synthesis of Lecithin from Phosphatidic Acid

Given the availability of a phosphatidic acid of desired fatty acid composition, optical activity, etc., monoesterification with a suitable choline salt can be readily effected, without resort to protecting groups, by any of several types of condensing agents, e.g., carbodiimides, trichloroacetonitrile, and sulfonyl chlorides. Since flexible procedures for the synthesis of almost any desired type of phosphatidic acid are known (see Section II, A), this combination of reactions probably represents the best general purpose method for the synthesis of phosphatidylcholines. This type of synthesis is also potentially very useful for the preparation of choline-labeled lecithins.

In a recent procedure described by Aneja and Chadha,[22] triisopropylbenzenesulfonyl chloride is used as the condensing agent. The authors claimed that lecithins prepared by this procedure have a specific and molecular rotation about 10% higher than those synthesized by other methods, and they suggested that their procedure gives compounds of greater optical purity (Note 1, Section II,D). The procedure is illustrated by the synthesis of L-α-distearoyl lecithin.

L-α-*Distearoyl Lecithin*

1,2-Distearoyl-*sn*-glycerol-3-phosphoric acid (110 mg, 0.16 mmole) and choline acetate (55 mg, 0.33 mmole; see Note 2, Section II,D) were mixed together and dried (over P_2O_5) overnight under

[18] J. S. Chen and P. G. Barton, *Can. J. Biochem.* **48**, 585 (1970).

[19] M. Kates, C. E. Parks, B. Palameta, and C. N. Joo, *Can. J. Biochem.* **49**, 275 (1971).

[20] H. van der Neut, J. H. Uhlenbroek, and P. E. Verkade, *Rec. Trav. Chim. Pays-Bas* **72**, 365 (1953).

[21] H. Eibl, *Biochemistry* **9**, 423 (1970).

[22] R. Aneja and J. S. Chadha, *Biochim. Biophys. Acta* **248**, 455 (1971).

vacuum (0.1 mm). Triisopropylbenzenesulfonyl chloride (140 mg, 0.45 mmole) and anhydrous pyridine (approximately 5 ml) were added and the mixture, protected from moist atmosphere, was stirred continuously for 1 hour at 70°, followed by stirring at room temperature for 4 hours. The choline acetate remaining undissolved was removed by filtration and the filtrate treated with water (1 ml). The solvents were removed under vacuum on a rotary evaporator and the residue extracted with chloroform. The chloroform extract was evaporated to dryness, the residual solid dissolved in chloroform–methanol–water (65:25:4 by volume; approximately 150 ml) and poured dropwise through a mixed bed ion-exchange column containing Amberlites IRC-50 (H$^+$) and IR-45 (OH$^-$) (15 g each). The eluate was evaporated to dryness under vacuum. The residue was dissolved in a small volume of benzene–methanol (7:3 by volume) and applied a column of neutral alumina (10 g; M. Woelm, Germany, activity grade 1) and the column was eluted with benzene–methanol (7:3 by volume). After a forerun (rejected), fractions containing phosphatidylcholine only, and then mixtures of phosphatidylcholine and lysophosphatidylcholine (deacylation on alumina column) were obtained. The fractions containing phosphatidylcholine only were combined and on evaporation to dryness gave a residue (74 mg, 60%) which gave a single spot, R_f approximately 0.35, on thin-layer chromatography on silica gel G plates developed with chloroform–methanol–water (65:25:4 by volume). For analyses, the sample was crystallized from hot dioxan, filtered, and washed with anhydrous ether. The white solid, m.p. 234–35°, $[\alpha]_D^{20}$ +6.95° [c 3; chloroform–methanol (1:1 by volume)], M$_D$ +54.9°.

2. Partial Synthesis of Lecithin from sn-Glycerol-3-phosphorylcholine

This is one of the most useful methods for the synthesis of lecithins of any desired fatty acid composition, but the simplest version, employing free glycerophosphorylcholine (GPC) is of very recent vintage.

The procedure depends on the efficient selective alcoholysis of egg lecithin to L-α-GPC, plus its reacylation by a modification of the procedure given above for the acylation of α-glycerophosphate. This procedure, coupled with a phospholipase A$_2$ deacylation followed by reacylation of the free glycerol β-ol, can also provide an indirect route to mixed-acid lecithins.[23]

Since this chapter is limited to purely chemical methods of synthesis, only such steps will be discussed below.

[23] E. Cubero Robles and D. van den Berg, *Biochim. Biophys. Acta* **187**, 520 (1969).

Dioleyl L-α-*Lecithin*

L-α-*Glycerophosphorylcholine.* The methanolysis procedure of Brockerhoff and Yurkowski[24] appears to be the most facile method for the deacylation of lecithin.

Purified egg yolk lecithin (3 g) was dissolved in 30 ml of diethyl ether and the solution was cleared by centrifugation. Then 3 ml of 1.0 M methanolic tetrabutylammonium hydroxide solution (K and K Laboratories, Plainview, New York) was added. After 1 hour at room temperature, the solvent was decanted from the glassy precipitate of GPC and the vessel rinsed with ether. In preparation for the following analyses, the material was dissolved in 1 ml of methanol and reprecipitated with 30 ml of ether. There was no loss of weight when the GPC was kept in a vacuum over phosphorus pentoxide for 2 days. The analyses indicated that the GPC contained 1 mole of water. The yield was 882 mg, or 84% of theory on the basis of the total phosphorus of the crude lecithin. The specific rotation of the preparation was $[\alpha]_D^{19}$ −2.84 (c 8.8, water).

Dioleyl L-α-*Lecithin* (*1,2 Dioleyl sn-Glycerol-3-phosphorylcholine*). The procedure of Cubero Robles and van den Berg,[23] given below, represents an adaptation of the fatty anhydride acylation method of Lapidot *et al.*[16]

L-α-Glycerophosphorylcholine (10 mmoles) and sodium oleate (20 mmoles) in 200 ml of methanol was evaporated to dryness *in vacuo* at 60°. The powder was dried overnight under vacuum over P_2O_5, and transferred to a 250-ml round-bottomed flask. Oleic anhydride (40 mmoles) was added, the flask was closed under vacuum and placed in an oil (or glycerol) bath at 70° for 48 hours while being rotated to ensure good contact between the anhydride and the powder. The reaction mixture was dissolved in chloroform (100 ml) and directly applied to a silicic acid column (700 g adsorbent). The lipids were separated by gradient-elution chromatography, using chloroform with gradually increasing methanol content. Two cylinders of equal diameter were interconnected by a Teflon tube (of about 2 mm diam), and another tube was connected to the top of the column. The mixing cylinder, provided with a mechanical stirrer, was filled with 1.5 liters of chloroform and the second cylinder with the same amount of chloroform–methanol (1:9 by volume). Fractions of approximately 17 ml were collected by means of an LKB fraction collector; the initial flow rate was 4.1 ml/minute. A total of 170 fractions was monitored by thin-layer chromatography. The

[24] H. Brockerhoff and M. Yurkowski, *Can. J. Biochem.* **43,** 1777 (1965).

lecithin peak appeared between fractions 36 and 90 (corresponding to 36 and 62% methanol–chloroform, respectively). A small amount of lysolecithin was detected in fractions 110–126, but it was not recovered. The lecithin fractions were combined and weighed. The yield was 7.05 mmole dioleyl lecithin (70.5% of theory).

Ether Lecithins

In this section the synthesis of ether lecithins, other than plasmalogens, is discussed. The synthesis of 1′-alkenyl ether phospholipids presents special problems and therefore are treated in a special section below (Section II,O).

Ether-containing glycerophospholipids can, of course, have 1-alkyl-2-acyl, 1-acyl-2-alkyl, or 1,2-dialkyl structures. Those phospholipids containing a 2-alkyl group are usually prepared by phosphorylation of a 3-hydroxy group. Although 1,2-dialkyl-3-iodoglycerides are readily prepared,[25] the stability of the ether groups allows glycerol 1,2-diethers to be manipulated without danger of positional migration. The reactivity of β-halo ethers is also less than that of the corresponding β-halo esters.

The synthesis of glycerol mono- and diethers are the initial synthetic problem in the preparation of the corresponding phospholipids and are thus considered in turn.

Glycerol-α-monoethers

rac-Glycerol-1-ethers or *sn*-glycerol-3-ethers are readily prepared by alkylation of the corresponding isopropylideneglycerol. According to Baumann and Mangold,[26] the best alkylating agents are the alkyl methanesulfonates, prepared by their procedure as follows.

cis,cis-9,12-Octadecadienyl Methanesulfonate. In a 500-ml three-necked flask equipped with a mechanical stirrer, inlet and outlet tubes for purified nitrogen, and a dropping funnel (connected with the nitrogen inlet tube) 23.0 g (86 mmoles) of *cis,cis*-9,12-octadecadienol is dissolved in 80 ml of absolute pyridine (kept dry by storing over solid potassium hydroxide). The solution is chilled in an ice bath and 15.0 g (130 mmoles) of methanesulfonyl chloride are added dropwise during 1 hour. The ice bath is removed and stirring is continued for another 5 hours at room temperature.

[25] A. F. Rosenthal, G. M. Kosolapoff, and R. P. Geyer, *Rec. Trav. Chim. Pays-Bas* **83**, 1273 (1964).

[26] W. J. Baumann and H. K. Mangold, *J. Org. Chem.* **29**, 3055 (1964).

The almost colorless slurry is dissolved by adding 150 ml of oxygen-free water and 200 ml of ether with stirring and cooling, and the mixture is transferred to a 1-liter funnel. The yield of methanesulfonate depends greatly upon the rapidity and the manner in which the extraction is done. After separation the water layer is kept in an ice bath. The ether phase is extracted consecutively with 50 ml of water, 2 N sulfuric acid (until acidic), 50 ml of water, 1% potassium carbonate solution (until neutral or basic), and 50 ml of water. The organic layer is dried over anhydrous sodium sulfate.

The original water phase and the other basic water layers are combined and treated with 200 ml of ether. After washing the ether extract with 50 ml of water, it is used to extract the original acidic layers (add 2 N H_2SO_4, if basic). The organic phase is then treated with 50 ml of water, 1% potassium carbonate solution, and 50 ml of water, and dried with the original extract over sodium sulfate.

After removing the solvent in a rotatory evaporator, the residue is dissolved in 120 ml of absolute ethanol and recrystallized at $-50°$. The long needles of methanesulfonate are collected at the same temperature on a chilled Büchner funnel. The product is transferred into a flask and dried first by a nitrogen stream, then under vacuum, yielding 23.7 g (79%), m.p. $-5°$.

cis,cis-9,12-Octadecadienyl Glyceryl-1-ether. In a 300-ml three-necked flask fitted with a reflux condenser, magnetic stirrer, nitrogen inlet and outlet tubes, and a dropping funnel (connected with the nitrogen inlet tube) are placed 80 ml of dry benzene and 1.6 g (40 mmoles) of clean pieces of potassium. The mixture is refluxed for 2 hours, then 5.3 g (40 mmoles) of 1,2-*O*-isopropylideneglycerol is added dropwise and refluxing is continued for another 4 hours. *cis,cis*-9,12-Octadecadienyl methanesulfonate (12.2 g, 35 mmoles), dissolved in 20 ml of dry benzene, is added dropwise and the reaction is continued for 15 hours. The flask is chilled in a water bath, moist ether and oxygen-free water are added, and the extraction is carried out in a 1-liter separatory funnel under nitrogen using approximately 250 ml of water and 150 ml of ether. After a second extraction with 150 ml of ether, the combined ether layers are washed with water and dried over anhydrous potassium carbonate. The solvent is removed by distillation in a rotary evaporator under vacuum. The residue, almost pure *cis,cis*-9,12-octadecadienyl 2,3-*O*-isopropylideneglyceryl-1-ether, is hydrolyzed by refluxing in a solution of 5 ml of concentrated hydrochloric acid in 50 ml of methanol. After cooling, the resultant product is treated with 100 ml of water and the solution is extracted with 250 and 150 ml of ether. The ether extract is washed consecutively with small amounts of water, 1% potassium carbonate solution, and water, and then dried over anhydrous sodium sulfate. Evaporation and recrystallization of the residue from 70 ml of low boiling hydrocarbon (Skellysolve F) at −50° and filtration over a chilled Büchner funnel yields 8.2 g (68%) of glyceryl ether, m.p. 8°.

The preparation of other ethers is performed similarly, altering only the crystallization temperature to that appropriate to the melting point of the individual compound.[26]

The optically active glycerol ethers thus prepared are of the sn-3 variety, which is the opposite configuration of the commonly occurring glycerol α-monoethers (e.g., batyl alcohol). Using an approach developed originally by Lands and Zschocke,[27] Chacko and Hanahan[28] devised a procedure for the conversion of long-chain sn-glycerol-3-ethers into the corresponding sn-1-ethers. The method depends upon the inversion at C-2 in the ether ditosylate during displacement of tosylate by acetate.

The procedure is illustrated by the conversion of sn-glycerol-3-octodecyl ether to sn-glycerol-1-octadecyl ether (batyl alcohol).

sn-Glycerol-1-octadecyl Ether

sn-Glycerol-1,2-ditosylate-3-octadecyl Ether. To a solution of 13.4 g (37.5 mmoles) of sn-glycerol-3-octadecyl ether in a 50-ml anhydrous pyridine were added 21.5 g (112 mmoles) of p-toluenesulfonyl chloride in 50 ml of anhydrous pyridine with constant stirring and the contents stirred at room temperature for 2 days. The reaction mixture was diluted with 300 ml of ether, washed several times with 1 M H_2SO_4 (cold) until the washings were acidic, then with 5% $NaHCO_3$, and finally with water. The ether solution, after drying over anhydrous Na_2SO_4, was evaporated to give 24.2 g of pale yellow oil. Thin-layer chromatography showed that besides the major desired component, there were two small spots, one running behind (batyl alcohol monotosylate) and the other one ahead (possibly p-toluenesulfonic acid) of the desired batyl alcohol ditosylate. For purification, 12.6 g of the crude material were placed on a 450-g column of silicic acid (Mallinckrodt, 100 mesh suitable for chromatography) and eluted initially with 15% ether in light petroleum (b.p. 30–60°). The first 1.5 liters of solvent removed fast running (thin-layer chromatography) component and the next 3 liters of solvent eluted the desired ditosylate in pure form. Evaporation of the

[27] W. E. M. Lands and A. Zschocke, *J. Lipid Res.* **6**, 324 (1965).
[28] G. K. Chacko and D. J. Hanahan, *Biochim. Biophys. Acta* **164**, 252 (1968).

solvent gave the product. Yield: 10.1 g (79% of theory), $[\alpha]_D^{22}$ +0.6° (c 75 mg/ml in hexane).

sn-Glycerol-1-octadecyl Ether. A mixture of 5 g (7.3 mmoles) of the ditosylate and about 3 g of freshly fused potassium acetate in 60 ml of absolute ethanol was refluxed with stirring under anhydrous conditions for 42 hours. This preparation was evaporated *in vacuo* to a white solid residue which was dissolved in water and extracted with chloroform. The chloroform solution was washed with water, dried over Na_2SO_4, and evaporated to yield a solid white residue. On thin-layer chromatography, a mixture of the following substances was observed: batyl alcohol diacetate, monoacetates, plus small amounts of unreacted batyl alcohol, which resulted from solvolysis.

The residue was hydrolyzed with NaOH in 95% ethanol by maintaining the mixture in a water bath at 50–60° for 30 minutes. The reaction mixture was neutralized with chloroform, washed with water, dried over Na_2SO_4, and evaporated to dryness. The residue was crystallized from hexane at room temperature to give the product. Yield: 2.3 g 90% of theory), m.p. 71–72°. It migrated as a single component on thin-layer chromatography; $[\alpha]_D^{22}$ −16.2° as isopropylidene derivative (c 46 mg/ml in hexane).

The *sn*-glycerol-1-monoether can then be tritylated to give an *sn*-glycerol-1-monoalkyl-3-trityl diether. This versatile intermediate can be acylated to give a 1-alkyl-2-acylglycerol after ditritylation or similarly alkylated to a mixed-alkyl glycerol-1,2-diether (see below). The first step is illustrated in the procedure of Chacko and Hanahan[28] below.

sn-Glycerol-1-cis-9'-octadecenyl-3-trityl Ether. A mixture of 1.75 g (5.1 mmoles) of *sn*-glycerol-1-cis-9'-octadecenyl ether and 3.0 g (10.8 mmoles) of triphenylchloromethane in 20 ml of dry pyridine was stirred for 24 hours at room temperature with 200 ml of diethyl ether. The ether-soluble fraction was washed successively with 0.25 M H_2SO_4 (ice cold), water, 5% $NaHCO_3$, and water. The ether phase was dried over anhydrous $NaSO_4$ and the solvent then evaporated to yield an oil, 3.1 g. This product was chromatographed on a 40-g SilicAR CC-7 column using the following solvent system in increasing order of polarity: hexane, hexane–benzene (1:1, v/v), benzene, chloroform and ether. The desired product, eluted with benzene and chloroform, weighed 2.72 g. Yield: 92% of theory, $[\alpha]_D^{22}$ 3.1° (c 43 mg/ml in hexane).

Chacko and Hanahan[28] also devised a practical method for detritylation of a 1-alkyl-2-acylglycerol without hydrogenation and without acids or bases which would bring about acyl migration (Note 3, Section II,D):

2-Stearoyl Selachyl Alcohol

1-O-cis-9'-Octadecenyl-2-O-octadecanoyl-3-O-tritylglycerol. To 3.0 g (5.1 mmoles) of *sn*-glycerol-1-*cis*-9'-octadecenyl-3-trityl diether in 20 ml of anhydrous ethanol-free chloroform was added 1 ml (12 mmoles) of anhydrous pyridine and 2 ml (6 mmoles) of stearoyl chloride. The solution was mixed well by swirling and kept in the dark at room temperature for 2 days. This mixture was extracted with diethyl ether and the extract washed successively with cold 0.25 M H_2SO_4, water, 5% $NaHCO_3$, and water. The ether phase was dried over anhydrous Na_2SO_4 and on evaporation yielded 4.1 g of yellow oil. Thin-layer chromatography showed one major component which, in hexane–ether–acetic acid (90:10:1 by volume) migrated faster than triolein. The entire sample in light petroleum (b.p. 30-60°) was chromatographed on a 150-g column of Silic-AR CC-7 prepared in light petroleum (b.p. 30–60°). The desired product, which showed a single spot on thin-layer chromatography, was eluted with 2 liters of 2.5% diethyl ether in light petroleum (b.p. 30–60°). Upon removal of the solvent, 3.0 g of a colorless oil was obtained. Yield: 89% of theory, $[\alpha]_D^{22}$ +5.00° (*c* 56 mg/ml in hexane).

1-O-cis-9'-Octadecenyl-2-octadecanoylglycerol (*2-Stearoyl selachyl Alcohol*). A mixture of 3.8 g (4.5 mmoles) of the diether ester and 1.2 g (18 mmoles) of boric acid powder were mixed with 15 ml of triethyl borate in a 50-ml round-bottomed flask, fitted with a vacuum adaptor and a stirrer. The contents were heated, while stirring, to 100° in an oil bath. The mixture cleared in approximately 5 minutes. Vacuum was now applied (by water pump) for removal of solvent and the mixture stirred at 100–150° for an additional 15 minutes, by which time the color of the solution turned yellow. The yellow viscous residue was cooled and dissolved in diethyl ether and the latter washed 3 times with equal volumes of water. Evaporation of the ether solvent yielded 3.9 g of an oil. Thin-layer chromatography of this mixture in hexane–ether–acetic acid (70:30:1 by volume) showed four components, i.e., the unhydrolyzed trityl derivative, triphenyl carbinol, and the 1,2- and 1,3-ether ester.

The oil was dissolved in light petroleum (b.p. 30-60°) and passed through a 200-g SilicAR CC-7 column packed in light petroleum (b.p. 30–60°) and the unhydrolyzed trityl derivative (1.7 g) was recovered by eluting the column with 1000 ml of 2.5% ether in hexane. Next 10% ether in hexane (3500 ml) eluted, first, triphenyl carbinol and then the 1,3 ether ester (0.45 g). Finally, 25% ether in hexane (2 liters) eluted the desired 1,2-ether ester (1.6 g). Another treatment of the unhydrolyzed trityl ether with boric acid, followed by work-up and column chromatography as described above, gave 0.15 g of 1,3-ether ester and 0.45 g of 1,2-ether ester. In both preparations the 1,2-ether ester was slightly contaminated with the 3 isomer. This was further purified by rechromatography on SilicAR CC-7 with the same solvents as above immediately before analysis as well as for use in the synthesis of the monoether phospholipids. Characteristics of the two isomers, recovered in 97% yield, are for 2-stearoyl-selachyl alcohol, yield 75%, $[\alpha]_D^{22}$ $-3.9°$ (c 45 mg/ml in hexane) and for 3-stearoyl selachyl alcohol, yield 22%, $[\alpha]_D^{22}$ $-0.7°$ (66 mg/ml in hexane; m.p. 43.0–43.5°) (recrystallized from hexane).

Glycerol Diethers

The glycerol α-alkyl α'-trityl, racemic or of either optical configuration, can be O-alkylated to give, after detritylation, a mixed alkyl glycerol 1,2-diether. This type of synthesis has been investigated extensively by Kates and co-workers;[29,30] however, these authors employed alkyl bromides as alkylating agents, which may be inferior to the methanesulfonates.[31] Whether this is actually the case or not, however, does not appear to have been the subject of any systematic comparative investigation as yet. In any case it seems undeniable that for those alkyl groups of which no bromide is commercially available (primarily unsaturated moieties) the preparation of the mesylates from the alcohols is significantly simpler (1 step)[26] than the preparation of the bromides (2 steps)[30] which have been reported.

An example of the mesylate alkylation procedure of Baumann and Mangold[32] is given below. Although the product prepared in this particular case is saturated and racemic, the O-alkylation of trityl-substituted glycerols is equally well adapted to the synthesis of optically active and unsaturated[33] glycerol-1,2-diethers.

2-Hexadecyloxy-3-octadecyloxy-1-propanol

In a 250-ml three-necked flask, fitted with water separation head, reflux condenser, dropping funnel, calcium chloride tubes, magnetic stirrer, and heating mantle, were placed 3.0 g of powdered potassium hydroxide, 80 ml of xylene, and 3.5 g (6 mmoles) of rac-glycerol 1-trityl-3-octadecyl ether. The mixture was refluxed for 1 hour to

[29] M. Kates, T. H. Chan, and N. Z. Stanacev, _Biochemistry_ **2**, 394 (1963).

[30] B. Palameta and M. Kates, _Biochemistry_ **5**, 618 (1966).

[31] W. J. Baumann, _in_ "Ether Lipids: Chemistry and Biology" (F. Snyder, ed.), p. 55. Academic Press, New York, 1972.

[32] W. J. Baumann and H. K. Mangold, _J. Org. Chem._ **31**, 498 (1966).

[33] P. J. Thomas and J. H. Law, _J. Lipid Res._ **7**, 453 (1966).

remove water by azeotropic distillation. Hexadecyl methanesulfonate (2.1 g, 6.6 mmoles) dissolved in 20 ml of xylene was added dropwise, and refluxing was continued for 6–8 hours. After removing about 50 ml of xylene by distillation, cooling, and addition of 100 ml of water and 150 ml of ether, the water phase was extracted twice with 100 ml of ether. Drying the organic phase over anhydrous potassium carbonate and evaporation yielded 2-hexadecyloxy-3-octadecyloxy-1-trityloxypropane.

This product was dissolved in 100 ml of 95% methanol, a stream of hydrogen chloride was led through the vigorously stirred reaction mixture, and refluxing was continued for 5–6 hours. The solvent was removed by distillation, 100 ml of water and 200 ml of ether were added, and the phases were separated in a 1-liter funnel. After a second extraction with 100 ml of ether, the combined organic layers were washed consecutively with 50 ml of water, 1% potassium carbonate solution (until basic), and 50 ml of water, and then dried over anhydrous sodium sulfate. The solvent was evaporated. The residue was taken up in 80 ml of Skellysolve F, and the precipitate of triphenylcarbinol formed on storing the solution in the refrigerator was filtered off and washed with a small amount of ice-cold Skellysolve F. Crystallization from the filtrate at freezer temperature (−30°), separation on a Büchner funnel, and recrystallization first from ethanol, then from Skellysolve F, yielded 3.0 g (88%) of product, m.p. 58.5–59°.

In the case of glycerol dialkyl ethers containing the same R groups, the synthetic problem is, in different aspects, at once easier and more difficult. The simplest case is that in which racemic compounds are desired; here, DL-1-trityl glycerol can be prepared directly from a tritylhalide and glycerol, the product alkylated, and finally detrylated with acid, making this relatively simple procedure fully applicable to the synthesis of saturated or unsaturated glycerol-1,2-diethers.[33]

When optically active derivatives are desired, however, serious limitations become obvious at once. If diethers of the sn-3-glycerophosphorylcholine series (the natural lecithin configuration) are desired, the corresponding sn-glycerol-3-trityl ether would have to be used; but this intermediate cannot be prepared by tritylation of sn-glycerol-1,2-isopropylidene ketal followed by acid-catalyzed deacetonation since the latter process also brings about detrylation. The 3-benzyl ether of sn-glycerol-1,2-isopropylidene ketal can readily be prepared and deacetonated to sn-glycerol-3-benzyl ether,[29] but then the sn-glycerol-1,2-diethers prepared from it must be completely saturated because the protective benzyl group must be removed by hydrogenolysis.

For the enantiomeric sn-glycerol-2,3-diether series, however, the corresponding sn-glycerol-1-trityl ether *can* be synthesized, paradoxically from the readily prepared sn-glycerol-1,2-isopropylidene ketal by a clever but rather roundabout procedure involving interchange of 1 and 3 substituents.[34] Thus, this optical series of sn-glycerol-2,3-diethers can be prepared without the "saturation" limitation[30]; this gives glycerol diethers of the configuration found in the glycerol diethers of certain halophilic bacteria.[35]

Theoretically, sn-glycerol-2,3-isopropylidene ketal could be similarly used to produce sn-glycerol-1,2-diethers, but as the starting ketal must be prepared from the commercially unavailable L-mannitol or by 1,3 interchange from sn-glycerol-1,2-isopropylidene ketal,[34] the whole synthesis then becomes so long as to be quite impractical. The best procedure for the synthesis of sn-glycerol-1,2-di(unsaturated) ethers is therefore the stepwise O-alkylation following inversion of Chacko and Hanahan,[28] which allows mixed R groups as well as similar R groups to be introduced (see above). Hardly a more extreme example exists in synthetic lipid chemistry where two such widely different methods of choice must be employed to prepare exactly the same type of compound when only the saturation of the alkyl groups is varied (Note 4, Section II,D).

The procedure of Kates *et al.*[29] given below, allows the facile synthesis of sn-glycerol-1,2-diethers containing identical saturated alkyl moieties:

3-O-Benzyl sn-Glycerol. A mixture of 19.1 g (0.145 mole) of D-acetone glycerol, 38 g (0.3 mole) of benzyl chloride, and 17 g of powdered potassium hydroxide in 100 ml of dry benzene was refluxed with stirring for 16 hours. The water formed was removed by means of a phase-separating head. The cooled mixture was diluted with benzene (100 ml) and washed successively with water, 0.1 N hydrochloric acid, 2.5% potassium bicarbonate, and again water. The benzene phase was dried over sodium sulfate and the solvent removed under reduced pressure. Distillation of the residue yielded unreacted benzyl chloride, boiling below 100° per 0.5 mm, and D-acetone glycerol benzyl ether b.p. 107–8° per 0.5 mm; yield, 28.0 g (87%). After acidic hydrolysis of the isopropylidene group 22.8 g (84% overall yield) of the 3-O-benzyl sn-glycerol was obtained; b.p. 142–146° per 0.7 mm, $[\alpha]_D^{21}$ +5.5° (pure liquid).

1,2-Di-O-Hexadecyl-3-O-benzyl sn-Glycerol. A mixture of 3.46 g (0.02 mole) of 3-O-benzyl sn-glycerol, 24.4 g (0.08 mole) of 1-bromohexadecane, and 4.5 g of powdered potassium hydroxide in 50 ml

[34] E. Baer and H. O. L. Fischer, *J. Amer. Chem. Soc.* **67**, 944 (1945).

[35] M. Kates, B. Palameta, and L. S. Yengoyan, *Biochemistry* **4**, 1595 (1965).

of dry benzene was refluxed with stirring for 16 hours, the water formed being removed with a phase-separating head. The cooled mixture was diluted with 50 ml of ethyl ether and washed successively with water, 1 N hydrochloric acid, 2.5% potassium bicarbonate solution, and finally with water. The benzene solution was dried over sodium sulfate and the solvent evaporated under reduced pressure. The residue was subjected to distillation *in vacuo* to remove unreacted hexadecyl bromide (14.2 g) boiling between 120–135° per 0.6 mm (bath, 155–165°). The residual oil (7.3 g, 58%) could not be crystallized and was not further purified (Note 5, Section II,D).

1,2-Di-O-hexadecyl sn-Glycerol. A solution of the above benzyl ether (7.3 g) in 50 ml of ethyl acetate was shaken with 0.5 g of freshly prepared palladium on charcoal catalyst in an atmosphere of hydrogen at room temperature and an initial pressure of about 40 cm of water. After 3–4 hours 240 ml of hydrogen was consumed (calculated for 1 mole, 280 ml). Slower rates of hydrogenolysis were found to result from incomplete removal of the alkyl bromide. In such cases, the catalyst was filtered off and hydrogenation resumed with fresh catalyst. When this did not suffice, the benzyl ether was again subjected to distillation as described above and again hydrogenated. After completion of the hydrogenolysis, the catalyst was removed by filtration and washed with chloroform, and the combined filtrates were evaporated to dryness under reduced pressure. The solidified residue was dried *in vacuo* in a desiccator (weight 6.0 g, 55% overall yield). Crystallization from 40 ml of ethyl acetate or from 50 ml of chloroform–methanol (1:4, v/v) at 5° yielded 3.4 g of D-α,β-di-O-hexadecyl glycerol with melting point 48.0–49.5°. Recrystallization from ethyl acetate raised the melting point to 48.5–49.5°, $[\alpha]_D^{21}$ $-7.5°$ (c 7.5 in chloroform); $[M]_D$ $-40.6°$.

Diether Lecithin

The phosphorus and choline functions can, of course, be introduced into glycerol-1,2-diethers by a variety of methods. The simplest and most versatile appears to be that introduced by Hirt and Berchtold[36] for ester lecithins in which the phosphodiester is introduced via β-bromoethylphosphoryl dichloride. Following hydrolysis of the unreacted chloride moiety, the quaternary ammonium function is introduced by means of trimethylamine (Note 6, Section II,D).

This route was improved by Eibl *et al.*[37] and adapted to the synthesis

[36] R. Hirt and R. Berchtold, *Pharmacol. Acta Helv.* **33**, 349 (1958).
[37] H. Eibl, D. Arnold, H. U. Weltzien, and O. Westphal, *Justus Liebigs Ann. Chem.* **709**, 226 (1967).

of ether as well as ester lecithins. The procedure is illustrated below.

β-Bromoethylphosphoryl Dichloride. The following method is given by Hirt and Berchtold[36]:

> To a mixture of freshly distilled phosphorus oxychloride (62 ml) in absolute carbon tetrachloride (50 ml) were added 62 g of 2-bromoethanol dropwise with stirring at 20°. The mixture was allowed to stand overnight and then heated to boiling for 2 hours to expel hydrogen chloride. The mixture was fractionated *in vacuo.* Yield: 70–77 g, b.p. 110–115° per 12 mm.

The general procedure given by Eibl *et al.*[37] can be used with long-chain α,β-diglycerides, α,α′-diglycerides, or glycerol-1,2-diethers.

DL-*Glycerol-1,2-dihexadecyl Ether-3-phosphorylcholine*

> Triethylamine (1.25 g, 12.5 mmoles) and 1.25 g (5 mmoles) of β-bromoethylphosphoryl dichloride were added to absolute chloroform (15 ml) at 0°. While stirring and cooling in an ice-water bath, a solution of 1.9 g (3.5 mmoles) of DL-glycerol-1,2-dihexadecyl ether in chloroform (15 ml) was added dropwise. The mixture was allowed to stir for 6 hours at room temperature and then at 40° for 12 hours. The dark solution was cooled at 0° and the phosphoromonochloridate was hydrolyzed by addition of 15 ml 0.1 KCl with stirring. After 1 hour 25 ml of methanol were added and the aqueous phase made to pH 3 with concentrated HCl. The mixture was shaken and the residue in the organic phase was dried over P_2O_5 in high vacuum.

> For the introduction of the trimethylammonium group the β-bromoethyl phosphodiester in butanone (chloroform, acetone, or, in the case of ethers, methanol can also be used) was treated with trimethylamine; for example, when a solution in butanone (50 ml) and trimethylamine (10 ml) was warmed for 12 hours at 55° and then cooled to 0°, the diether lecithin bromide crystallized and was filtered, washed with acetone, water, and again acetone. For complete purification of the product, it was stirred in 90% aqueous methanol with 1 g of silver acetate for 30 minutes and chromatographed, if necessary, on silica. The yield of pure diether lecithin, recrystallized from butanone, was about 60%.

D. Notes

1. This point is important in evaluating the purity of standard phospholipids. We have, however, noted in our laboratory considerable differences between optical rotations taken on optical vs. photoelectric polar-

imeters.[38] For a definitive settlement of the question, what would be required is to perform the synthesis of the same individual phospholipid by a number of different methods simultaneously from the same batch of starting material (e.g., *sn*-glycerol-1,2-isopropylidene ketal), and to observe the optical rotations of the products using the identical instrument.

2. Prepared by adding an excess of acetic acid to choline hydroxide, evaporating off the excess acetic acid, and drying under vacuum. Another recommended choline salt used in similar condensations in our laboratory and others[39] is the *p*-toluenesulfonate prepared from methyl tosylate and dimethylaminoethanol.[40]

3. Another and perhaps simpler method of detritylation is that used by Buchnea[41] for the synthesis of mixed acid *sn*-glycerol-1,2- and -2,3-diesters.

4. A more modern and practical method for the possible preparation of *sn*-glycerol-1- and 3-trityl ethers has recently been introduced (see Section III,C,1 and Pfeiffer *et al.*[42]) which involves a peculiar but much less tedious interchange of substituents. This newer procedure is not as facile or flexible for the synthesis of monoethers or ether esters as that of Chacko and Hanahan,[28] but it may compete with the latter to produce diethers containing identical R groups.

5. In our own experience with this type of reaction using dioxan as solvent[43,44] relatively little unreacted halide is obtained at the conclusion of the *O*-alkylation. Instead, the major byproduct is the symmetrical ether ROR, which in this case would be dihexadecyl ether. This does not affect the formation of the desired product, however, although the inert ether may be a less objectionable by-product when a subsequent hydrogenation is performed. The comparative conditions of alkylation under which ROR vs. RBr is the major byproduct remain to be fully elucidated. One of the factors affecting completeness of reaction of RBr is the state of subdivision of the KOH. In our experience the quickest and simplest way to prepare powdered KOH of high reactivity is by grinding dry KOH pellets in a dry, covered Waring blender for a few minutes.

[38] A. F. Rosenthal and M. Pousada, *Lipids* **4**, 37 (1969).
[39] P. P. M. Bonsen, G. J. Burbach-Westerhuis, G. H. de Haas, and L. L. M. van Deenen, *Chem. Phys. Lipids* **8**, 199 (1972).
[40] A. F. Rosenthal, *J. Lipid Res.* **7**, 779 (1966).
[41] D. Buchnea, *Lipids* **6**, 734 (1971).
[42] F. R. Pfeiffer, C. K. Miao, and J. A. Weisbach, *J. Org. Chem.* **35**, 221 (1970).
[43] A. F. Rosenthal, *J. Chem. Soc., London* p. 7345 (1965).
[44] A. F. Rosenthal, L. Vargas, and S. C. H. Han, *Biochim. Biophys. Acta* **260**, 369 (1972).

Another peculiarity of this type of reaction is the specificity of KOH; we have attempted to substitute NaOH, LiOH, and NaH with little or no success.

6. The great advantage of this type of synthesis is that, since only one phosphoric ester group must be formed at a time, an excess of β-bromoethylphosphoryl dichloride can be used, ensuring a relatively smooth reaction with minimal phosphotriester formation. Furthermore, since only a phosphorochloridate serves as the "protecting" group, the method is completely adaptable to the synthesis of unsaturated compounds. Our experience with syntheses using β-bromoethylphosphoryl dichloride (for the preparation of β-lecithins, cholesterylphosphoryl-choline, and straight-chain phosphorylcholines) has produced uniformly good results.

7. *Related compounds.* The large variety of phosphate lecithin analogues which have been synthesized are too numerous to list here. The interested reader is referred to the recent paper of Bonsen et al.,[39] as well as the reviews of van Deenen and de Haas[1] and Slotboom and Bonsen.[3]

E. Synthesis of Phosphatidylethanolamine

The methods for the synthesis of phosphatidylethanolamines are generally analogous to those for the synthesis of lecithins, although with certain important differences. One of the main points of difference is the necessity for protecting the unsubstituted amino function of the intermediates in the synthesis of phosphatidylethanolamine. Another is the relative impracticality of preparing the amino lipid from glycerophosphorylethanolamine because of the much greater difficulty of preparing this intermediate in a pure state from practical starting materials. Accordingly, in this section no synthesis of this type is described. The syntheses of the diester phosphatidylethanolamines discussed below employ either phosphatidic acid or iodohydrin diesters as starting materials.

1. Synthesis of Phosphatidylethanolamine from Phosphatidic Acid

There are two essential problems in this type of synthesis: employing (1) a suitable N-protecting group for the ethanolamine, and (2) an appropriate condensing agent for the esterification.

Obviously, protection of the amino function should be accomplished via a moiety which can be removed with complete selectivity, i.e., unmasking of the protecting group should not be accompanied by ester hydrolysis or alteration of unsaturated chains. The activating (condensing) agent, one of the by now classic phosphodiester forming type (carbodiimides, trichloroacetonitrile, sulfonyl chlorides, etc.), should be chosen to give high yields and a pure product.

In the recent procedure of Aneja *et al.*[45] given below, *N*-tritylethanolamine is condensed with a phosphatidic acid by means of 2,4,6-triisopropylbenzenesulfonyl chloride:

Dipalmitoyl sn-Glycerol-3-phosphorylethanolamine

N-Trityl-O-(1,2-distearoyl-sn-glycero-3-phosphoryl)ethanolamine. Anhydrous *N*-tritylethanolamine (135 mg, 0.44 mmole, Note 1, Section II,F) and trisopropylbenzenesulfonyl chloride (200 mg, 0.66 mmole) in anhydrous pyridine (approximately 5 ml) was stirred for 0.5 hour at room temperature and 1,2-distearoyl-*sn*-glycero-3-phosphoric acid (150 mg, 0.22 mmole) added. The reaction mixture was warmed (60°) briefly (approximately 1 minute) to dissolve the phosphatidic acid. The solution was stirred at room temperature and continued for another 1½ hours. Water (approximately 1 ml) was added and the solvents were removed on a rotary evaporator and finally evaporated to dryness under reduced pressure. The residue was chromatographed on SilicAR CC-7 (40 g) impregnated with triethylamine; elution with chloroform containing 0.5% triethylamine (250 ml) followed by elution with chloroform–methanol (19:1, v/v) also containing 0.5% triethylamine (approximately 200 ml) gave the product, *N*-tritylphosphatidylethanolamine, dried over P_2O_5 (0.5 mm, 20 hours) (196 mg, 0.2 mmole, 93%), and crystallized from cold (−15°) acetone, m.p. 113–114°, $[\alpha]_D^{22}$ +6.3° (c 2.0).

The removal of the trityl group by these authors was accomplished in this particular case by hydrogenolysis, which makes the procedure unavailable for the synthesis of unsaturated compounds. The trityl group can also be removed by very mild acid hydrolysis according to Billimoria and Lewis,[46] whose procedure for the homologous dipalmitoyl compound is given below (see also Note 2, Section II,F):

Detritylation with glacial acetic acid–water (9:1; 5 ml) was performed at reflux temperature for 3 minutes. The solution when cooled and set aside overnight at room temperature deposited crystals. The mixture was diluted with acetone; crystals of the phosphatidylethanolamine (99.5%) were filtered off; m.p. 210–211°, $[\alpha]_D^{25.0}$ +6.2°.

[45] R. Aneja, J. S. Chadha, and A. P. Davies, *Biochim. Biophys. Acta* **218**, 102 (1970).
[46] J. D. Billimoria and K. O. Lewis, *J. Chem. Soc., London* p. 1404 (1968).

2. Synthesis of Phosphatidylethanolamine from Glycerol-α-iodohydrin Diesters

The relative ease of preparation of specific diesters of glycerol-α-iodohydrin containing saturated or unsaturated fatty acids (see above) allows the synthesis of phosphatidylethanolamines of almost any desired structure. What is also required, of course, is a silver salt of a substituted phosphorylethanolamine containing easily removed protective groups.

The reagent introduced by Daemen,[47] silver tert-butyl (N-tert-butyl-oxycarbonyl-2-aminoethyl) phosphate, appears to be the most useful salt of this type so far described, since it allows simultaneous removal of both phosphoester and amine protecting groups under conditions which do not affect unsaturation or ester groups. The preparation of this reagent, and its use in the synthesis of an optically active mixed-acid, partially unsaturated phosphatidylethanolamine, is given below.

Silver tert-Butyl-(N-tert-butyloxycarbonyl-2-aminoethyl) Phosphate

N-tert-Butyloxycarbonylethanolamine.[48]

A solution of 15.25 g (0.25 mole of ethanolamine in 300 ml of water was vigorously shaken for 10 minutes with 10.1 g (0.25 mole) of finely powdered magnesium oxide. After addition of 71.4 g (0.5 mole) of freshly distilled tert-butyloxycarbonyl-azide (Aldrich Chemical Co.) in 300 ml of dioxan, the suspension was stirred for 24 hours at 50°. After completion of the reaction the magnesium oxide was centrifuged and the supernatant layer concentrated to a volume of about 300 ml. The solution was diluted with 100 ml of water and extracted 4 times with 100 ml of ethyl acetate. The combined extracts were dried over magnesium sulfate. After evaporation of the solvent, the faintly colored product was subjected to distillation in a high vacuum, thus affording 36.1 g (90%) of tert-butyloxycarbonylaminoethanol, a thick colorless oil, b.p. 0.3 mm; 92°, n_D^{20} 1.4525.

Silver tert-Butyl(N-tert-butyloxycarbonyl-2-aminoethyl) Phosphate.[47] 1.75 g of dicyclohexylammonium tert-butyl phosphate (Note 3, Section II, F) (5 mmoles) was suspended in 25 ml of dry pyridine and 2.4 g of N-tert-butyloxycarbonylethanolamine (15 mmoles) and 7.2 g of trichloroacetonitrile (50 mmoles) were added. The mixture was kept in a closed vessel at 90–100° for 4 hours. The

[47] F. J. M. Daemen, *Chem. Phys. Lipids* **1**, 476 (1967).
[48] F. J. M. Daemen, G. H. de Haas, and L. L. M. van Deenen, *Rec. Trav. Chim. Pays-Bas* **82**, 487 (1963).

resulting brown reaction mixture was diluted with 100 ml of water and continuously extracted with ether for 1 hour. To the aqueous layer 2 ml of cyclohexylamine were added and the solvent was removed in a rotating evaporator below 40°. The residue was once recrystallized from dimethylformamide, containing 2% of cyclohexylamine. The precipitate was collected at 4°, washed with acetone and ether, yielding 1.2 g (60%) of the cyclohexylammonium salt with m.p. 185–187°. A column of Amberlite IR 120 H$^+$ (15 g wet weight) was prepared in 50% ethanol and converted into the Ag$^+$ form by percolating a solution of 15g of silver nitrate in 50% ethanol. After washing out the adhering electrolyte, 1.2 g of the cyclohexylammonium salt was applied on the column and eluted with 50% ethanol (100 ml). The eluate was concentrated and the residue recrystallized from absolute ethanol. The silver salt was obtained as a pure white, light-sensitive powder with m.p. 158–160°. Yield: 1.06 g (95%). The conversion was carried out in dim red light.

1-Palmitoyl-2-oleyl-sn-glycerol-3-phosphorylethanolamine

1-Palmitoyl-2-oleyl-sn-glycerol-3-iodohydrin was prepared analogously to other mixed-acid glycerol iodohydrins (see Section II,A,1). Silver tert-butyl(N-tert-butyloxycarbonyl-2-aminoethyl) phosphate (1.5, 2.7 mmoles) and the iodohydrin diester (2.2 g, 3.1 mmoles) were introduced into 200 ml of freshly distilled, anhydrous benzene and the suspension was boiled under reflux in a nitrogenous atmosphere for 2 hours. After removal of the precipitated silver iodide the reaction mixture was washed with potassium bicarbonate solution and water. After drying on sodium sulfate, 2.4 g of a colorless oil was obtained. This oil, mainly consisting of phosphoric acid triester, was dissolved in 100 ml of anhydrous ether and dry hydrogen chloride was led through this solution at 0° for 2 hours. The solvent was evaporated *in vacuo* below 40° and the remaining solid, containing the hydrogen chloride adduct of the phosphatidylethanolammonium chloride, was dissolved in 50 ml of a mixture of ether, ethanol, and water (4:2:1 v/v) and percolated over a column of Amberlite IR 45 (OH$^-$). After evaporation, the residue was crystallized once from chloroform–acetone to yield 1.24 g (55%) of the desired product; m.p. 194–196° $[\alpha]_D^{22}$ +6.35° (c 3 in chloroform).

3. Ether Phosphatidylethanolamines

The synthesis of ether phosphatidylethanolamines is completely analogous to the synthesis of the corresponding lecithins. The same glycerol ethers are employed as starting materials; both ether esters[28] and diethers[28,33] have been employed. In the procedure given below, the phosphorylethanolamine function is introduced in one step by esterification with 2-phthalimidoethylphosphoryl dichloride, followed by hydrolysis of the chloride and hydrazinolysis of the phthalyl group.

1,2-Di-(9-cis-octadecenyloxy)-rac-glycerol-3-(2'-aminoethyl) Phosphate

2-Phthalimidoethylphosphoryl Dichloride. The synthesis of this reagent was described by Hirt and Berchtold.[49]

β-Hydroxyethylphthalimide (60 g), 250 ml of absolute benzene, and 132 g of phosphorus oxychloride were heated under reflux for 4 hours with exclusion of moisture. Excess benzene was removed *in vacuo* at 60° and the remaining POCl₃ was distilled off. The residual red syrupy mass was dissolved in 500 ml of absolute ether and cooled to 0°; after 2–3 hours the product crystallized as a white powder, which was rapidly filtered and washed with ether. The yield was 72 g of a white powder, m.p. 72–73°, which quickly decomposes in moist air with the evolution of hydrogen chloride.

1,2-Di(9-cis-octadecenyloxy)-rac-glycerol-3-(2'-aminoethyl) Phosphate. Although in this specific example a racemic product containing two identical ether chains is prepared, the method is not limited only to substances of this type, provided that the requisite glycerol ether is available. The procedure is that of Thomas and Law.[33]

The glyceryl diether (0.5 g) was treated in 100 ml of dry chloroform with 2-phthalimidoethyl dichlorophosphate (3 g) and pyridine (1.6 g). The solution was allowed to stand 48 hours at room temperature; the chloroform was then removed *in vacuo* and the residue was dissolved in wet ether and allowed to stand 24 hours. The ether solution was then washed with 20 ml each of water, 2.5% hydrochloric acid, twice again with water, and dried over anhydrous sodium sulfate. The solution was filtered and the ether was evaporated *in vacuo*. The white solid residue was dissolved in 100 ml of 99% ethanol; 15 ml of 10% hydrazine in ethanol were added in small portions, with cooling, and the solution was then refluxed for 2 hours. After the solution was again cool, 10 ml of 20% hydrochloric acid was added, and the solution was allowed to stand 2 hours at room of the diacyl compound. This fraction was dissolved in benzene and temperature.

The solvent was removed *in vacuo*, and the residue was taken up in 150 ml of chloroform–methanol 2:1. The insoluble material

[49] R. Hirt and R. Berchtold, *Helv. Chim. Acta* **40**, 1928 (1957).

was removed by filtration, and the solution was washed with 30 ml of 1% sodium chloride. The upper layer was discarded, and the lower layer was again washed with 20 ml of 1% sodium chloride. The product, after evaporation of the chloroform, was purified by chromatography on DEAE-cellulose. The sample (500 mg) was applied to a tightly packed, 35 × 20 cm column in chloroform–methanol 7:1, fraction 1 was eluted with 250 ml of chloroform–methanol 7:1 and fractions 2 and 3 with 250 ml of chloroform–methanol 7:3. Fraction 3 contained 440 mg of material. This gave a single ninhydrin-positive spot on silicic acid TLC (Adsorbosil-1, Applied Science Laboratories, Inc., State College, Pa.) running with or slightly ahead of diacyl phosphatidylethanolamine when developed in chloroform–methanol 2:1. The R_f of the diether cephalin is virtually the same as that of the diacyl compound. This fraction was dissolved in benzene and lyophilized to give the final product, a fluffy snow-white powder, in 73% yield from the glyceryl diether. The diether glycerophosphatide is soluble in chloroform, ethanol, ether, and benzene, and slightly soluble in hexane. It does not exhibit a sharp melting point, but softens at around 200°.

F. Notes

1. N-Tritylethanolamine is prepared simply by 1:1 molar reaction of trityl chloride with ethanolamine, since N-tritylation is much more rapid than O-tritylation. A recent procedure, that of Butskus et al.[50] is the following (see also Note 2):

A mixture of 2.4 g of ethanolamine and 5.6 g of triphenylchloromethane in 15 ml of pyridine was boiled for 5 hours. Water was gradually added to the reaction mixture, precipitating the reaction product, which was filtered off and treated with hot ethanol. The ethanolic solution was filtered and water was added to precipitate the desired compound. Yield: 5.1 g (84.1%), m.p. 75–76°.

2. The acid-catalyzed detritylation is invariably accompanied by a small amount of lysophosphatidylethanolamine formation.[51] Thus, other derivatives of ethanolamine may be preferable in this reaction, e.g., N-tert-butyloxycarbonylethanolamine (readily removable with hydrogen chloride at 0°)[52,53] or N-(β,β,β-trichloroethoxycarbonyl)ethanolamine[54]

[50] P. Butskus, R. Saboniene, and D. Lemesiene, Zh. Org. Khim. 6, 1984 (1970).
[51] R. Aneja, J. S. Chadha, and J. A. Knaggs, Biochim. Biophys. Acta 187, 579 (1969).
[52] I. Barzilay and Y. Lapidot, Chem. Phys. Lipids 3, 280 (1969).
[53] S. Rakhit, J. F. Bagli, and R. Deghenghi, Can. J. Chem. 47, 2906 (1969).
[54] F. R. Pfeiffer, S. R. Cohen, and J. A. Weisbach, J. Org. Chem. 34, 2795 (1969).

(removable with zinc and acetic acid at room temperature). One of the most easily demasked derivatives is N-(o-nitrobenzenesulfenyl)ethanolamine[55]; it is prepared simply from ethanolamine and o-nitrobenzenesulfenyl chloride, which is available commercially (Aldrich Chemical Co.). The present method of *Aneja et al.*[45] was selected for inclusion in this section because 2,4,6-triisopropylbenzenesulfonyl chloride is probably superior to dicyclohexylcarbodiimide[52] or trichloroacetonitrile[53] in effecting this condensation.[45]

3. Cramer *et al.*[56] gave a relatively simple preparation of mono-*tert*-butyl phosphate, which is summarized as follows: To a mixture of anhydrous phosphoric acid (3.9 g), triethylamine (8.1 g), acetonitrile (30 ml), and a few drops of water was added, during 2 hours with stirring, a mixture of 6.3 g of *tert*-butanol and 17.3 g of trichloroacetonitrile. After additional stirring, the mixture was diluted to 300 ml with acetone and 30 ml of cyclohexylamine added. The mixture was cooled and filtered; the product was recrystallized from boiling ethanolcyclohexylamine–water (90:10:1) to give 5.4 g of the biscyclohexylamine salt of mono-*tert*-butyl phosphate.

4. *Related compounds.* The synthesis of N-methylphosphatidylethanolamine and N,N-dimethylphosphatidylethanolamine was also accomplished by Aneja *et al.*[45] from the corresponding phosphatidic acids in what appears to be the simplest preparation of these intermediate methylated lipids. For syntheses of N-acylphosphatidylethanolamines as well as other analogs, the reader is again referred to the review of Slotboom and Bonsen.[3]

G. Synthesis of Phosphatidylserine

1. Synthesis of Phosphatidylserine from Phosphatidic Acid

When glycerol iodohydrin diesters are used as starting materials for the preparation of phosphatidylserine, a rather complex silver salt is then required for introduction of the phosphate and serine functions.[57] Fortunately, the recent work of Aneja and co-workers,[45] which allows the synthesis of phosphatidylserine by condensation of phosphatidic acid with an amino- and carboxy-protected serine, should allow the synthesis to proceed by an appreciably simpler route.

As discussed above, such phosphatidic acid condensations are not in-

[55] I. Barzilay and Y. Lapidot, *Chem. Phys. Lipids* **7**, 93 (1971).
[56] F. Cramer, W. Rittersdorf, and W. Boehm, *Justus Liebigs Ann. Chem.* **654**, 180 (1962).
[57] G. H. de Haas, H. van Zutphen, P. P. M. Bonsen, and L. L. M. van Deenen, *Rec. Trav. Chim. Pays-Bas* **83**, 99 (1964).

herently limited only to "same-acid," racemic, or saturated compounds, but in the particular procedure of Aneja et al.,[45] given below, unsaturated compounds cannot be employed because of the specific protecting groups for the serine chosen by these workers (see Note 1, Section II,H).

1,2-Distearoyl sn-Glycerol-3-phosphoryl-DL-serine

O-(1,2-Distearoyl-sn-glycero-3-phosphoryl)-N-carbobenzoxy-DL-serine Benzyl Ester. Dried 1,2-distearoyl-sn-glycero-3-phosphoric acid (197 mg, 0.28 mmole), dry N-carbobenzoxy-DL-serine benzyl ester (Note 1, Section II, H; 172 mg, 0.52 mmole), and triisopropylbenzensulfonyl chloride (255 mg, 0.84 mmole) in anhydrous pyridine (approximately 5 ml) were warmed (bath temperature 70°) until a clear solution was obtained (approximately 1 minute) and then stirred at room temperature for 3 hours. Ethanol-free dry chloroform (approximately 3 ml) was added and stirring continued for another 1 hour; water (2 ml) was then added and the solvents were removed on a rotary evaporator. The residue was extracted with diethyl ether and the diethyl ether–soluble material chromatographed on Silic-AR CC-7 (15 g) and eluted with chloroform–methanol (19:1 and then 9:1, v/v). The fraction containing the product on evaporation gave mainly the desired compound (233 mg, 1.23 mmoles, 83%). For purification by crystallization it was dissolved in warm acetone (5 ml) and cooled to room temperature. The cloudy solution obtained was filtered and the filtrate on cooling (−15°) deposited the title compound, m.p. 44°, $[\alpha]_D^{22}$ +3.7° (c 1.4).

1,2-Distearoyl-sn-glycero-3-phosphoryl-DL-serine. A solution of the N-carbobenzoxy benzyl ester (150 mg, 0.15 mmole) in glacial acetic acid (approximately 20 ml) was hydrogenated at 25 lb/inch² in the presence of palladium catalyst (200 mg, 5% on carbon). After the absorption of hydrogen was complete, the mixture was diluted with chloroform and methanol and filtered through a bed of Hyflo. The filtrate was evaporated and the residue washed with anhydrous diethyl ether. The residual compound was pure phosphatidylserine (thin-layer chromatography). It was crystallized from hot glacial acetic acid (approximately 10 ml), filtered, and washed with diethyl ether. The compound was freed from solvents by continuous evacuation for 1 hour. The dry compound, m.p. 159–160°, was insoluble in common organic solvents but soluble in hot dioxan and hot acetic acid, and could be crystallized from both these solvents. It dissolved in chloroform–acetic acid on heating and accordingly a rotation was recorded in this system, $[\alpha]_D^{55}$ +7.4° (c 0.5; chloroform, acetic acid (9:1, v/v)).

2. Synthesis of Phosphatidylserine from Substituted Glycerophosphorylserine

Ordinarily phosphorus oxychloride is not the method of choice for the introduction of the phosphate moiety into phospholipids since it usually leads to considerable pyrophosphate or other byproduct formation, but Shvets and co-workers[58] have described such a procedure which appears to give an unsaturated phosphatidylserine in good yield. The difficultly prepared unsaturated 1,2-diglycerides are not used as the starting materials. Instead, 1,2-isopropylideneglycerol is treated successively with phosphorus oxychloride and the N-phthalyserine; hydrolysis of the ketal and with phosphomonochloride groups is accomplished simultaneously to give glycerol-3-phosphoryl-N-phthalylserine as the barium salt after neutralization with barium hydroxide. Reaction with an excess of fatty acid chloride then gives the desired phosphatidylserine. The authors also claimed that this esterification can even be run in stepwise sequence, giving a mixed-acid phosphatidylserine. The procedure of Shvets[58] is as follows.

1,2-Dilinoleyl rac-Glycero-3-phosphorylserine

Barium Salt of α-Glyceroylphosphoryl-N-phthaloylserine. To 2.1 g of α,β-isopropylideneglycerine (b.p. 92–94° at 20 mm, n_D^{20} 1.4381) was added with stirring 3.0 g of phosphorus oxychloride and 2.41 ml of quinoline at 0°. The reaction was left for 1 hour at 25° and then the temperature was lowered at 0° and 4.4 g of N-phthaloyl-serine (mp. 149.5–150.5°) in 100 ml of dimethylformamide with 5 ml of pyridine were added. The reaction mixture was left at 18–20° for 20 hours. Then 1.5 ml of water were added and stirred for 1 hour. Dimethylformamide was removed under vacuum (20 mm). The residue was dissolved in water, filtered, acidified with 3% hydrochloric acid to pH 1.5–2, and left for 3.5 hours at 18–20°. Impurities were removed from the aqueous solution which was extracted with chloroform (3 × 100 ml). Water was removed under vacuum (20 mm). The residue was dissolved in 100 ml of methanol and filtered; the methanol was evaporated. To the material dissolved in 75 ml of water was added saturated aqueous barium hydroxide solution to pH 8–9. The reaction mixture was left for 12 hours at 18–20°. Excess barium ion was precipitated with carbon dioxide.

[58] V. I. Shvets, M. K. Petrova, G. A. Kazenova, and N. A. Preobrazhenskii, *Zh. Obshch. Khim.* **37,** 1454 (1967).

The solution was filtered; chloride was removed by treating the reaction solution with stirring over 2 hours with 8.7 g of silver carbonate obtained from silver nitrate and sodium carbonate. The solid was filtered and hydrogen sulfide was passed through the solution. The sulfides were filtered off. Water was removed in vacuum (20 mm). The material obtained was treated with 150 ml of methanol. The solid was removed and the solvent was evaporated. This operation was repeated 3 times. The material obtained was dried for 1.5 hour at 0.1 mm and 40–50°. The yield was 8.2 g (79.5%).

α-(α′,β-Dilinoleoyl)glycerylphosphoryl-N-phthaloylserine. To 2.0 g of the barium salt in 150 ml of dimethylformamide and 2.8 ml of pyridine was added 12.0 g of linoleoyl chloride (b.p. 150–152° at 0.5 mm). The reaction mixture was kept at 50° for 48 hours and then poured into 400 ml of 5% hydrochloric acid. It was extracted with chloroform (4 × 40 ml). The organic layer was washed with saturated aqueous sodium bicarbonate solution (2 × 20 ml), dried over sodium sulfate, and evaporated. The residue (12.0 g) in 40 ml of chloroform was chromatographed over 120 g of silicic acid dried for 4 hours at 120°. The final product was eluted with 120 ml of a mixture of chloroform and methanol (4:1). The solvents were evaporated and the material was dried at 0.15 mm and 20–25° for 2 hours. The waxy material was very soluble in chloroform, ether, benzene, or methanol and formed an emulsion with water. The homogeneity of the material was substantiated by thin-layer chromatography on silicic acid (gypsum) in the systems: petroleum ether (b.p. 40–60°)–ether (1:1) (material at the origin), and diisobutyl ketone–acetic acid–water (8:5:1) (R_f 0.86). Spots of the material were detected on the chromatograms in all cases by spraying the plates with 10% sulfuric acid and subsequent charring at 200–250°. Yield: 3.1 g (88.9%).

1,2-Dilinoleyl-rac-glycero-3-phosphorylserine. To 3.0 g of the phthalyl phosphatidylserine in 150 ml of methanol was added 0.25 ml of hydrazine hydrate at 20°. The temperature of the reaction was increased to 60° for 2.5 hours. The hot solution was filtered and acidified with glacial acetic acid (0.5 ml). It was cooled to 5°. Impurities were filtered and the filtrate was concentrated under vacuum (10–15 mm). The residue was dissolved in 15 ml of chloroform, cooled to 5°, filtered, and evaporated. The material obtained (2.1 g) in 20 ml of chloroform was chromatographed on 30 g of dry silicic acid activated for 4 hours at 120°; phosphatidylserine was eluted with 60 ml of a mixture of chloroform and methanol (4:1). The solvents were evaporated and the residue was dried at

0.15 mm and 25–30° for 2.5 hours. The waxy material was very soluble in benzene, chloroform, ether, and methanol and gave a stable emulsion in water. The homogeneity of the compound was shown by thin-layer chromatography on silicic acid in mixed diisobutyl ketone–acetic acid–water 8:5:1, R_f 0.8. Yield: 0.97 g (69.2%).

H. Notes

1. N-Carbobenzoxyserine benzyl ester was prepared according to Baer and Maurukas.[59] Synthetic directions for this protected serine are not included here for two reasons: They are quite long and detailed, and the original source is readily available; and the derivative is now a rather obsolete one, requiring hydrogenation for removal of the protecting groups. A better compound for this purpose might be, e.g., N-tert-butoxycarbonylserine or N-tert-butoxycarbonylserine tert-butyl ester, which should allow the ready synthesis of phosphatidylserine containing unsaturated groups. From the published literature it is not even altogether certain that the carboxy group must be protected during this condensation. If this turns out ultimately to be unnecessary, the probability of a very facile synthesis of phosphatidylserine from N-tert-butoxycarbonylserine (commercially available) is indicated.

2. *Related compounds.* The number of synthetic phosphatidylserine analogues known is not large. Noteworthy, however, is a recent synthesis of phosphatidylthreonine by Moore and Szelke,[60] which employs both a $POCl_3$ phosphorylation of isopropylideneglycerol step and a sulfonyl chloride–activated condensation.

I. Synthesis of Phosphatidylglycerol

Of the several reported syntheses of this phospholipid, the procedure of Bonsen *et al.*[61] appears to be the most general, allowing the preparation of saturated or unsaturated mixed-acid or same-acid phosphatidylglycerol of any configuration. Actually, the configurational problems associated with the synthesis of phosphatidylglycerol are very appreciable because the unacylated glycerol moiety of the natural lipid has the difficulty synthesized sn-1′ configuration (i.e., sn-1,2-diacylglycerophosphoryl-sn-1′-glycerol). Thus, 2,3-isopropylidene-sn-glycerol (L-acetone-glycerol) must be employed as a starting material.

[59] E. Baer and J. Maurukas, *J. Biol. Chem.* 212, 25 (1955).

[60] J. W. Moore and M. Szelke, *Tetrahedron Lett.* 50, 4423 (1970).

[61] P. P. M. Bonsen, G. H. de Haas, and L. L. M. van Deenen, *Chem. Phys. Lipids* 1, 33 (1966).

The procedure of Bonsen and co-workers,[61] given below, illustrates the versatility of the P-O-benzyl moiety a protective group, since it can be removed not only by hydrogenation but also quite selectively by highly nucleophilic reagents (in this case iodide ion).

1-Oleyl-2-palmitoyl-sn-glycerol-3-phosphoryl-sn-1'-glycerol

2,3-Isopropylidene-sn-glycerol-1-dibenzylphosphate. 32.5 g of dibenzyl phosphoric acid (Note 1, Section II,C) and 12.0 g of dicyclohexyl carbodiimide were dissolved in chloroform to give tetrabenzylpyrophosphate. After 1 hour a solution containing 8.8 g of 2,3-isopropylideneglycerol (Note 2, Section II,C) and 7.95 g of imidazole in chloroform was added dropwise. The mixture was stirred for 48 hours at room temperature. After filtration of dicyclohexylurea the solvent was evaporated, the residue dissolved in ether and washed in succession with an ice-cold 10% hydrochloric acid and 4 times with water. The ethereal solution was dried over sodium sulfate and evaporated *in vacuo*, giving the product, a colorless viscous oil, in a yield of 55–65%. A small amount of the substance was chromatographed over silica with benzene–ether as eluents for elemental analysis and optical rotation. $[\alpha]_D^{20}$ +3.65° (*c* 10 in chloroform).

Barium Salt of 2.3-Isopropylidene-sn-glycerol-(benzyl) Phosphate. Debenzylation of the foregoing product was effected by barium iodide in dry acetone at boiling temperature for 4 hours. The precipitated barium salt was filtered off and washed several times with acetone. The product revealed one distinct spot on thin-layer chromatograms. Yield: 93%, m.p. over 300°, $[\alpha]_D^{20}$ +1.27° (*c* 10 in water).

Silver Salt of 2,3-Isopropylidene-glycerol-1-(benzyl) Phosphate. In the dark an aqueous solution of an equivalent amount of silver sulfate was added to a solution of the barium salt in water. After centrifugation of barium sulfate, the clear solution was evaporated *in vacuo*. The silver salt was obtained as a white powder in a yield of 100%; m.p. 194–198°, $[\alpha]_D^{20}$ +1.40° (*c* 10 in water).

1-Oleoyl-2-palmitoyl-glycerol-3-(benzyl)phosphoryl-1'-(2', 3'-isopropylidene)glycerol. 4.1 g of 1-oleoyl-2-palmitoyl-glycerol iodohydrin (see Section II,A) was reacted with 3.1 g (30% excess) of the silver salt in toluene at a bath temperature of 100°. After 1 hour the mixture was cooled, the silver iodide removed by centrifugation, and the solution evaporated *in vacuo*. The residue was dissolved in pentane and washed subsequently with a bicarbonate solution and water.

After drying over sodium sulfate the solution was evaporated *in vacuo*. The yellowish oily residue was chromatographed over silica with 5% ether in benzene, giving the product as a colorless oil in a yield of 75%; $[\alpha]_D^{20}$ +0.75° (*c* 10 in chloroform).

Sodium Salt of 1-Oleoyl-2-palmitoyl-glycerol-3-phosphoryl-1′-glycerol. 0.6 g of the isopropylidene derivative was dissolved in 10 ml of trimethylborate, 1.5 g of boric acid was added and dissolved by refluxing on a water bath for 5 minutes. The solvent was evaporated and the residue heated at 100° for 10 minutes with a rotary evaporator. After cooling the product was dissolved in ether and washed with a saturated sodium chloride solution. After drying over sodium sulfate and evaporation of the solvent, the product was purified by chromatography over silica with chloroform–methanol (85:15 v/v) as eluent. The pure phosphatidylglycerol obtained was crystallized as the sodium salt from chloroform acetone, giving a white powder in a yield of 85%; m.p. 176–179°, $[\alpha]_D^{20}$ +1.02 (*c* 10 in chloroform).

J. Notes

1. Dibenzylphosphoric acid is now available commercially but is rather expensive. It may also be prepared by oxidation of the cheaper dibenzyl phosphite according to the procedure[15] already described (Section II,A) for oxidation of di-*tert*-butyl phosphite. Another and even simpler method of preparation of dibenzyl phosphate as a sodium salt is by heating the moderately priced tribenzyl phosphate (Eastman) with sodium iodide in dry acetone. The precipitated and washed sodium salt in a small amount of water gives a precipitate of silver dibenzyl phosphate on treatment with aqueous silver nitrate.

2. 2,3-Isopropylidene-*sn*-glycerol may be prepared from L-mannitol in the same way as its enantiomer is prepared from D-mannitol. Unfortunately, L-mannitol is not available commercially; thus, it must first be synthesized from L-arabinose, a good grade of which is available from Aldrich Chemical Co.

L-Arabinose is first converted to L-mannose according to Sowden and Fischer[62]:

L-*Mannose.* A suspension of 30 g of L-arabinose in 110 ml of nitromethane and 150 ml of methanol was shaken for 18 hours at room temperature with a solution of 6.9 g of sodium in 200 ml of

[62] J. C. Sowden and H. O. L. Fischer, *J. Amer. Chem. Soc.* **69**, 1963 (1947).

methanol. The resulting amorphous sodium salts of the nitroalcohols, after drying in high vacuum over phosphorus pentoxide, weighed 52 g and contained 9.4% of sodium. The sodium salts were dissolved in 200 ml of cold water and the resulting solution was added dropwise at room temperature to a stirred solution containing 40 ml of sulfuric acid and 50 ml of water. The resulting solution was neutralized to congo red with solid sodium carbonate, treated with decolorizing carbon and filtered. The pale amber solution was then made just alkaline to litmus with sodium bicarbonate solution and again acid to litmus with a few drops of acetic acid. A solution of 20 ml of phenylhydrazine in 40 ml of 25% acetic acid was then added. The precipitation of L-mannose phenylhydrazone began after a few minutes. After standing overnight, the hydrazone was collected and washed successively with water, 60% ethanol, absolute ethanol, and acetone. The colorless product weighed 12.2 g (22.6%) and melted at 186–188°.

Four grams of the hydrazone was refluxed for 2½ hours with a solution containing 50 ml of water, 10 ml of ethanol, 5 ml of benzaldehyde, and 0.5 g of benzoic acid. The cooled solution was decanted from benzaldehyde phenylhydrazone, extracted 3 times with chloroform, decolorized with carbon, and concentrated at reduced pressure. The resulting syrup crystallized readily. Filtration with absolute ethanol then yielded 2.2 g (82.4%) of L-mannose, m.p. 128–132°, and $[\alpha]_D^{25}$ $-14.5°$ (equil.) in water, c 3.4.

The above procedure can be simplified by adding the dry amorphous sodium nitroalcohols directly to the aqueous sulfuric acid. Thus, when the sodium salts obtained from 5 g of D-arabinose and nitromethane were added, with stirring, to 25 ml of 40% (by weight) sulfuric acid, there was obtained, after neutralization and addition of phenylhydrazine, 1.9 g (21%) of D-mannose phenylhydrazone.

L-Mannose is then converted to L-mannitol according to the procedure of Vargha et al.[63]:

L-*Mannitol.* An L-mannose solution (from 72.4 g L-mannose phenylhydrazone; 450 ml) was made alkaline (10 N NaOH) and to it was added sodium borohydride (11 g) in water (500 ml). After 4 hours the mixture was carefully made acidic with acetic acid and evaporated to dryness in vacuum. The residue was heated 1 hour in 6% HCl in methanol (1200 ml), when a clear solution was formed. After standing overnight at 0° the crystallized L-mannitol was filtered off and washed with a little methanol. Yield: 15 g of compound of

[63] L. Vargha, L. Toldy, and E. Kasztreiner, *Acta Chim.* (*Budapest*) **19**, 295 (1959).

m.p. 164–166°. From the mother liquor evaporated to 200 ml was obtained another 29 g of product, m.p. 155–156°, which was recrystallized from ethanol to give 21.6 g of pure L-mannitol. Combined yield: 36.6 g.

It seems likely that 2,3-isopropylidene-sn-glycerol could be prepared from 1-trityl-sn-glycerol-2,3-carbonate (see Baer and Fischer[34] and Pfeiffer et al.[42]), but this has not been completely investigated in detail for every step.

3. Related compounds. The synthesis of diether analogues of phosphatidylglycerol containing phytanyl groups has been reported by Joo and Kates[64]; such a substance is found in certain bacteria.[65] Amino acid[66] and glucosaminyl[67,68] esters of phosphatidylglycerol have been synthesized by van Deenen and co-workers. The closely related phosphatidylglycerophosphate has been synthesized by Bonsen and de Haas[69]; the corresponding phytanyl diether phospholipid has also been synthesized by Joo and Kates.[70]

K. Synthesis of Diphosphatidylglycerol (Cardiolipin)

Although the 1,3-(diphosphatidyl)glycerol structure has by no means been established for cardiolipins from all natural sources, it is believed to be the structure of the most common substances of this group. In the synthesis of diphosphatidylglycerol reported by de Haas and van Deenen,[71] a silver monobenzyl phosphate of an sn-1,2-diglyceride (72 moles) is reacted with 1,3-diiodo-2-tert-butoxypropane to give a protected cardiolipin derivative. After demasking of the phosphate and alcohol groups the cardiolipin was obtained in good yield; since it was optically active, it almost certainly has the configuration of the natural substance.[10] The synthesis gave a mixed-acid partially unsaturated product, for which it is well adapted.

[64] C. N. Joo and M. Kates, Biochim. Biophys. Acta 152, 800 (1968).

[65] M. Kates, B. Palameta, C. N. Joo, D. J. Kushner, and N. E. Gibbon, Biochemistry 5, 4092 (1966).

[66] P. P. M. Bonsen, G. H. de Haas, and L. L. M. van Deenen, Biochim. Biophys. Acta 106, 93 (1965).

[67] P. P. M. Bonsen, G. H. de Haas, and L. L. M. van Deenen, Biochemistry 6, 1114 (1967).

[68] H. M. Verheij, P. P. M. Bonsen, and L. L. M. van Deenen, Chem. Phys. Lipids 6, 46 (1971).

[69] P. P. M. Bonsen and G. H. de Haas, Chem. Phys. Lipids 1, 100 (1967).

[70] C. N. Joo and M. Kates, Biochim. Biophys. Acta 176, 278 (1969).

[71] G. H. de Haas and L. L. M. van Deenen, Rec. Trav. Chim. Pays-Bas 84, 436 (1965).

1,3-Bis-(1-stearoyl-2-oleyl-sn-glycerol-3-phosphoryl)glycerol

α,γ-Diiodoglycerol. α,γ-Diiodoglycerol was prepared either from α,γ-di-*p*-toluenesulfonylglycerol or from α,γ-dichloropropanol by reaction with sodium iodide in acetone. Boiling point 70° per 0.004 mm Hg; n_D^{22} 1.6650. Conversion of the foregoing compound into the corresponding *tert*-butyl ether derivative was carried out in a dry methylene chloride solution with isobutylene and a trace of concentrated sulfuric acid. The product could not be obtained in analytically pure form. Incomplete conversion into the *tert*-butyl ether derivative and/or a slight decomposition of a highly labile substance during working-up resulted in a product which upon thin-layer chromatography appeared to be contaminated with a small amount of the starting material (α,γ-di-iodoglycerol).

Dibenzyl β-Oleoyl-γ-stearoyl-L-α-glycerol Phosphate. A solution of 32.0 g (0.044 mole) of β-oleoyl-γ-stearoyl-L-α-iodopropanediol (see Section II,A for general preparation of mixed-acid iodohydrin diesters) in 350 ml of anhydrous benzene was mixed with 21.0 g (0.55 mole) of silver dibenzyl phosphate (see Note 1, Section II, J) and the mixture was refluxed in the dark with vigorous stirring for 3 hours. After removal of the precipitated silver iodide, the solution was dried *in vacuo* and the residue was crystallized from petroleum ether (b.p. 40–60°) at a temperature of −20°. The dibenzyl compound weighed 31.0 g (yields of several batches varied from 80 to 95%); $[α]_D^{20}$ +4.2° in chloroform (*c* 10). The existence of unstable modifications of this compound was shown by several transition points in the melting curve and by a significant increase of the melting point after storage.

Silver Benzyl β-Oleoyl-γ-stearoyl-L-α-glycerol Phosphate.[12] Debenzylation of 39.0 g of dibenzyl ester was carried out by boiling for 3 hours with 9.0 g of dry sodium iodide in 300 ml of acetone in the presence of a trace of mercury and in a nitrogen atmosphere. After storage at −20° the pale yellow precipitate formed was filtered off. This crude sodium salt was purified by low temperature crystallization from anhydrous acetone. The recovered colorless monobenzyl sodium salt weighed 32.3 g (91%) and was immediately converted into the corresponding free acid by treatment with 230 g of Amberlite IR-120 (H⁺) in 600 ml of boiling acetone for 3 hours. The crude material was purified by crystallization from ethanol to yield 25.5 g 80%) of monobenzyl β-oleoyl-γ-stearoyl-L-α-glycerol hydrogenphosphate, colorless crystals with m.p. 22–23°, $[α]_D^{20}$ +2.6 in chloroform (*c* 10).

$\alpha'\gamma$-$Bis[benzyl$-(β'-$oleoyl$-γ'-$stearoyl$-L-α'-$glyceroyl$)$phosphoryl$]-$glycerol\ tert$-$Butyl\ Ether$. The silver salt and the di-iodoglycerol derivative were reacted in a molar ratio of 2.2:1 in dry toluene at 100°. Thin layer chromatography analysis showed an optimal reaction time of 60 minutes, after which small amounts of contaminants (mainly mono-substituted products) were removed by chromatography on silica using ether-hexane mixtures are eluent. The product was obtained as a colorless syrup in a yield of 65% ($[\alpha]_D^{20}$ +2.15° in chloroform, c 9).

$Dibarium\ Salt\ of\ \alpha,\gamma$-$Bis$($\beta'$-$oleyl$-$\gamma'$-$stearoyl$-L-$\alpha'$-$glyceroyl$-$phosphoryl$)$glycerol\ tert$-$Butyl\ Ether$. A solution of 2.40 g of the dibenzyl ester in 25 ml of dry acetone was heated for 5 hours at 60–65° in the presence of 0.83 g of dry barium iodide. After chilling, the precipitated dibarium salt was isolated and purified by crystallization from ether–acetone. Yield: 2.19 g (94%), m.p. 201–204°, $[\alpha]_D^{20}$ +1.66° in chloroform (c 11).

$Dibarium\ Salt\ of\ \alpha,\gamma$-$Bis$($\beta'$-$oleoyl$-$\gamma'$-$stearoyl$-L-$\alpha'$-$glyceroyl$-$phosphoryl$)$glycerol$. The foregoing barium salt dissolved in anhydrous chloroform was de-$tert$-butylated by treatment with a rapid stream of dry hydrogen chloride at 0°. After 4 hours chloroform and excess of hydrogen chloride were removed $in\ vacuo$ and the residue was dissolved in a 1:1 mixture of ether and methanol. The crude barium salt (was precipitated by the addition of a solution of barium acetate in aqueous methanol. Purification of the barium salt was carried out by crystallization from chloroform–acetone and ether–ethanol followed by very rapid chromatography on a silicic acid column. Yield: 69%, colorless crystals, m.p. 192–194, $[\alpha]_D^{20}$ —6.25° in chloroform (c 6).

L. Notes

$Related\ compounds$. Very few diphosphatidylglycerol analogues of well-defined structure have been synthesized. Noteworthy is a synthesis of the closely related diphosphatidyl(β-acyl)glycerol by de Haas and van Deenen.[72]

M. Synthesis of Phosphatidylinositol

Phosphatidylinositol presents a number of interesting but difficult challenges to the synthetic chemist. The stereochemistry alone is a formidable problem, because in addition to the usual asymmetry of the glyc-

[72] G. H. de Haas and L. L. M. van Deenen, $Rec.\ Trav.\ Chim.\ Pays$-Bas **82,** 1163 (1963).

erol moiety there exists the asymmetry of the inositol group. The latter produces difficulties because, like glycerol, inositol itself is optically inactive but its 1-phosphate (the moiety present in phosphatidylglycerol) is optically active. Thus, a synthetic asymmetrically substituted inositol derivative must be employed for the synthesis of a phosphatidylinositol containing an optically active inositol moiety, the natural configuration being sn-myoinositol-1-phosphoryl-3'-(sn-glycerol-1',2'-diacylate), in the nomenclature of Klyashchitskii and co-workers.[73]

Optical resolution is required to prepare such an optically active myoinositol intermediate. This was recently accomplished by Klyashchitskii et al.,[74,75] who employed the resolved intermediates in a total synthesis of completely optically active dipalmitoylphosphatidylinositol.[76] Their procedure is not, however, given in detail here, because the resolution sequence is quite complex and the protecting groups chosen (benzyl and phenyl) for the phosphorylation make the route inapplicable to the synthesis of unsaturated compounds.

The synthesis of unsaturated phosphatidylinositol was reported by Molotkovskii and Bergelson[77] and is given below. These authors employed an optically active glycerol derivative but a racemic myoinositol moiety, introduced via an acetylated inositolphosphate benzyl ester. Considering the work of these two laboratories, the total synthesis of a completely optically active unsaturated phosphatidylinositol identical in structure with the natural substance appears imminent, but it is certain to be quite lengthy.

Molotkovsky and Bergelson[77] employed as a key intermediate rac-2,3,4,5,6- (or rac-1,2,4,5,6- as it is also called) pentaacetylmyoinositol, which was first synthesized by Angyal and co-workers[78] by means of an unusual acetolysis reaction from rac-1,4,5,6-tetra-O-benzylmyoinositol. The first intermediate, rac-1,2,-O-cyclohexylidenemyoinositol was prepared by Angyal et al.[79]:

[73] B. A. Klyashchitskii, V. I. Shvets, and N. A. Preobrazhenskii, Chem. Phys. Lipids 3, 393 (1969).

[74] B. A. Klyashchitskii, G. D. Strakhova, V. I. Shvets, S. D. Sokolov, and N. A. Preobrazhenskii, Dokl. Akad. Nauk SSSR 185, 594 (1969).

[75] B. A. Klyashchitskii, V. V. Pimenova, V. I. Shvets, and N. A. Preobrazhenskii, Zh. Obshch. Khim. 39, 1653 (1969).

[76] B. A. Klyashchitskii, E. G. Shelvakova, V. V. Pimenova, V. I. Shvets, R. P. Yevstigneeva, and N. A. Preobrazhenskii, Zh. Obshch. Khim. 41, 1386 (1971).

[77] Y. G. Molotkovskii and L. D. Bergelson, Dokl. Akad. Nauk SSSR 198, 461 (1971).

[78] S. J. Angyal, M. H. Randall, and M. E. Tate, J. Chem. Soc., London p. 919 (1967).

[79] S. J. Angyal, G. C. Irving, D. Rutherford, and M. E. Tate, J. Chem. Soc., London p. 6662 (1965).

rac-2,3,4,5,6-Pentaacetyl-O-myoinositol

($\pm$)-1,2-O-Cyclohexylidenemyoinositol. In a flask fitted with an efficient stirrer, a wide condenser and a Dean and Stark separator connected to the flask by a tube at least 2 cm in diameter, finely powdered myoinositol (50 g) is heated with purified cyclohexanone (500 ml) and benzene (130 ml) until no more water distills. Toluene-p-sulfonic acid (0.5 g of monohydrate) is then added and the solution is boiled vigorously with rapid stirring until the separation of water becomes slow and nearly all the inositol has dissolved (about 4 hours). The success of the reaction at this stage is indicated by the disappearance of the solid; the amount of water collected in the separator (usually about 20 ml) is not a true indication since it exceeds the theoretical, owing to the inevitable self-condensation of cyclohexanone. The solution is cooled and decanted from the solid which is washed with benzene (50 ml), more benzene (200 ml), light petroleum (b.p. 60–80°) (250 ml), ethanol (50 ml), and cyclohexyli-denemyoinositol (0.5–1 g for seeding) are added, in this order, to the solution. Hydrolysis of the tri- and diketals begins as soon as the ethanol is added; the cyclohexylidene groups are removed as cyclohexanone diethyl ketal. The monoketal is readily hydrolyzed further to myoinositol if it remains in solution; hence, it is essential to remove it from solution as soon as possible. This aim is achieved by generous inoculation and stirring; a copious precipitate appears within 10 minutes. After storage at 0° overnight, triethylamine (1 ml) is added with stirring, the solution is filtered, and the solid (ca. 66 g) is well washed with benzene. To remove the myoinositol, the solid is extracted with boiling ethanol (1000 ml) containing triethyl-amine (1 ml) to neutralize any acid which may be present; if the reaction is successful only a small amount of myoinositol remains undissolved. On cooling, 1,2-O-cyclohexylidenemyoinositol (ca. 54 g, 74%), m.p. 181–183° (with transition at 158–160°), crystallizes; it usually contains a trace of myoinositol. To obtain the compound free from myoinositol, it is recrystallized from three parts of water.

Myinositol 1,4,5,6-tetra-O-benzyl ether was prepared by Angyal and Tate[80]:

1,4,5,6-Tetra-O-benzyl-2,3,-O-cyclohexylidenemyoinositol. 1,2-O-Cyclohexylidenemyoinositol (17.5 g) was vigorously stirred with powdered potassium hydroxide (105 g) and benzyl chloride (175 ml) on a steam bath for 19 hours. The mixture was steam-distilled (to remove benzyl chloride) and was then extracted with benzene

[80] S. J. Angyal and M. E. Tate, J. Chem. Soc., London p. 6949 (1965).

(3 × 100 ml); the extract was washed with water (100 ml), dried (MgSO₄), and evaporated to dryness. The residue, after dissolution in boiling methanol (400 ml), deposited the tetrabenzyl ether (32.0 g, 74%) as prisms, m.p. 80–84°; recrystallization raised the m.p. to 85–87°.

1,4,5,6-Tetra-O-benzylmyoinositol. The foregoing monoketal (7.74 g) was heated on a steam bath with glacial acetic acid (160 ml) and water (40 ml) for 2.5 hours. The mixture was evaporated, methanol was added to the residue and again evaporated. The crystalline residue was crystallized from methanol (25 ml) to give needles (5.72 g, 84%), m.p. 114–115°; after another crystallization from methanol or benzene, the tetra-O-benzylmyoinositol melted at 115°. Occasionally, a second allotropic modification, m.p. 127°, crystallized from the crude reaction mixture.

Finally, the pentaacetylinositol was prepared by Angyal *et al.*[78]:

Acetolysis of 1,4,5,6-Tetra-O-benzylmyoinositol. After standing at 0° for 31 hours, a solution of the above compound (0.6 g) in the acetolysis mixture (3 ml of 200:1 cold acetic anhydride, 67% perchloric acid) was poured into ice water containing sodium hydrogen carbonate. The solid which precipitated was crystallized from ethanol to give 1,2,4,5,6-penta-O-acetyl-3-O-benzylmyoinositol (0.45 g, 83%), m.p. 158–160°. Two crystallizations raised the m.p. to 167–168°.

1,2,4,5,6-Penta-O-acetylmyoinositol. A rapid stream of hydrogen was passed through a mixture of 1,2,4,5,6-penta-O-acetyl-3-O-benzylmyoinositol (2.0 g), glacial acetic acid (50 ml), and 10% palladium chloride on carbon (2.0 g); thin-layer chromatography showed that hydrogenolysis was complete in 45 minutes. The catalyst was removed by centrifugation and washed with acetic acid (2 × 20 ml). The combined liquors were evaporated under reduced pressure, and the residue was triturated with water (4 ml) whereupon it crystallized. Filtration and washing with ice-cold water gave 1,2,4,5,6-penta-O-acetylmyoinositol (1.24 g, 76%), m.p. 158–160°. Crystallization from water (20 ml) gave needles, m.p. 161–164°.

The synthetic procedure of Molotkovskii and Bergelson[77] employs phosphorus oxychloride and then benzyl alcohol for the preparation of the pentaacetylinositol dibenzylphosphate, which could not be obtained directly from the inositol derivative and dibenzylphosphoryl chloride. Monodebenzyation followed by reaction of the silver salt of the pentaacetylinositol phosphate monobenzyl ester with a mixed-acid partially unsaturated *sn*-glycerol-3-iodohydrin diester gave the protected phosphatidylinositol. Iodide debenzylation gave the phosphodiester; removal of the acetyl groups was accomplished with hydrazine at ambient temperature to give, finally, the unsaturated phosphatidylinositol. This order of removal of the protecting groups appears to meet the objections of Gent et al.,[81] who stressed the great tendency of unprotected myoinositol phosphotriesters to undergo positional migration. The detailed synthetic procedure is given by Molotkovskii and Bergelson[77] as follows.

1-(1'-Lauroyl-2'-oleyl-sn-glycerol-3'-phosphoryl)-rac-myoinositol

rac-1-Dibenzylphosphoryl-2,3,4,5,6-penta-O-acetylmyoinositol. To a solution of 0.72 g of $POCl_3$ in 1 ml of absolute chloroform was added, during 30 minutes at room temperature, a solution of 1.8 g of *rac*-2,3,4,5,6-penta-O-acetylmyoinositol and 0.65 g of quinoline in 5 ml of dry chloroform. The mixture was allowed to stand for 24 hours, cooled to 5°, and to it was added during 30 minutes a solution of 1.1 g of benzyl alcohol and 1 ml of pyridine in 2 ml of dry chloroform. After standing for 5 hours at 0–5°, the mixture was diluted with 50 ml of ether and 50 ml of ethyl acetate, and the solution was washed successively with 1 N H_2SO_4, water, saturated $KHCO_3$ solution, water, and saline, after which it dried with Na_2SO_4, filtered, and the solvent evaporated. The residue was recrystallized from ether–petroleum ether mixture to give 1.16 g (39% yield) of the product, which was chromatographically homogenous (by TLC in the system 2:1 chloroform–ethyl acetate). After further recrystallization from methanol–water and ether–petroleum ether the substance melted at 147–149°.

rac-2,3,4,5,6-Penta-O-acetylmyoinositol-1-benzylphosphate Silver Salt. A solution of 1.8 g of the phosphotriester and 450 mg of NaI in 10 ml of acetone was heated for 2½ hours at 80° in a sealed tube. The mixture was cooled to −5° and filtered, and the residue was dissolved in 7 ml of 19:1 chloroform–methanol. The solution was evaporated to half its initial volume, 15 ml of acetone were added, and the mixture was left overnight at −5°. The precipitate was collected on a filter, washed with cold acetone, and dried over P_2O_5, to give 1.4 g (87% yield) of sodium *rac*-2,3,4,5,6-penta-O-acetylmyoinositol-1-benzylphosphate, m.p. 264–265°, which was chromatographically homogeneous.

To a solution of 1.52 g of this sodium salt in 7 ml of methanol was added a solution of 0.45 g of $AgNO_3$ in 1 ml of water and 3 ml of methanol, followed by 20 ml of acetone. The mixture was left at 0° overnight, and the silver salt which precipitated was collected

[81] P. A. Gent, R. Gigg, and C. D. Warren, *Tetrahedron Lett.* p. 2575 (1970).

on the filter. Yield: 0.47 g of white powder, m.p. 246–249° (dec.). A further 0.15 g separated out of the filtrate on standing; total yield 93%.

rac-1-(1'-Lauroyl-2'-oleoyl-sn-glyceryl-3'-benzylphosphoryl)-2,3,4, 5,6-penta-O-acetylmyoinositol. A mixture of 700 mg of 1-lauroyl-2-oleoyl-*sn*-glycerol-3-iodohydrin (see Section II,A) with 700 mg of silver salt in 5 ml of anhydrous toluene was heated at boiling point for 12 hours, in the dark and with vigorous stirring, filtered to remove the AgI precipitate formed, and evaporated. The residue was chromatographed on a column of 120 g of silica gel, applying gradient elution in the system benzene–ethyl acetate. As a control, we applied TLC in the solvent system benzene–dioxane (4:1). Using benzene with 20–25% of ethyl acetate we eluted 640 mg (56%) of triphosphate from the column, in the form of a white semisolid mass, $[\alpha]_D +2.2$ (c 9.5, dioxane).

rac-1-(1'-Lauroyl-2'-oleoyl-sn-glyceryl-3'-phosphoryl) myoinositol (Phosphatidylinositol). The phosphotriester (120 mg) was debenzylated by treatment with NaI in acetone at 80° for 2.5 hours in a sealed tube, evaporated *in vacuo* to dryness; a chloroform solution of the residue was applied to a column packed with 4 g of silica gel. The column was washed with 20 ml of chloroform, which was followed by 15 ml of 4:1 chloroform–methanol mixture, which eluted *rac*-1-(1'-lauroyl-2'-oleoyl-*sn*-glycerol-3'-phosphoryl)-2,3,4,5,6-penta-*O*-acetylmyoinositol. This was dissolved in 5 ml of 95% ethanol, and 40 mg of hydrazine hydrate were added. The mixture was left overnight at room temperature, the pH was adjusted to 4 by addition of acetic acid, and the solution was evaporated to dryness *in vacuo*. The residue was examined by TLC, in the solvent system chloroform–methanol–water–7 N NH_4OH (130:70:7:7), and found to contain three major components—phosphatidylinositol and two less polar substances—which were not further examined. The mixture was chromatographed on a column containing 15 g of silica gel, which was eluted with a gradient of chloroform with a 9:1 methanol–20% NH_4OH mixture (control TLC was done). In order to avoid the risk of deacylation of the product, the time of elution was kept at 2–2.5 hours, using a micropump for forcing the eluant through the column. The eluate was evaporated to dryness *in vacuo* at 20°, to give 40 mg (44%) of phosphatidylinositol ammonium salt, which was crystallized from chloroform–acetone at 0°, and dried at 20° per 0.05 mm, over P_2O_5. The product was a white powder, $[\alpha]_D +2.3°$ (c 0.3, chloroform). It sinters at about 110° and melts at 180–195° with decomposition (Note 1, Section II,N).

N. Notes

1. According to the authors, the synthetic product could not be separated by TLC from natural phosphatidylinositol isolated from yeast in neutral, acidic, or basic solvent systems.

2. *Related compounds.* The synthesis of phosphatidylglucose was reported by Verheij *et al.*[82] and by Volkova *et al.*[83] A phosphatidylinositol aminoethyl phosphotriester isomer has been synthesized by Preobrazhenskii's group.[84]

O. Synthesis of Plasmalogens

The chemical synthesis of plasmalogens presents a unique problem compared to other glycerophosphatides since the diglyceride portion of the molecule contains a distinctly different moiety, the acid-labile 1-alkenyl ether group. Aside from the initial problem of formation of this moiety and then of performing the remainder of the synthesis without disturbing it, the preparation introduces a new stereochemical difficulty; the natural substance has the *cis* configuration at the enol ether group, while presently available synthetic methods produce *cis-trans* mixtures of varying composition.

The chemical synthesis of plasmalogens, because of these complexities, has not progressed as far as most of the other syntheses discussed in this chapter. The procedures are currently undergoing rapid refinement in several laboratories and more definitive syntheses can be expected in the near future.

The enol ether–containing intermediate employed for the plasmalogen syntheses described below is a glycerol 1'-alkenyl ether. Since the enol ether linkage is acid-labile but base-stable, the starting material employed for its preparation is the base-labile glycerol 1,2-cyclic carbonate. For the preparation of the isomer of natural optical configuration the "L-glycerol" derivative *sn*-glycerol-2,3-cyclic carbonate is required. Two recent methods of preparation of this substance are (1) by means of an inversion at C-2 of *sn*-glycerol-3-benzyl ether (prepared from 1,2-isopropylidene-*sn*-glycerol), followed by reaction with ethyl carbonate to give *sn*-glycerol-1-benzyl ether-2,3-cyclocarbonate; hydrogenation then removes the protective benzyl group; (2) by means of a peculiar cycliza-

[82] H. M. Verheij,. P. F. Smith, P. P. M. Bonsen, and L. L. M. van Deenen, *Biochim. Biophys. Acta* **218**, 97 (1970).

[83] L. V. Volkova, M. G. Luchinskaya, H. M. Karimova, and R. P. Yevstigneeva, *Zh. Org. Khim.* **42**, 1405 (1972).

[84] A. V. Lukyanov, A. I. Lyutik, V. I. Shvets, and N. A. Preobrazhenskii, *Zh. Org. Khim.* **36**, 1029 (1965).

tion–elimination reaction of 1-trityl-3-(2′,2′,2′-trichloroethyl)-sn-glycerol when it is warmed with a tertiary amine. The resulting sn-glycerol-1-trityl ether-2,3-cyclocarbonate was not actually converted to sn-glycerol-2,3-cyclocarbonate because in that particular synthesis (not of plasmalogens) the latter was not desired, but it seems certain that it could be prepared by hydrogenation under the same conditions employed with the corresponding 1-benzyl ether or by some mild acid hydrolysis procedure.

The first route is due to Fedorova and co-workers[85] while the second was developed by Pfeiffer et al.[42] Because sn-glycerol-2,3-carbonate is a very valuable synthetic intermediate, both synthetic routes are detailed below even though the hydrogenolysis (or hydrolysis) of the trityl ether in the latter route was not described; nonetheless, it is doubtlessly the simpler procedure. The remainder of the entire synthesis is rather complex and will be discussed in the following sections.

1. sn-Glycerol-2,3-cyclocarbonate

By Inversion from sn-Glycerol-3-benzyl Ether.[85]

1,2-Di-p-TOLUENESULFONYL-3-O-BENZYL-sn-GLYCEROL. To a solution of 54 g of tosyl chloride in 90 ml of pyridine cooled to 0° was added a solution of 25 g of 3-O-benzyl-sn-glycerol (see "Glycerol Diethers" in Section II,C,3) in 23 ml of pyridine and the solution was stirred at 18–20° for 36 hours. Ether (700 ml) and water (200 ml) were then added to the reaction mixture. The ether layer was washed with 1 N sulfuric acid, 5% sodium bicarbonate, and dried over magnesium sulfate. Evaporation of the solution gave an oil which crystallized from ether at −20°. Yield: 53.2 g (79%), m.p. 54–55°.

1-O-BENZYL-sn-GLYCEROL.   1,2-Di-p-tosyl-sn-glycerol-3-benzyl ether (46.5 g) in 50 ml of dimethylformamide was heated at 140° for 1 hour with 37 g of freshly fused potassium acetate. The reaction mixture was cooled to 18–20°, the precipitate filtered off, the filtrate concentrated, and the residue extracted with chloroform. After evaporation of the solvent the residue was heated with sodium hydroxide (7 g) in methanol (30 ml) for 30 minutes at 50°. The reaction mixture was then neutralized with hydrochloric acid, the precipitate filtered, and the methanol removed. The residue was extracted with chloroform, washed with water, dried over magnesium sulfate, concentrated, and distilled. Yield: 10.5 g (60.5%), b_4 135–136°, $[\alpha]_D^{20}$ −5.9° (neat).

[85] G. N. Fedorova, G. A. Serebrennikova, A. G. Yefimova, and R. P. Yevstigneeva, *Zh. Org. Khim.* **7**, 957 (1971).

sn-GLYCEROL-1-BENZYL ETHER-2,3-CYCLOCARBONATE. 1-O-Benzyl-sn-glycerol in 25 ml of diethyl carbonate was refluxed for 8 hours in the presence of 110 mg of sodium bicarbonate while the alcohol evolved was continuously distilled off. After the usual work-up and removal of excess diethyl carbonate the product was obtained as a viscous oil. Yield: 11.4 g (95%). Low temperature crystallization from methanol gave a crystalline product, m.p. 30–31°, $[\alpha]_D^{20}$ —25° (neat, supercooled).

sn-GLYCEROL-2,3-CYCLOCARBONATE. sn-Glycerol-1-benzyl ether-2,3-cyclocarbonate (11.2 g) in 40 ml of glacial acetic acid was hydrogenated in the presence of 1 g 10% palladium on charcoal at 20° and atmospheric pressure. After the usual work-up the product was obtained as a viscous oil. Yield: 6.53 g (99%), $[\alpha]_D^{20}$ +11°.

From sn-Glycerol-3-(2′,2′,2′-trichloroethylcarbonate).[42]

sn-GLYCEROL-1-TRITYL ETHER-*2,3*-CYCLIC CARBONATE. The preparation of sn-glycerol-3-trichloroethyl carbonate is described in Section III,C,1. A mixture of 2.67 g (0.01 mole) of the crude carbonate (prepared from HCl hydrolysis of its isopropylidene derivative, Section III,C,1), 2.78 g (0.01 mole) of trityl chloride, 10 ml of dry pyridine, and 15 ml of dry chloroform was stirred at 60° for 18 hours. The mixture was diluted with ethyl acetate and washed with dilute HCl, water, 5% NaHCO$_3$, and water. After being dried the solvent was evaporated and the residue was crystallized from acetone–petroleum ether to give 2.27 g (63%) of sn-glycerol-1-trityl ether-2,3-cyclocarbonate, m.p. 216–217°. Two recrystallizations from acetone afforded an analytically pure product, m.p. 218–220°, $[\alpha]_D^{25}$ +18.9° (c 1.67 in chloroform).

2. sn-Glycerol-cis-1′-alkenyl Ethers

Of the several methods which have been proposed for the formation of the 1-alkenyl ether group of plasmalogens in its natural *cis* configuration, that of Gigg and Gigg[86] seems the most definitive. Its virtues are that the enol ether moiety is formed unequivocally in the 1′ position and that, although a 1′-*cis-trans* mixture inevitably is formed, the former (natural) isomer predominates. In this route a fatty aldehyde dialkyl acetal is reacted with glycerol-2,3-cyclocarbonate; an exchange reaction occurs to produce a di-(glycerol-2,3-carbonate)-1-acetal. Treatment with an acid chloride produces a 1′-chloro-1-alkoxyglycerol-2,3-carbonate.

[86] J. Gigg and R. Gigg, *J. Chem. Soc., London* p. 16 (1968).

Dehydrohalogenation of the reactive α-halo ether occurs simply on heating with a tertiary amine to give the *cis-trans*-1'-alkenyl ether from which the base-sensitive carbonate group is cleaved by alkali. At some stage after formation of the enol ether moiety, the *cis* and *trans* isomers are separated on silver nitrate–impregnated silica gel.

The preparation of a long-chain acetal is illustrated in the procedure of Gigg and Gigg[86]:

> *Octadecanal Dimethyl Acetal* (Note 1, Section II,P). Sodium hydrogen carbonate (43 g) and *n*-octadecyl methanesulfonate (see "Glycerol α-Monoethers" in Section II,C,3) (56 g, prepared from *n*-octadecanol) in dimethyl sulfoxide (225 ml) were stirred vigorously at 150° for 20 minutes, cooled, and poured into ice water. The precipitated solid was collected and washed thoroughly with water. After drying, the crude aldehyde was heated under reflux in dry methanol (500 ml) containing toluene-*p*-sulfonic acid (2 g) for 3 hours. The acid was neutralized by stirring with an excess of potassium carbonate, the solvent evaporated, and the residue extracted with ether. The extract was dried and the solvent evaporated to give the crude dimethyl acetal (42 g). Thin-layer chromatography [ether–light petroleum (1:3) as mobile phase] showed the acetal (R_f 0.9) and traces of olefin (R_f 0.95), alcohol (R_f 0.3), and methanesulfonate (R_f 0.46). The product was filtered through an alumina column (8 × 10 cm) in ether–light petroleum (1:5) to give the dimethyl acetal (30 g) still contaminated with the trace of olefin but free from alcohol and methanesulfonate.

The plasmalogens are, of course, inherently unsaturated structures, so that methods for their synthesis would permit other points of unsaturation in the long-chain groups. By using the method of Gigg and Gigg[86] described above it is also possible to introduce such unsaturation before the enol ether moiety is formed. In the recent modification of the procedure by Federova *et al.*[85] such a synthesis is actually accomplished. The final removal of the carbonate group in this modification is performed on a basic alumina column rather than by alkali.

sn-Glycerol-1-cis,cis-(1'9'-octadecadienyl) Ether[85]

> *Di-(sn-glycerol-2,3-cyclocarbonate) Acetal of Oleyl Aldehyde.* *sn*-Glycerol-2,3-cyclocarbonate (2 g) and 2.9 g of oleic aldehyde diethyl acetal in 40 ml of benzene were heated under reflux for 2 hours in the presence of a catalytic amount of *p*-toluenesulfonic acid with azeotropic removal of the ethanol formed. The reaction mixture was washed with saturated aqueous sodium bicarbonate and dried with sodium sulfate. After evaporation of the solution the residue

was subjected to chromatographic purification on silicic acid. The product was eluted with ether. Yield: 3.28 g (80%), m.p. 31–32° $[\alpha]_D^{20}$ +13.6°, R_f 0.52 (activity III alumina in ether–acetone 47:3).

1-O-(Octadecadien-1′,9′-yl)-sn-glycerol-2,3-cyclocarbonate. The above product (1.9 g) in 3 ml of benzene and a catalytic quantity of pyridine was heated at 40° with 0.6 ml of thionyl chloride. After excess thionyl chloride was removed *in vacuo*, the residue was heated with 19 ml of dry triethylamine for 10 minutes at 70°. The reaction mixture was diluted with anhydrous ether, triethylamine hydrochloride was filtered off, and the filtrate evaporated. The residue was purified by chromatography on 20 g of silicic acid. The product was eluted with a mixture of petroleum ether and ether (4:1). Yield: 0.99 g (69.3%), R_f 0.6 (*cis,cis*) and 0.7 (*trans,cis*) on silica gel in ether.

Separation of cis,cis-1-O-(Octadecadien-1′,9′-yl) and trans,cis-1-O-(Octadecadien-1,9-yl)-sn-glycerol-2,3-cyclocarbonate. The isomeric mixture (0.69 g) was subjected to chromatography on a column of silica gel (6 g) impregnated with silver nitrate (4:1). The *trans,cis* isomer was eluted with a mixture of petroleum ether and ether (3:1). Yield: 0.11 g; R_f 0.6 (silica gel impregnated with silver nitrate; in ether).

cis,cis-1-O-(Octadecadien-1′,9′-yl)-*sn*-glycerol-2,3-cyclocarbonate was eluted with ether, giving the product as a viscous oil. Yield: 0.425 g (79.5%), R_f 0 5 (silica gel impregnated with silver nitrate; in ether), $[\alpha]_D^{20}$ −11.8°.

cis,cis-1-O-(Octadecadien-1′,9′-yl)-sn-glycerol. cis,cis-1-O-(Octadecadien-1′,9′-yl)-*sn*-glycerol-2,3-cyclocarbonate (0.120 g) was applied to a column of alumina (activity III–IV) and eluted with ether over 2 hours. Yield: 0.109 g (98%), R_f 0.4 (alumina activity IV, ether–acetone, 47:2), R_f 0.5 (silica gel impregnated with silver nitrate, ether–acetone, 41:1); $[\alpha]_D^{20}$ −1.9°.

3. Plasmalogens

The remainder of the synthetic route follows conventional lines for unsaturated glycerophospholipids with the additional proviso that no acidic reagent can be used in any step as a result of the lability of enol ethers. In the procedure of Serebrennikova *et al.*,[87] the glycerol-1-(1′-alkenyl) ether was reacted with 1 mole of tosyl chloride to produce the 3-tosylate. Acylation with a fatty acid chloride followed by reaction with

[87] G. A. Serebrennikova, V. I. Titov, and N. A. Preobrazhenskii, *Zh. Org. Khim.* **5**, 550 (1969).

sodium iodide gave the glycerol-1-enol ether-2-acylate-3-iodohydrin; silver dibenzylphosphate then yielded the 3-dibenzylphosphate. In a continuation of the work, Serebrennikova and co-workers[88] replaced one benzyl group by a silver salt by successive treatment with sodium iodide and silver nitrate. Reaction with dimethylaminoethyl chloride then gave the dimethylamino-plasmalogen benzyl ester, which was finally freed from its protective benzyl group by reaction with sodium iodide. A catalytic quantity of pyridine was necessary in each step to safeguard the enol ether group. Thus, a plasmalogen intermediate between phosphatidalethanolamine and phosphatidalcholine was prepared. In this example a racemic plasmalogen was synthesized, but the procedure is completely applicable to optically active compounds.

1-O-(Dodecen-1'-yl)-2-stearoyl-rac-glycerol-3-dibenzylphosphate[87]

1-O-(Dodecen-1'-yl)-3-O-tosylglycerol. To a solution of 1.0 g of 1-O-(dodecen-1'-yl) glycerol in 7.5 ml of chloroform and 0.75 g of pyridine was added a solution of tosyl chloride (0.74 g) in chloroform (2.5 ml) with stirring at 0°, which was continued for 20 hours. The reaction mixture was then diluted at 0–5° with 30 ml of water and 15 ml of chloroform. The chloroform extract was washed with water (4 × 20 ml) and dried with sodium sulfate. After evaporation of the solution, the residue was chromatographed on 20 g of silicic acid. Benzene removed low polarity impurities; a mixture of benzene–ether (95:5) eluted the product. Yield: 0.82 g (50.8%), R_f 0.26 (activity IV alumina, benzene–ether 3:2).

1-O-(Dodecen-1'-yl)-2-stearoylglycerol-3-O-tosylate. To a solution of 0.8 g of 1-O-(dodecen-1'-yl)-3-O-tosylglycerol in 8 ml of chloroform was added at 0° 0.7 g of pyridine and a solution of 0.72 g of stearoyl chloride in chloroform (8 ml). The reaction mixture was kept for 26 hours at 18–20°, diluted with ether (30 ml), washed with 5% aqueous potassium carbonate (2 × 20 ml), and dried with sodium sulfate. The residue obtained after evaporation of the solvent (1.08 g) was used for the reaction with sodium iodide. R_f 0.56 (activity IV alumina, petroleum ether–ether 3:1).

1-O-(Dodecen-1'-yl)-2-O-stearoyl-3-iodopropanediol (1,2). To a solution of 0.7 g of the tosylate in 20 ml of acetone were added 0.1 g of pyridine and 3 g of sodium iodide, and the solution heated under reflux for 6 hours. After cooling to 18–20° the reaction mixture was diluted with 20 ml of ether and 50 ml of water, the ethereal layer

[88] G. A. Serebrennikova, P. L. Ovechkin, I. B. Vtorov, and N. A. Preobrazhenskii, *Zh. Org. Khim.* **5**, 546 (1969).

separated, and the aqueous layer extracted with ether (3 × 10 ml). The combined extract was washed with 10% sodium thiosulfate and dried with sodium sulfate. The solvent was removed, and the residue chromatographed on 10 g of silicic acid. The substance was eluted with a mixture of ether and benzene, 2:1. Yield: 0.37 g (46.3%), m.p. 27.5—29°, R_f 0.71 (activity IV alumina, petroleum ether–ether, 6:1).

1-O-(Dodecen-1-yl)-2-O-stearoylglycerol-3-dibenzylphosphate. To a solution of 0.2 g of the iodo compound in 2.5 ml of xylene and 0.025 g of pyridine was added 0.18 g of silver dibenzylphosphate (freshly prepared, m.p. 218–220°) and the mixture heated under reflux in the dark for 30 minutes. After cooling to 18–20° the silver salts were filtered off and washed with 10 ml of ether. The combined filtrate and washings were evaporated in vacuum and the residue was chromatographed on 4 g of silicic acid. A mixture of petroleum ether–ether removed low polarity impurities and the product was eluted with petroleum ether–ether (5:2). Yield: 0.135 g (54.1%), m.p. 35–36°, R_f 0.62 (silica gel without binder, benzene–ether, 3:1).

The rest of the synthetic sequence was performed by Serebrennikova and co-workers[ss] with the 1-(hexadecen-1'-yl)-2-stearoyl homologue (prepared by a different method).

1-O-(Dodecen-1'-yl)-2-stearoyl-rac-glycerol-3-(N,N-dimethylaminoethyl)phosphate[ss]

Silver Salt of 1-O-(Hexadecen-1'-yl)-2-stearoylglycerol-3-mono-benzylphosphate. A solution of 0.40 g of 1-O-(hexadecen-1'-yl)-2-O-stearoylglycerol-3-dibenzylphosphate, 0.016 of sodium iodide in 8 ml of acetone, and 0.01 ml of pyridine was heated under reflux for 3 hours. The sodium salt which precipitated on cooling to −5° was filtered, washed with 10 ml of cold ether, and dried in vacuum. Yield: 0.24 g (64.5%), m.p. 64.5–66°, R_f 0.52 (on silica gel without binder in acetone–chloroform 3:1). To a warm (40°) solution of the sodium salt in 12 ml of acetone was added a warm solution of 0.049 g of silver nitrate in a mixture of 1.4 ml of water and 3 ml of acetone. After standing for 2 hours at 0°, the silver salt was filtered, washed with 15 ml of cold acetone, and dried in vacuum (0.15 mm). Yield: 0.173 g (77.9%), m.p. 73.5–75°.

1-O-(Hexadecen-1'-yl)-2-stearoylglycerol-3-(2''-N,N-dimethyl-aminoethyl)benzylphosphate. A solution of 0.173 g of the silver salt and 0.033 g of N,N-dimethyl-β-chloroethylamine (from the hydrochloride) in 6 ml of benzene was stirred vigorously for 2 hours at 80–85°. The reaction mixture was cooled, filtered from silver chloride, which was washed with 5 ml of benzene, and the combined filtrate and washings dried in vacuum. The residue was chromatographed on 4 g of silicic acid. Low polarity impurities were removed with chloroform, and the product was eluted with chloroform–acetone, 95:5. Yield: 0.15 g (90.9%), m.p. 17–18.2°, R_f 0.25 (on silica gel without binder in acetone–chloroform, 2:15).

1-O-(Hexadecen-1'-yl)-2-stearoylglycerol-3-(N,N-dimethylaminoethyl)phosphate, Phosphatidaldimethylaminoethanol. A solution of 0.3 g of the benzyl ester and 0.076 g of sodium iodide in 8 ml of acetone and 0.01 ml of pyridine was heated under reflux for 3 hours. On cooling the reaction mixture to −5° the sodium salt precipitated; it was filtered and washed with cold acetone (10 ml) and dried in vacuum (0.2 mm). Yield: 0.20 g (70.9%), m.p. 138–139.3°, R_f 0.25 (on silica gel without binder in ether–chloroform–methanol, 10:2:8).

P. Notes

1. A few saturated fatty aldehydes are obtainable commercially, but they are in the form of a trimer from which the free aldehyde must still be obtained (see Section II,S). The bisulfite adducts are also obtainable in a few cases (e.g., from Aldrich Chemical Co.). Some dimethyl acetals are available from supply houses specializing in lipid derivatives (e.g., Supelco, Applied Science Labs) but they are prohibitively expensive for synthetic purposes.

2. *Related compounds.* Slotboom *et al.*,[89] in earlier work, reported a partial synthesis of the *trans* isomer of phosphatidalcholine. Among the more recent synthetic analogues of natural plasmalogens reported is an acetal phosphatidalcholine by Chacko and Perkins,[90] which can also be formed from natural lysophosphatidal plasmalogen.[91]

Q. Synthesis of Lysoglycerophospholipids

The chemical synthesis of lysophosphatides has received a certain amount of attention despite the fact that enzymatic hydrolysis of the fully acylated lipids remains the simpler and much more frequently used method of preparation of the lyso compounds. The extreme lability of monoacylated phospholipids, especially of the 2-acyl structure, to acyl rearrangements under the influence of even very weakly basic or acidic

[89] A. J. Slotboom, G. H. de Haas, and L. L. M. van Deenen, *Chem. Phys. Lipids* **1**, 192 (1967).
[90] G. K. Chacko and E. G. Perkins, *J. Org. Chem.* **32**, 1623 (1967).
[91] M. F. Frosolono, A. Kisic, and M. M. Rapport, *J. Org. Chem.* **32**, 3998 (1967).

conditions greatly limits the synthetic procedures which can be employed. The only well-established procedure for demasking the free-OH protecting groups without inducing acyl migration is hydrogenation in neutral solvents, which limits the procedure to saturated structures. Thus, the purely chemical synthesis of unsaturated lysophosphatides remains an intriguing and largely unsolved problem, particularly in the case of 2-acyl structures.

By contrast, the specific enzymatic degradation of diacyl phospholipids to pure 1-acyl or 2-acyl lysophospholipids is well established. Pancreatic lipase, which specifically removes only the 1-acyl group, and venom phospholipase A₂, removing only the 2-acyl moiety, are the enzymatic reagents of choice. It is of particular note for the purification of the chemically synthesized lysolipids that even silicic acid chromatography brings about acyl migration.[92]

The two procedures given in detail below represent, respectively, methods for the chemical synthesis of 1-acyl-lysolecithin and 2-acyl-lysophosphatidylethanolamine. They illustrate the general approach to the formation of these isomers which has been taken by most investigators in this area. With the single theoretically possible exception noted below, the procedures are adaptable only to saturated lysophosphatides, but they may be used to prepare any desired optical form of these isomers.

DL-*1-Stearoyl-3-lysolecithin*

The synthesis of DL-1-stearoyl-3-lysolecithin by Slotboom *et al.*[93,94] illustrates the use of the benzyl ether group in protecting the glycerol 2 position.

rac-Stearoyl-2-O-benzylglycerol-3-iodohydrin[94]

α-p-Toluenesulfonyl-(3-O-benzyl)glycerol. To a solution of 11.38 g (0.062 mole) of *β-O*-benzylglycerol (Note 1, Section II,R) in 40 ml of dry pyridine were added 11.82 g (0.062 mole) of *p*-toluenesulfonyl chloride dissolved in 40 ml of dry pyridine dropwise with stirring during 2 hours at 0°. After 24 hours at room temperature the reaction mixture was diluted with ether and the ethereal solution washed successively with ice-cold 0.5 *N* sulfuric acid (3 times), ice-cold water (1 time), ice-cold 5% potassium bicarbonate solution (1

[92] G. H. de Haas and L. L. M. van Deenen, *Biochim. Biophys. Acta* **106**, 315 (1965).

[93] A. J. Slotboom, G. H. de Haas, and L. L. M. van Deenen, *Chem. Phys. Lipids* **1**, 317 (1967).

[94] A. J. Slotboom, G. H. de Haas, and L. L. M. van Deenen, *Rec. Trav. Chim. Pays-Bas* **82**, 469 (1963).

time), and ice-cold water (2 times). The separated ether layer was dried over anhydrous sodium sulfate. A precipitate formed at $-12°$; it was filtered off and proved to be α,γ-di-p-toluenesulfonyl-β-O-benzylglycerol, m.p. 110–112°. The filtrate was brought to dryness *in vacuo* to yield 14.6 g (70%) of α-p-toluenesulfonyl-β-O-benzyl-glycerol as a pale yellow oil containing traces of α,γ-ditosylbenzyl-glycerol and unreacted β-benzylglycerol. The product was used in the next step without further purification.

α-p-Toluenesulfonyl-β-O-benzyl-γ-stearoylglycerol. To a solution of 9.55 g (0.028 mole) of the foregoing crude oil and 2.21 g (0.028 mole) of anhydrous pyridine in 60 ml of chloroform (freshly distilled from phosphorus pentoxide) was added dropwise with stirring at 0° a solution of 8.47 g (0.028 mole) of freshly distilled stearoyl chloride in 50 ml of absolute chloroform. After storage for 30 hours at room temperature ether was added and the solution was washed several times with ice-cold 0.5 *N* sulfuric acid and then with water until neutral. The separated ether layer was dried over anhydrous sodium sulfate and then stored at 0°. After removal of a small amount of stearic acid by filtration, some methanol was added and the solution stored at $-12°$ overnight. The white crystalline material was filtered off. Yield: 13.6 g (80%), m.p. 44–46°, after recrystallization from ether–methanol.

rac-γ-Stearoyl-β-O-benzylglycerol-α-iodohydrin. 8.0 g (0.013 mole) of the foregoing compound and 2.85 g (0.019 mole) of dry sodium iodide dissolved in 110 ml of anhydrous acetone were heated at 110° for 24 hours. After removal of the sodium p-toluenesulfonate (100%) the solution was evaporated *in vacuo*. The residue, dissolved in ether, was washed with 5% sodium thiosulfate solution, then with water, and dried over anhydrous sodium sulfate. The filtered ether solution was evaporated to dryness *in vacuo* and the residue was crystallized from ether–methanol to give 5.5 g (75%) of product, m.p. 41–42.5°.

(rac)-1-Stearoyl-2-O-benzylglycerol-3-(dibenzyl)phosphate.[93] A magnetically stirred suspension of 8.0 g of (rac)-1-stearoyl-2-O-benzylglycerol iodohydrin and 5.86 g of silver dibenzylphosphate in 140 ml of absolute toluene was refluxed in the dark for 5½ hours. After cooling to room temperature the precipitated silver iodide was centrifuged and washed twice with toluene. The combined toluene solutions were washed with a 5% potassium bicarbonate solution and twice with water. Drying of the organic phase over sodium sulfate and evaporating *in vacuo* gave 9.4 g of a colorless oil (94%). Thin-layer chromatography showed only one spot.

Sodium Salt of (rac)-1-Stearoyl-2-O-benzylglycerol-3-(benzyl)-phosphate. 9.4 g of (rac)-1-stearoyl-2-O-benzylglycerol-3-(dibenzyl)-phosphate and 2.98 g of anhydrous sodium iodide were refluxed for 2 hours in 60 ml of absolute acetone. After cooling to —15° the precipitate was centrifuged and recrystallized from chloroform–acetone at —15°. 7.45 g (87%) of the sodium salt were isolated as a colorless solid with m.p. 88.5–89.9°. Thin-layer chromatography in chloroform–methanol–ammonia (70:20:1.5, v/v/v) revealed a pure compound to be present.

Silver Salt of (rac)-1-Stearoyl-2-O-benzylglycerol-3-(benzyl) phosphate. To a magnetically stirred solution of 7.26 g of the sodium salt of (rac)-1-stearoyl-2-O-benzylglycerol-3-(benzyl)phosphate in 125 ml of tetrahydrofuran and 50 ml of 15% acetone in distilled water was slowly added in the dark a solution of 1.93 g of silver nitrate in 30 ml of distilled water. After addition of more acetone the solution was cooled to 0° and the colorless precipitate (4.75 g, m.p. 75–77.5°) was centrifuged. Addition of more cold acetone to the filtrate yielded a second crop of silver salt (2.15 g, m.p. 72.5–74.5°). The total yield amounted to 84%.

(rac)-1-Stearoyl-2-O-benzylglycerol-3-phosphoryl-(N,N-di-methyl)ethanolamine. 5.15 g of the foregoing silver salt and 5.41 g of 2-bromoethyl-N,N-dimethylamine picrate (Note 2, Section II,R) dissolved in 350 ml of absolute toluene were refluxed in the dark for 2 hours with magnetic stirring. The reaction mixture, cooled to 25°, was centrifuged, and the precipitate washed twice with toluene. The combined filtrate and washings were evaporated thoroughly *in vacuo* and the residue was dissolved in 250 ml of absolute acetone. Two grams of anhydrous lithium bromide were added and the mixture heated at 60° for 2 hours. The acetone solution was diluted to 1 liter with a mixture of methanol and water (9:1, v/v). Removal of inorganic material was carried out by passing this solution over an ion-exchange column containing equal amounts of Amberlite IRC-50 (H⁺) and IR-45 (OH⁻). After washing the column the eluate was evaporated to dryness *in vacuo*. The residue was chromatographed on a silicic acid column with chloroform–methanol mixtures containing up to 30% of methanol. After crystallization from chloroform–methanol 2.64 g (62%) of (rac)-1-stearoyl-2-O-benzylglycerol-3-phosphoryl-(N,N-dimethyl)ethanolamine (m.p. 163–169°) was obtained. Thin-layer chromatography in chloroform–methanol–water (65:25:4, v/v/v) showed that a pure compound was present.

(rac)-1-Stearoyl-2-O-benzylglycerol-3-phosphorylcholine. To a solution of 2.05 g of the dimethylamino compound in 21 ml of metha-

nol and 12 ml of tetrahydrofuran were added 0.71 g of methyl iodide
and 0.47 g of cyclohexylamine in 6 ml of methanol. The reaction
mixture was left at room temperature in the dark. Thin-layer chromatographic control (chloroform–methanol–water 65:25:4, v/v/v)
showed that the reaction was complete in 4 hours. Eighty milliliters
of ice-cold 0.5 N sulfuric acid were added and the mixture was extracted twice with 90 ml of chloroform. The combined chloroform
layers were extracted twice with 50% methanol and the resulting
chloroform phase was evaporated *in vacuo*. The residue, dissolved
in a mixture of 30 ml of chloroform, 80 ml of methanol, and 10 ml
of water, was treated with silver oxide (prepared from 2 g of silver
nitrate and 0.5 g of sodium hydroxide) for 15 minutes. After centrifugation the clear supernatant was evaporated *in vacuo* and the residue
chromatographed on a silicic acid column with chloroform–methanol
mixtures containing up to 60% of methanol. Crystallization from
chloroform–acetone yielded 1.3 g (60%) of a colorless solid, melting
at   225–227°.   Thin-layer   chromatography   in chloroform–methanol–water (65:25:4, v/v/v) gave one spot.

   (*rac*)-*1-Stearoylglycerol-3-phosphorylcholine*. 0.87 g of the foregoing product was hydrogenolyzed in absolute ethanol over a palladium catalyst for 2 hour at just above atmospheric pressure. After
crystallization from chloroform–acetone at −15° 0.7 g (*rac*)-1-
stearoylglycerol-3-phosphorylcholine (93%) was obtained with m.p.
246–252°; the product was chromatographically pure.

2-O-Palmitoyl-sn-glycerol-3-(2'-aminoethyl Hydrogen Phosphate)

   The synthesis of this 2-acyl-lysophosphatidylethanolamine by Billimoria and Lewis[46] illustrates the utility of the trityl protecting group
in the synthesis of lysolipids. A peculiarity of the experimental procedure
given is that detailed directions, except for the final product, are given
only for the synthesis of DL-intermediates. The physical properties of
some of the corresponding L compounds are briefly given at the end of
the appropriate section.

   It is claimed by these authors that if the removal of the trityl groups
in the final step is accomplished with refluxing 90% aqueous acetic acid
for 10 minutes rather than by hydrogenation, the pure 1-palmitoyl isomer
is obtained. If this is the case, it should be possible to employ this procedure to give an unsaturated 1-acyl lysophosphatidylethanolamine.

   3-Idodo-1-O-trityl-DL-propane-1,2-diol. 3-Iodo-DL-propane-1,2-diol
(8.1 g) was added to a stirred solution of dry pyridine (3.2 g)
in dry chloroform (10.0 ml) kept at 0°. A solution of trityl chloride
(11.2 g) in dry chloroform (50.0 ml) was added dropwise during

1 hour. After further stirring and cooling for 1 hour the solution was allowed to warm to room temperature then set aside for 3 days. It was diluted with chloroform and poured into water; the chloroform layer was washed with water until neutral (pH 7), dried (Na_2SO_4), and evaporated under reduced pressure to yield the trityl iodohydrin as a light yellow oil (17.3 g, 97.1%) which gradually solidified. Recrystallization from benzene–light petroleum (b.p. 40–60°) gave needles, m.p. 82–84°.

3-Iodo-2-O-palmitoyl-1-O-trityl-DL-propane-1,2-diol. To the crude iodide (17.3 g) dissolved in dry chloroform (50.0 ml) was added dry pyridine (3.2 g). The solution was cooled in ice, then palmitoyl chloride (11.1 g) in dry chloroform (20.0 ml) was added dropwise with stirring. The mixture was stirred for a further 1 hour at 0° and set aside for 3 days at room temperature. The chloroform layer was washed with water until neutral (pH 7), dried (Na_2SO_4), and evaporated under reduced pressure to leave an oil. The oil (from chloroform–methanol at −20°) gave the 2-palmitoyl compound as a crystalline solid (18.6 g, 70.0% that melted to an oil between 4° and room temperature; $n_D^{22.0}$ 1.54390.

3-Iodo-2-O-palmitoyl-1-O-trityl-L-propane-1,2-diol. Under conditions similar to those described for the racemic compounds, 3-iodo-L-propane-1,2-diol (4.05 g) was converted into the optically active trityl iodohydrin (8.8 g, 99.0%), m.p. 66–67°, $[\alpha]_D^{23.0}$ − 6.54°, identical on TLC with the racemic form.

2-O-Palmitoyl-1-O-trityl-DL-glycerol-3-(dibenzylphosphate). A solution of the above compound (22.0 g) in anhydrous benzene (250 ml) was stirred with silver dibenzylphosphate (13.64 g) under reflux for 4 hours. When cool, the benzene solution was filtered and evaporated to dryness to give the phosphate triester as a light yellow oil (24.1 g, 89.8%). Recrystallization from methanol at −20° gave a white solid which melted at room temperature to an oil.

2-O-Palmitoyl-1-O-trityl-DL-glycerol-3-(silver benzylphosphate). A solution of the crude phosphate triester (22.9 g) in dry acetone (300 ml) containing dry sodium iodide (4.54 g) was heated under reflux under anhydrous conditions for 4 hours while being stirred vigorously to disperse a few drops of added mercury. The cooled solution was stored at 4° overnight and subsequently filtered. The acetone mother liquors were treated with triethylamine (19.3 ml) then, after 2 hours, evaporated to dryness under reduced pressure at 40°. The residue was extracted with dry ether; the extract was evaporated to give the sodium salt as a yellow oil. This oil, dissolved in acetone (600 ml), was treated with a solution of silver nitrate

(4.68 g) in aqueous acetone (50%, 60.0 ml). The acetone was evaporated from the cloudy solution at 40° to leave a mixture of the oily silver salt and water. The silver salt was extracted with chloroform and the organic layer was dried (Na₂SO₄). Evaporation of the solution *in vacuo* gave the product as a pale yellow oil (22.1 g, 94.6%) which was used without further purification. Precipitation of a sample from ether–methanol gave a pale yellow oil that set to a vitreous solid when dried *in vacuo* at 56°.

*2-O-Palmitoyl-1-O-trityl-*DL*-glycerol-3-(Benzyl-2-tritylaminoethylphosphate).* The silver salt dissolved in anhydrous benzene (150 ml) was stirred in the dark under reflux while a solution of 2-tritylaminoethyl iodide (Note 3, Section II,R) (4.86 g) in anhydrous benzene (80.0 ml) was added dropwise during 2 hours. Stirring under reflux was continued for a further 2½ hours, after which the cooled solution was filtered and evaporated *in vacuo* to give the phosphate triester (12.0 g, 99.1%). Several precipitations of a sample of this oil from methanol at −20° gave the pure triester.

A somewhat different route was employed for the synthesis of the optically active compound. Silver benzyl-2-tritylaminoethylphosphate (Note 4, Section II,R) was prepared and reacted directly with 1-O-trityl-2-palmitoyl-*sn*-glycerol-3-iodohydrin to give 2-O-palmitoyl-1-O-trityl-*sn*-glycerol-3-benzyl-2′-tritylaminoethylphosphate.

*2-O-Palmitoyl-*L*-glycerol-1-(2′-aminoethyl Hydrogen Phosphate).* The crude fully protected phosphatide was hydrogenolyzed over 10% palladium–charcoal in absolute ethanol (100 ml). The product (from hot absolute ethanol) gave crystals of the lysophosphatide (0.088 g, 12.3%), m.p. 220° (sinters 160–180°), indistinguishable on TLC (silica gel, chloroform–methanol–water, 65:25:4) from the 1-palmitoyl isomer. The material was quite soluble in cold chloroform, in sharp contrast to the marked insolubility of the 1-acyl isomer.

R. Notes

1. 2-O-Benzylglycerol can be prepared by the benzylation of 1,3-benzylideneglycerol followed by acid hydrolysis of the protective cyclic acetal. Although the *cis* or higher melting isomer of 1,3-benzylideneglycerol has usually been employed in such reactions, it is likely that both *cis* and *trans* isomers can serve as starting materials. Briefly, one such synthesis is as follows:

1,3-Benzylideneglycerol.[95] A continuous stream of air was drawn

[95] N. Baggett, J. S. Brimacombe, A. B. Foster, M. Stacey, and D. H. Whiffen, *J. Chem. Soc., London* p. 2574 (1960).

through a mixture of benzaldehyde (200 g), glycerol (220 g), and concentrated sulfuric acid (10 drops) at 95°. Benzene (275 ml) was added and the water (25 ml, *ca.* 70%) was removed azeotropically. The cooled mixture was seeded with *cis*-1,3-*O*-benzylideneglycerol and stored at 0° for 2 days. The crystalline product was collected, dissolved in benzene and washed with aqueous ammonia, and fractionally crystallized from benzene petroleum ether to yield a slightly impure *cis* isomer (111 g, 32.5%), m.p. 70–78°, and *cis-trans* mixtures, m.p. 54–60°; The isomers could be separated by elution from activity III alumina (benzene–ether 3:2 and more polar mixtures).

2-O-Benzylglycerol.[96] Finely powdered potassium hydroxide (470 g) was added to a solution of 1,3-benzylideneglycerol (250 g) in benzyl chloride (2 liters). The mixture was stirred vigorously for 1 hour at 80–90°, cooled, washed with water, and dried. Excess benzyl chloride was removed by distillation *in vacuo*. The residue was recrystallized from benzene–petroleum ether to give 278 g of 1,3-*O*-benzylidene-2-*O*-benzylglycerol, m.p. 77–78°. This product was added to a solution of concentrated sulfuric acid (80 ml) in 1 liter of 40% aqueous ethanol, and the mixture heated under reflux for 6 hours. The mixture was then steam-distilled to remove benzaladehyde; the nonvolatile residue was extracted with ether, the ether solution dried and evaporated *in vacuo* to give 178 g of 2-*O*-benzylglycerol, m.p. 37–39°.

2. 2-Bromoethyl-*N,N*-dimethylamine hydrobromide can be prepared from dimethylaminoethanol and hydrobromic acid according to the procedure given in *Organic Syntheses.*[97] The picrate salt,[93] m.p. 143.5–145.5°, was prepared from the hydrobromide and picric acid.

Slotboom *et al.*[93] also prepared DL-1-*O*-benzyl-2-stearoylglycerol-3-phosphorylcholine monobenzyl ester, without the necessity for methyl iodide quaternization, directly from 2-bromoethyl (trimethyl) ammonium picrate and the corresponding silver salt. There seems to be no reason why this could not be accomplished equally well with the 1-stearoyl-2-benzyl isomer. 2-Bromoethyl(trimethyl)ammonium bromide is an inexpensive commercial chemical (e.g., Aldrich Chemical Co.) from which the picrate can be easily prepared.

3. 2-Tritylaminoethyl iodide is prepared[46] in two steps as follows:

2-Iodoethylamine Hydrodide. Ethanolamine (47.0 g) was added to hydriodic acid (d 1.7; 300 ml) at 0°. The mixture was slowly distilled to half its volume and left at 4° overnight; the precipitate was then filtered off. The crystals were washed free of iodine with

[96] A. J. E. Porck and B. M. Craig, *Can. J. Chem.* **33,** 1286 (1955).

[97] F. Cortese, *Org. Syn., Collect.* **2,** 91 (1946).

chloroform on the filter pad and gave plates of amine hydroiodide
(142.0 g, 61.0%), m.p. 198–199° (from ethanol–chloroform).

2-Tritylaminoethyl Iodide. To 2-iodoethylamine hydroiodide
(42.5 g) stirred under dry chloroform (60.0 ml) at 0° a solution of
trityl chloride (39.6 g) and triethylamine (28.7 g) in dry chloroform
(120 ml) was added dropwise. After 1 hour at 0°, the solution was
allowed to warm to room temperature and set aside overnight. The
mixture was filtered, diluted with chloroform (to 500 ml) and washed
with water (3 × 300 ml). The chloroform layer was dried (NaSO$_4$
and evaporated to leave an oil which gave the product (40.8 g,
69.5%). Recrystallization from nitromethane (80°) gave pale yellow
microneedles, m.p. 98–100°.

4. Silver benzyl 2-tritylaminoethyl phosphate is prepared in two steps
from 2-trimethylaminoethyl iodide in two steps[46]:

Dibenzyl 2-Tritylaminoethyl Phosphate. Dry silver dibenzylphosphate (14.5 g) was stirred in boiling anhydrous benzene in the
dark while 2-tritylaminoethyl iodide (12.0 g) in dry benzene (60.0
ml) was added dropwise during 2 hours. The mixture was stirred
under reflux for a further 5 hours then cooled and filtered. Evaporation of the benzene solution *in vacuo* left the phosphate triester as
a pale yellow oil (17.0 g). It was dissolved in ether and light petroleum (b.p. 40–60°) was added until the solution was turbid. The solution, at room temperature, deposited a yellow oil (3.0 g) from which
the colorless supernatant was decanted. At 4° the supernatant gave
crystals (10.0 g, 61.1%) of phosphate triester which gave needles,
m.p. 50–51° (from chloroform–methanol at −20°).

Silver Benzyl 2-Tritylaminoethyl Phosphate. The phosphate triester was dissolved in dry acetone (80.0 ml) with dry barium iodide
(4.34 g), and the solution was heated under reflux for 4½ hours.
When cool it was clarified by filtration, evaporated to a small bulk,
and diluted with ether (900 ml). The yellow precipitate (10.0 g)
was filtered off and dissolved in hot ethanol (100 ml). The cooled
solution yielded crystals which were then suspended in hot ethanol
(100 ml). Colorless crystals (8.0 g, 83.3%) of the barium salt were
filtered from the cooled suspension. Recrystallization from acetone–ether gave crystals, m.p. 148° (sintering at 144°). Equal portions
(3.6 g) of the barium salt were dissolved in separate portions of hot
acetone (1200 ml). To each cooled solution was added, with swirling
and in the dark, a cooled aqueous solution of silver sulfate [1.04
g in water (160 ml)]. The solutions were set aside in the dark for
45 minutes then filtered. The combined filtrates were evaporated at
40° under reduced pressure until amost all the acetone had been re-

moved. The remaining heterogeneous mixture was stored at 4° for 1 hour and the pure silver salt (5.4 g, 70.0%) was then filtered off; it gave needles, m.p. 192–194° from benzene–light petroleum (b.p. 40–60°).

5. *Related compounds.* Optical and positional isomers of lysolecithins were synthesized by de Haas and van Deenen.[98] An interesting synthesis of a D-α-lysolecithin via 1,2,5,6-D-mannitol tetrabenzyl ether was reported by Eibl *et al.*[99] A synthesis of DL-α- and β-ether lysolecithins was reported by Arnold *et al.*,[100] while O-methylated and acetylated lysolecithins were prepared by Weltzien and Westphal.[101]

S. Synthesis of Sphingomyelin

Unlike the cases of the previously discussed phospholipids, the synthesis of sphingomyelin is almost entirely the work of one laboratory, that of Shapiro and co-workers. These investigators prepared spingomyelin in racemic and D and L forms. Since natural sphingomyelin has the D-*erythro* configuration, emphasis was placed on this form, but L and *threo* compounds were prepared as well with essentially no alteration of the synthetic route. Although sphingomyelins containing no unsaturated fatty acid amides were reported, there seems little reason to doubt that these could be prepared by the same procedure.

Since the synthetic route employed is rather complex, it is given in schematic form (Fig. 2). The reader is referred to the original papers (see below) for additional discussion about the individual steps in the reaction sequence; only the experimental details are given below. Many of these details are given in the authors' work in terms of a family of compounds, but where possible the N-palmitoyl compound will be described as the model for the group.

*trans-erythro-*D-*Sphingomyelin*

The synthesis of the first six intermediates in the sequence was reported by Shapiro *et al.*[102]:

      trans-2-Hexadecenoic Acid. Forty-three grams of myristaldehyde (Note 1, Section II,T) were added to a solution of 24 g of malonic

[98] G. H. de Haas and L. L. M. van Deenen, *Biochim. Biophys. Acta* **106**, 315 (1965).

[99] H. Eibl, O. Westphal, H. van den Bosch, and L. L. M. van Deenen, *Justus Liebigs Ann. Chem.* **738**, 161 (1970).

[100] D. Arnold, H. U. Weltzien, and O. Wesphal, *Justus Liebigs Ann. Chem.* **709**, 234 (1967).

[101] H. A. Weltzien and O. Westphal, *Justus Liebigs Ann. Chem.* **709**, 240 (1967).

[102] D. Shapiro, H. Segal, and H. M. Flowers, *J. Amer. Chem. Soc.* **80**, 1194 (1958).

acid in 50 cc of pyridine at a temperature not exceeding 35°. After addition of 1.5 cc of piperidine the mixture was warmed for 1 hour at 50–55° and for 3 hours at 80–90°. The reaction product was poured into ice water, 50 cc of concentrated hydrochloric acid were added, and the mixture was extracted with ether. The oily residue was taken up with 75 cc of petroleum ether (60–80°) and cooled overnight. Yield: 35 g, m.p. 48–49°.

trans-2-Hexadecenoyl Chloride. To a warm solution of 57 g of hexadecenoic acid in 90 cc of dry petroleum ether were added with stirring, over a period of 15 minutes, 40 cc of thionyl chloride (purified by distillation over beeswax). Refluxing was continued for 4 hours. After distilling off the solvent and the excess of thionyl chloride *in vacuo,* the latter was removed as completely as possible by distilling the residue twice with 50 cc portions of petroleum ether; b.p. 145–148° (0.05 mm). Yield: 55 g. The material was redistilled for further use. The chloride distilled at 145–148° (0.05 mm) as a slightly yellow liquid, n_D^{25} 1.4644.

Ethyl-2-acetyl-3-oxo-4-octadecenoate. To a suspension of 8.46 g of powdered sodium in 1250 cc of ether were added 56 g of ethyl acetoacetate and the mixture was stirred for about 4 hours at room temperature until all sodium had reacted. The suspension was then cooled to 5° and 94 g of pure hexadecenoyl chloride were added with stirring during 2–3 minutes (Note 2, Section II,T). Stirring was continued at room temperature for 16–18 hours. The slightly turbid solution was then poured into water and enough ice to maintain the temperature at about +5°, and 125 cc of 20% sulfuric acid were added. The ether layer was washed to neutral, dried, and evaporated. The remaining oil was taken up with 2 volumes of ethanol, cooled while scratching with a rod until crystallization set in, and left overnight in the refrigerator. The ester was filtered, washed with a little cold ethanol and recrystallized from 800 cc of ethanol. Yield: 80 g, m.p. 34–35.5°.

Ethyl-2,3-dioxo-4-octadecenoate-2-phenylhydrazone. A standard diazo solution was prepared from 28 g of aniline, 100 cc of hydrochloric acid (specific gravity 1.19), 330 cc of water, and a solution of 21.5 g of sodium nitrite in 40 cc of water. This solution amounted to 500 cc and was partly neutralized with 36 g of sodium carbonate (dissolved in 360 cc of water) at −5° before use.

A solution of 14.4 g of the above ester in 1 liter of ethanol was cooled to 12°, then 32 cc of sodium acetate solution (45 g of sodium acetate in 50 cc of water) and 20 g of ammonium chloride were added, followed immediately by 120 cc of the neutralized diazo solu-

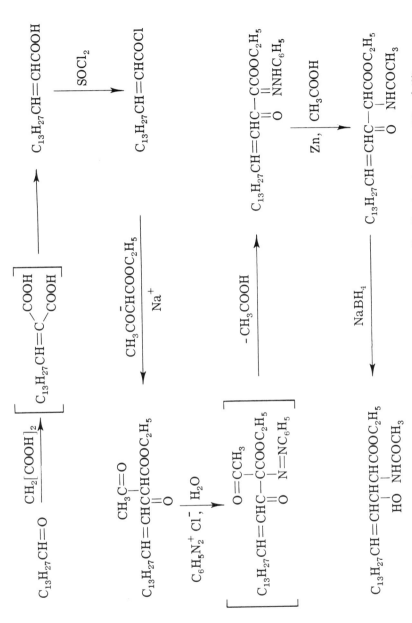

Fig. 2. Scheme for synthesis of sphingomyelin (see text for details). Continued on pages 496 and 497.

$$C_{13}H_{27}CH=CHCHCHCOOC_2H_5$$
$$\text{HO}\quad \text{NHCOCH}_3$$

$$\xrightarrow[\substack{C_2H_5O-C-C_6H_5 \\ \parallel \\ NH}]{HCl-C_2H_5OH;}}$$

$$C_{13}H_{27}CH=CHCH-CHCH_2OH$$
$$O-C=N-C_6H_5$$

$$\xrightarrow{LiAlH_4}$$

$$C_{13}H_{27}CH=CHCH-CHCH_2OH$$
$$O-C=N-C_6H_5$$

$$\xrightarrow[]{H_2SO_4,}$$

$$C_{13}H_{27}CH=CHC-CHCH_2OH$$
$$H\quad NH_2$$
$$O=C-C_6H_5$$

$$\xrightarrow[\substack{\text{then} \\ ROCl, NaOCCH_3}]{\text{resolve via} \\ \text{tartrate salts;}}$$

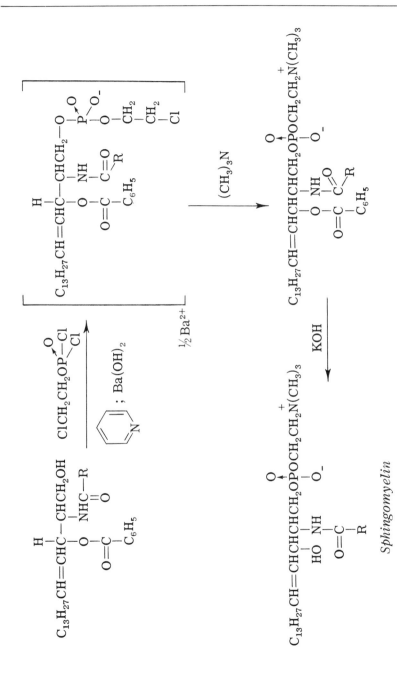

Fig. 2. For legend see page 495.

tion. The latter was added during 2–3 minutes with vigorous stirring. After 30 minutes, 100 cc of ether were added in a thin stream. A flocculent yellow precipitate was soon formed, and vigorous stirring was continued for 1 hour at 8–10°. After cooling overnight in a refrigerator, the product was filtered, washed with cold 70% ethanol, then with cold water and dried *in vacuo* over calcium chloride. Yield: 14–15 g, m.p. 37–39° (Note 3, Section II,T). The phenylhydrazone thus prepared, in most cases, was pure enough for the next reaction. It can be crystallized from petroleum ether. The product is not stable and should be kept in the cold.

DL-*Ethyl-2-acetamido-3-oxo-4-octadecenoate*. A solution of 11.25 g of the phenylhydrazone in 100 cc of glacial acetic acid was added dropwise, with good stirring, over a period of 75–90 minutes to a suspension of 15 g of zinc powder in 60 cc of glacial acetic acid and 25 cc of acetic anhydride, the temperature being maintained at 20–22°. After the addition, stirring was continued until the yellow color of the solution had completely disappeared. The zinc was filtered off, washed with glacial acetic acid, and the filtrate poured into ice water. The oily precipitate solidified after shaking for a few minutes. The product was collected, washed with cold water, and dried over calcium chloride. Yield: 9.3 g. After recrystallization from methanol, it melted at 63–65°.

DL-*Ethyl-erythro-2-acetamido-3-hydroxy-4-octadecenoate*. A solution of 10 g of the acetamido ester in 400 cc of methanol was treated at 10–15° with a solution of 0.5 g of sodium borohydride in 10 cc of water to which 4 drops of *N* sodium hydroxide solution had been added. The solution was kept at this temperature for 30 minutes, poured into a mixture of 300 cc of ice water and 300 cc of saturated sodium chloride solution, and extracted twice with ether. After washing several times with water to remove the methanol, the ether was dried and evaporated *in vacuo*. The oily residue was recrystallized twice or thrice from 8–10 volumes of petroleum ether (60–80°), the first time in the cold, then preferably at 27–30°. Yield of the *erythro* isomer 4.8 g, m.p. 64–67°.

The following two intermediates were prepared by Shapiro *et al.*[103]:

Ethyl-erythro-2-amino-3-hydroxy-4-octadecenoate. A solution of DL-ethyl-*erythro*-2-acetamido-3-hydroxy-4-octadecenoate (38 g) in 15% absolute alcoholic hydrochloric acid (300 ml) was boiled under reflux for 1½ hours. After evaporation of the solvent, excess of dry ether was added to the cooled paste, the precipitate filtered and

[103] D. Shapiro, H. M. Flowers, and S. Spector-Shefer, *J. Amer. Chem. Soc.* **81**, 4360 (1959).

washed thoroughly with dry ether. A single recrystallization from ethyl acetate gave 20.1 g of m.p. 110–112°, identical with an authentic sample. The ester hydrochloride thus obtained was hand stirred with a mixture of ether and an excess of 10% sodium carbonate solution to which a little methanol had been added. After a few minutes the liberated amino ester dissolved completely in the ether layer which was washed and concentrated to a low melting solid. When recrystallized from n-hexane, it gave 13.2 g (41%) of m.p. 63–65°. A second crystallization raised the m.p. to 64–65°

cis-2-Phenyl-4-hydroxymethyl-5-(1'-pentadecenyl)-2-oxazoline. A solution of the above ester (13 g) and ethyl iminobenzolate hydrochloride (Note 4, Section II,T) (9 g) in dry chloroform (100 ml) was boiled under reflux for 3 hours. After a few minutes, ammonium chloride began to precipitate. To the cooled mixture, dry ether (200 ml) was added, and the precipitate removed by filtration. The filtrate was evaporated and the crude ester reduced as follows[104]: A stirred suspension of lithium aluminum hydride (2.6 g) in dry ether (260 ml) was boiled for 15 minutes and cooled in ice. To this was added, dropwise, over a period of 20 minutes, a solution of the crude ester in ether (220 ml). The mixture was stirred for a further 20 minutes at room temperature (20°), then boiled for 10 minutes. After cooling at 0°, ethyl acetate (10 ml) was added, dropwise, followed by the rapid addition of N-hydrochloric acid (130 ml) and 25% acetic acid (87.5 ml). The bulky white precipitate was dissolved by adding more ether and a little chloroform and methanol to facilitate separation. The organic layer was washed successively with ice water, sodium carbonate solution until pH 9, and again with water, and concentrated in vacuo to a white solid. Crystallization from five parts of ethyl acetate gave 8.7 g (59%) of product m.p. 98–99°, which remained constant on further crystallization.

Hydrolysis of the oxazoline ring and resolution of the 3-O-benzyl-DL-sphingosine thus obtained was reported by Shapiro and Flowers,[105] who also acylated the product to a 3-O-benzyl-D-ceramide:

3-O-Benzoyl-DL-sphingosine Sulfate. A solution of the substituted oxazoline (10 g) in tetrahydrofuran (150 ml) and 3 N sulfuric acid (30 ml) was allowed to stand at room temperature for 15–18 hours. The tetrahydrofuran solution was diluted with ice water. On cooling the mixture for a short time, the sulfate separated quantitatively. It was washed thoroughly with water and cold dilute methanol, dried

[104] D. Shapiro, H. M. Flowers, and S. Spector-Shefer, J. Amer. Chem. Soc. 81, 3743 (1959).
[105] D. Shapiro and H. M. Flowers, J. Amer. Chem. Soc. 83, 3327 (1961).

and recrystallized from 70 parts of 90% methanol. Yield: 10.3 g, m.p. 148–149°.

Resolution of 3-O-Benzoylsphingosine. A warm solution of the DL-sulfate (7.5 g) in 50% tetrahydrofuran (160 ml) was added with light swirling to a solution prepared from L-tartaric acid (2.5 g), distilled water (95 ml), and 0.38 N barium hydroxide solution (45 ml). The mixture was warmed to boiling, and hot ethanol (800 ml) was added. The warm suspension was filtered immediately by gravity and the filtrate was evaporated *in vacuo.* The wet product was taken up several times with small portions of isopropyl alcohol which was distilled off to remove the excess water. The dry salt thus obtained was recrystallized 2–3 times from ethanol (280 ml) at room temperature and yielded 2 g of mp 121–122° and $[\alpha]_D^{26}$ $-14°$ (*c* 1.2, in methanol). Melting point and specific rotation remained constant on further crystallization. To obtain the D isomer, 3 N sulfuric acid (10 ml) was added to the filtrate, and the sulfate which precipitated by addition of ice water was filtered and washed thoroughly with water. The dry product was converted to the D-tartrate as described above, and the salt was recrystallized 2–3 times from ethanol (200–240 ml depending on the room temperature). Yield: 1.6 g of m.p. 125–127°, $[\alpha]_D^{25}$ $+16.9°$ (*c* 1.3, in methanol).

3-O-Benzoyl-D-ceramides. The tartrate was dissolved in tetrahydrofuran (15 volumes); 1 N acetic acid (5 volumes) was added. To the rapidly stirred mixture were added simultaneously, during 30–40 minutes, 50% sodium acetate solution (200 ml) and an equivalent amount of the acid chloride dissolved in 5–10 volumes of dry ether. After stirring for 2–3 hours, the ethereal extract was treated with sodium bicarbonate, washed, dried, and concentrated *in vacuo.* The residue was recrystallized from methanol and, if necessary, a second time from *n*-hexane.

D-*erythro-N*-Palmitoyl-3-*O*-benzoyl-D-sphingosine had m.p. 88–99°, $[\alpha]_D^{25}$ $+13.0$ (*c* 1.05 in chloroform).

The introduction of the phosphorus function was reported by Shapiro and Flowers[106] by using the general procedure of the Hirt and Berchtold[36] (see Section II,C):

β-Chloroethylphosphates of 3-O-Benzoylceramides. To a stirred solution of β-chloroethylphosphoryl dichloride (Note 5, Section II, T) (0.0066 mole) in dry chloroform (10 ml) cooled to $-10°$ was added dropwise dry pyridine (0.0066 mole). After the addition of the benzyol ceramide (0.0066 mole) dissolved in dry chloroform (20

[106] D. Shapiro and H. M. Flowers, *J. Amer. Chem. Soc.* **84**, 1047 (1962).

ml) had been completed in 5–10 minutes, the temperature was maintained at 0 to +5° for 4–5 hours. The clear reaction mixture was transferred to a vigorously stirred solution of 5% barium hydroxide (in slight excess), and the temperature was allowed to rise slowly to 20°. After 30 minutes, the mixture was recooled to 10° and cold ether was added slowly. Stirring was continued for 1 hour at a final temperature of 20–25°. The upper layer was separated, washed several times with water, and left overnight in the refrigerator. The bulky precipitate which separated was washed with a little cold acetone and ether and air-dried. It was found advantageous to cool the solution to −10° for about 30 minutes before filtering off the precipitated barium salt. The barium salts isolated were decomposed by shaking with a mixture of ether and dilute hydrochloric acid, and the residue obtained on evaporation of the ether solution was crystallized from methanol at room temperature, giving 50–60% yields of the β-chloroethylphosphates.

D-*erthythro*-N-Palmitoyl-3-O-benzoyl-ceramide-1-O-(2′-chloroethylphosphoric acid) had a m.p. of 76–78° and $[\alpha]_D^{25}$ +3.6° (chloroform).

Quaternization and removal of the protective benzoyl group was performed according to the procedure of Shapiro et al.[103]:

D-*erythro*-N-*Palmitoyl-sphingomyelin*. The barium salt of the above chloroethylphosphate (1.6 g) in dry benzene (6 ml) and trimethylamine (5 ml) was warmed in a sealed tube at 60° for 4 days. The solvent was evaporated *in vacuo*, the residue dissolved in warm methanol (40 ml), and the solution allowed to stand at room temperature for 2 hours. A small precipitate was removed, and the filtrate treated for 4 hours with 2 N sodium hydroxide solution (2 ml). To the gelatinous mass was added N methanolic hydrochloric acid (5 ml) followed by 2 N aqueous hydrochloric acid (5 ml) and acetone (100 ml). After cooling for 30 minutes, the precipitate was filtered and washed with a mixture of equal volumes of 70% methanol and acetone. To completely remove the barium ions, it was redissolved in methanol (30 ml) and treated again with hydrochloric acid and acetone. The product thus obtained weighed 0.9 g after drying over phosphorus pentoxide. It was dissolved in methanol (100 ml), the solution filtered and, after addition of distilled water (5 ml), passed over a column of Amberlite IRA-45. After evaporation of the solvent, the dry residue (0.7–0.8 g) was recrystallized first from methanol and acetone and then from butyl acetate giving a slightly hygroscopic powder of m.p. 215–217°, $[\alpha]_D^{25}$ +6.1 (1:1 chloroform–methanol).

T. Notes

1. Baer and Sarma,[107] who repeated the entire synthesis up to the preparation of the 3-*O*-benzoyl-D-ceramide, have made several modifications and/or corrections in the experimental details. The following is one significant point:

Myristaldehyde, available commercially (e.g., Aldrich Chemical Co.), is mainly trimer and must be depolymerized in order to undergo the initial condensation with malonic acid. The best way of accomplishing this is by distillation before use at atmospheric pressure from a short-necked distilling flask connected to an air condenser, using an open flame for heating.

2. Baer and Sarma[107] made the following modification: The ethereal solution of *trans*-octadecenoyl chloride must be added slowly to the ethereal suspension of sodioacetoacetic ester over 30 minutes at 0°. Before recrystallization of the ethyl 2-acetyl-3-oxo-4-octadecenoate, it should be freed from ethyl acetoacetate and volatile byproducts by keeping it for 3 hours at 50° and 0.01 mm.

3. Baer and Sarma[107] recommended that the ethyl-2,3-dioxo-4-octadecenoate-2-phenylhydrazone be filtered off as rapidly as possible, 30 minutes after addition of ether at the end of the reaction to avoid isomerization of the phenylhydrazone to ethyl-1,2,3,4-tetrahydro-1-phenyl-6-tridocylpyrazin-4-one-3-carboxylate.

4. Ethyl iminobenzoate hydrochloride (ethyl benzimidate hydrochloride), m.p. 119–120° is available from Pfaltz and Bauer, Flushing, N.Y. It may also be prepared by passing dry hydrogen chloride into a solution of benzonitrile in absolute ethanol with cooling.

5. 2-Chloroethylphosphoryl dichloride may be prepard from 2-chloroethanol by the method used for the 2-bromoethyl compound (Section II,C). Undoubtedly 2-bromoethylphosphoryl dichloride can be used with at least equal success in this reaction, but no physical constants are available for the intermediates thus prepared.

6. *Related compounds*. The L- and DL-*erythro*-sphingomyelins were prepared by Shapiro and Flowers[106] along with a few *threo* isomers. The closely related dihydrosphingomyelins were also synthesized.[104]

III. Synthesis of Phospholipid Analogues Containing C–P Bonds

This section deals primarily with phospholipid analogues containing the phosphonic acid $(-C-P(O)(OH)O-)$ group. Synthetic interest in these compounds has developed within the past decade from three major

---

[107] E. Baer and G. R. Sarma, *Can. J. Biochem.* **47**, 609 (1969).

directions: (1) the discovery of certain members of this class as natural constituents from biologic sources, (2) their possible use as inhibitors of enzymes which utilize phospholipids as substrates, and (3) their potential use as analogues of phospholipids in physicochemical and model biologic systems.

The synthesis of phosphonic acid analogues has not, however, received the large amount of attention accorded the phosphate phospholipids, which has resulted in gradual refinements in the synthesis of the latter compounds so that a variety of specific structural features (optical activity, ethers, unsaturation, etc.) can be included as desired. Thus, for many types of phosphonate lipids, syntheses have not as yet been developed which are readily adaptable to the preparation of compounds containing certain of these special structural features.

Since, with the exception of phosphatidic acids, phospholipids are phosphodiesters, their phosphonate analogues may have a C–P linkage on either the base or "glycerol" side of the molecule. Only substances of the former type have so far been found in nature, but substances of the latter type often show more interesting biochemical properties.

A. Synthesis of C–P Analogues of Phosphatidic Acids

1. Diester Phosphonate Analogues

The C–P bond in the phosphatidic acid analogues must, of course, be on the "glycerol" chain. The only syntheses so far reported of the diester phosphonic acid analogues of phosphatidic acids involve the acylation of 2,3-dihydroxypropylphosphonic acid according to the procedure described by Lapidot et al.[16] for the synthesis of phosphatidic acids. The initial problem is, therefore, the synthesis of the requisite dihydroxypropylphosphonic acid.

In the procedure of Baer and Basu[108] the optically active forms of the acid are prepared by Arbuzov reaction of the corresponding isopropylidene glycerol-α-iodohydrin with triethyl phosphite followed by acid hydrolysis.

Barium L-Dihydroxypropylphosphonate

Acetone L-*Dihydroxypropylphosphonic Acid Diethyl Ester.* Into a two-necked 500-ml round flask equipped with a Liebig condenser and gas-inlet tube were placed 24.2 g (0.10 mole) of acetone α-iodo-L-propylene glycol and 100 g (0.6 mole) of triethyl phosphite, and

[103] E. Baer and H. Basu, *Can. J. Biochem.* **47**, 955 (1969).

the mixture was kept for 24 hours at 120–125° (oil bath temperature). To facilitate the removal of ethyl iodide, nitrogen was passed
through the apparatus and the temperature of the condenser water
was kept at 55° ($\pm 5°$). The reaction mixture was distilled at a
pressure of 18–13 mm Hg, collecting the fraction boiling from
135–155° at 13 mm Hg (bath temperature 170°). This fraction on
redistillation gave 17.2 g (68.2% of theory) of pure acetone L-dihydroxypropylphosphonic acid diethyl ester; b.p. 143–144° (9–10 mm
Hg), n_D^{28} 1.4330, $[\alpha]_D^{25}$ +20.5° in anhydrous ethanol–free chloroform (c 6.2).

L-*Dihydroxypropylphosphonic Acid Diethyl Ester.* A mixture of
5.04 g (0.02 mole) of acetone L-dihydroxypropylphosphonic acid
diethyl ester and 100 ml of 0.05 N sulfuric acid was stirred vigorously at room temperature for 24 hours. At the end of this period,
50 ml of distilled water and 75 ml of Rexin AG5 (OH⁻ form) were
added, and the stirring was continued for 30 minutes. The resin was
filtered off and washed with 150 ml of water, and the combined filtrates were evaporated under reduced pressure at a bath temperature
of 35–40°. The remaining material, a viscous oil, was dissolved in
50 ml of anhydrous ethanol, the solution was filtered if necessary,
and the filtrate was evaporated under reduced pressure at 35–40°.
The residue (weighing 4.11 g) on distilling at low pressure gave 2.70
g (63.7% of theory) of L-dihydroxypropylphosphonic acid diethyl
ester; b.p. 123–124° at 0.05 mm Hg, n_D^{22} 1.4538, $[\alpha]^{25}$ −12.2° in
anhydrous ethanol (c 4.1).

Barium L-*Dihydroxypropylphosphonate.* A solution of 5.04 g
(20.0 mmoles) of acetone L-dihydroxypropylphosphonic acid diethyl
ester in 120 ml of 2.0 N sulfuric acid was boiled under reflux (temperature of the boiling mixture 102°) for 48 hours. After cooling the
solution to room temperature and adding 120 ml of distilled water
and 45 g of barium carbonate, the mixture was stirred vigorously
for 10 minutes. It was then made slightly alkaline to phenolphthalein
by dropwise addition of an aqueous solution of barium hydroxide.
The excess of barium hydroxide was neutralized with gaseous carbon
dioxide. The insoluble barium salts were removed by centrifugation,
and the supernatant solution was concentrated under reduced pressure and a bath temperature of 35–40° to a volume of 45 ml. The
concentrate on standing at room temperature (23°) spontaneously
formed a crystalline precipitate. After standing overnight, the mother
liquor was decanted and the precipitate was washed with 30 ml of
99% ethanol. The barium salt on drying over phosphorus pentoxide
at the temperature of boiling xylene and at a pressure of 0.01 mm

Hg weighed 3.72 g. Addition of the wash-alcohol to the mother liquor and drying of the precipitate as described above gave further 0.54 g of barium salt. For further purification, the barium salt (4.26 g) was suspended with stirring in 90 ml of distilled water for 2 hours and the insoluble material was removed by centrifugation. The aqueous solution was concentrated under reduced pressure to a volume of 20 ml, and the barium salt was precipitated by the addition of 10 ml of 99% ethanol. It was dried for 4 hours over phosphorus pentoxide at the temperature of boiling xylene and a pressure of 0.01 mm Hg. The anhydrous barium salt of L-dihydroxypropyl-phosphonic acid weighed 3.33 g (57.2% of theoretical yield); $[\alpha]_D$ —3.7° in 2 N hydrochloric acid (c 4.4).

If only the racemic form of the acid is required, a different and much simpler procedure can be employed[109]:

Dilithium 2,3-Dihydroxypropylphosphonate. To 35.6 g (0.200 mole) of diethyl allylphosphonate (Note 1, Section III,B) were added 1 ml of 88% formic acid and 35 ml of 30% hydrogen peroxide. The mixture was warmed to 40° for a few minutes until its temperature rose above 42°. It was removed from the warming bath and intermittently immersed in a bath at 25° to keep the temperature at 40–45°. This exothermic phase lasted approximately 2 hours. The reaction was completed by keeping the mixture at 40–45° overnight. The liquid was evaporated to a clear, colorless syrup (bath temperature, 70–80°; water pump) which was then heated on a steambath for 30 minutes. Hot, saturated aqueous lithium hydroxide was added slowly until a few drops of 1% phenolphthalein became red. Twenty milliliters more of saturated lithium hydroxide were added, and after a few minutes a white precipitate began to form. The mixture was next heated in a steam autoclave at 120° for 5 hours at the end of which time it became quite solidified. Enough water to make a total of 250 ml was added and the mixture heated to boiling and filtered hot. Forty milliliters more of hot saturated lithium hydroxide were added and the liquid returned to the autoclave at 120° for 7.5 hours.

The precipitate on the funnel was washed 3 times with 95% ethanol, then with ether and air dried. A second crop was isolated in the same way from the autoclaved filtrate. The combined yield of analytically pure dilithium-2,3-dihydroxypropylphosphonate was 19.0 g (62.5%). The substance is a fluffy, white, nonhygroscopic powder. It is readily soluble in water, but like the lithium, calcium, and barium salts of the corresponding α-glycerophosphoric acid, it is less soluble in hot than in cold water.

[109] A. F. Rosenthal and R. P. Geyer, *J. Amer. Chem. Soc.* **80**, 5240 (1958).

The acylation of the phosphonic acid (as the pyridinium salt) described by Baer and Basu[110] is so similar to the original acylation of glycerophosphate[16] above that it is given only in summary form below:

Dipalmitoyl-L-dihydroxypropylphosphonic Acid. The acylation of the pyridinium salt of L-dihydroxypropylphosphonic acid by palmitic anhydride was carried out as in the procedure above of Lapidot *et al.*[16] The material was converted in chloroform–methanol–water (130:50:8) to the acid form by stirring with Amberlite IR-120 (H[+]). Pyridine was added, the solution was kept 16 hours at room temperature, and the solution again treated with an excess of Amberlite IR-120 (H[+]). After evaporation of the filtrate to dryness, the residue was dissolved in chloroform and chromatographed on a silicic acid column, which was washed initially with chloroform. The product was eluted with chloroform–methanol (9:1), the eluate evaporated, and the product crystallized from absolute ethanol. Yield: 25%, m.p. 83–84° (slight sintering at 81°); $[\alpha]_D$ +1.4° in chloroform–methanol (4:1, *c* 4.6).

2. Diether Phosphonate Analogues of Phosphatidic Acids

Dihydroxypropylphosphonic acid, because it contains ionizable groups, cannot be directly alkylated by the KOH-mesylate (or halide) procedure. No procedure has yet been described for the synthesis of optically active or unsaturated compounds of this type, but the synthesis described below[25] is readily adapted to the preparation of "mixed-alkyl" derivatives.

In this synthesis, an allyl alkyl ether is used as the starting material. The second ether group and primary halide are added together in one step by a modified hypohalite addition in the presence of zinc oxide (Note 2, Section II,B). Reaction with triethyl phosphite followed by hydrolysis of the phosphonate ester groups yields the desired diether phosphatidic acid analog. The compound whose synthesis is described below is a good inhibitor of phosphatidate phosphatase[111]:

rac-2-Hexadecoxy-3-octadecoxypropylphosphonic Acid

2-Hexadecyloxy-3-octadecyloxyiodopropane. Freshly prepared allyl octadecyl ether (Note 3, Section, III,B) b.p. 157–158° at 0.45 mm (80.0 g, 0.258 mole) was dissolved in 750 ml of anhydrous 1,2-dimethoxyethane. Iodine (150 g), zinc oxide (60 g), and cetyl alcohol (150 g) were added and the mixture stirred at 50–55° for 24 hours.

[110] E. Baer and H. Basu, *Can. J. Biochem.* **48**, 1010 (1970).
[111] A. F. Rosenthal and M. Pousada, *Biochim. Biophys. Acta* **125**, 265 (1966).

Most of the solvent was evaporated *in vacuo* and the viscous residue was taken up in hexane and washed with 10% sodium thiosulfate until both layers were colorless, then with water and sodium sulfate solution. The hexane layer was dried with magnesium sulfate and evaporated. The residual oil was treated with an equal volume of ether and cooled to 20–22°. Fifteen hundred milliliters of anhydrous methanol at the same temperature were added slowly; the granular white product precipitated at once. It was filtered off with a minimum of cooling and washed with methanol (20–22°). The yield was 132.1 g (75.5%) m.p. 27–28.5°. After two recrystallizations from ethyl acetate–acetone the m.p. was 28–29°. The product is readily soluble in nonpolar solvents but poorly soluble in acetone, ethanol, and dimethylformamide.

rac-2-Hexadecoxy-3-octadecoxypropylphosphonic Acid. 2-Hexadecyloxy-3-octadecyloxyiodopropane (30.0 g, 0.0442 mole) dissolved in 250 ml of triethyl phosphite was heated in a static nitrogen atmosphere at 120° in an apparatus which allowed escape of ethyl iodide as formed. After 40 hours no iodine remained in the reaction mixture. The cooled reaction mixture was poured into 5% aqueous hydrochloric acid (1.5 liters) and extracted with hexane. The hexane extract was washed twice with 5% hydrochloric acid, then with 10% ammonium sulfate solution and dried over magnesium sulfate. After evaporation of the solvent the residual oil was triturated with cold acetonitrile (150 ml) and kept overnight at 0–5°. After filtration the yellowish solid was washed with ice-cold acetonitrile and air-dried. The weight of crude product at this point was 28 g. The crude product was dissolved in about 500 ml of warm propionic acid and 8 *M* hydrochloric acid (75 ml) added. The mixture was heated under reflux for 16 hours. As much liquid as possible was removed *in vacuo* and the residual material was extracted with warm chloroform and water. The chloroform layer was dried over MgSO$_4$ and evaporated. The product was crystallized from hexane. The fine white granular product had a m.p. of 74.5–75.5°; yields from the iodo-diether are about 40%.

B. Notes

1. Diethyl allylphosphonate[112] b.p.$_{10}$ 58–60°, is prepared in excellent yield by heating allyl bromide with an equivalent amount of triethyl phosphite in a bath at 160° overnight and distilling the product.

[112] A. E. Arbuzov and A. I. Razumov, *Izv. Akad. Nauk SSSR, Otd. Khim. Nauk* p. 714 (1951).

2. Mercuric oxide (a more common reagent for hypohalite additions) and cadmium oxide are equally effective in this reaction but offer no advantages at all over the cheaper and less toxic zinc oxide. This is one of the very few hypohalite additions known in which a long-chain alcohol is one of the reagents used. Presumably, beginning with an allyl *ester* a mixed 2-ether 3-ester propylphosphonate could be prepared, provided special techniques (e.g., Rabinowitz[113]) are used to remove the phosphonate ester groups without affecting the carboxylic ester. This approach has not yet been described in practice, however.

3. Allyl octadecyl ether is easily prepared by a modification of the method of Stegerhoek and Verkade[114] as follows: To a solution of 30 g of sodium in 350 ml of anhydrous allyl alcohol was added 1-bromo-octadecane (120 g) and the mixture was stirred at 55–60° for 20 hours. The excess allyl alcohol was removed *in vacuo* and the mixture diluted with ether (500 ml) and extracted with water several times. The ether solution was dried and evaporated, and the residue distilled. The product, b.p.$_{0.45}$ 157–158°, solidifies at room temperature. Yields are about 75–80%.

4. *Related compounds.* A medium-chain diester analogue has been reported by van Deenen and co-workers.[39] Isosteric[43] and long-short chain[25] diether analogues were reported by Rosenthal, who also described a "hydrocarbon-chain" analogue.[40] The synthesis of the isosteric analogue is described in Section III,E.

C. Synthesis of C–P Analogues of Phosphatidylethanolamine

1. *Glycerol Derivatives*

Members of this group were historically the first lipids containing C–P bonds whose hydrolysis products were found in nature, although the actual isolation of the intact phosphatidylethanolamine analogue occurred after some other C–P lipids had been isolated and characterized.

Syntheses of those analogues reported employ 1,2-diglycerides as starting materials rather than 3-iododiglycerides. It should be emphasized that there is no inherent reason why the more versatile and stable iodo compounds could not be used, but to date the requisite silver salts have not been prepared and thus no such synthesis has yet been reported.

Therefore, the initial synthetic problem is the preparation of 1,2-diglycerides. The recent procedure of Pfeiffer and co-workers[42] is a rela-

[113] R. Rabinowitz, *J. Org. Chem.* **28**, 2975 (1963).
[114] L. J. Stegerhoek and P. E. Verkade, *Rec. Trav. Chim. Pays-Bas* **75**, 143 (1956).

tively simple approach to the synthesis of saturated or unsaturated *sn*-1,2- or *sn*-2,3-diglycerides. The method of Buchnea[41] allows the preparation of mixed-acid diglycerides of either configuration, but it is too long a synthesis unless such compounds are specifically required. Although none of the synthetic approaches to phosphonate analogues of phosphatidylethanolamine has been sufficiently recent to utilize either of these improved procedures, they are to be recommended to future workers in the area. The synthesis of *sn*-1,2-diglycerides by Pfeiffer is therefore given below.

1,2-Isopropylidene-sn-glycerol-3-β,β,β-trichloroethylcarbonate. A solution of 200 g (0.943 mole) of β,β,β-trichloroethyl chloroformate (Aldrich Chemical Co.) in 100 ml of $CHCl_3$ (distilled from P_2O_5) was added dropwise to an ice-cold mixture of 124.6 g (0.942 mole) of 1,2-isopropylidene-*sn*-glycerol (see Section II,A), 50 ml of dry pyridine, and 100 ml of $CHCl_3$. The solution was stirred at room temperature for 18 hours, diluted with Et_2O (800 ml), and washed successively with dilute HCl, H_2O, 5% $NaHCO_3$, and H_2O. The dried (Na_2SO_4) organic extract was concentrated and distilled to give 252 g (87%) of the colorless, syrupy product; b.p. 140–145° (0.25 mm), $[\alpha]_D^{25}$ −1.5° (*c* 0.87, $CHCl_3$).

sn-Glycerol-3-β,β,β-trichloroethylcarbonate. A mixture of 126 g (0.41 mole) of the above product, 150 ml of Et_2O, 40 ml of MeOH, and 40 ml of 3 *N* HCl was stirred at room temperature overnight. The solvents were evaporated at 40° at H_2O aspirator pressure, the residue was extracted with EtOAc, and the organic phase was washed with brine 5 times. After being dried (Na_2SO_4), the solvent was evaporated and the residue was azeotroped several times with C_6H_6 at 40°. The yield was quantitative. Analysis of the crude product as the derived trifluoroacetate[27] showed a mixture of 94% of the desired product and 6% of *sn*-glycerol-2,3-carbonate. This mixture was used without further purification in the acylation reactions described below and can be stored at 0° for several months. The pure material is a colorless oil; $[\alpha]_D^{25}$ −1.31° (*c* 1.0, $CHCl_3$).

1,2-Diacyl-sn-glycerol-3-(β,β,β-trichloroethyl)carbonates. The appropriate acid chloride (0.12 mole) in $CHCl_3$ (75 ml, distilled from P_2O_5) was added dropwise with stirring to a cold (0°) solution of the crude glycerol trichloroethylcarbonate (0.06 mole), 14 ml of anhydrous pyridine, and 75 ml of $CHCl_3$. The reaction mixture was then stirred at 25° for 24 hours, diluted with Et_2O (500 ml), and washed with dilute HCl, H_2O, 5% $NaHCO_3$, and H_2O. The dried (Na_2SO_4) solution was concentrated, dissolved in a minimum amount of 8:1 Et_2O–petroleum ether (b.p. 30–60°), and applied to a column (65 × 6.5 cm) packed with 800 g of Florisil (100–200 mesh). The column was developed with increasing concentrations of Et_2O in petroleum ether, and was monitored by TLC; the desired compounds had an R_f of *ca.* 0.75–0.85 on silica gel G (0.25 mm) using a system of 3:1 cyclohexane–EtOAc. The distearoylglycerol trichloroethylcarbonate had a m.p. of 56–57°, $[\alpha]_D^{25}$ −1.7° (*c* 0.88 in $CHCl_3$) and −1.8° (*c* 1.49 in $CHCl_3$), respectively.

sn-Glycerol-1,2-diacylates. The above compounds (0.03 mole) were dissolved in a mixture of HOAc (30 ml) and Et_2O (20 ml) and cooled in an ice bath, and 25 g of activated zinc (Note 1, Section III,D) were added. The suspension was stirred at 20–25° for 1–2 hours or until the indicated complete conversion to the desired product. After dilution with 4:1 Et_2O–$CHCl_3$ (300 ml), the inorganics were filtered and the filter cake was washed with additional solvents. The filtrate was washed with H_2O 3 times, 5% $NaHCO_3$, and brine. After being dried (Na_2SO_4), the solvents were evaporated at 30°. Thin-layer chromatography of the isolated diglycerides on silica gel G impregnated with 10% H_3BO_3 with hexane–Et_2O (3:2) or $CHCl_3$–Me_2CO (96:4) showed little or no 1,3-diglyceride in the crude products. The analytical specimens were prepared by column chromatography over acid-washed Florisil containing 10% of H_3BO_3 and elution with hexane–ether mixture or by thick layer chromatography on plates coated with 1 mm of silica gel G mixed with 10–15% of H_3BO_3 and using $CHCl_3$–MeCO (96:4) as the moving phase. 1,2-Disteaoryl-*sn*-glycerol, m.p. 73–74.5°, had $[\alpha]_D^{25}$ −2.6° (*c* 1). The dioleyl and dilinoleyl glycerols, which are oils, had $[\alpha]_D^{25}$ −1.91° (*c* 6.2) and −2.2° (*c* 1). Rotations were taken in chloroform.

The preparation of the phosphatidylethanolamine analogue requires 2-phthalimidoethylphonic acid from which an acid chloride is prepared. Reaction with a 1,2-diglyceride gave the *N*-phthalyl phosphonate lipid, which is treated with hydrazine to give the phosphatidylethanolamine analogue.

The procedures of Baer and Stanacev[115] and of Rosenthal and Pousada[116] for the dipalmitoyl compound give virtually identical results. In both methods the major loss occurs in the hydrazinolysis step, probably by lyso-lipid formation. Presumably, *N*-protecting groups which can be removed by more selective methods would give higher yields. They would, however, have to prove themselves capable of withstanding the rather vigorous conditions necessary to produce the phosphonic acid

[115] E. Baer and N. Z. Stanacev, *J. Biol. Chem.* **239**, 3209 (1964).

[116] A. F. Rosenthal and M. Pousada, *Rec. Trav. Chim. Pays-Bas* **84**, 833 (1965).

moiety; it is problematical whether any of the more recent protecting groups could fulfill this condition.

The procedure of Rosenthal and Pousada[116] is given below:

2-Phthalimidoethylphosphonic Acid

2-Bromoethylphthalimide (39 g, 0.15 mole) and triethyl phosphite (26 g, 0.16 mole) were reacted together at 160° for 4½ hours and the product was extracted with 4 × 25 ml of hexane. The hexane washings were discarded, and the insoluble oily residue was thoroughly evaporated *in vacuo* to remove small amounts of hexane. The viscous oil was heated at 130–140° for 4 hours while a moderate stream of hydrogen bromide was passed through it. After the mixture had cooled isopropanol (175 ml) was added, and the mixture was heated to boiling until the solid material had almost completely dissolved. The hot solution was filtered rapidly through a celite mat which was then washed with two 35-ml portions of hot isopropanol. The filtrate and washings were again heated almost to boiling and vigorously stirred mechanically while 440 ml of hot hexane were added rapidly. The mixture was cooled to 5° overnight, filtered, and washed with cold 2:1 hexane–isopropanol and then with hexane. The product obtained by drying the precipitate *in vacuo* was almost pure. Yield: 17.2 g (44% from bromoethylphthalimide), m. 197.5–199.5°. Another hexane–isopropanol purification gave an analytically pure product of m.p. 198.5–199.5°.

1,2-Dipalmitoyl-sn-glycerol-3-(2'-aminoethyl)phosphonate

To a magnetically stirred mixture of phthalimidoethylphosphonic acid (4.55 g, 0.018 mole) and thionyl chloride (45 ml) were added 3 drops of dimethylformamide. Within a few minutes a homogeneous solution was formed; to ensure complete reaction the mixture was stirred overnight. Excess thionyl chloride was evaporated off thoroughly *in vacuo* and the residue was dissolved in a mixture of 70 ml of chloroform (freshly distilled from P_2O_5) and 4.5 ml (0.055 mole) of anhydrous pyridine.

To the vigorously stirred acid chloride solution at 5° was added a solution of 5.08 g (0.0089 mole) of D-α,β-dipalmitin ($[\alpha]_D^{25}$ −2.8°) in 53 ml of anhydrous ethanol-free chloroform dropwise over 1 hour. The mixture was stirred at 5° for 30 minutes more and at room temperature for 1 hour. To the mixture was added 450 ml of ether, the solution was cooled to 5°, and shaken with 750 ml of ice-cold 0.5 N HCl. The ether solution was washed twice with half-saturated aqueous ammonium sulfate and with 20% aqueous ethanol, dried with magnesium and sodium sulfates, and finally evaporated *in vacuo* at 40°.

The residue weighed 6.8 g; it was dissolved in warm hexane, filtered, and allowed to crystallize at 20°. The product at this point had a m.p. of 60–62° and traces of impurities. Without further purification the product was dissolved in 133 ml of 2-methoxyethanol (freshly distilled from potassium borohydride) and neutralized with 0.1 N NaOH (53.7 ml, phenolphthalein indicator). Hydrazine hydrate (99%; 0.4 ml, 0.008 mole) was added, and the mixture was stirred at 37° overnight. Ether and water were removed at the water pump (bath, 40°) as thoroughly as possible and the residue was freed of traces of volatile material in a high vacuum at 40° for 1½ hours. The residue was dissolved in 450 ml of freshly distilled tetrahydrofuran at 40° and while stirring rapidly 16.5 g of Amberlite IRC-50 were added. Stirring was continued for 30 minutes, and the solution filtered and then evaporated at 45° and dried *in vacuo* overnight; weight, 5.50 g of product.

The crude phosphono-cephalin in 200 ml of warm chloroform was applied to a 4×30 cm column of silicic acid (205 g, Mallinkrodt AR, 100 mesh) and 25 g of celite which had just been thoroughly washed with methanol and then chloroform. Chloroform (600 ml) eluted only phosphorus-free impurities; 10% methanol in chloroform (1200 ml) eluted all the unreacted starting material. The product was eluted with 25% methanol in chloroform (1000 ml); it weighed 1.90 g (35% of material put onto column or 33% from D-α,β-dipalmitin). The product was given a final purification by precipitation from warm chloroform with methanol (10 volumes) at 20° followed by crystallization from chloroform at 5°. The product after drying (48 hours) at 40° over P_2O_5 at 0.5 mm had a m.p. of 180–181° and $[\alpha]_D^{25}$ +6.30° (chloroform; c 1 mg/ml); it was chromatographically pure.

2. *Dihydroxypropylphosphonic Acid Derivatives*

The synthesis of phosphatidylethanolamine analogues of this class requires the corresponding phosphonic acid which is then condensed with an N-protected ethanolamine. Removal of the protecting group then gives the desired lipid analogue.

DL-2-*Aminoethyl 2-Octadecoxy-3-octadecoxypropylphosphonate*

The synthesis of this diether analog, described by Rosenthal and Pousada,[116] is given below:

2-Hexadecyloxy-3-octadecyloxypropylphosphonic acid (see Section III,A) (1.26 g, 0.0020 mole), dry pyridine (12 ml), 2-hydroxyethylphthalimide (0.600 g, 0.0031 mole), and trichloroacetonitrile (2.0 ml) were warmed at 65–70° for 12 hours. Most of the pyridine was removed *in vacui*, and 2 ml of trifluoroacetic acid and then 3 volumes of acetonitrile were added. The precipitated product was filtered and washed with acetonitrile. It was recrystallized from acetone at 20°. Yield: 1.35 g (84%), m.p. 50–51°. 810 mg (0.0010 mole) of the above product were dissolved in 20 ml of warm isopropanol; hydrazine hydrate (85%; 0.5 ml) was added and the mixture was warmed at 65° for an hour and then for 20 minutes on a steam bath. The copious precipitate which formed contained all the reaction products. It was filtered off, washed with isopropanol, and ground well with a solution of 10% glacial acetic acid in chloroform. The solution was filtered through a celite mat, which was then washed well with chloroform; the filtrate and washings were evaporated *in vacuo*. The residue was redissolved in a little chloroform, and the solution again was filtered and evaporated. The product was redissolved in 3 ml of chloroform, acetone (15 ml) was added, and the mixture was allowed to stand for several hours at 2°. The product, which was almost pure, was filtered off and dried *in vacuo*. Yield: 600 mg (89%), m.p. 168–171° after softening at 100° and becoming brown above 135°. The product was freed of a trace of the starting phthalimido acid by recrystallization first from chloroform–methanol and then from isopropanol–trichlorethylene at 5°. The analytically pure cephalin analogue softened at 105°, turned brown at 140°, and melted sharply at 169–170°.

D. Notes

1. Activated zinc is prepared by the method of Baer and Buchnea[117] as follows:

To 150 g of zinc dust in a 1-liter glass beaker were added 200 ml of 2 *N* hydrochloric acid, and the mixture was stirred vigorously for 1 minute at room temperature. Upon addition of 400 ml of distilled water the stirring was continued for another minute. The zinc was allowed to settle and the acid was decanted. The procedure was repeated twice more, each time a fresh 200-ml portion of 2 *N* hydrochloric acid being used. The zinc then was suspended with stirring in 500 ml of distilled water, allowed to settle, and the wash water was free from chlorine ions. The zinc was washed with two portions of 99% ethanol, followed by two por-

[117] E. Baer and D. Buchnea, *J. Biol. Chem.* **230**, 447 (1958).

tions of ether. The zinc, which loses its activity on standing or in contact with air, is prepared for immediate use only and is kept covered with ether until it is transferred to the reaction vessel.

2. *Related compounds.* (a) Glycerol derivatives: An unsaturated diester analogue was reported by Turner et al.[118] Saturated diether analogues were reported by Baer and Stanacev.[119] Saturated and unsaturated ether ester analogues were reported by Baer and Basu[120] and by Chacko and Hanahan,[121] respectively. *N*-Methyl analogues were described by Baer and Pavanaram[122] and *N,N*-dimethyl analogues by Baer and Rao.[123] Ethylene glycol analogues were synthesized by Baer and Basu.[124]

(b) Dihydroxypropylphosponic acid derivatives: An isosteric diether analogue was synthesized by Rosenthal,[40] who also reported the synthesis of a "hydrocarbon chain" analogue.[40]

E. Synthesis of Lecithin Analogues Containing C–P Bonds

1. Glycerol Derivatives

The synthesis of 1,2-diacylglycerol-3-(2′-trimethylammoniumethyl)-phosphonate by Baer and Stanacev[125] is similar in principle to the synthesis of the corresponding phosphatidylethanolamine analogue. The acid chloride of bromoethylphosphonic acid was reacted with a 1,2-diglyceride and the product was reacted with trimethylamine under anhydrous conditions to give the lecithin analogue. As in the case of phosphatidylethanolamine analogue, the synthetic route is adaptable to the preparation of unsaturated compounds, although no diesters of this type were described. A somewhat different, one-step approach by Rosenthal and Pousada[38] employs an acid chloride of 2-trimethylammoniumethylphosphonic acid, which is reacted with the diglyceride to give the lecithin analogue. Although both methods give roughly comparable yields, the procedure of Baer and Stanacev[125] is described below since it gives a more easily purified product:

2-Bromoethylphosphonic Acid Monoanilinium Salt. A solution of 39.0 g of diethyl 2-bromoethylphosphonate (Aldrich Chemical Co.)

[118] D. L. Turner, M. J. Silver, R. R. Holburn, E. Baczynski, and A. B. Brown, *Lipids* 3, 234 (1968).
[119] E. Baer and N. Z. Stanacev, *J. Biol. Chem.* 240, 44 (1965).
[120] E. Baer and H. Basu, *Can. J. Biochem.* 50, 988 (1972).
[121] G. K. Chacko and D. J. Hanahan, *Biochim. Biophys. Acta* 176, 190 (1969).
[122] E. Baer and S. K. Pavanaram, *Can. J. Biochem.* 48, 979 (1970).
[123] E. Baer and K. V. J. Rao, *Can. J. Biochem.* 48, 184 (1970).
[124] E. Baer and H. Basu, *Can. J. Biochem.* 46, 1279 (1968).
[125] E. Baer and N. Z. Stanacev, *J. Biol. Chem.* 240, 3754 (1965).

in 200 ml of concentrated hydrobromic acid (48%) was kept in an oil bath at 95–100° for 20 hours. The reaction mixture was evaporated under reduced pressure (10 mm Hg) from a bath the temperature of which was gradually raised to 60°. The remaining viscous, yellow oil was dissolved in 50 ml of 99% ethanol, and aniline was gradually added to the solution until no further precipitate was formed. The precipitate was filtered off with suction, washed on the filter with 5 ml of cold 99% ethanol, and dried at room temperature over phosphorus pentoxide in a vacuum of 10 mm Hg for several hours. This material, weighing 41.2 g (91.8% of theory), on recrystallization from 700 ml of boiling 99% ethanol gave 29.3 g (65.3% of theory) of analytically pure 2-bromoethylphosphonic acid monoanilinium salt; m.p. 150–151.5° with decomposition, with sintering at 132°. The anilinium salt is soluble at room temperature in water. acetone, and 50% aqueous ethanol, and moderately soluble in warm (50°) anhydrous ethanol.

2-Bromoethylphosphonic Acid. A solution of 50.0 g of pure monoanilinium salt of 2-bromoethylphosphonic acid in 1 liter of a warm (50°) mixture of ethanol and water (1:1, v/v) was passed 3 times through a column (32 × 340 mm) containing 300 ml of Amberlite IR-120 (H⁺ form). The column was washed with 200 ml of the same solvent mixture, and the combined effluents were evaporated to dryness under reduced pressure at a bath temperature of 40°. The solid residue was dissolved in 500 ml of boiling chloroform (reflux condenser), and the solution was set aside for crystallization at room temperature. After 4 hours, the 2-bromoethylphosphonic acid was filtered off with suction and dried at room temperature over phosphorus pentoxide in a vacuum of 0.02 mm Hg. Yield was 25.84 g (77.2% of theory); m.p. 93–95°. The substance is soluble at room temperature in water, methanol, ethanol, acetone, and benzene, moderately soluble in boiling chloroform, and insoluble in petroleum ether (b.p. 30–60°).

1,2-Dipalmitoyl-sn-Glycerol-3-(2'-trimethylammoniumethyl)phosphonate

Dipalmitoyl-L-α-glycerol-(2-bromoethyl)phosphonate. Preparation of 2-bromoethylphosphonic acid monochloride is as follows. A solution of 3.78 g (20 mmoles) of thoroughly dried 2-bromoethylphosphonic acid and 4.16 g (20 mmoles) of phosphorus pentachloride in 50 ml of anhydrous ethanol-free chloroform was prepared in a 250-ml round flask provided with a reflux condenser protected by a calcium chloride tube. It was heated for 10 minutes under gentle reflux. The solution was evaporated to dryness under *anhydrous conditions* at a pressure of about 10 mm Hg (bath temperature, 30–35°), and the residue was kept in a vacuum of 0.1 mm Hg at 50° for 30 minutes.

Phosphonylation was done as follows: The 2-bromoethylphosphonic acid monochloride was dissolved in 50 ml of anhydrous ethanol-free chloroform; to the solution, cooled in a water bath at 10°, a solution of 5.69 g (10 mmoles) of D-α,β-dipalmitin and 3.0 ml (22 moles) of anhydrous triethylamine in 50 ml of anhydrous chloroform was added under anhydrous conditions and with stirring over a period of 1 hour. The reaction mixture was allowed to stand at room temperature for 20 hours. At the end of this period, 1 ml of water was added, and the mixture was stirred vigorously for 2 hours. The solvent and excess of triethylamine were distilled off under reduced pressure at a bath temperature of 30–35°, and the residue was dissolved in a mixture of 100 ml of chloroform, 50 ml of methanol, and 5 ml of water. The solution was stirred for 2 hours with 20 ml of Amberlite IR-120 (H$^+$). The Amberlite was removed by filtration, washed with a small volume of the solvent mixture, and the combined filtrates were evaporated to dryness under reduced pressure at 30–35°. The remaining solid material was dissolved in 60 ml of boiling acetone, and the solution was kept overnight at 6°. The precipitate was filtered with suction and washed with 5 ml of cold acetone. The material was recrystallized in the same manner from 40 ml of acetone and dried over phosphorus pentoxide in a vacuum of 0.2 mm Hg at 35°. The dipalmitoyl-L-α-glycerol-(2-bromoethyl)-phosphonate, a white crystalline substance, weighed 3.88 g (52.5% of theory), m.p. 54–57°, $[\alpha]_D^{23}$ +3.84° in chloroform (*c* 10). The substance is readily soluble at room temperature in chloroform, sparingly soluble in 95% ethanol, acetone, and insoluble in water, methanol, and dimethylformamide.

Dipalmitoyl-L-α-glycerol-(2-trimethylammoniumethyl)phosphonate. A suspension of 2.9 g of dipalmitoylglyceryl bromoethylphosphonate in a mixture of 50 ml of anhydrous dimethylformamide and 15 ml of trimethylamine was kept in a pressure bottle at 50–55°. The suspension cleared on warming and after 1 hour showed the first signs of, the formation of a new precipitate. At the end of the third day, the reaction mixture was cooled to room temperature and filtered with suction. The solid material was resuspended in 50 ml of warm (50°) dimethylformamide; the suspension was filtered while still warm, and the material on the filter was washed with 15 ml of warm dimethylformamide. The crude phosphonolipid, after drying for several hours over potassium hydroxide at room temperature in a vacuum of 0.02 mm Hg, weighed 1.75 g. It was dissolved in a mixture of 25 ml of chloroform, 25 ml of methanol, and 2.5 ml of water; 5 g of silver carbonate were added to the solution, and the mixture was stirred vigorously for 1 hour. The silver salts were removed by filtration, and the filtrate was evaporated to dryness under reduced pressure at 30–35° bath temperature. The remaining material was dissolved in 20 ml of chloroform and reprecipitated by the addition of 70 ml of acetone. The recovered material was treated once more in the same manner. The precipitate, weighing 1.3 g, was dissolved in a mixture of 25 ml of chloroform, 25 ml of methanol, and 2.5 ml of water; 10 ml of Amberlite IR-120 (H$^+$ form) were added to the solution, and the mixture was stirred for 1 hour. The amberlite was removed by filtration, washed with a small amount of the solvent mixture, and the combined filtrates were evaporated to dryness under reduced pressure at 30–35°. The residue was dissolved in 20 ml of chloroform and reprecipitated by the addition of 200 ml of acetone. The pecipitate was dried over phosphorus pentoxide in a vacuum of 0.02 mm Hg at 56° (boiling acetone) for 6 hours. The analytically pure dipalmitoyl-L-α-glycerol(2-trimethyl-ammoniumethyl)phosphonate weighed 1.06 g (36.7% of theory); m.p. 195–196.5°, with sintering at 193.5°; $[\alpha]_D^{23}$ +75° in a mixture of chloroform and methanol, 3:1 (c 10), M_D +55.4°. At room temperature, the substance is fairly soluble in methanol, slightly soluble in chloroform, and sparingly soluble in acetone and 99% ethanol, and insoluble in water, ether, and petroleum ether.

2. Choline Alkylphosphonate Derivatives

Phosphonate lecithins containing the C–P group on the alkyl moiety analogous to glycerol have been synthesized by condensation of a choline salt with the parent phosphonic acid. Most of the reported compounds of this class have been designed for maximum hydrolytic stability and are thus of the diether or hydrocarbon type.

The isosteric diether lecithin analogue whose synthesis is described below is a powerful inhibitor of C. perfringens phospholipase C.[126] The synthesis of the requisite dihydroxybutylphosphonic acid diether reported by Rosenthal[43] was accomplished by a straightforward route beginning with 3-buten-1-ol, which was successively tritylated, hydroxylated, and etherified. Detritylation, bromination of the hydroxyl group, and reaction

[126] A. F. Rosenthal and M. Pousada, Biochim. Biophys. Acta 164, 226 (1968).

with sodium diethyl phosphite gave the phosphonate diester from which the phosphonic acid was obtained by acid hydrolysis. The synthetic route is not well adapted to optically active isomers, but unsaturated derivatives (not yet reported) should be preparable thereby.

3,4-Dioctadecoxybutylphosphonic Acid

Trityl But-3-en-1-yl Ether. But-3-en-1-ol (K and K Laboratories, Plainview, N.Y.) (5.0 g, 0.069 mole) in pyridine (25 ml) was added to a suspension of trityl bromide (22.3 g, 0.069 mole) in anhydrous ethyl acetate (175 ml). After stirring overnight at 40° the solvents were evaporated and the yellow residual oil taken up in ether and washed several times with water; the ether layer was dried over magnesium sulfate and evaporated. Hexane (100 ml) was added and the solution was kept at 2° and filtered from a small precipitate of tritanol. The oil obtained by evaporation of the filtrate crystallized from aqueous acetonitrile at 2° to yield the product (16.3 g.), m.p. 53–54°; a further 2.3 g were obtained by recrystallizing the residue from the evaporated filtrate (total yield 85.5%). The substance was chromatographically homogeneous (R_f 0.90 in 19:1 hexane–ethyl acetate).

3,4-Dihydroxybutyl Trityl Ether. Potassium permanganate (15.8 g, 0.100 mole) in water (500 ml) was added to a solution of but-3-en-1-yl trityl ether (31.6 g, 0.100 mole) in acetone (3.21) dropwise during 9½ hours at 5 ± 1°. The product crystallized at 3° after filtration and evaporation of the filtrate to about 11. Recrystallization from aqueous acetonitrile yielded the product (31.2 g, 89.5%), m.p. 124.5–126°. Further recrystallization raised the m.p. to 129–130°. The product was chromatographically homogeneous (R_f 0.70 in ethyl acetate).

3,4-Dioctadecyloxybutanol. A mixture of 3,4-dihydroxybutyl trityl ether (1.05 g, 0.0030 mole), powdered dry potassium hydroxide (6.0 g), 1-bromooctadecane (7.0 g, 0.021 mole), calcium hydride (2.0 g) and dioxan (35 ml) was stirred just below reflux for 16 hours. The mixture was filtered through celite and the filtrate evaporated. The residue was dissolved in warm 3:1 acetone–ethyl acetate (50 ml), kept at 21° overnight, and filtered. Evaporation of the filtrate gave a white powder (1.58 g, 62%) consisting of almost pure triether but containing a small amount of dioctadecyl ether (R_f 0.44 and 0.63, respectively, in 19:1 hexane–ethyl acetate).

The triether (1.000 g, 0.0017 mole) in propanol (15 ml) and 12 N hydrochloric acid (7 ml) was refluxed for 3 hours with vigorous magnetic stirring. The residue obtained after thorough evaporation of the mixture was precipitated with acetonitrile and the product filtered off. Reprecipitation with acetonitrile removed traces of trityl compounds and left fairly pure 3,4-dioctadecyloxybutanol (700 mg, 97%) (R_f 0.05 in 19:1 hexane–ethyl acetate) containing a small quantity of dioctadecyl ether.

3,4-Dioctadecyloxybutylphosphonic Acid. Fairly pure 3,4-diocta-decyloxybutanol (5.00 g, 0.0082 mole) was dissolved in warm 1,2-dimethoxyethane and added in two portions with shaking to dibro-motriphenoxyphosphorane (10 g) (Note 1, Section III,F). The homogeneous yellow solution was allowed to stand for 45 minutes at room temperature and the solvent evaporated. To the residue was added acetonitrile (60 ml) and the granular product filtered off. The 1-bromo-3,4-dioctadecyloxybutane (5.50 g, 99%) (R_f 0.48 in 19:1 hexane–ethyl acetate) was fairly pure but contained a small amount of dioctadecyl ether and traces of 3,4-dioctadecyloxybutanol and phosphorus compounds. The bromide was used directly in the Michaelis-Becker reaction.

A solution of sodium diethyl phosphite was prepared from de-oiled sodium hydride (1.50 g) and diethyl hydrogen phosphite (30 ml) in anhydrous peroxide-free 1,2-dimethoxyethane (200 ml). 1-Bromo-3,4-dioctadecyloxybutane (3.00 g, 0.0464 mole) was added and the mixture heated under reflux for 4.5 hours, evaporated, and the residue extracted with ether and washed several times with water and dilute hydrochloric acid. The dried ether solution was evapo-rated and the residue was dissolved in propionic acid (85 ml), 8 N hydrochloric acid (15 ml) was added, and the residue heated under reflux for 16 hours. The product was obtained by diluting the mixture with cold water (1 liter), filtering, and recrystallizing from ace-tone–hexane. Yield of product, m.p. 76–78°, was 1.62 g (53%). Re-crystallization from isopropanol–acetonitrile and from chloro-form–acetone yielded an analytically pure product, m.p. 77.5–78.5°.

3,4-Dioctadecoxybutylphosphonylcholine

The synthesis of the isosteric diether phosphonate lecithin described by Rosenthal[40] employs trichloroacetonitrile as condensing agent with choline *p*-toluenesulfonate (salt):

3,4-Dioctadecoxybutylphosphonylcholine. 3,4-Dioctadecoxylbutyl-phosphonic acid (674 mg, 1.00 mmole) and choline *p*-toluenesulfonate (2.0 g) (Note 2, Section III,F) were dissolved in a mixture of pyridine (15 ml) and trichloroacetonitrile (4 ml) at 50°, and the solution was kept at this temperature for 48 hours. About two-thirds of the volatile material was carefully removed *in vacuo* and the prod-uct was precipitated by addition of acetonitrile (50 ml). Filtration and washing with acetonitrile yielded a tan product, which was dissolved in tetrahydrofuran–water 7:3 at 30° (100 ml) and passed through a short column of Amberlite MB-3 previously equilibrated with the same solvent (flow rate about 5 ml/minute). The resin was washed with a further 500 ml of aqueous tetrahydrofuran and the combined eluates were taken to dryness *in vacuo*.

The decolorized product was recrystallized twice from hexane (50 to 18°) and then from trichloroethylene–acetone (*ca.* 1:2). The final yield of white crystalline material was 490 mg (63%). The isosteric lecithin softens slightly above 120°, becomes partially transparent above 150°, browns slightly at 190–195°, and finally melts sharply at 200.5–201.5°.

3. Phosphinate Analogues of Lecithin

The phosphinate group contains two carbon atoms atached to a single phosphorus; recently, several lecithin analogues containing this moiety have become available by synthesis. Although compounds of this type have not so far been found in nature, some of them have displayed valuable biochemical and physicochemical properties; for example, a diether lecithin analogue containing the C–P–C group was the first completely nonhydrolyzable phospholipid (other than phosphatidic acid) analogue known; it is a potent inhibitor of phospholipase C[127] and phospholipase A.[128] In addition, it has been employed in a liposome preparation to show that complement-mediated damage to cellular membranes does not depend upon lecithin hydrolysis[129]; the compound has also proved valuable in helping to elucidate some of the features of phospholipid–cholesterol interactions in model membranes.[130,131]

For these reasons the synthesis of the specific compound used in these studies

$$C_{18}H_{37}OCH_2CH(OC_{16}H_{33})CH_2—P \overset{O}{\underset{O—}{\nearrow}}—CH_2CH_2\overset{+}{N}(CH_3)_3$$

[127] A. F. Rosenthal, S. V. Chodsky, and S. C. H. Han, *Biochim. Biophys. Acta* **187**, 385 (1969).

[128] A. F. Rosenthal and S. C. H. Han, *Biochim. Biophys. Acta* **218**, 213 (1970).

[129] S. C. Kinsky, P. P. M. Bonsen, C. B. Kinsky, L. L. M. van Deenen, and A. F. Rosenthal, *Biochim. Biophys. Acta* **233**, 815 (1971).

[130] R. Bittman and L. Blau, *Biochemistry* **11**, 4831 (1972).

[131] B. de Kruyff, R. A. Demel, A. J. Slotboom, L. L. M. van Deenen, and A. F. Rosenthal, *Biochim. Biophys. Acta.* **307**, 1 (1973).

is described in this section. The presence of the second C–P group requires that a very different, and more difficult, synthetic route must be employed than that used in the synthesis of any of the phosphonate analogues described above.

The key intermediate, analogous to a trialkyl phosphite, is an appropriate dialkyl alkylphosphonite ($(RO)_2PR'$) which can react with a 2,3-dialkoxy-1-halopropane to produce the carbon–phosphorus skeleton of the lecithin analogue. Such alkylphosphonites are not commercially available but must be freshly synthesized just before use; they react avidly with atmospheric oxygen as well as moisture, and thus are very difficult to store. The intermediates in the synthesis of the phosphonite display the same properties as well.

The trivalent phosphorus compound employed was diisopropyl allylphosphonite. Reaction with 2-hexadecoxy-3-octadecoxyiodopropane gave the dialkoxypropyl(allyl)phosphinate; hydroxylation of the allyl group with osmate-periodate gave the corresponding phosphorus-substituted acetaldehyde. Reduction with borohydride, mesylation of the ethanol derivative, and reaction with dimethylamine gave the dimethylaminoethyl compound. Quaternization with methyl iodide and hydrolysis finally gave the diether phosphinate lecithin. The synthesis is not adaptable to unsaturated derivatives, but if an optically active halodiether were available, the optically active phosphinate lecithins could be prepared by this route.

The procedure of Rosenthal and co-workers[127] for this synthesis is described below. First, however, is described the synthesis of diisopropyl allylphosphonite by Razumov et al.[132] with additional details indicated where appropriate by the present author.

Diisopropyl Allylphosphonite

The preparation of the phosphonite must be accomplished with minimal storage of any intermediate (no longer than overnight under nitrogen) and completed within approximately 4 days. All synthetic operations must be carried out under dry nitrogen; all reagents must be anhydrous. It is essential that oil baths rather than heating mantles be employed for all heating operations, especially including distillations. Much better yields are obtained when the preparation is done on a large rather than a small scale.

    Allyldichlorophosphine. Into a 3-liter, four-necked round-bottomed flask, fitted with a mechanical stirrer and a seal, making it possible to run the stirring under vacuum, a reflux condenser, drop-

[132] A. I. Razumov, B. G. Liorber, M. B. Gazizov, and Z. M. Khammatova, *Zh. Obshch. Khim.* **34**, 1851 (1964).

ping funnel and a thermometer, were charged 267 g of aluminum chloride and 550 g of phosphorus trichloride, after which the mixture was stirred vigorously. Then 76.5 g of allyl chloride was added in drops to the suspension at 40–50°. When necessary, the heating was continued until all of the aluminum chloride had dissolved. The complex was cooled, dissolved in 350 ml of methylene chloride, and 1 liter of dibutyl phthalate was added with cooling and stirring at first, then at a gradually increased rate at room temperature so that the methylene chloride stayed below its boiling point. 120 g (50% excess) of finely divided antimony were added, after which the mixture was heated under reflux for 4 hours at 80–90° (in the bath). After this the methylene chloride was distilled off at atmospheric pressure, and the mixture of phosphorus trichloride and allyldichlorophosphine *in vacuo*, gradually raising the bath temperature up to 135–140°. The distillate was collected in a well-cooled receiver. The product was carefully redistilled using an efficient fractionating column. The pure dichlorophosphine had b.p.$_{46}$ 51–52° d_{20} 1.2325. Yield: about 35%.

Diisopropyl Allylphosphonite. To a solution of 51 g of isopropanol and 86 g of triethylamine in 750 ml of pentane was added dropwise a solution of 61 g of allyldichlorophosphine in about 60 ml of pentane with stirring at —5 to —10° during about 2 hours. The mixture was heated under reflux for 2½ hours; it was then cooled to room temperature and carefully filtered in a nitrogen atmosphere (in a dry box or glove bag). The precipitate was washed twice with pentane. Most of the solvent was carefully distilled off at atmospheric pressure and the residue distilled *in vacuo* using an efficient fractionating column; b.p.$_{15}$ 66–67°, d_{20} 0.9031. Yield: about 55%.

DL-*2-Hexadecoxy-3-octadecoxypropyl[2-(trimethylammonium)-ethyl]phosphinate*

The procedure of Rosenthal *et al.*[127] is given below:

DL-*Isopropyl 2-Hexadecoxy-3-octadecoxypropyl(allyl)phosphinate.* 2-Hexadecoxy-3-octadecoxyiodopropane (see Section III,A) (22.4 g, 0.033 mole) was dissolved in 15 g (0.08 mole) of freshly distilled diisopropyl allylphosphonite with a trace of hydroquinone under a static N_2 atmosphere and was kept at 120–122° for 40 hours. The mixture was dissolved in hexane, washed with HCl and then water; the hexane phase was dried over $MgSO_4$, filtered, and finally evaporated *in vacuo*. The product was precipitated with cold acetonitrile, filtered, washed with acetonitrile, and dried *in vacuo*. The crude product (22.2 g) in methylene chloride (200 ml) was applied to a

38-mm diam column containing 300 g of SilicAR CC-7 (Mallinkrodt Chemical Works). Methylene chloride (1 liter) and 4% ethanol in methylene chloride (1.5 liters) eluted low polarity impurities. The product was eluted with 10% ethanol in methylene chloride (1.5 liters); on evaporation of the solvent it weighed 10.1 g. Since it retained a small amount of impurity, it was rechromatographed in the same way to yield 9.6 g (42%) of pure product (m.p. 31–32°) after crystallization from chloroform–acetone. This consisted of two approximately equal components at R_f's 0.58 and 0.64 in 6% ethanol in methylene chloride or R_f's 0.67 and 0.77 in formic acid–ethanol–methylene chloride (1:2:47 by volume). These are apparently the two racemic diastereomers (asymmetric P as well as C-2). In various preparations 2–3 g of additional product could be obtained by repeated rechromatography of the impure intermediate fractions eluted with 10% ethanol in methylene chloride.

DL-*Isopropyl 2-Hexadecoxy-3-octadecoxypropyl (2'-acetaldo)phosphinate.* Isopropyl 2-hexadecoxy-3-octadecoxypropyl(allyl)phosphinate (1.50 g, 0.0022 mole) was dissolved in 50 ml of absolute ethanol. To this was added, with vigorous stirring, a mixture of NaIO$_4$ (1.50 g) in 50 ml of water plus 400 ml of ethanol, followed immediately by a freshly prepared solution of OsO$_4$ (85 mg) in absolute ethanol (50 ml). The mixture was stirred for 2 hours at room temperature and then was evaporated *in vacuo* (35–37° bath) until almost all the ethanol had been removed. The mixture was extracted with chloroform; the combined chloroform extracts were dried over MgSO$_4$ and filtered; the solvent was removed *in vacuo*. The crude product weighed 1.47 g and was shown by thin-layer chromatography to consist of several spots. The infrared spectrum differed from that of the allyphosphinate principally in showing a strong, sharp aldehyde band at 1710 cm^{-1}. Further purification was not attempted; instead, the aldehyde was reduced at once.

DL-*Isopropyl 2-Hexadecoxy-3-octadecoxypropyl(2'-hydroxyethyl)phosphinate.* To the crude aldehyde (1.47 g) in absolute ethanol (50 ml) was added sodium borohydride (500 mg); the mixture was stirred for 16 hours at room temperature. It was cooled at 5° and 12 M HCl was carefully added dropwise until no more H$_2$ was evolved. After concentration almost to dryness *in vacuo* (bath temperature, 40°), the reaction mixture was extracted with chloroform and water. The chloroform layer was washed twice with water, dried with MgSO$_4$, filtered, and evaporated to dryness *in vacuo* (bath, 40°). The crude alcohol weighed 1.29 g and showed several spots on thin layer chromatography.

DL-*Isopropyl 2-Hexadecoxy-3-octadecoxypropyl(2'-mesyloxy-ethyl)phosphinate.* DL-Isopropyl-2-hexadecoxy-3-octadecoxypropyl-(2'-mesyloxyethyl)phosphinate was prepared by dissolving the crude alcohol (1.29 g) in dry pyridine (25 ml), cooling to 0–5°, and adding methanesulfonyl chloride dropwise (3 ml) during 10 minutes with vigorous stirring. After stirring in the cold for 20 minutes, the reaction was completed by stirring at room temperature for 30 minutes. To the reaction mixture, again at 0–5°, was added ether (100 ml) and then 50 ml water dropwise. The mixture was transferred to a separatory funnel with another 25 ml of ether, and the aqueous phase was extracted twice with 50 ml portions of ether. The combined ether extract was washed successively with 2×100 ml water, 1×100 ml 1 M aqueous H_2SO_4, 1×100 ml water, 1×100 ml 3% aqueous Na_2CO_3, and finally $1 \times$ ml water. The ether solution was dried with $MgSO_4$, filtered, and evaporated *in vacuo* (bath, 40°). The product weighed 1.12 g; there were two predominant spots, R_f's 0.78 and 0.86 in addition to several much smaller components, on thin-layer chromatography in 8% ethanol in methylene chloride.

DL-*Isopropyl 2-Hexadecoxy-3-octadecoxypropyl[2-(trimethylam-monium)ethyl]phosphinate Iodide.* To 580 mg of the crude sulfonic ester in 30 ml of tetrahydrofuran were added 10 ml of 40% aqueous dimethylamine and 8 ml of water. The mixture was stirred at room temperature overnight and concentrated *in vacuo* (bath, 40°). The residue was dissolved in ether (50 ml) and swirled with a solution of 4 g K_2CO_2 in 10 ml of 0.1 M aqueous NaOH without vigorous shaking. The layers were allowed to separate completely before the aqueous layer was removed. To the ethereal solution of isopropyl 2-hexadecoxy-3-octadecoxypropyl-(2'-dimethylaminoethyl)phosphinate were added 5 ml of methyl iodide, and the solution was kept in the dark at room temperature for 4 days. The mixture was cooled to about 17° for a few hours and filtered. The precipitate was washed thoroughly with minimal quantities of cold ether. After drying *in vacuo*, the yield was 446 mg; m.p., 166–167° after crystallization from chloroform–acetone. The overall yield from the allylphosphinate was thus 47%.

DL-*2-Hexadecoxy-3-octadecoxypropyl[2'-(trimethylammonium)-ethyl]phosphinate.* To 175 mg of the methiodide were added 40 ml of acetic acid and 5 ml of peroxide-free 1,2-dimethoxyethane and 4 ml of 6 M hydrochloric acid. The mixture was warmed to 70–75° forming a clear solution, and was kept at this temperature for 24 hours. It was evaporated almost to dryness *in vacuo*, precipitated with a little cold acetonitrile, and filtered. The crude yield at this point was about 73 or 34% overall from the allylphosphinate. The precipitate was dissolved in 1 ml of chloroform, 3 drops of pyridine were added, and the zwitterionic product was again precipitated with cold acetonitrile, filtered, and dried *in vacuo*. Recrystallization from chloroform–acetone twice yielded an analytically pure white microcrystalline product (92 mg), m.p. 202–203° (decomposition).

F. Notes

1. Dibromotriphenoxyphosphorane is prepared by dropwise addition of a solution of bromine in anhydrous carbon tetrachloride to very slightly more than an equivalent quantity of triphenyl phosphite in carbon tetrachloride. The orange-yellow product, which precipitates immediately, is very moisture-sensitive and is therefore best isolated simply by evaporation of the solvent rather than by filtration.

2. The choline salt of *p*-toluenesulfonic acid is prepared by allowing methyl *p*-toluenesulfonate to react overnight with an equivalent quantity of 2-dimethylaminoethanol in dry ether. The precipitated product is filtered and recrystallized from anhydrous acetone. The hygroscopic salt must be stored in a dessicator or otherwise protected from moisture.[40]

3. *Related compounds.* (a) Glycerol derivatives: saturated diether and unsaturated ether-ester phosphonate lecithin analogues were reported by Baer and Rao[133] and Chacko and Hanahan,[121] respectively. Glycol analogues were reported by Baer and Robinson.[134] (b) Choline alkylphosphonate derivatives: the synthesis of medium-chain diester analogues was reported by Bonsen and co-workers.[39] The synthesis of diether propylphosphonate and hydrocarbon-chain propylphosphonate lecithin analogues was reported by Rosenthal and Pousada[38] and Rosenthal,[40] respectively. (c) Phosphinate lecithins: *rac*-diether analogues isosteric on either[135] or both[136] sides of the phosphorus function have been prepared by Rosenthal and Chodsky, as was a simple alkylphosphinyl analogue.[135] Optically active diether phosphinate analogues, partly and fully isosteric with natural lecithin were synthesized by Rosenthal and co-workers.[44]

G. Sphingomyelin Analogues

Phosphonic acid–containing sphingolipids have been isolated from natural sources, but it is curious that all, or almost all, of these substances

[133] E. Baer and K. V. J. Rao, *Lipids* 1, 291 (1966).
[134] E. Baer and R. Robinson, *Can. J. Biochem.* 46, 1273 (1968).
[135] A. F. Rosenthal and S. V. Chodsky, *J. Lipid Res.* 12, 277 (1971).
[136] A. F. Rosenthal and S. V. Chodsky, *Biochim. Biophys. Acta* 239, 248 (1971).

have been 2-aminoethylphosphonate derivatives in spite of the fact that phosphate sphingomyelins are phosphorylcholine derivatives.

A sphingomyelin analogue of the naturally occurring type, *erythro-N*-acyl-D-sphingosine 2-aminoethylphosphonate, was synthesized by Baer and Sarma,[107] who also synthesized the corresponding L-sphingosine derivative. The starting *erythro*-3-*O*-benzoyl-*N*-acylsphingosines were prepared by the long synthetic route of Shapiro and co-workers (see Section II,S). The remainder of the synthesis was similar to that employed for the corresponding phosphatidylethanolamine analogues.

erythro-N-Palmitoyl-D-Sphingosyl-1-(2'-Aminoethyl)phosphonate

To an ice-cold solution of (2-phthalimidoethyl)phosphonic acid chloride in 40 ml of anhydrous and ethanol-free chloroform, prepared from 0.715 g (2.8 mmoles) of (2-phthalimidoethyl)phosphonic acid and 0.583 g (2.8 mmoles) of phosphorus pentachloride (Note 1, Section III,H), was added with stirring and under anhydrous conditions in the course of 30 minutes a solution of 0.900 g (1.4 mmoles) of *erythro-N*-palmitoyl-3-*O*-benzoyl-D-sphingosine (Section II,S) and 1.40 ml (10.3 mmoles) of anhydrous triethylamine in 50 ml of chloroform. The reaction mixture was kept at room temperature for 48 hours. At the end of this time, 1.0 ml of triethylamine and 1.5 ml of distilled water were added and the mixture was stirred vigorously for 2 hours. It was then evaporated to dryness under reduced pressure at a bath temperature of 30°. To the residue were added 15 ml of benzene, and the benzene was distilled off under reduced pressure. This procedure was repeated twice more with 15 ml portions of benzene. The remaining material was extracted with four 30-ml portions of anhydrous ether, and the combined ether extracts were evaporated to dryness under reduced pressure at 20°. The residue was dissolved in 75 ml of a mixture of chloroform–methanol–water (5:4:1, v/v/v); to the solution were added 12 g of Amberlite IR-120 (H⁺ form), and the mixture was shaken vigorously for 2 hours. The Amberlite was filtered off and the filtrate was evaporated to dryness under reduced pressure. The remaining material was triturated with 20 ml of distilled water, and the mixture was separated by centrifugation. The procedure was repeated with two 20-ml portions of water and 10 ml of ice-cold methanol. The residue was dissolved in 12 ml of warm (70°) glacial acetic acid, and the solution was kept overnight at room temperature (24°). The precipitate, separated by centrifugation, was triturated with 2 ml of cold glacial acetic acid and two 25-ml portions of water, followed by 10 ml of ice-cold methanol. The chromatographically pure *erythro-N*-palmitoyl-3-*O*-benzoyl-D-sphingosyl-1-(2-phthalimidoethyl)phosphonate, on drying over potassium hydroxide at room temperature and a pressure of 0.04 mm Hg, weighed 0.755 g (61.4% of theoretical yield), m.p. 119–121°, R_f 0.82 (silica gel H; chloroform–methanol–water, 65:25:4 v/v/v).

erythro-N-Palmitoyl-3-O-benzoyl-D-sphingosyl-1-(2'-aminoethyl) phosphonate. In a 250-ml three-necked, round-bottomed flash fitted with a reflux condenser and gas inlet tube were placed 0.670 g (0.76 mmole) of 70 ml of 90% ethanol, and 3.5 ml of a 1.0% solution of hydrazine (anhydrous) in 90% ethanol, and the contents of the flask were heated to gentle reflux for 4 hours while passing nitrogen through the apparatus. At the end of the first and second hour, 3.5 ml and 2.7 ml, respectively, of the hydrazine solution were added. The solution was then concentrated under reduced pressure in an atmosphere of nitrogen and a bath temperature of 35–40° to an approximate volume of 10 ml. The concentrate was diluted with 2 ml of distilled water and kept overnight at 6°. The precipitate was filtered off with suction, washed thoroughly with water, and dissolved in 100 ml of a mixture of chloroform–methanol–water (5:5:1, v/v/v). To the solution were added 10 g of Amberlite IR-120 (H⁺ form), and the mixture was stirred for 1 hour. The resin was filtered off, washed with two 25-ml portions of the solvent mixture, and the combined filtrates were concentrated under reduced pressure at 35–40° to a volume of approximately 10 ml. To the concentrate were added 10 ml of water, and the mixture was kept overnight at 6°. The precipitate was collected by centrifugation, triturated with distilled water, and dried at room temperature over calcium chloride at a pressure of 8–10 mm Hg. The crude hydrazinolysis product, weighing 0.500 g, was dissolved in 5 ml of chloroform, the solution was passed through a column (2.5 × 25 cm) of silicic acid, and the material was eluted successively with 200 ml of chloroform (eluate 1), 200 ml of a mixture of chloroform and methanol (98:2 v/v, eluate 2), and 250 ml of chloroform and methanol (80:20 v/v, eluate 3). Evaporation under reduced pressure of eluates 2 and 3 at 35–40°, and drying the residues at room temperature over phosphorus pentoxide at 0.02 mm Hg, gave 0.180 g of the original phthaloyl compound (eluate 2) and 0.290 g of *erythro-N*-palmitoyl-3-*O*-benzoyl-D-sphingosyl-1-(2'-aminoethyl)phosphonate (eluate 3). The latter, on recrystallization from 3.5 ml of methanol, gave 0.250 g of chromatographically pure material. The starting material (0.180 g) recovered from eluate 2, on reprocessing as described above, gave a further 0.100 g of the desired compound. The total yield of pure compound was 0.350 g (61.3% of theory), m.p. 161.5–163.0°, R_f 0.52 (silica gel H, chloroform–methanol–water, 65:25:4 (v/v/v).

erythro-N-Palmitoyl-D-sphingosyl-1-(2'-aminoethyl)phosphonate.

To a solution of 0.300 g (0.4 mmole) of *erythro*-N-palmitoyl-3-O-benzoyl-D-sphingosyl-1-(2-aminoethyl)phosphonate in 22 ml of warm (35°) methanol was added 0.42 ml of a 2 N sodium hydroxide solution. The mixture was kept with stirring for 30 minutes at 35° and for 24 hours at room temperature (24°). The solution was then made slightly acidic with 9.5 ml of 0.1 N hydrochloric acid, and the suspension was concentrated under reduced pressure from a bath at 30° to about one-half of its original volume. The mixture was separated by centrifugation, and the precipitate was washed thoroughly with water (until the latter was free of hydrochloric acid) followed by two 25-ml portions of acetone. The dry material was dissolved in 7 ml of a mixture of chloroform and methanol (4:1, v/v) and reprecipitated with 40 ml of acetone. This process was repeated once more. The final precipitate was dried at room temperature over phosphorus pentoxide at a pressure of 8–10 mm Hg. The pure *erythro*-N-palmitoyl-D-sphingosyl-1-(2'-aminoethyl)phosphonate weighed 0.180 g (69.8% of theory), m.p. 172–173° (slight sintering at 91°), $[\alpha]_D^{22}$ +25.1° [4] in chloroform–methanol, 4:1 v/v (c 1.2).

H. Notes

1. Phthalimidoethylphosphonyl monochloride is prepared by the procedure of Baer and Sarma[137] as follows:

A suspension of 1.53 g (6.0 mmoles) of finely powdered, sieved (150 mesh), and thoroughly dried 2-phthalimidoethylphosphonic acid in a solution of 1.25 g (6.0 mmoles) of phosphorus pentachloride in 25 ml of anhydrous ethanol-free chloroform was heated at gentle reflux under anhydrous conditions (condenser protected by a calcium chloride tube) until all the solid material had dissolved. This usually requires from 10 to 15 minutes, depending on the fineness of the particles of the phosphonic acid. When all the solid material had dissolved, the solution was refluxed for a further 10 minutes, but no longer. The chloroform and phosphorus oxychloride were distilled off under reduced pressure at a bath temperature of 30°, with adequate protection against moisture, and the last traces of phosphorus oxychloride were removed by keeping the material at a pressure of 0.1 mm at 40° for 1 hour.

2. *Related compounds.* Besides the *erythro*-L-sphingomyelin analogue mentioned above, Baer and Pal[138] have reported the synthesis of the corresponding *erythro*-DL-dihydroceramide aminoethylphosphonate.

[137] E. Baer and G. R. Sarma, *Can. J. Biochem.* **43**, 1353 (1965).
[138] E. Baer and B. C. Pal, *Can. J. Physiol. Pharmacol.* **46**, 525 (1968).

Acknowledgments

Part of this work was supported by NIH grant No. AM-07699. Thanks are given to Miss Patricia Goldman and Mrs. Regina McArdle for their competent secretarial assistance.

Throughout this chapter, quotations from *Biochemistry, J. Amer. Chem. Soc.,* and *J. Org. Chem.* are reprinted with permission. Copyright by the American Chemical Society.

[49] Separation of Sphingosine Bases by Chromatography on Columns of Silica Gel[1]

By Yechezkel Barenholz and Shimon Gatt

The three main sphingosine bases of animal tissues are sphingosine ($CH_3-(CH_2)_{12}-CH=CH-CHOH-CHNH_2-CH_2OH$) and dihydrosphingosine ($CH_3-(CH_2)_{14}-CHOH-CHNH_2-CH_2OH$), while in plants phytosphingosine ($CH_3-(CH_2)_{13}-CHOH-CHOH-CHNH_2-CH_2OH$) predominates. Similar bases with smaller or greater chain length have been described.[2] These bases are released from the phospho- or glycosphingolipids by acid or alkaline hydrolysis.[3] Such hydrolyses yield mixtures of the above bases as well as several derivatives which are formed as a result of the chemical reaction.[3] The reaction medium must therefore be subjected to a purification step in which the individual bases are separated from the above mixture. In this procedure silica gel chromatography is used to separate, on a preparative scale, the pure bases from acid hydrolysis reaction mixtures. This is here exemplified by two experiments. In the first, pure sphingosine and dihydrosphingosine were obtained from the hydrolyzate of bovine spinal cord sphingolipids. In the second, phytosphingosine and dihydrosphingosine were isolated from the hydrolyzate of the acetylated bases secreted into the growth medium of the yeast *Hansenula ciferri*.

Materials

Bases of Spinal Cord

Bovine spinal cord lipids were obtained from Armour and Co. and stored for several years in the dry state at room temperature. These were hydrolyzed using 5 N aqueous HCl–methanol, 2:8.[3] The content of the ampule was extracted 5 times with 0.5 volume each of petroleum ether

[1] Y. Barenholz and S. Gatt, *Biochim. Biophys. Acta* **152**, 790 (1968).

[2] K. A. Karlsson, *Chem. Phys. Lipids* **5**, 6 (1970).

[3] R. G. Gaver and C. C. Sweeley, *J. Amer. Oil Chem. Soc.* **42**, 294 (1965).

(b.p. 60–80°) to remove fatty acids. The pH was adjusted to 12 with 15 N KOH; chloroform and water were then added to a final mixture of chloroform–methanol–water, 8:4:3.[4] After shaking, the phases were separated and the lower phase was washed twice with equal volumes of a mixture of methanol–water–chloroform, 94:96:6.[4] The lower phase was evaporated *in vacuo*.

Bases of *Hansenula ciferri*

The yeast was grown and the bases were isolated from the growth medium as described by Barenholz and Gatt.[5]

Silica Gel

For the separation of sphingosine and dihydrosphingosine, silica gel (0.05–0.2 mm) was purchased from Merck, Darmstraat, Germany. For the separation of phytosphingosine and dihydrosphingosine, silica gel S was purchased from Riedel de Haen, Hanover, Germany (catalog No. 31614).

Analytical Procedures

Quantitative estimation of the bases was made using the method of Lauter and Trams.[6] Thin-layer chromatography on silica gel G was done using a mixture of chloroform–methanol–2 N NH$_4$OH, 77:23:2.3 (modified from Sambasirarao and McCluer[7]). Gas–liquid chromatography of the bases was done using the trimethylsilyl derivatives[8] or the aldehydes obtained after oxidation with periodate.[9]

Silica Gel Chromatography

Method 1. Separation of Sphingosine and Dihydrosphingosine. A hydrolyzate of spinal cord lipids which contained 117 μmoles of sphingosine and 68 μmoles of dihydrosphingosine was mixed with 32 μmoles of bases of *H. ciferri* (mostly phytosphingosine). These bases were dissolved in 2 ml of a mixture of chloroform–methanol–2 M NH$_4$OH, 90:10:1

[4] J. Folch, M. Lees, and G. H. Sloan-Stanley, *J. Biol. Chem.* **226**, 497 (1957).
[5] Y. Barenholz and S. Gatt, *Biochemistry* **7**, 2603 (1968).
[6] J. L. Lauter and E. G. Trams, *J. Lipid Res.* **3**, 136 (1962).
[7] K. Sambasivarao and R. H. McCluer, *J. Lipid Res.* **4**, 136 (1963).
[8] H. E. Carter and R. C. Gaver, *J. Lipid Res.* **8**, 391 (1967).
[9] C. C. Sweeley and E. A. Moscatelli, *J. Lipid Res.* **1**, 40 (1959).

(v/v/v). Fifty grams of silica gel (Merck, 0.05–0.2 mm) were suspended in chloroform–methanol, 1:1, packed by gravity into a glass column (diameter, 2.2 cm) and washed with chloroform till transparent; final column height, 30 cm. The solution of the bases was applied onto the column, and the bases were eluted with a discontinuous gradient of increasing concentrations of ammoniacal methanol in chloroform as follows: A lower mixing chamber contained 300 ml of chloroform–methanol–2 M NH$_4$OH (90:10:1) and an upper reservoir had 600 ml of chloroform–methanol–2 M NH$_4$OH (50,50:5, v/v/v). When all 600 ml were delivered from the upper reservoir, it was again filled with 500 ml of chloroform–methanol–2 M NH$_4$OH (30:70:7, v/v/v). Ten milliliter fractions were collected. Storage of the bases in the alkaline medium produced other products. Chloroform, methanol, and water were therefore added, without undue delay, to each tube to a final ratio of chloroform–methanol–water, about 8:4:3. The tubes were mixed and the upper phases were removed. Aliquots of the lower phases were chromatographed on thin-layer plates of silica gel, and the bases were visualized by iodine vapors or by spraying with ninhydrin or with sulfuric acid. Figure 1 shows several peaks: Fractions 4–20 contained methyl ethers and other nonidentified compounds; fractions 33–42 contained 95 μmoles of pure sphingosine; fractions 43–47 had 23 μmoles of a mixture of sphingosine and dihydrosphingosine; fractions 45–58 had 40 μmoles of pure dihydrosphingosine; fractions 60–62 had 5 μmoles of a mixture of dihydrosphingosine and phytosphingosine; fractions 63–67 had 8 μmoles of pure phytosphingosine; and fractions 68–85 had about 1–2 μmoles of unidentified, ninhydrin-

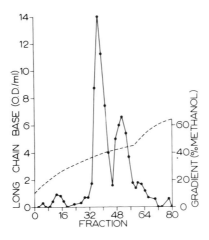

Fig. 1

positive compounds which migrated on the silica gel plates with an R_f of about 0.9.

This column therefore resulted in a good yield of pure sphingosine (80%) and dihydrosphingosine (60%), but in a rather poor yield of phytosphingosine (25%).

Method 2. Separation of Phytosphingosine and Dihydrosphingosine. The mixtures of bases obtained from the growth medium of *H. ciferri*[5] contained 260 μmoles of ^{3}H-labeled phytosphingosine, 40 μmoles of ^{3}H-labeled dihydrosphingosine, and about 30 mg of unidentified compounds. It was dissolved in 2 ml of chloroform–methanol, 95:5 (v/v), and applied to a column (33 × 2.2 cm) of 60 g silica gel S (Riedel de Haen) prepared as described in the former experiment. The column was first eluted with 500 ml of chloroform–methanol, 9:1, then with two successive continuous gradients as follows: The lower mixing chamber contained 300 ml of chloroform–methanol–2 M NH$_4$OH (90:10:1 v/v/v) and the upper reservoir had 400 ml of chloroform–methanol–2 M NH$_4$OH (75:25:2.5). After taking all 700 ml through the column, a second gradient was applied. The lower mixing chamber contained 300 ml of chloroform–methanol–2 M NH$_4$OH (70:30:3), and the upper reservoir had 400 ml of chloroform–methanol–2 M NH$_4$OH (55:45:4.5). Twelve milliliter fractions were collected. Chloroform, methanol, and water were added as in Fig. 1 and the phases were separated. Aliquots of the lower phases were taken for determination of radioactivity, for thin-layer chromatography, for quantitation,[6] and for gas–liquid chromatography. Figure 2 shows the results of this separation. Fractions 6–12 contained fast moving compounds, possibly methyl ethers, anhydro base and other unidentified compounds. Dihydrosphingosine (contaminated with some phytosphingosine) was eluted as a shallow peak in fractions 60–100. Pure phytosphingosine (over 200 μmoles) was eluted as a sharp peak in fractions 125–135.

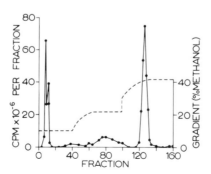

Fig. 2.

Remarks

1. The main advantage of the above-mentioned methods is in the fact that the free bases are used and the separation, with a good quantitative recovery, requires only one passage through a silica gel column.

2. Although separation of other long-chain bases was not tested the method might be adapted, for this purpose, by choosing the type of silica gel and suitable gradient.

3. Two previously described methods utilized the N-succinyl derivatives[10] or the acetonyl- or p-nitrobenzene derivatives[11] of the bases. In either case, the blocking agent must be removed after the separation.

4. Method 2 did not provide a good separation between sphingosine and dihydrosphingosine, the two main bases of animal tissue. It might however be utilized for crude extracts and partial hydrolyzates as follows. The lipids other than the long-chain bases are eluted with chloroform–methanol mixtures wihout ammonia. The bases are then eluted with mixtures of chloroform–methanol–ammonia and rechromatographed by Method 1.

5. Method 2 was employed successfully for separation of up to 3 g of phytosphingosine, using more silica gel. With Method 1, separation was somewhat poorer when the quantities of silica gel and bases were increased.

Acknowledgment

This work was supported in part by a U.S. Public Health Service Grant (NB-02967).

[10] J. B. Wittenberg, *J. Biol. Chem.* **216**, 379 (1955).
[11] B. Weiss and R. L. Stiller, *Lipids* **5**, 782 (1970).

[50] Chemical Synthesis of Choline-Labeled Lecithins and Sphingomyelins

By WILHELM STOFFEL

Principle

Synthetic or naturally occurring phosphatidylcholines or sphingomyelins are demethylated to the corresponding phosphatidyl-N,N-

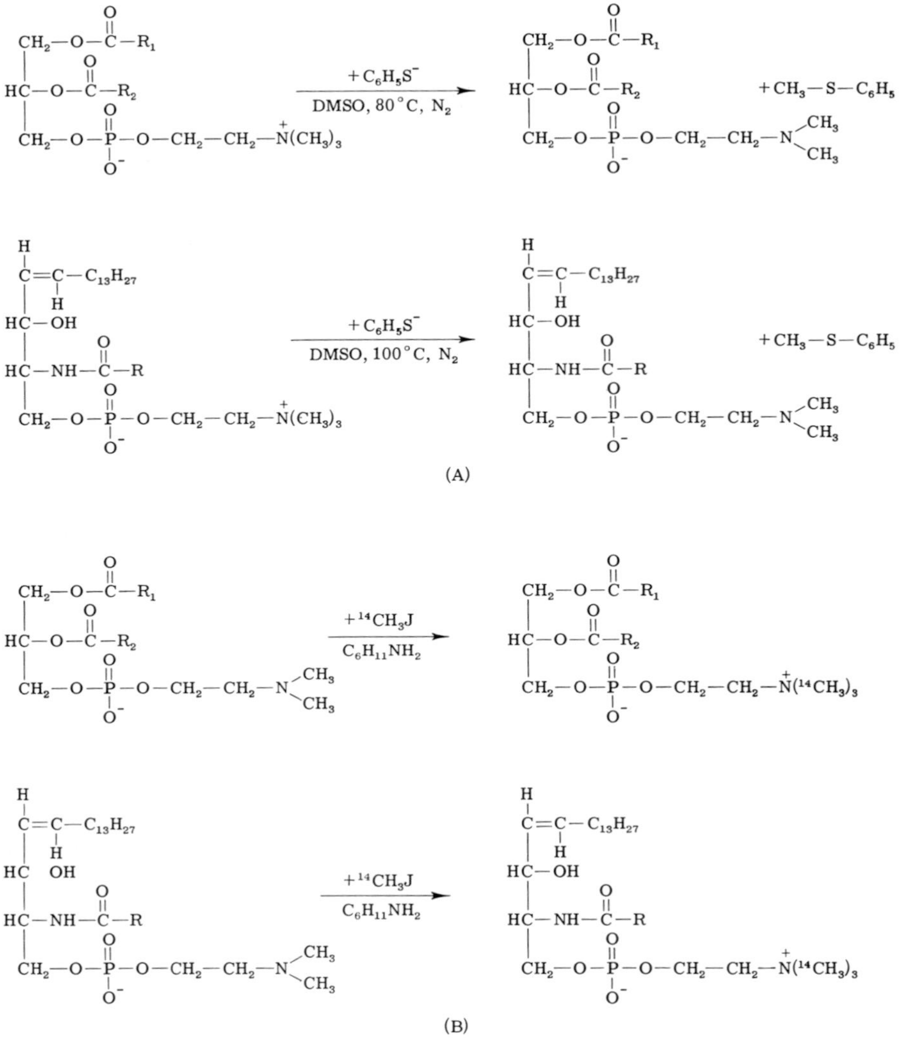

(A)

(B)

FIG. 1. Reaction schemes: (a) demethylation and (B) methylation with labeled CH₃I.

dimethylethanolamines or ceramide-1-phosphoryl-N,N-dimethylethanol-amines with sodium benzenethiolate. The purified N,N-dimethyl compounds are quaternized in high yield with [^{14}C]- or [^{13}C]methyliodide to the respective lecithins and sphingomyelins labeled in the choline moiety.[1] The reaction scheme given in Fig. 1 summarizes the two steps. The procedure allows the convenient chemical preparation of choline-labeled lecithins and sphingomyelins of high specific activity regardless of the degree of unsaturation. The procedure is simple, rapid, and inexpensive as compared to the biosynthetic preparation of choline-labeled lecithins and sphingomyelins. The latter can hardly be obtained in high specific activity by biosynthetic procedures.

Procedures

1. Preparation of Sodium Benzene Thiolate

Sodium benzene thiolate was prepared essentially according to Jenden et al.[2] with the following modifications: equivalent amounts of sodium (17 g, 0.75 mole) and thiophenol (81 g, 0.75 mole) were refluxed in 100 ml of absolute methanol for 1 hour. Two-hundred milliliters of toluene were added. The sodium salt crystallized from the clear solution on cooling at −10°. The crystalline precipitation was collected on a Büchner funnel, washed with cold toluene and petroleum ether (30–60°), and stored in a desiccator over P_2O_5.

2. Demethylation Procedures

a. *Phosphatidyl-N,N-dimethylethanolamine.* Ten grams (12.8 mmoles) phosphatidylcholine are dissolved in 250 ml freshly distilled dimethylsulfoxide and 3.9 g (25.6 mmoles) sodium benzene thiolate are added to the solution. The mixture was heated under a stream of nitrogen at 80° with stirring under exclusion of air moisture. The progress of the reaction is followed by thin-layer chromatography (silica gel H, solvent system: chloroform–methanol–water, 65:25:4). After 2½ hours additional sodium benzene thiolate (1 g, 6 mmoles) is added. The reaction is completed in general after 5 hours. Byproducts are traces of free fatty acids and lysolecithin. The cooled solution is poured on ice-cold 1 N HCl, and the reaction product extracted with five 50-ml portions of chloroform. The combined extracts are dried over Na_2SO_4, the solvent evaporated under vacuum, and the yellow residue dissolved in chloroform. (Traces of DMSO do not interfere with the following purification procedure.) The phosphatidyl-N,N-dimethylethanolamine is purified by silicic acid column chromatography (4 × 70 cm; approximately 350 g SiO_2) apply-

[1] W. Stoffel, D. LeKim, and T. S. Tschung, *Hoppe Seyler's Z. Physiol. Chem.* **352**, 1058 (1971).
[2] D. J. Jenden, I. Hanin, and S. I. Lamb, *Anal. Chem.* **40**, 125 (1968).

ing increasing concentrations of methanol in chloroform (1:20, 1:9, 1:5, 1:4, 1:3) and collecting the sample with an automatic fraction collector. Yield: 7.0 g (9.1 mmoles) 72% of theory.

b. *Ceramide 1-Phosphoryl-N,N-dimethylethanolamine.* Eight-hundred milligrams (1 mmole) sphingomyelin is dissolved in 20 ml of dimethysulfoxide. Five millimoles sodium thiophenolate are added and the mixture heated with stirring at 100° under nitrogen for a period of 6 hours three millimoles sodium thiophenolate are added after 3 hours. The demethylation is controlled by thin-layer chromatography (silica gel H, solvent system: chloroform–methanol–10% NH_4OH, 60:35:8). The demethylated product is isolated as described in Section 2,a and separated from contaminations by silicic acid chromatography by increasing concentrations of methanol in chloroform. Yield: 0.7–0.8 mmole, 70–80% of theory.

c. *Convenient Demethylation Procedure for Phosphatidylcholine and Spingomyelin.* One mmole of phosphatidycholine and 6 mmoles of 1,4-diazabicyclo-(2,2,2)-octane dissolved in 20 ml dimethyl formamide are refluxed under a stream of nitrogen for 3 to 6 hours. The progress of the reaction is followed by thin-layer chromatography, and the isolation and purification of the products are carried out as described under Sections *2a* and *2b*, respectively. Almost no lysolecithin is formed, and the ceramide 1-phosphoryl-*N,N*-dimethyl-ethanolamine can be used for methylation without further purification.

3. Methylation Procedure

a. [*N*-$^{14}CH_3$]*Choline Lecithin.* Two-hundred milligrams (0.25 mmole) phosphatidyl-*N,N*-dimethylcholine are dissolved in 5 ml absolute methanol in a Teflon-lined screw cap tube. Sixty microliters (0.52 mmole) freshly distilled cyclohexylamine and 80 μl (25 μmoles) $^{14}CH_3I$ are added, the tube flushed with nitrogen, and tightly closed. The mixture is left at room temperature in the dark for 18 hours, time sufficient for the completion of the reaction. This can be controlled by thin-layer chromatography (silica gel H, solvent system: $CHCl_3$–CH_3OH–H_2O, 65:25:4). If some unreacted dimethyl compound is still present, additional cold CH_3I should be added and the reaction continued. After the completion of the reaction 25 ml ether are added and the ether solution extracted twice with 5% $Na_2S_2O_3$, cold 2 N HCl, and water. The combined aqueous phases are reextracted twice with 10 ml chloroform, and the combined ether–chloroform extracts dried over Na_2SO_4. The solvents are evaporated completely, the residue dissolved in a few milliliters of chloroform, and applied to a silicic acid column (2.5 × 30 cm, 50 g SiO_2).

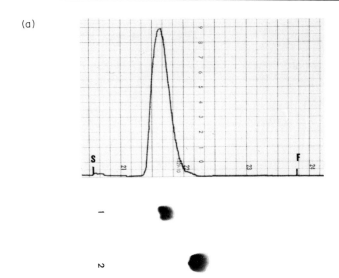

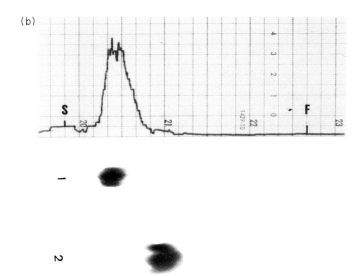

FIG. 2. Thin-layer chromatogram and radio scan of (a) 1. [CH₃-¹⁴C]choline sphingomyelin, 2. ceramide 1-phosphoryl-*N,N*-dimethylethanolamine. Solvent system: chloroform–methanol–10% NH₄OH, 60:35:8. (b) 1. [CH₃-¹⁴C]choline phosphatidylcholine, 2. phosphatidyl-*N,N*-dimethylethanolamine. Solvent system: chloroform–methanol–water 65:25:4.

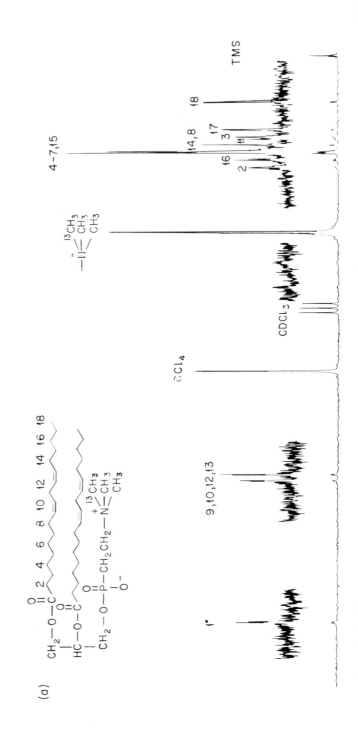

(a)

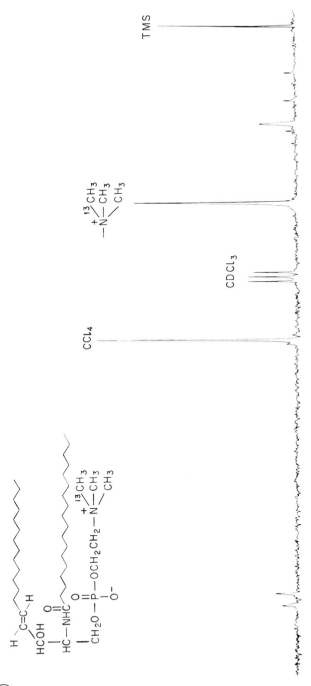

FIG. 3. ^{13}C-Nuclear magnetic resonance spectrum of $[N$-$^{13}C]$choline-labeled (a) phosphatidylcholine and (b) sphingomyelin.

Traces of lysolecithin are separated from the main product, which is pure labeled lecithin, by elution with increasing concentrations of methanol in chloroform (1:2, 2:3, 1:1, 3:1 M/C). Yield: 160 mg (0.20 mmole) pure [N-$^{14}CH_3$]choline lecithin. Additional 25–30 mg of labeled lysolecithin are recovered.

   b. [N-$^{14}CH_3$]*Choline Sphingomyelin.* A solution of 200 mg (0.275 mmole) ceramide-1-phosphoryl-N,N-dimethylethanolamine, 30 μl (0.35 mmole) cyclohexylamine, 90 μl (0.28 mmole) $^{14}CH_3I$ in 5 ml dry methanol in a tightly sealed Teflon-lined screw cap tube is stored at room temperature in the dark for approximately 18 hours. After this period of time complete methylation to sphingomyelin occurs. The solvent is evaporated under vacuum, the residue dissolved in chloroform, and the chloroform solution is washed twice with 5% $Na_2S_2O_3$, 2 N HCl and water. Centrifugation of the mixture for phase separation might be necessary. The organic phase is dried over Na_2SO_4 and the solvent evaporated under vacuum. Thin-layer chromatography of the reaction product reveals a complete methylation, and a further purification of the labeled sphingomyelin is not required in most cases (silica gel H, solvent system: $CHCl_3$–CH_3OH–10% NH_4OH, 60:35:8). If, however, a purification is demanded by the result of the thin-layer chromatographic analysis, silicic acid column chromatography as described in Section 3,a for the purification of lecithin or preparative thin-layer chromatography is the method of choice. Sphingomyelin can be quantitatively eluted from silica gel with a mixture of chloroform–methanol–water, 1:1:0.1.

Characterization of Products

   The purity of the demethylation products and the labeled phosphatidylcholine and sphingomyelin was tested by thin-layer chromatography, Fig. 2. The methylated compounds proved to be radiochemically pure as demonstrated by the radio scan of the thin-layer chromatograms indicated in these figures. The NMR spectrum of phosphatidyl-N,N-di-

methylethanolamine exhibits the signals characteristic for the $-N\begin{smallmatrix}\diagup CH_3 \\ \diagdown CH_3\end{smallmatrix}$

group at 2.9, CH_2—N at 3.3, —CH_2—O—P— at 4.4, $-CH_2-\overset{\overset{\displaystyle O}{\|}}{C}-O$ at

2.3, and $-CH_2-O-\overset{\overset{\displaystyle O}{\|}}{C}$ at 4.2 ppm and was identical with that of a synthetic reference compound.[3] The NMR spectrum of demethylated sphingomyelin showed the following typical protons: $-N\begin{smallmatrix} \nearrow CH_3 \\ \searrow CH_3 \end{smallmatrix}$ at 2.8,

$-NH-\overset{\overset{\textstyle O}{\|}}{C}$ at 7.3, and CH_2-O-P- at 4.0–4.2 ppm. Typical fragments are present in the mass spectra of phosphatidyl-N,N-dimethylethanolamine and ceramide 1-phosphoryl-N,N-dimethylethanolamine at m/e 72 $CH_2-CH_2-N(CH_3)_2$; m/e 58 $CH_2=N(CH_3)_2$; m/e 169 $HO-CH_2-CH_2-O-P(OH)_2-O-CH_2-CH_2$; the deacylated tris(trimethylsilyl)-sn-glycerol-3-phosphoryl-N,N-dimethylethanolamine at m/e 58, 71, 72, 555 (M-15), 369 (M-90), and 356 (M-103).

Phospholipase C from *B. cereus*[4] releases labeled phosphorylcholine completely with hydrolysis of the choline-labeled lecithin and sphingomyelin which on paper electrophoresis [acetic acid (150 ml), formic acid (50 ml), water (800 ml), pH 1.9, 10 V/cm, 3 hours] yielded one single radioactive band identical with the test substance. Carbon-13 magnetic resonance spectroscopy of [N-^{13}C]choline lecithin and sphingomyelin with 90% natural abundance of carbon-13 exhibits the characteristic chemical shift to 54 ppm[5] (Fig. 3).

[3] D. Chapman and A. Morrison, *J. Biol. Chem.* **241**, 5044 (1966).

[4] F. Haverkate and L. L. M. van Deenen, *Biochim. Biophys. Acta* **106**, 78 (1965).

[5] W. Stoffel, O. Zierenberg, and B. D. Tunggal, *Hoppe Seyler's Z. Physiol. Chem.* **354**, 1962 (1972).

[51] The Preparation of Tay-Sachs Ganglioside Specifically Labeled in Either the *N*-Acetylneuraminosyl or *N*-Acetylgalactosaminyl Portion of the Molecule

By JOHN F. TALLMAN, EDWIN H. KOLODNY, and ROSCOE O. BRADY

Introduction

Knowledge of the pathways of metabolism of Tay-Sachs ganglioside, N-acetylgalactosaminyl-(N-acetylneuraminosyl)-galactosylglucosylceramide (G$_{M2}$) is essential for an understanding of the pathogenesis of Tay-Sachs disease. The enzymatic hydrolysis of G$_{M2}$ *in vitro* has been difficult to demonstrate in the past because of the lack of an assay with sufficient sensitivity to detect the small amounts of reaction products which result. However, through specific labeling of G_{M2} with radioactivity as described in this method the minimum detectable level of enzymatic activity is greatly reduced.

Reagents

Chloroform and methanol, 1:1 (v/v)
          2:1 (v/v)
"TUP-KCl," chloroform, methanol, and 0.1 KCl 3:48:47 (v/v/v)
"TUP-H$_2$O," chloroform, methanol, and water 3:48:47 (v/v/v)
Sodium methoxide, dissolve 0.1 g of sodium in 5 ml of methanol (absolute)
Methanolic HCl, 2 N
Chloroform, methanol, and water, 60:30:4.5 (v/v/v)
Chloroform, methanol, and water, 60:35:8 (v/v/v)
Chloroform, methanol, and water, 60:40:10 (v/v/v)
n-Propanol and water 7:3 (v/v)
H$_2$SO$_4$ (aq) 0.05 N
Ba(OH)$_2$(aq) 1%
n-Propanol and water 9:1 (v/v)
Chloroform, methanol, and water 65:25:4 (v/v/v)
0.25 M sucrose
0.2 M sodium cacodylate-HCl pH 7.3
Sodium deoxycholate, 3%
Uridine diphosphate N-acetyl-D-[1-^{14}C]galactosamine, New England Nuclear Co. >40 mCi/mM
N-Acetyl-D-mannosamine [^{3}H(G)] 2–10 mCi/mM
Phosphodiesterase and alkaline phosphatase, Worthington Biochemical Co.
Ether
Acetone
NaCl, 0.9%
Acetic acid, 0.5%
Anasil S., Analabs, Inc., Hamden, Conn.
Hyflo Supercel, Fisher Chemical Co.

N-Acetylneuraminic Acid Labeled G_{M2}

The method employed to prepared [^{3}H]NeuAc-labeled G_{M2} has been described previously.[1] We have since modified and simplified the proce-

---

[1] E. H. Kolodny, J. Kanfer, J. M. Quirk, and R. O. Brady, *J. Lipid Res.* **11**, 144 (1970).

dure so that much better yields are obtained. These innovations are included in this contribution. Generally labeled [³H]N-acetyl-D-mannosamine, 5.06 mCi/mmole (New England Nuclear Co., Boston, Mass.) is used in this preparation. This distribution of label throughout the molecule minimizes the possibility that studies on the enzymatic hydrolysis of G_{M2} labeled in the acetic acid moiety of N-acetylneuraminic acid may indicate only deacetylation.

Preparation of Mixed Labeled Brain Gangliosides. Fifty eight-day-old Sprague Dawley rats are used for each preparation. Each rat is injected intracerebrally with 10 μl of an aqueous solution of labeled N-acetyl-D-[³H]mannosamine containing 5×10^7 cpm. The injection is made with a 10-μl Hamilton syringe at a point slightly anterior to the intrauricular line and lateral to the midline. The rats are returned to their mothers for 48 hours. The rats are decapitated, their brains are removed, and an acetone powder of the pooled brains is prepared.[2] The gangliosides are extracted by refluxing for 1 hour with 20 ml/g of chloroform–methanol (C–M), 2:1. The suspension is filtered, and the residue is extracted for 30 minutes with 10 ml/g of C–M, 1:1, and again filtered. The filtrates were combined and sufficient chloroform added to bring the composition of the filtrate of 2:1 in C–M. This mixture is partitioned according to Folch *et al.*[3] with 0.1 M KCl (0.2 volume), TUP–KCl (0.4 volume), and TUP–H$_2$O (0.4 volume). The aqueous phase is dialyzed overnight against distilled water and the retentate is lyophilized. The residue is dissolved in absolute methanol and the phospholipids are transesterified with sodium methoxide,[4] the mixture is neutralized with methanolic HCl, dissolved in C–M, 2:1, repartitioned, and dialyzed. The retentate is passed over a cation exchange resin (Dowex 50 H⁺ form, 100–200 mesh) and the material in the effluent solution is lyophilized. A white powder is obtained consisting of a mixture of gangliosides which are dissolved in a small volume of distilled water. About 50 mg of mixed labeled gangliosides are obtained from 50 g (wet weight) of the brain.

Conversion to G_{M1}. A portion of the mixed ganglioside solution is treated with *V. cholera* neuraminidase (500 IU/ml, Behring Diagnostics, Inc., Woodbury, N.Y.). This solution already contains 0.050 M sodium acetate buffer pH 5.5 in 0.9 g/dl NaCl and 0.1 g/dl CaCl$_2$. One milliliter of neuraminidase solution is added to each 15 mg of mixed labeled gangliosides and the mixture is incubated for 48 hours at 37° with additions of 0.5 ml of neuraminidase at 12, 24, and 36 hours. The conversion of mixed gangliosides to G_{M1} is quantitative with this neuraminidase

[2] J. Kanfer, Vol. 14 [62].
[3] J. Folch, M. Lees, and G. H. Sloane-Stanley, *J. Biol. Chem.* **226**, 497 (1957).
[4] G. V. Marinetti, *Biochemistry* **1**, 350 (1962).

treatment. The incubation mixture is lyophilized, redissolved in 5 ml C–M–H$_2$O (60:30:4.5), and filtered. The filter paper is washed with 5 ml of the same solvent and the filtrates are combined. Five grams of Sephadex G-25 (superfine) are washed with 15 ml of C–M–H$_2$O (60:30:4.5) and the ganglioside is applied to this column. The column is eluted with C–M–H$_2$O (60:30:4.5) and ganglioside radioactivity appears in the initial portion of the effluent.[5] An aliquot is removed for examination by thin-layer chromatography and the remainder (almost entirely [³H]NeuAc-G$_{M1}$) is dried under N$_2$ and taken up in H$_2$O.

Removal of Terminal Galactose from G_{M1}. The terminal molecule of galactose in G$_{M1}$ is susceptible to enzymatic hydrolysis by a highly active β-galactosidase from rat liver.[6,7] An acetone powder of this pellet II[7] is prepared and the β-galactosidase is extracted from the powder by trituration with 5 volumes of distilled water. The suspension is periodically stirred for 1 hour at 0°C then centrifuged at 25,000 g for 20 minutes. The supernate contains the β-galactosidase. The enzyme is stable indefinitely when stored at the temperature of liquid N$_2$. The conversion of [³H]G$_{M1}$ to [³H]G$_{M2}$ is accomplished by incubating 15 mg portions of [³H]G$_{M1}$ with 2 ml of the lysosomal extract (4–6 mg of protein) in 0.1 M potassium acetate buffer (pH 5.0) in a final volume of 2.5 ml for 5 hours at 37°. The incubated mixture is lyophilized. C–M–H$_2$O (60:30:4.5) is added and the suspension filtered and passed over a column of Sephadex G-25 as described above. The released galactose is separated from ganglioside reaction products by this procedure. Usually about 30–50% conversion of G$_{M1}$ to G$_{M2}$ is obtained. [³H]G$_{M2}$ is separated from unreacted [³H]G$_{M1}$ by preparative thin-layer chromatography on 20 × 40 cm silica gel G plates. The solvent system used is C–M–H$_2$O (60:35:8). Authentic standards are used to identify the migration of compounds. The thin-layer plate is scraped, transferred to a small column, and the [³H]G$_{M2}$ is eluted from the silica gel with 500 ml of C–M–H$_2$O (60:40:10). The volume of solvent is reduced under N$_2$, the [³H]G$_{M2}$ dialyzed, and the retentate is lyophilized. The white powder is taken up in C–M–H$_2$O (60:30:4.5) and treated as before with Sephadex G-25. A yield of between 10 and 15 mg of G$_{M2}$ is obtained.

Identity of the Product and Distribution of the Label. The identity of the purified [³H]NeuAc-G$_{M2}$ is established by chromatography with authentic G$_{M2}$ in two solvent systems, *n*-propanol-H$_2$O (7:3) and C–M–2.5 N ammonia (60:35:8). The plates are scanned for radioactivity

[5] M. A. Wells and J. C. Dittmer, *Biochemistry* **2**, 1259 (1963).
[6] F. A. Cumar, J. F. Tallman, and R. O. Brady, *J. Biol. Chem.* **247**, 2322 (1972).
[7] H. Ragab, C. Beck, C. Dillard, and A. L. Tappel, *Biochim. Biophys. Acta* **148**, 501 (1967).

and then the portion of the plate containing the unlabeled standard is sprayed with resorcinol reagent.[8] The remainder of the plate is protected with a pane of glass during this procedure. More than 95% of the radio-activity usually comigrates with the G_{M2} standard.

In order to demonstrate the distribution of label in the gangliosides, several methods of hydrolysis are utilized. To liberate the sialic acid, mild acid hydrolysis (0.05 N H_2SO_4, 80°, 90 minutes) is employed. Approximately 30 nmoles (based on sialic acid content) of mixed ganglioside (18,000 cpm) are heated in a total volume of 0.6 ml. In order to correct for the amount of sialic acid destroyed in the hydrolysis procedure, an equal amount of unlabeled sialic acid is processed in the same manner as control. After hydrolysis, the solutions are adjusted to pH 4.0 with a 1% (w/v) solution of $Ba(OH)_2$. The $BaSO_4$ precipitate is removed by centrifugation and washed with 1 ml of distilled water. The combined supernatants are dialyzed against distilled water and the dialysate is lyophilized. The residue is analyzed for N-acetylneuraminic acid,[9] N-acetylhexosamine,[10] and radioactivity. Another aliquot is separated by thin-layer chromatography into its components using the system devised by Gal[11] employing n-propanol–water 9:1. In our hands, all of the radioactivity was confined to the region of the N-acetylneuraminic acid standard; none was found in the area of the N-acetylhexosamine. The retentate from the dialysis is lyophilized and the residue taken up in C–M 2:1 and partitioned. The lower phase lipids, including such hydrolysis products as ceramide trihexoside, ceramide lactoside, and ceramide glucose, were chromatographed in a solvent system consisting of C–M–H$_2$O (65:25:4).[12] No radioactivity was found either by scanning or by direct counting of the gel after spraying with ammonium bisulfate.

Extensive hydrolysis of [³H]G_{M2} was carried out[13] and glucose, galactose, and N-acetylgalactosamine were identified by thin-layer chromatography.[11] N-acetylneuraminic acid is largely destroyed by this treatment. No radioactivity should be associated with these neutral sugars either on scanning or direct counting of the gels. Determination of the quantity of hexose to hexosamine[10] indicated a ratio of approximately 2:1.

The specific activity is determined by taking duplicate samples at two different concentrations for determination of radioactivity and quantitation of N-acetylneuraminic acid.[8] The specific activity of our

[8] L. Svennerholm, *Biochim. Biophys. Acta* **24**, 604 (1957).

[9] L. Warren, *J. Biol. Chem.* **234**, 1971 (1959).

[10] R. Spiro, Vol. 8 [1].

[11] A. Gal, *Anal. Biochem.* **24**, 452 (1968).

[12] R. O. Brady, J. Kanfer, and D. Shapiro, *J. Biol. Chem.* **240**, 39 (1965).

[13] J. J. Gallai-Hatchard and G. M. Gray, *Biochim. Biophys Acta* **116**, 532 (1966).

synthesized $[^3H]G_{M2}$, which is based on the amount of radioactivity per mole of N-acetylneuraminic acid, was $5-7 \times 10^5$ cpm/μmole.

N-Acetylgalactosamine Labeled G_{M2}

Labeled G_{M2} was synthesized enzymatically from uridine diphosphate-N-[1-^{14}C]acetylgalactosamine and N-acetylneuraminosylgalactosylglucosylceramide (G_{M3}).[14]

$$\text{UDP-}[^{14}\text{C}]\text{GalNAc} + \text{NeuAc-GalGlcCer} \xrightarrow[\text{galactosaminyltransferase}]{\text{hematoside: } N\text{-acetyl-}}$$

$$\text{Gal-}[^{14}\text{C}]\text{NAc-(NeuAc)GalGlcCer} + \text{UDP}$$

Preparation of G_{M3}. Hematoside (G_{M3}) is prepared from dog erythrocytes according to the procedure of Yamakawa *et al.*[15] with minor improvements.[16]

Whole heparinized dog blood (4L) is centrifuged to remove serum. The pellet is resuspended in an equal volume of 0.9% saline and recentrifuged. The pellet is then suspended in 0.5% acetic acid (2.5 volume/pellet volume) and the cells are allowed to hemolyze overnight at 4°C. The pellet is washed 4 times with H_2O (2.5 volume/pellet volume) and the remaining stroma is lyophilized. The dried red cell stroma is successively extracted with acetone (10 volume/pellet volume), ether 5 volume/pellet volume), and C–M, 1:1 (10 volume/pellet volume). Both the ether and C–M extracts contain G_{M3}. These solutions are dried and the powder redissolved in a minimal volume of C–M, 1:1. The resuspended lipid extract is slowly added dropwise into 250 ml of cold (−10°C) acetone. The insoluble material is again resuspended in C–M, 1:1, and the procedure repeated. The insoluble material from this second acetone powder is dissolved in about 50 ml of C–M, 2:1, and partitioned according to Folch *et al.*[3] using TUP–H_2O (see above). Seven washes remove most of the G_{M3} to the upper phase. The upper phases are combined and dialyzed. The retentate is lyophilized. This crude G_{M3} preparation is further purified by column chromatography.

Twenty grams of Hyflo Super Cel is dried overnight at 105°C along with 40 g of AnaSil S. A slurry is prepared with 200 ml of chloroform and poured into a column of dimensions 2×55 cm. The upper phase is·resuspended in chloroform and placed onto the column. The column is successively eluted with 200 ml of C–M in the proportions, 10:0, 8:2,

[14] J. M. Quirk, J. F. Tallman, and R. O. Brady, *J. Label. Compounds* **8**, 483 (1972).
[15] T. Yamakawa R. Irie, and M. Iwanaga, *J. Biochem. (Tokyo)* **48**, 490 (1960).
[16] F. A. Cumar, R. O. Brady, E. H. Kolodny, V. W. McFarland, and P. T. Mora, *Proc. Nat. Acad. Sci. U.S.* **67**, 757 (1970).

6:4, 5:5, 4:6, and 2:8. Fractions containing G_{M3}, followed by monitoring a small part of the effluent by thin-layer chromatography (as carried out above), are combined and the volume is reduced under vacuum. Preparative thin-layer chromatography of G_{M3} containing fractions is performed as for [^{3}H]AcNeu-G_{M2}; G_{M3} areas are noted by exposure of the plate to I_2 vapor. These areas are eluted and purified by Sephadex G-25 from the gel contaminants as described above. A yield of around 250 mg of G_{M3} is obtained.

Conversion of G_{M3} to G_{M2}. Brains from 15-day-old Sprague Dawley rats are homogenized with 8 volumes of 0.25 M sucrose in a glass homogenizer. The homogenate is centrifuged at 9000 g for 20 minutes. The supernatant is centrifuged at 105,000 g for 45 minutes to obtain a light mitochondrial microsomal pellet. This pellet is resuspended in 0.2 M sodium cacodylate HCl buffer, pH 7.3 (0.7 ml per rat brain). Sodium deoxycholate (3%) is added to a final concentration of 0.18%. The suspension is homogenized and centrifuged as above. The supernate is utilized as the source of the enzyme.[17] All procedures were carried out at 0–5°. The incubation mixtures contain 2.4 μmoles of G_{M3}, 5 μCi of UDP-N-[1-^{14}C]acetylgalactosamine (0.117 μmole), 10 mg of Triton X-100, 50 μmoles of $MnCl_2$, and the enzyme (10 mg of protein) in a total volume of 2.5 ml. Incubations are carried out for 2 hours at 37°.

Purification of Labeled G_{M2}. The reaction is stopped by the addition of 20 volumes of C–M (2:1) and the suspension is filtered to remove denatured protein. The filter paper is washed with C–M (2:1) and the wash is combined with the filtrate. The volume is reduced under vacuum and the remaining solution dialyzed against 1000 volumes of distilled water. The retentate is lyophilized.

The dried powder is taken up in 8 ml of C–M–H_2O (60:30:4.5) and applied to a 1.5 × 10 cm column of superfine Sephadex G-25. The column was eluted with 60 ml of the same solvent. The labeled ganglioside appears in the initial fractions (30 ml).

The ganglioside solution is treated with alkaline phosphatase and phosphodiesterase and dialyzed.[2] The retenate is lyophilized and the G_{M2} is purified by preparative thin-layer chromatography on 500 mm silica gel G plates in an ascending solvent system of C–M–2.5 N ammonia (60:35:8). The identity of the product is verified by cochromatography with authentic standard in adjacent lanes which are sprayed with resorcinol reagent after the plate has been scanned for radioactivity and radioactive areas scraped from the plate.[8] The ganglioside is eluted from these scrapings by 500 ml of C–M–H_2O, 60:40:10. The volume of the eluate

[17] F. A. Cumar, P. H. Fishman, and R. O. Brady, *J. Biol. Chem.* **246**, 5075 (1971).

is reduced under vacuum and dialyzed. The retentate is lyophilized. Non-lipid material is removed by passing the ganglioside over Sephadex G-25 column as described for NeuAc[^{3}H].

Identity of the Product and Distribution of the Label. Homogeneity of the isolated compound is established by chromatography in other solvent systems with appropriate standards as described above. Acid hydrolysis of the molecule[17] as described for [^{3}H]NeuAc-G_{M2} and treatment with mammalian neuraminidase should produce the expected products (labeled *N*-acetylgalactosaminylgalactosylglucoseceramide) from the G_{M2}.[18] The specific activity of each ganglioside produced is based on the amount of radioactivity per mole of *N*-acetylneuraminic acid. The ratio of *N*-acetylneuraminic acid to *N*-acetylgalactosamine in the biosynthetically produced G_{M2} is unity.

The preparation of [^{14}C]G_{M2} using the soluble galactosaminyltransferase in a representative experiment resulted in an overall yield of 10.3% of purified ganglioside based on the starting quantity of UDP-*N*-[1-^{14}C] acetylgalactosamine. The specific activity of the synthesized [^{14}C]G_{M2} was 6.7×10^6 cpm/μmole.

Importance of Labeled Substrates

Using these two radioactive gangliosides as substrates, the fact was established that mammalian tissues contain two enzymes which are involved in the catabolism of G_{M2}. One enzyme catalyzes the hydrolysis of *N*-acetylneuraminic acid from G_{M2} yielding ceramidetrihexoside[18,19] the other enzyme catalyzes the cleavage of *N*-acetylgalactosamine yielding G_{M3}.[18] The catabolism of G_{M2} is impaired in the pathological condition in humans known as Tay-Sachs disease as a result of a drastic attenuation of G_{M2}-hexosaminidase activity.[20,21]

[18] J. F. Tallman and R. O. Brady, *J. Biol. Chem.* **247**, 7570 (1972).

[19] E. H. Kolodny, J. Kanfer, J. M. Quirk, and R. O. Brady, *J. Biol. Chem.* **246**, 1426 (1971).

[20] E. H. Kolodny, R. O. Brady, and B. W. Volk, *Biochem. Biophys. Res. Commun.* **37**, 526 (1969)

[21] J. F. Tallman, W. G. Johnson, and R. O. Brady, *J. Clin. Invest.* **51**, 2339 (1972).

[52] Removal of Water-Soluble Substances from Ganglioside Preparations

By Thomas P. Carter and Julian N. Kanfer

Investigations of the individual reactions responsible for ganglioside biosynthesis studied *in vitro* usually employ specific radioactive nucleotide sugar donors (e.g., uridine diphosphogalactose and cytidine monophospho-*N*-acetylneuraminic acid). A number of techniques have been used for the removal of these water-soluble precursors and other components from ganglioside containing preparations. Such methods include silicic acid[1] and Sephadex G-25 column chromatography,[2] paper chromatography,[3] high voltage paper electrophoresis,[3,4] dialysis,[5,6] and coprecipitation of gangliosides with bovine serum albumin by trichloroacetic acid.[7]

Principle. The "Folch" partitioning method is the procedure most commonly employed for the removal of water-soluble contaminants from a crude lipid mixture.[6] The final washed lower chloroform phase obtained in this method contains all of the lipids initially present except for the gangliosides. These sialic acid–containing glycosphingolipids are found in the upper or aqueous methanol phase. Retention of the gangliosides in the chloroform-rich lower phase is possible because of the formation of water-insoluble ganglioside–calcium complexes. A recent study reported on the stability constants and the nature of the complexes formed.[8] The actual structure of these complexes is not well characterized.

Applicability of Method. Although this experimental procedure was developed for *in vitro* ganglioside studies to remove labeled water-soluble precursors, specifically nucleotide sugars, this method can easily be modified and used for other purposes. This method can be used for *in vivo* studies by "scaling-up" the procedure, without decreasing the recovery of gangliosides. In addition, the procedure can be used for the isolation

[1] M. Holm and L. Svennerholm, *J. Neurochem.* **19**, 609 (1972).

[2] M. A. Wells and J. C. Dittmer, *Biochemistry* **2**, 1259 (1963).

[3] B. Kaufman, S. Basu, and S. Roseman, *in* "Inborn Disorders of Sphingolipid Metabolism" (S. M. Aronson and B. W. Volk, eds.), p. 193. Pergamon, Oxford, 1967.

[4] Kaufman, B., S. Basu, and S. Roseman, Vol. 8, p. 365.

[5] J. N. Kanfer and R. O. Brady, *in* "Inborn Disorders of Sphingolipid Metabolism" (S. M. Aronson and B. W. Volk, eds.), p. 187. Pergamon, Oxford, 1967.

[6] J. Folch-Pi, M. Lees, and G. H. Sloane-Stanley, *J. Biol. Chem.* **226**, 497 (1957).

[7] J. F. Tallman and R. O. Brady, *J. Biol. Chem.* **247**, 7570 (1972).

[8] J. P. Behr and J. M. Lehn, *FEBS Lett.* **31**, 297 (1973).

of small quantities of ganglioside, free of nonganglioside lipids or water-soluble contaminants, such as gangliosides derived from subcellular organelles or nonneural tissue. In the latter case, gangliosides contained in the washed, chloroform-rich lower phase are purified by column chromatography employing DEAE-cellulose,[9] DEAE-Sephadex,[10] or silicic acid.[1]

Materials and Reagents

A typical simulated incubation mixture: 1 mg of lyophilized bovine white matter (400 μg of protein) suspended in a solution containing 0.90% NaCl, 1.15% KCl, 2.11% KH_2PO_4, 1.85% $MgSO_4$, and 1.3% $NaHCO_3$ and buffered with 0.1 M Na_2HPO_4-NaH_2PO_4 (pH 7.4).

Glycine–HCl buffer, 0.2 M (pH 2.2) containing 100 mM $CaCl_2$

Ethylenediaminetetraacetic acid (tetrasodium salt), 0.1 M adjusted to pH 7.0 with 10 N HCl

Aqueous–methanol solution, chloroform–methanol–0.1 M EDTA (pH 7.0), 3:48:47 (v/v/v)

Chloroform–methanol, 1:1 (v/v)

Chloroform–methanol, 2:1 (v/v)

Bovine brain gangliosides and mixed nonganglioside lipid were isolated by conventional techniques.[6,9]

[^{3}H]G_{M1} (specific activity 2.0×10^6 cpm/μmole), labeled mono-sialoganglioside was prepared by the galactose oxidase-sodium [^{3}H]borohydride method.[11]

UDP-[^{14}C]glucose (specific activity 4.4×10^8 cpm/μmole) was obtained from New England Nuclear, Boston, Mass.

[^{3}H]N-acetylgalactosamine (specific activity 1.19×10^8 cpm/μmole) was prepared by the acetic anhydride method.[12]

Methanol, chloroform, hexane, and benzene were reagent grade.

Procedure

(A) To 200 μl of any incubation mixture in which gangliosides would be expected to be present, 6 ml of chloroform–methanol (2:1) are added, and the contents are mixed.

(B) The solution so obtained is vigorously extracted with 1.5 ml of 0.2 M glycine–HCl buffer, pH 2.2, containing 100 mM $CaCl_2$,

[9] C. C. Winterbourn, *J. Neurochem.* **18**, 1153 (1971).
[10] R. W., Ledeen, R. K. Yu, and L. F. Eng, *J. Neurochem.* **21**, 829 (1973).
[11] N. S. Radin, L. Hof, R. M. Bradley, and R. O. Brady, *Brain Res.* **14**, 497 (1969).
[12] P. Brunetti, G. W. Jourdian, and S. Roseman, *J. Biol. Chem.* **237**, 2447 (1962).

and is then centrifuged. The upper phase is removed and discarded.

(C) The chloroform phase plus any interfacial material is extracted 3 times as in step B above. Any material appearing at the interface is left with the final lower phase.

(D) Chloroform and methanol, 3.0 ml each, are added so that the resulting volume ratio of chloroform to methanol is 2:1, respectively.

(E) The mixture is extracted once with 2.4 ml 0.1 M EDTA, pH 7.0, and 4 times with 2.0 ml portions of the aqueous–methanol solution containing EDTA. The phases are separated by centrifugation and the upper layer removed and saved.

(F) The pooled upper phases are transferred to a 30-ml test tube and the contents are concentrated under reduced pressure with the aid of either hexane or benzene as an azeotropic agent.

(G) To the residue containing principally EDTA and gangliosides, 10 ml of chloroform–methanol (1:1) are added, the solution is vigorously mixed and passed through a fine-porosity sintered glass filter. The clear filtrate, containing gangliosides at this stage, is virtually free of other lipids and water-soluble contaminants and can be used for further detailed compositional analyses.

Validity of Procedure. When 200 μl of a simulated incubation mixture containing [^{3}H]G$_{M1}$ (11 nmoles) was subjected to steps A–D above, the recovery of gangliosides in the lower phase was always greater than 96%. With mixed nonradioactive bovine brain gangliosides, the recovery of sialic acid in the lower phase exceeded 90% for concentrations less than 500 nmoles/ml. In similar control experiments carried through steps A–D only, when either [^{3}H]N-acetylgalactosamine or UDP-[^{14}C]glucose was employed, less than 4% of these water-soluble precursors were found in the lower phase.

In experiments where mixed bovine brain gangliosides were subjected to the entire procedure of steps A–G, the final chloroform–methanol (1:1) filtrate always contained more than 85% of the lipid-bound sialic acid at sphingolipid concentrations less than 400 nmoles/ml. Since the initial ganglioside content of typical *in vitro* mixtures is usually less than 80 nmoles (4.0 $\times$ 10^{-4} M ganglioside), a ganglioside recovery of 85% is the lower limit. Carrier nonganglioside lipid dissolved in chloroform–methanol must be added at step A when the total lipid content of the individual incubation mixtures is less than 800 μg.

Section VIII
Preparation of Single Cells

[53] Isolation of Free Brown and White Fat Cells

By JOHN N. FAIN

Free fat cells can be readily isolated by incubation of white[1] or brown[2] adipose tissue with crude bacterial collagenase. The fat cells are readily separated from the other cells since they float to the top of the incubation medium after low-speed centrifugation. This results in a relatively pure preparation of a single cell type. Furthermore, the adipose tissue from several animals can be pooled and enough cells obtained to examine the effects of agents in paired experiments. In general, studies with isolated fat cells are also more convenient and reproducible than studies in perfused adipose tissue or segments of adipose tissue incubated *in vitro*.

Reagents

1. Crude bacterial collagenase from *Clostridium histolyticum:* The crude collagenase supplied by Worthington Biochemical Corporation has been adequate. The collagenase solution is usually prepared fresh daily but aliquots of a stock solution can be frozen and then thawed for each experiment.

2. Bovine fraction V albumin powder: Albumin supplied by Armour or Pentex (Miles) is satisfactory. Care must be exercised since the response of fat cells to catecholamines and insulin is sometimes very poor for cells isolated and incubated in certain lots of albumin. We have generally found that in acceptable lots of albumin there should be a good lipolytic response to 0.1 μM norepinephrine or low concentrations of theophylline (0.1–0.2 mM) which is inhibited by insulin (10–100 microunits/ml).

Defatted albumin is sometimes desired in studies with fat cells and some lots of albumin have an unacceptably high content of free fatty acids. The most convenient procedure to remove bound free fatty acids and certain other alcohol-soluble contaminants is by ethanol extraction of the albumin powder. We use a modification of the procedure published by Guillory and Racker[3] in which approximately 250 g of albumin powder are stirred overnight at 3°C with 35 liters of 95% ethanol. The ethanol is removed by filtration using a Büchner funnel. The extraction procedure is repeated and the ethanol removed from the moist albumin powder by lyophilization using a vacuum pump.

---

[1] M. Rodbell, *J. Biol. Chem.* **239**, 375 (1964).
[2] J. N. Fain, N. Reed, and R. Saperstein, *J. Biol. Chem.* **242**, 1887 (1967).
[3] R. J. Guillory and E. Racker, *Biochim. Biophys. Acta* **153**, 490 (1968).

3. Krebs-Ringer phosphate buffer: NaCl, 128 mM MgSO$_4$, 1.4 mM; CaCl$_2$, 1.4 mM; KCl, 5.2 mM; and Na$_2$HPO$_4$, 10 mM.

Procedure

The major problem in the isolation of hormonally sensitive fat cells is the variability seen in commercially available collagenase preparations. These problems cannot be circumvented as yet by the use of more highly purified preparations of collagenase since Kono[4] found that digestion of adipose tissue requires the presence of a proteolytic enzyme(s) as well as collagenase. We should eventually be able to use purified collagenase plus specific protease(s) for cell digestion. However, none of the commercially available proteases has been found suitable. Not only must suitable preparations contain both collagenase and at least one proteolytic enzyme, but also other tryptic-like proteolytic enzymes should be absent. It is known that exposure of fat cells to trypsin and many other proteolytic enzymes abolishes the response of these cells to insulin and glucagon.[5,6] Problems in the use of collagenase preparations which are contaminated with tryptic-like enzymes should be suspected when isolated fat cells show little stimulation of glucose metabolism by insulin (10–100 microunits/ml).

Some crude preparations of commercial collagenase may contain appreciable amounts of phospholipase C.[7] All of the radioactive choline taken up into adipose tissue lecithin during an incubation for 1 hour was lost during subsequent digestion with collagenase. Egg lipoprotein is considered to be a good substrate for phospholipase C, but in our laboratory we have not as yet seen any beneficial effects from including this as a possible preferential substrate of phospholipase C during digestion of fat cells.[8] The addition of preparations of phospholipase C from *Clostridium perfringens* to white[9] or brown[8] fat cells stimulated glucose oxidation at low concentrations while higher concentrations lysed cells. Rosenthal and Fain[8] did not observe any effect of egg lipoprotein on the stimulation of glucose metabolism by some *C. perfringens* preparations which suggested that the effects attributed to such preparations may result from contaminants. There is an unknown substance secreted by *C. perfringens* which is more potent than phospholipase C in stimulating glucose metab-

---

[4] T. Kono, *Biochim. Biophys. Acta* **178**, 397 (1969).
[5] T. Kono, *J. Biol. Chem.* **244**, 1772 (1969).
[6] J. N. Fain and S. C. Loken, *J. Biol. Chem.* **244**, 3500 (1969).
[7] P. Elsbach and M. Rizack, *Biochim. Biophys. Acta* **198**, 82 (1970).
[8] J. W. Rosenthal and J. N. Fain, *J. Biol. Chem.* **246**, 5888 (1971)
[9] M. Rodbell, *J. Biol. Chem.* **241**, 130 (1966).

olism.[8] The presence of a high level of basal glucose metabolism and a poor response to insulin may be the result of contamination of collagenase preparations with phospholipase C or other agents.

Another problem arises from the finding that fat cells must be incubated in buffer containing albumin as an acceptor to bind the fatty acids released during activation of triglyceride hydrolysis. Even in studies in which lipolysis is not measured, the presence of albumin results in an improved response of fat cells to hormones such as insulin.[10-12]

Our experience is that the available preparations of bovine fraction V albumin vary markedly. The best procedure is to test a given lot of albumin and, if it is suitable, obtain a sufficient amount of a given lot for all experiments which are contemplated. The problems are not circumvented by the use of crystalline bovine albumin, fraction V albumin isolated from other animal species or defatted albumin. We have found no simple procedure which will consistently remove impurities from kilogram amounts of albumin preparations.

Albumin has a remarkable ability to bind rather tightly a very large number of substances. Albumin preparations invariably contain a heavy metal (probably copper) contaminant which is required in order to see a stimulation of glucose metabolism by thiols. The cofactor required for thiol action cannot be completely removed by either dialysis or passage of albumin over Sephadex G-50 columns.[11] This contaminant can be removed if the albumin solution is exposed to a metal chelator such as o-phenanthroline and then chromatographed on columns of Sephadex G-50.[11]

For the isolation of white fat cells, white adipose tissue is obtained from the appropriate source and kept at 37°. The tissue is rinsed, cut into pieces weighing about 200 mg, and any connective tissue present is removed. Isolation of white fat cells from most animal species is readily accomplished except with ruminants where the cells often clump. This may result from the difficulty in keeping the triglycerides of ruminant fat cells in the liquid state. The pieces of adipose tissue are rinsed with buffer containing albumin and added to 1 ounce polyethylene bottles containing buffer plus albumin and collagenase. Plastic 25 ml Erlenmeyer flasks are also suitable for digestion of adipose tissue.

Two milliliters of Krebs-Ringer phosphate buffer containing 4% albumin and 0.2–0.7 mg/ml of collagenase are added per gram of adipose tissue. The tissue is ordinarily incubated for about 45 minutes at 37° in a shaking water bath. Our experience with rat white adipose tissue

[10] T. Kono, *J. Biol. Chem.* **244**, 5777 (1969).
[11] M. P. Czech and J. N. Fain, *J. Biol. Chem.* **247**, 6218 (1972).
[12] J. Gliemann, *Diabetologia* **3**, 382 (1967).

has not indicated that shorter periods of digestion using greater amounts of collagenase resulted in superior results.

The phosphate buffer is made up fresh daily and the pH adjusted to 7.4 after addition of bovine fraction V albumin powder. The presence of glucose during the isolation of fat cells does not appear to affect the yield or hormonal response of fat cells and is therefore usually omitted from the medium used for tissue digestion.

Krebs-Ringer phosphate buffer with air as the gas phase is used for convenience since we have found that the response of fat cells to hormones is generally the same as for cells incubated in Krebs-Ringer bicarbonate buffer. Studies done with bicarbonate buffer must be done with 5% CO_2 in the gas phase since the final pH will be around 8.2 to 8.4 in its absence. There is a marked increase in basal glucose oxidation under such circumstances.[12]

Most studies on the isolation and incubation of fat cells have been done with cells isolated and incubated in water bath shakers equipped for reciprocal motion. Gliemann[12] has shown that shaking fat cells in 22 ml plastic vials at up to 120 cpm did not affect the hormonal response, but in cells isolated by shaking at 140 cpm there was an increase in basal glucose metabolism which suggests cell damage. Lysis of cells resulting from vigorous shaking is more of a problem when cells isolated from adipose tissue of adult humans and obese animals for these tissues tend to have large and fragile fat cells. Less damage can be obtained by using a water bath shaker equipped for orbital motion (Fermentation Design or New Brunswick Scientific).

After 45–60 minutes of incubation with collagenase in buffer containing albumin, the cells are filtered through nylon chiffon to remove undigested adipose and connective tissue. This is readily accomplished by cutting the top out of 1 ounce bottle caps, placing a piece of chiffon on top of the bottle, and then screwing on the top. The filtration of the cells is facilitated by squeezing the side of the polyethylene bottle.

The suspension of fat cells is added to plastic centrifuge tubes (12 ml) and centrifuged briefly. The time and force of centrifugation should be just sufficient to float the fat cells to the surface. Generally 10–15 seconds of centrifugation in an International clinical centrifuge with a horizontal head to hold 15 ml tubes is used. With large fat cells centrifugation can even be omitted. The stromal-vascular cells, erythrocytes, and other sedimented material are removed by aspiration using a Pasteur pipette attached to a water aspirator. The cells are rinsed with albumin buffer (5 ml/g of adipose tissue originally present), the cell suspension is centrifuged, the sedimented material and infranatant are removed, and the procedure repeated again. The cells are resuspended in medium for addition to incubation tubes or flasks. Further washing of fat cells has not been found beneficial in our laboratory. Gliemann[12] has pointed out that as little as 4 μg/ml of crude bacterial collagenase degraded insulin; thus, cells should be washed several times to remove collagenase. In his experiments the best response to insulin was seen in fat cells washed at least 4 times.

Each aliquot of cells added to incubation tubes should contain the same number of cells. The cell suspension is added to a 1-ounce plastic bottle, the cells swirled to insure a uniform suspension and then a 0.2–1-ml aliquot is quickly drawn into a piece of plastic tubing (i.d. of about 3 mm) attached to a 1-ml glass syringe mounted on a small ring stand (lubricated to maintain a tight fit). Cells can also be dispensed with automatic pipettes using plastic tips. The end of the tips should be cut off to provide an i.d. at the narrowest point of 2 mm or more to reduce breakage of fat cells. With this procedure it is possible to dispense aliquots of cells into 50 tubes within 3 minutes or less without cell breakage. Gliemann[12] has described a slightly different procedure in which the cells are kept in a uniform suspension by agitation with a magnetic stirrer.

For incubation volumes up to 1.5 ml, cells are conveniently incubated in 17 × 100 mm disposable plastic culture tubes (polystyrene or polypropylene). Larger volumes are incubated in 1 ounce bottles either in an upright position for volumes up to 3 ml or on their side for volumes from 3 to 10 ml. Occasionally plastic bottles have been found to be contaminated with glycerol. This is undesirable in studies in which lipolysis is based on the accumulation of glycerol in the medium. Rinsing plastic bottles with distilled water has been sufficient to remove contaminants found to date.

Cells are viable with respect to glucose metabolism for up to 4 hours, but after that time they generally have a marked increase in basal glucose metabolism.[12,13] There are a wide variety of agents which stimulate glucose oxidation by fat cells, and many of these agents at higher concentrations lyse fat cells.[14] Apparently mild membrane damage results in an increased rather than decreased basal glucose metabolism of fat cells.

Brown fat cells are prepared by the same procedure described for white fat cells with the following exceptions. The brown adipose tissue is carefully dissected to remove adhering muscle and white adipose tissue. The tissue is then minced into many small pieces with scissors and incubated for 45–60 minutes with twice as much collagenase as used for isolation of white fat cells. Every 15 minutes the contents are removed from

[13] D. J. Galton and J. N. Fain, *Biochem. J.* **98,** 557 (1966).
[14] J. N. Fain, *Pharmacol. Rev.* **25,** 67 (1973).

the shaker and shaken vigorously by hand for 10–20 seconds. The centrifugation of brown fat cells must be carried out at higher speeds than are required for white fat cells. Generally 1 minute at 400 g is sufficient.

The isolation of brown fat cells from the dorsal interscapular adipose tissue of rats kept at room temperature is readily accomplished. We have had difficulty in isolating brown fat cells from cold-acclimated rats because the cells did not contain enough lipid to float during centrifugation. Brown fat cells can also be isolated from the brown adipose tissue of hamsters.[15]

The viability of fat cells can be assessed by staining fat cells with acridine orange[16] and examining them under the fluorescence microscope. If there is a visible layer of fat on top of the fat cells this is usually an indication of cell lysis during isolation. Cell lysis occurs primarily as a result of the presence of toxic factors in the albumin and collagenase or, in the case of large fat cells, from vigorous shaking during isolation.

The amount of fat cells added per tube has been usually based on the triglyceride content which is readily determined by direct analysis[17] or by measurement of total fatty acid content after alkaline hydrolysis.[1,2] It is also possible to weigh aliquots of the fat cells to determine the amount added per tube. In experiments in which one is comparing the response of fat cells from animals under conditions which might affect the lipid content per cell the data should be expressed per number of cells. This can be readily determined by counting fat cells in a siliconized cell counter used for experiments with blood cells.[12,16]

The response of fat cells as compared to intact adipose tissue has been most carefully studied with respect to glucose metabolism and the response to insulin.[18,19] Fat cells are more sensitive than intact adipose tissue to low concentrations of insulin.[18] This may result from uncovering of insulin receptors during isolation of fat cells, less insulin degrading activity of cells, or greater ease of insulin access to its receptor. The stimulation of glucose metabolism by insulin in cells is similar to that seen with intact pieces of adipose tissue except for a larger stimulation of lactate formation by insulin in cells.[18,19] The basal glucose metabolism is increased to a greater extent in cells than in tissue by raising the extracellular glucose concentration[18] which may indicate a greater leakiness of fat cells under some conditions.

If dilute suspensions of fat cells have been incubated for a period

[15] S. B. Prusiner, B. Cannon, and O. Lindberg, *Eur. J. Biochem.* **6**, 15 (1968).
[16] E. Lorch and G. Rentsch, *Diabetologia* **5**, 356 (1969).
[17] F. Snyder and N. Stephens, *Biochim. Biophys. Acta* **34**, 244 (1959).
[18] J. Gliemann, *Acta Physiol. Scand.* **72**, 481 (1968).
[19] J. N. Fain and L. Rosenberg, *Diabetes* **21**, 414 (1972).

of time with one agent there is a loss of responsiveness to hormones if the cells are transferred to fresh medium. One way to bypass this problem is to include the first agent in the medium used for digestion of cells and isolating the cells after periods of up to 2 hours. The cells can then be washed and incubated for periods of up to 2 hours with different agents.[20] The variability of the results for quadruplicate tubes from one lot of fat cells is far less than the variability for an experiment which is replicated 4 times. It is most misleading to present data as the mean ± standard error of quadruplicate tubes from a selected experiment. Such values give little indication of the reproducibility of results. The most reliable results are obtained by paired comparisons involving experiments replicated 3 times or more on as many different days unless the differences seen are fairly large.

[20] J. N. Fain, A. Dodd, and L. Novak, *Metab., Clin. Exp.* **20**, 109 (1971).

[54] Isolation of Specific Brain Cells

By Shirley E. Poduslo and William T. Norton

The isolation of neurons and glia from brain is a problem which has challenged neuroscientists for many years. Biochemists studying the nervous system have been forced to work either with whole tissue, carefully selected topographic areas, or hand-dissected neurons and clumps of surrounding neuropil. However, any gross dissection of the brain will yield a heterogeneous population of cells. The microdissection techniques are restricted to certain large and possibly highly specialized or atypical cells, cannot furnish pure glial cells, and, of course, limit studies to very small samples. Ideally, then, one desires a procedure that can efficiently and benignly resolve this tissue into its component cells so that their individual properties can be determined.

The incredible complexity of the microarchitecture of the brain, composed as it is of several cell types, each with multiple branching processes compactly intertwined, made the problem of bulk isolation of pure cell types seem insoluble until recently. The first major effort in this direction was carried out by Korey and co-workers[1] in 1956, who attempted to isolate glia from white matter. Although the cell separation was probably

[1] S. R. Korey, M. Orchen, and M. Brotz, *J. Neuropathol. Exp. Neurol.* **17**, 430 (1958).

crude, the basic techniques used by these investigators have been continued in most of the subsequent investigations.

After this first attempt there was a lapse of 10 years before this problem was pursued by others. Since 1967 there have been four major procedures developed for the bulk isolation of brain cells that show promise for a variety of investigations. We have only had extensive experience with our own methods and these are the only ones which will be described in detail. The other three procedures will be described briefly, but it is not the purpose of this chapter to comment extensively on the comparative merits of the different procedures. All four seem to be reasonable approaches toward the isolation of pure cell types, but at present no one method could be described as being totally satisfactory for all purposes.

The problems of isolating pure cell fractions in bulk fall into three main categories: (a) the selection of a medium which will maintain the integrity of the cells, (b) the disruption of tissue through screens (with or without prior treatment) to dissociate it into single cells, and (c) selection of a centrifugation procedure to separate the different cell types. All media used so far incorporate a high molecular weight material; either Ficoll, polyvinylpyrrolidone, or albumin. The mode of action of these polymers is not clear but they appear to lessen cell breakage. The screening procedures are all similar and involve forcing the tissue through nylon bolting cloth or stainless steel screens of successively smaller meshes. In some cases the tissue is used as soon as collected[2,3]; in others it is incubated first with[4] or without[5] enzymes. The centrifugation procedures developed have varied considerably, involving step gradients of sucrose or sucrose and Ficoll, or continuous gradients in zonal rotors. The g forces used in these procedures have also varied over a wide range.

The first of the newer procedures was developed by Rose.[2] About 20 rat cortices (12 g) were placed in a bag of 110 mesh (130 μ apertures) nylon bolting cloth and the tissue was gently "teased" through the bag, by stroking with a glass rod, into a medium of 10% Ficoll, 100 mM KCl, and 10 mM potassium phosphate buffer, pH 7.4. The suspension was screened a second time through a stainless steel screen of 40 μ pore size. The cell suspension was placed on a step gradient of 30% Ficoll in the KCl-phosphate buffer medium over 1.45 M sucrose. The tubes were centrifuged either at 125,000 g for 45 minutes or 53,000 g for 120 minutes. The layer on 30% Ficoll was the glial rich layer (now termed the

[2] S. P. R. Rose, Biochem. J. 102, 33 (1967).
[3] O. Z. Sellinger, J. M. Azcurra, D. E. Johnson, W. G. Ohlsson, and Z. Lodin, Nature (London), New Biol. 230, 253 (1971).
[4] W. T. Norton and S. E. Poduslo, Science 167, 1144 (1970).
[5] C. Blomstrand and A. Hamberger, J. Neurochem. 17, 1187 (1970).

"neuropil layer"), and the layer on 1.45 M sucrose was the neuronal layer. The yield of neurons was about 1.2×10^6 cells/cortex or 2.7 mg dry weight.

More recently this procedure has been modified.[6] After disruption of the tissue the cell suspension was filtered through a column of glass beads to remove capillaries. It was then separated by centrifugation as before.

The procedure developed by Hamberger and colleagues was initially described as utilizing the disruption of fresh tissue,[7] but was subsequently modified to include a preincubation step before disruption.[5] The final method is best described in one of the later papers.[8] Rabbit cortex was chopped mechanically and incubated for 30 minutes at 37° under O_2, in a medium containing 35 mM tris-chloride, pH 7.6, 120 mM NaCl, 5 mM sodium phosphate buffer, pH 7.6, 20 mM glucose, 2.5 mM MgCl$_2$, 2.5 mM ADP, and 2% Ficoll. This medium is quite different from the original one of 0.32 M sucrose, tris buffer, EDTA, and Ficoll.[7] After incubation the tissue was disrupted by passing it through a nylon mesh of 1 mm openings glued to a plastic syringe. The suspension was then diluted with 0.32 M sucrose, 100 mM NaCl, 10 mM tris buffer, pH 7.6, and screened through successively smaller nylon meshes. The suspension was centrifuged at 150 g to pellet the cells and the cell pellet was suspended in 20% Ficoll made up in the sucrose–salt–tris medium. The cell suspension was placed in the center of the gradient having beneath it 30% Ficoll over 40% sucrose and above it 15% Ficoll and 10% Ficoll. These Ficoll solutions were made up in the sucrose–salt medium. In some cases three layers were present above the suspension: 15, 12.5, and 10% Ficoll. The tubes were centrifuged 120 minutes at 55,000 g. The neuronal fraction was collected in the interface between 40% sucrose and 30% Ficoll and the glial fraction between 15 and 12.5% Ficoll. The yield from one rabbit cortex (900 mg protein) was 6 mg glial protein and 4 mg neuronal protein.

The third major procedure has been devised by Sellinger and co-workers.[3,9] Rat cortices were minced in a hypotonic medium consisting of 7.5% of polyvinylpyrrolidone (PVP), 10 mM CaCl$_2$, and 1% bovine serum albumin (BSA). The mince was disrupted by passing it successively through nylon bolting cloth of 333, 110, and 73 μ openings attached to a cut-off plastic syringe. The suspension was layered on a two-step gradient of 1.0 M sucrose plus 1% BSA over 1.75 M sucrose plus 1% BSA. The tubes were centrifuged at 75,000 g for 30 minutes. The neuronal perikarya formed a pellet. The glia on 1.75 M sucrose were suspended

[6] S. P. R. Rose and A. K. Sinha, *Life Sci. Part II* **9**, 907 (1970).
[7] C. Blomstrand and A. Hamberger, *J. Neurochem.* **16**, 1401 (1969).
[8] A. Hamberger, O. Eriksson, and K. Norrby, *Exp. Cell Res.* **67**, 380 (1971).
[9] D. E. Johnson and O. Z. Sellinger, *J. Neurochem.* **18**, 1445 (1971).

in a medium of 7.5% PVP, 5% Ficoll, and 1% BSA and passed through a mesh of 73 μ. The crude glial suspension was relayered onto a gradient of 30% Ficoll over 1.2 M sucrose over 1.65 M sucrose, all containing 1% BSA. This was centrifuged at 20,000 rpm/30 minutes. Neurons and capillaries were pelleted once again. The glia collected on 1.65 M sucrose. The glia were resuspended in 0.32 M sucrose and placed on a third gradient of 1.3 M sucrose over 1.65 M sucrose. This was centrifuged at 5,000 rpm/20 minutes. The glia collected on the 1.65 M sucrose interface. Yields of about 1.6 mg protein/g cortex were obtained in the neuronal fraction while 0.063 mg protein/g cortex were obtained in the glial fraction. The yield of neurons decreased in direct proportion to the age of the animal.

Several other procedures, and modifications of the main procedures, have appeared, some of which deserve comment. Satake and Abe[10] have isolated neurons from tissue pretreated with a mixture of acetone, glycerol, and water. A variation of this basic procedure was developed by Freysz et al.[11] for preparing both neurons and glia. Satake and co-workers[12] have also developed another procedure using a salt–Ficoll medium for the isolation of large multipolar cells from pig brain stem. Flangas and Bowman[13] claim to be able to isolate neurons and several fractions of glia by zonal centrifugation of a cell suspension prepared initially by disruption of brain through a tissue press in the absence of medium.

Oligodendroglia were first isolated by Fewster et al.[14] from bovine and rat white matter using a modification of the technique of Rose.[2] Takahashi et al.[15] have also isolated neuroglia from white matter using a Ficoll–CaCl$_2$ medium for disruption similar to that devised by Flangas and Bowman.[13] Hemminki and Holmila[16] separated neurons and glia by customary gradient techniques from a cell dispersion prepared by collagenase-hyaluronidase treatment. The method of Norton and Poduslo[4] has been modified by Giorgi[17] who used zonal centrifugation for the cell fractionation. Iqbal and Tellez-Nagel[18] have used a combination of the procedures of Norton and Poduslo[4] and Blomstrand and Hamberger[5] for isolating cells from postmortem tissue or from rat brain after incubation without enzymes.

[10] M. Satake and S. Abe, J. Biochem. (Tokyo) 59, 72 (1966).
[11] L. Freysz, R. Bieth, C. Judes, M. Sensenbrenner, M. Jacob, and P. Mandel, J. Neurochem. 15, 307 (1968).
[12] M. Satake, S. Hasegawa, S. Abe, and R. Tanaka, Brain Res. 11, 246 (1968).
[13] A. L. Flangas and R. E. Bowman, Science 161, 1025 (1968).
[14] M. E. Fewster, A. B. Scheibel, and J. F. Mead, Brain Res. 6, 401 (1967).
[15] Y. Takahashi, S. Hsu, and S. Honma, Brain Res. 23, 284 (1970).
[16] K. Hemminki and E. Holmila, Acta Physiol. Scand. 82, 135 (1971).
[17] P. P. Giorgi, Exp. Cell Res. 68, 273 (1971).
[18] K. Iqbal and I. Tellez-Nagel, Brain Res. 45, 296 (1972).

Several reviews of the principal procedures and the results obtained by them have appeared.[19-23] In the remainder of this chapter we describe in detail the methods we have developed for isolating cells and show their adaptability and the possible problems that may arise during their use.

Reagents

The suppliers of the chemicals we use are listed, but this should not be taken to mean other sources would not be adequate.

1. Glucose (Fisher Scientific Co.)
2. Sucrose (Fisher Scientific Co.)
3. Fructose (Sigma)
4. Bovine serum albumin, Cohn Fraction V (Sigma)
5. Trypsin, $2\times$ crystallized salt free (Nutritional Biochemicals)
6. Calf serum (Gibco)

Apparatus

1. Nylon monofilament screen cloth, ASTM 100 mesh (149 μ openings) (Tobler, Ernst and Traber, 420 Saw Mill River Road, Elmsford, N.Y.)
2. Stainless steel screens, ASTM 200 mesh (having 74 μ openings). Special order from Newark Wire Cloth Co., ordered through any laboratory supply company.
3. Centrifuge with a swinging bucket rotor capable of 3300 g (we use the Spinco SW-27 rotor).
4. Centrifuge for large-scale work-up. We use the Sorvall RC-3 with the HG-4 rotor which takes both 650 and 250 ml bottles as well as smaller tubes.
5. A phase microscope is essential for monitoring the fractions.

Solutions

The solutions mentioned in this chapter, and used throughout the cell isolation procedures are defined below.

1. *Buffer:* KH_2PO_4:NaOH, pH 6.0. For the stock solution (100 mM)

[19] S. P. R. Rose, *in* "Applied Neurochemistry" (A. N. Davison and J. Dobbing, eds.), p. 332. Davis, Philadelphia, Pennsylvania, 1968.
[20] S. P. R. Rose, *in* "Handbook of Neurochemistry" (A. Lajtha, ed.), Vol. II, p. 183. Plenum, New York, 1969.
[21] P. V. Johnston and B. I. Roots, *Int. Rev. Cytol.* **29**, 265 (1970).
[22] B. I. Roots and P. V. Johnston, *in* "Research Methods in Neurochemistry" (N. Marks and R. Rodnight, eds.), Vol. I, p. 3. Plenum, New York, 1972.
[23] S. E. Poduslo and W. T. Norton, *in* "Research Methods in Neurochemistry" (N. Marks and R. Rodnight, eds.), Vol. I, p. 19. Plenum, New York, 1972.

usually 20 ml 0.4 M KH$_2$PO$_4$ + 3 ml 0.4 N NaOH is diluted to 80 ml with H$_2$O. This is diluted 1:10 in the medium. All pH adjustments of the solutions are made at room temperature and are made using either 0.4 N NaOH or 0.4 M KH$_2$PO$_4$.

2. *Medium:* 5% glucose, 5% fructose, 1% albumin, 10 mM buffer, pH 6.0, specific gravity 1.046. The pH is adjusted to 6.0 by adding approximately 0.3–0.4 ml 0.4 N NaOH/100 ml medium. (Note that the buffer concentration in this medium is given incorrectly in Norton and Poduslo.[4]

3. *Buffered calf serum:* 20 ml serum plus 2 ml 100 mM buffer + 4 ml 0.4 M KH$_2$PO$_4$, pH 6.0.

4. *Trypsin:* 1% (1g/100 ml medium) for neuron and astrocyte isolation. 0.1% (0.1 g/100 ml medium) for oligodendroglia isolation.

The trypsin solution is prepared just before it is used. Trypsin is dissolved in the medium and filtered if the solution is hazy. The pH of the solution must be checked and readjusted to a pH of 6.0 if necessary.

5. *Sucrose solutions, pH 6.0, for gradients:*
   0.9 M, specific gravity 1.154
   1.35 M, specific gravity 1.204
   1.4 M, specific gravity 1.210
   1.55 M, specific gravity 1.227
   2.0 M, specific gravity 1.286

All of the sucrose solutions are dissolved in medium and the pH is adjusted to 6.0. Usually 0.01 ml 0.4 N NaOH/10 ml sucrose solution is required. All of the sucrose solutions mentioned in this paper are those made up in medium.

We recommend that all solutions be made fresh daily to minimize bacterial contamination and to give greater consistency. The use of frozen or old solutions can give rise to different cell distributions in the gradients. All solutions, with the exception of the trypsin, must be chilled to 0–4° before use, and all procedures, with the exception of the trypsinization step, are performed at this temperature. It is critical that all solutions be adjusted to pH 5.95–6.05. The highest yield of intact cells is obtained if the preparation is carried through without interruption.

The isolations can be done on a small scale or on a large scale for each cell type, and these variations are given. The large-scale preparations tend to become "messy" and less efficient. It is suggested that one start on a small scale and perfect the technique before attempting a large-scale procedure. Each preparation should be monitored by phase microscopy.

The method for isolation of neurons and astrocytes has been developed for rat whole brain trimmed of cerebellum. The oligodendroglial isolation uses calf corpus callosum and centrum semiovale. It is suggested that one use these tissues to repeat the procedures before adapting the method to other animals. Briefly, the procedure is as follows. The tissue is dissected, minced in medium, and incubated with trypsin at 37° for 30–90 minutes. The trypsinized tissue is cooled, treated with cold calf serum to inactivate the trypsin and then washed several times to eliminate the trypsin. The tissue pieces are then put through screens to obtain a cell suspension. The suspension is layered onto discontinuous sucrose gradients and centrifuged at fairly low speed (3300 g) for 10–20 minutes. The resulting layers are collected, pooled, diluted slowly with medium, and spun down gently. Astrocytes are purified on a second discontinuous sucrose gradient.

I. The Isolation of Neurons and Astrocytes on a Small Scale[4] (for 1–6 Rats)

Solutions Required

Medium: 800 ml

Sucrose in medium: 150 ml 0.9 M, and 50 ml each of 1.35 M, 1.4 M, 1.55 M, and 2.0 M

Trypsin: 60 ml of a 1% solution in medium (see however the comment on this solution in the *Precautions* section)

Buffered calf serum: 12 ml

One to six rats are decapitated, the brains quickly removed, and the cerebellums discarded. The tissue is placed in cold medium in ice until it is all collected. The brains are finely minced in a small amount of medium on a polyethylene plate. Either a razor blade or Stadie-Riggs blade can be used. Care must be taken to keep the tissue cold during the mincing. The minced tissue is then transferred to the trypsin solution and incubated with shaking at 37° for 90 minutes. Incubation has been carried out both in air and in 100% oxygen. Cells prepared after incubation in oxygen appear to respire better. After 90 minutes the mixture is cooled in ice. Twelve milliliters of cold buffered calf serum are added. The mixture is centrifuged at 800 rpm (140 g) for 4 minutes. The cloudy supernatant is discarded and the tissue pellet is washed twice by suspending it in 20 ml of cold medium and centrifuging at low speed as before.

The washed tissue pieces are suspended in cold medium for screening. A piece of nylon screen cloth is placed over the top of a Hirsch funnel so that it does not touch the perforated plate. The funnel is set up in a vacuum filtration device having a tube or beaker which is packed in ice to collect the cell suspension after it is screened. A convenient apparatus is the Fisher Filtrator. The nylon mesh on the funnel is moistened with medium and a slight vacuum started to hold the nylon in place. The tissue pieces are placed on the nylon in medium and are quickly and gently stroked through the nylon with a blunt glass rod. For 1–6 young rat brains this should take less than a minute. The nylon is rinsed carefully with a few milliliters of medium. The underside of the nylon is scraped clean with a spatula and this is also added to the suspension. Only slight vacuum is needed to facilitate this process. Too much vacuum causes the medium to foam.

The crude suspension is then put through the stainless steel screen. This screen is also placed on the funnel in the vacuum filtrator set up. It is wet with medium and a minimum amount of vacuum applied. The crude suspension is passed quickly through the screen at least 2 times. If the tissue is heavily myelinated, a third time may be necessary to insure complete dissociation of the tissue. The cell suspension is finally adjusted to 10 ml/g tissue with medium.

The discontinuous sucrose gradients are made in the 39-ml tubes of the Spinco SW-27 rotor. Ten milliliters of cell suspension are layered onto previously made gradients consisting of, from the top down, 14 ml of 0.9 M sucrose, 5 ml of 1.35 M sucrose, 5 ml of 1.55 M sucrose, and 5 ml of 2.0 M sucrose. For 6 rats, 6 tubes are used. The tubes are centrifuged at 5000 rpm (3300 g) for 10 minutes. Four layers are formed which are removed carefully with a Pasteur pipette. The top whitish layer, which consists of the 0.9 M sucrose-suspension interface and much of the 0.9 M sucrose layer, contains myelin, processes, small particles, and some cells and is discarded. The layer on 1.35 M sucrose is the crude glial layer and is reserved for further purification. The layer on 1.55 M sucrose contains a mixed population of neurons, astrocytes, oligodendrocytes, and capillaries, and is discarded. The neuronal perikarya form a layer on 2.0 M sucrose and require no further purification. There is usually little or no pellet.

The crude glial layer (on 1.35 M sucrose) is further purified on a second gradient. This layer is diluted slowly with medium to 25 ml (it must be diluted at least 1:1) and layered onto a two-step gradient of 5 ml of 0.9 M sucrose over 5 ml of 1.4 M sucrose. It is possible to combine two glial layers from the first gradient centrifugation step for purification on one gradient tube. These tubes are centrifuged at 5000 rpm (3300 g), but for 20 minutes this time. The purified astrocytes collect at the 1.4–0.9 M sucrose interface. Contaminating cells are pelleted through 1.4 M sucrose.

Both the neurons and the astrocytes can be concentrated by diluting them slowly with cold medium at least fivefold (an aliquot can be taken here for a cell count) and centrifuging them at 1500 rpm (630 g) for 10 minutes.

II. The Isolation of Neurons and Astrocytes on a Large Scale (for 50 Rat Brains)

Solutions Required

Medium: 4800 ml
Sucrose in medium: 1000 ml of 0.9 M, 400 ml each of 1.35 M and 1.55 M, and 200 ml each of 1.4 M and 2.0 M
Trypsin: 400 ml of a 1% solution in medium (see comment in *Precautions* section)
Buffered calf serum: 80 ml

The volumes of solutions given are for processing 50 rat brains at one time. Of course fewer may be used with appropriate modifications. The rats are killed, one or two at a time, the brains removed, and the cerebellums discarded as before. The brains are placed in medium in ice until all of the tissue is collected. The tissue is finely minced with a razor blade or Stadie-Riggs blade as before. The mince is added to the filtered trypsin solution and incubated with shaking at 37° for 90 minutes. The incubated tissue is then cooled, 80 ml of buffered serum are added, the tissue centrifuged at 800 rpm (140 g) and washed twice with 50–100 ml of medium.

To screen this large amount of tissue, a larger (8–10 inch diam) funnel must be used. It is sometimes difficult to get an even vacuum through the nylon mesh when it is just placed over a funnel, so the nylon is stretched tightly across the top and taped to the sides of the funnel. Then the vacuum is checked after the nylon is moistened with medium. The tissue pieces are taken up in 0.9 M sucrose in medium rather than just medium alone. The tissue is stroked quickly through the nylon into a beaker in ice. This crude suspension is then passed through the stainless steel screen at least 3 times to completely dissociate the tissue.

The cell suspension is diluted to 400 ml with 0.9 M sucrose. The sucrose gradients are now made in the 250-ml polycarbonate bottles which fit the Sorvall RC-3 HG-4 rotor. Portions (100 ml each) of the cell suspension are layered on three-step gradients consisting of 75 ml of 1.35 M sucrose, 70 ml of 1.55 M sucrose, and 15 ml of 2.0 M sucrose. These gradients are made readily in a cold room by adding the sucrose solutions from separatory funnels. The bottles are balanced, tightly capped, and centrifuged at 3600 rpm (3400 g) for 15 minutes.

The appearance of the tubes is similar to those in the small-scale preparation, although the zones are not as discrete. As before the neurons collect on the 2.0 M sucrose layer, the mixed cell layer on 1.55 M sucrose, and the crude astrocyte layer on 1.35 M sucrose. Most of the 0.9 M layer is discarded. The layers can be removed using Pasteur pipettes or by gentle aspiration directly into iced bottles.

Unfortunately, it is difficult to purify the astrocytes further by using gradients in the Sorvall centrifuge. Instead the astrocytes are diluted to either 300 or 450 ml depending upon the collected volume of the layer (at least a 1:2 dilution with medium must be made). Using the 39-ml tubes of the Spinco SW-27 rotor, 25 ml of the crude glial suspension is layered over a two-step gradient of 7 ml of 0.9 M sucrose over 7 ml of 1.4 M sucrose. These tubes are centrifuged at 5000 rpm (3300 g) for 20 minutes. Two or three centrifugation runs must be made to process the total volume of crude glial suspension. The astrocyte layers on 1.4 M sucrose can then be combined, diluted slowly with medium to 550 ml, and concentrated by centrifugation in 150 ml tubes of the Sorvall RC-3 at 3600 rpm for 10 minutes. The neurons obtain on the first gradient can also be diluted and pelleted in exactly the same way.

III. The Isolation of Oligodendroglia on a Small Scale[24] (for 60 g of White Matter)

Solutions Required

Medium: 2000 ml
Sucrose in medium: 800 ml of 0.9 M, 150 ml each of 1.4 M and 1.55 M
Trypsin: 200 ml of 0.1% solution in medium
Buffered calf serum: 40 ml

Two calf brains are removed from the animals at the abattoir as quickly as possible, packed in ice, and transported to the laboratory. In our laboratory this takes 2–3 hours. White matter from the corpus callosum and centrum semiovale is dissected free of gray matter and placed in medium in ice until the dissection is completed. Usually 30–40 g wet weight of white matter can be obtained per brain. The white matter is very finely minced using either a Stadie-Riggs blade or a sharp chef's knife. Since mincing may easily take 5–10 minutes, the tissue on the polyethylene plate must be kept on ice during this procedure.

The minced tissue is then added to 200 ml of the filtered trypsin solution and incubated with shaking at 37° for 90 minutes. It should be noted that the trypsin concentration is 0.1% instead of 1%. After incubation the tissue is cooled, 40 ml of buffered serum are added, and the tissue is centrifuged at 800 rpm for 4 minutes in the Sorvall centrifuge. The tissue is then resuspended in 25–50 ml of medium 2 times more and recentrifuged.

For screening, the washed trypsinized tissue pieces are taken up in 0.9 M sucrose. The nylon is placed over a medium-sized (4 inch diam) funnel and moistened and vacuum is applied. The tissue is passed through the nylon using as much vacuum as is necessary and somewhat more vigorous stroking than required for rat brain. The suspension is filtered 3 times through the 200 mesh stainless steel screen.

The sucrose gradients are made in the 60-ml tubes of the Spinco SW-25.2 rotor. The gradient consists of 8 ml of 0.9 M sucrose, 8 ml of 1.4 M sucrose, and 8 ml of 1.55 M sucrose. The cell suspension is diluted to 420 ml with 0.9 M sucrose and 35 ml portions are layered onto each gradient. The gradients are centrifuged at 5000 rpm (3300 g) for 10 minutes. Four runs are necessary to process all of this tissue. The top layer floating on 0.9 M sucrose is a thick pasty layer of myelin, axons, etc. The layers on 0.9 and 1.4 M sucrose are discarded. These layers contain oligodendroglia with varying amounts of myelin attached to them, but they are heavily contaminated with extraneous myelin and processes in the 0.9 M layer and with endothelial cells on the 1.4 M layer. The main oligodendroglial layer is on and extends into the 1.55 M sucrose layer. These cell layers are collected, pooled, and diluted slowly with medium to 500 ml. An aliquot can be removed here for a cell count. The cells can be concentrated by centrifuging at 1500 rpm (630 g) for 10 minutes in the Sorvall centrifuge.

IV. The Isolation of Oligodendroglia on a Large Scale[24]
(for 200 g of White Matter)

Solutions Required

Medium: 4500 ml
Sucrose in medium: 1500 ml of 0.9 M, 600 ml of 1.4 M, and 500 ml of 1.55 M
Trypsin: 600 ml of 0.1% solution in medium
Buffered calf serum: 120 ml

Six calf brains can be processed in 1 day. About 200 g of white matter can be dissected from the corpus callosum and centrum semiovale of 6 calf brains, each weighing about 300 g. The tissue, which is kept in the medium in ice, is divided into 4 portions and each portion is minced thoroughly using a sharp chef's knife. The tissue is minced, turned over and minced again until finely divided. This procedure may take 30 min-

[24] S. E. Poduslo and W. T. Norton, *J. Neurochem.* **19**, 727 (1972).

utes. The tissue mince is then put into 600 ml of filtered trypsin solution and is incubated with shaking at 37° for 90 minutes.

The trypsinized tissue is then cooled and 40 ml of cold buffered calf serum are added. The mixture is centrifuged at 800 rpm (140 g) for 4 minutes in the Sorvall. The tissue is resuspended in 50–100 ml of medium and recentrifuged 2 times more. Efficient screening of this much tissue requires a large funnel. We use a two-piece plastic one of about 10 inches diam. Nylon is placed between the two parts of the funnel and taped tightly to the sides. The nylon is moistened with medium and vacuum applied. The tissue pieces are again taken up in 0.9 M sucrose and the whole tissue mixture is put onto the nylon. It is then stroked through using a glass rod, the bottom of a small beaker, or by hand. When most of the tissue is through, the nylon is rinsed and removed. The underside of the nylon is also scraped into the suspension. The crude suspension is then put through the stainless steel screen a minimum of 3 times.

The gradients are made in the 650-ml bottles which fit the HG-4 rotor of the Sorvall RC-3 centrifuge. The separatory funnels are again used to prepare the gradients in a cold room. One-hundred milliliters of 0.9 M sucrose is layered onto 140 ml of 1.4 M sucrose over 100 ml of 1.55 M sucrose. The cell suspension is adjusted to 1000 ml and 250 ml portions are layered onto each gradient. The bottles are balanced, capped, and centrifuged at 3600 rpm (3400 g) for 20 minutes.

The gradient zones are less clear, but the results are the same as for the small-scale preparation. There is a thick, pasty, myelin layer floating on top and a very hazy zone extending downward to the reddish layer on 1.4 M sucrose. All of this top portion of the gradient including the reddish layer is discarded. There are many oligodendroglia here, but again they are difficult to purify from the myelin and endothelial cells. The bottom layer including the 1.4–1.55 M sucrose interface is taken completely. If the dissection of the tissue was carefully done the pellet will also contain many oligodendroglia. However, if gray matter was included then neurons will contaminate the pellet. This oligodendroglial layer is diluted slowly to 200 ml with medium and concentrated by centrifugation in 150 ml tubes at 3600 rpm for 10 minutes in the Sorvall RC-3.

Precautions

If the procedures are followed strictly, good yields of morphologically intact cells are obtained. However, variable or poor yields can usually be traced to specific deviations from the procedure. We have found, both from our own experience and that of others, that intact astrocytes are the most difficult cell to obtain. Low yields of fragmented astrocytes can be caused by essentially any variation of the procedure, i.e., solutions not at pH 6.0, especially during incubation; less than 90 minutes incubation; too rapid dilution of cell layers; etc. Incubation for less than 90 minutes produces astrocytes still embedded in neuropil giving them a "ball-like" shape (see, e.g., Fig. 4 in Giorgi[17]).

Although the procedures for the isolation of neurons and astrocytes are given exactly as developed and still used successfully in New York, one of us (S.E.P.) has recently found that when the procedure was used in Baltimore, the amount of trypsin had to be reduced drastically. When the procedure described herein was followed, very granular neurons with swollen nuclei were obtained, and many free nuclei were found in the neuronal layer. The astrocyte fraction consisted mostly of broken processes with few cell bodies. When 0.1% trypsin was substituted for 1% trypsin, but without any other changes in the procedure, the method could be duplicated successfully. We are still unable to explain this discrepancy since several batches of trypsin were tried with the same results. This points up the rather fragile nature of such procedures and the absolute necessity of monitoring each preparation. The microscopic appearance of the cell layers should be carefully compared with our published photographs.[4,23,25] If they differ significantly then it is probable the procedure is not being carried out under optimal conditions.

If the tissue is heavily myelinated, it must be minced as fine as possible to facilitate later screening. The initial screening step should be done under continuous vacuum so the tissue can be sieved as quickly as possible. Also, the cell suspension should be prepared in 0.9 M sucrose in medium so that during centrifugation the myelin and myelinated axons will float upward while the cells migrate downward.

After incubation, it is very important to inactivate the trypsin completely. This is done by cooling the minced tissue and adding cold buffered calf serum. Washing the tissue pellet several times insures removal of the trypsin. Once the tissue is cooled, it must remain so during the isolation. Cell contact with sucrose should also be kept to a minimum. Because of their large size the cells segregate in the gradient quickly at low speeds. Since the gradient media are hypertonic and the cells are osmometers, no attempt is made to reach density equilibrium in this step. Rather separation is achieved by a combination rate-zonal technique. The use of higher speeds or longer centrifugation times will cause the cells to migrate to higher density zones, and eventually different cell types will again become mixed.

Dilution of the hypertonic sucrose solutions so that the cells can be

[25] C. S. Raine, S. E. Poduslo, and W. T. Norton, *Brain Res.* **27**, 11 (1971).

concentrated must be done very gradually since the cells are susceptible to fragmentation when the medium is changed. The addition of small aliquots of medium with slow mixing seems the most favorable procedure. Only low-speed centrifugation is required to pellet the cells.

General Comments

Before this method was finally adopted, many other approaches were attempted with varying degrees of negative results. Tissue disruption by either mechanical or enzymatic means or various combinations of both was evaluated. In our initial attempts at mechanical dissociation alone, poor neuronal suspensions could be obtained from young animals, but no intact astrocytes were ever seen. Various hydrolytic enzymes besides trypsin (i.e., neuraminidase, hyaluronidase, papain, pronase, crude pancreatin, and various mixtures of these) were tried with mostly negative results. Initial trials with trypsin involved the use of sucrose or 10–15% glucose solutions. Morphologically intact neurons and astrocytes were produced with the high glucose concentrations. However, the cells did not take up oxygen nor did brain slices in the same medium. The present medium was adopted after many empirical trials. This medium in a combination with trypsin produced the highest yield of intact cells which still respired to some degree and which excluded dyes impermeable to living cells.

One of the strongest criticisms of this method has been the use of trypsin to assist in tissue disruption. It is probable that trypsin acts on the intercellular glycoproteins reducing cell-to-cell adhesion. It has been noted by others that mechanical disruption is greatly facilitated by prior incubation[5,8,18] or postmortem autolysis.[18] In these cases it is reasonable to assume that intercellular adhesion is being lessened by endogenous proteolytic action. In our hands it did not seem possible to obtain intact glia from rat brains of widely varying stages of maturity without some type of prior incubation. Although the concentrations of trypsin seem high and times of trypsinization long, the following factors must be considered: the pH of the medium is not optimal for trypsin activity; there is considerable albumin in the mixture; and the trypsinization is carried out on tissue pieces, not single cells. Thus the enzyme is not being used efficiently. Even so this incubation in our particular medium is necessary and enables the tissue to be disrupted without disrupting the cells. A further indication that the action of trypsin in this medium is not as deleterious as one might think is that myelin isolated from the 0.9 M sucrose layer of the cell isolation procedure has the same lipid and protein content as that isolated by conventional methods. Yet myelin has at least one protein (basic protein) which is very labile to trypsin. It is also possible to observe actively motile ciliated ependymal cells in the mixed cell layer on 1.55 M sucrose after a 30–60-minute trypsin incubation.

We have avoided pelleting either the initial cell suspension or the separated cells until the end of the procedure. Centrifuging and resuspending cells increases fragmentation and lowers cell yields. Some investigators who find the presence of albumin in the final cell pellet undesirable, have washed out the albumin by suspending the cells in medium made without albumin and recentrifuging at low speed.

These cell pellets must be handled carefully since they are subject to deterioration on storage. Storing washed pellets in the freezer, or frozen-dried cells in a desiccator, will cause a breakdown of lipid by autoxidation.[26] Protein degradation also occurs, possibly by the same mechanism. Thus such constituents must be extracted from fresh cells. The first lipid to oxidize is the plasmalogen form of ethanolaminephosphatides producing lyso compounds.

These isolated cells can be used as a starting point for subcellular fractionation. We have obtained both a plasma membrane fraction and a "myelin" fraction from isolated oligodendroglia.[27]

Adaptations

This method has been used successfully to isolate neurons and astrocytes from mouse, dog, and cat brain. No modifications were needed. However, cells from brains of different animals or cells from specific areas of the brain may not necessarily all have the same density nor be of the same size as in rat or calf brain. Thus the sucrose gradients should be modified accordingly to accommodate any differences; for example, preliminary experiments show that this general method can be used to isolate neurons from human brain tissue obtained at autopsy (the amount of trypsin must be reduced to allow for any proteolysis that has occurred). Neurons from mature adults have high concentrations of lipofuchsin which enables them to collect on a sucrose layer less dense than 2 M. In our laboratory oligodenroglia have only been isolated from calf white matter. It seems entirely feasible, however, that oligodendroglia could be isolated from any species with only slight modifications of the procedure given here.

If only neurons are desired, a 30-minute trypsin incubation is sufficient for their isolation. Cells isolated under these conditions may have

[26] W. T. Norton and S. E. Poduslo, *J. Lipid Res.* 27, 11 (1971).
[27] S. E. Poduslo and W. T. Norton, *Trans. Amer. Soc. Neurochem.* 4, 123 (1973)

longer processes than those obtained after a 90-minute incubation. Successful isolation of astrocytes and oligodendroglia, however, requires the full 90 minute incubation.

If one desires to isolate cells from the CNS of species different from the ones mentioned, we can recommend the following approach. Trypsinize the tissue as in the small-scale procedure I. If the tissue is well-myelinated, prepare the cell suspension in 0.9 M sucrose in medium rather than in medium alone. If the desired cells are very large (i.e., anterior horn neurons) a larger mesh screen (100 mesh) might be used for the final screening instead of the 200 mesh screen. Choosing a proper step-gradient can be facilitated by initial centrifugations on continuous sucrose gradients (0.9–2.0 M sucrose) at the recommended speeds and times. The numbers of each cell type can be determined by phase microscopy throughout the gradient and the type and degree of contamination evaluated. It should then be possible with a minimum of trials to design a step-gradient which combines the greatest yield of cells with the least contamination.

It is possible to eliminate contamination by procedures other than density gradient centrifugation. The large neurons from rabbit spinal cord collect on a 1.7 M sucrose layer but are contaminated with many oligodendroglia. Differential centrifugation of this layer results in considerable improvement in neuronal purity. We have recently been reinvestigating the oligodendroglial separation[28]: If one uses a gradient consisting of 0.9 M sucrose, 1.15 M sucrose, and 1.80 M sucrose instead of 0.9 M sucrose, 1.4 M sucrose, and 1.55 M sucrose, an increased yield of cells is obtained on the 1.80 M sucrose layer. However, this layer is badly contaminated with capillaries. These can be almost completely removed by passing the cell layer through a bed of glass beads of 50–100 μ diam as recommended by Rose and Sinha.[6] However, if neurons are present in the cell suspension, they may contaminate the oligodendroglial layer.

Chemistry and Morphology

Neurons, astrocytes, and oligodendroglia differ strikingly from one another when examined by phase microscopy.[4,25] The neurons are large cells having a large nucleus, a single nucleolus, and abundant cytoplasm. The astrocytes have much smaller perikarya with a prominent nucleus, little visible cytoplasm but very highly branched processes. The oligodendroglia are the smallest cells, not much larger than erythrocytes. By light microscopy it is difficult to distinguish these cells from nuclei, except under oil (1000× magnification), where a small round nucleus surrounded

---

[28] T. Abe and W. T. Norton, in preparation.

by a rim of cytoplasm can be discerned. Many of the oligodendroglia have a small process.

It was essential to use a tissue culture technique to prepare these isolated cells for electron microscopy.[25] The usual methods of fixation used for tissue or subcellular fractions produced very fragmented cells. Since many people have had difficulty in obtaining good electron micrographs of isolated cells the technique will be repeated in some detail.

After sucrose density gradient centrifugation the cell layers were transferred to cellulose nitrate tubes in an ice bath and 0.01 volume of 50% glutaraldehyde was added to partially fix the cells. The suspension was then gradually diluted with 4 volumes of medium and centrifuged at 600 g for 10 minutes at 4°. The supernatant was discarded and the pellet was fixed with 2.5% glutaraldehyde in Millonig's buffer (0.05 M NaH_2PO_4–NaOH, pH 7.3, +0.015 M glucose) at 0° for 1 hour. The pellets were then post-fixed in 1% osmic acid in the same buffer for 30 minutes. The pellets were dehydrated by 5 minute treatments each of a graded series of ethyl alcohol (30, 50, 70, 90, 95, 100, and 100%). During the fixation and dehydration the pellet was not disturbed. Solutions were carefully layered over the pellet, kept in ice for the desired time, and then carefully removed. The pellets were removed as flat discs from the cellulose nitrate tubes and transferred to glass vials. They were then treated twice with propylene oxide, 10 minutes each, and infiltrated with a 1:1 mixture of propylene oxide:epon for 1 hour. They were left in epon overnight at 4° and finally placed in capsules with new epon which was polymerized at 45 and 60° over a 36-hour period. Specimens were later cut from the hardened plastic and reoriented end on and reembedded in gelatin capsules. Sectioning of specimens thus embedded permitted simultaneous examination of all strata within the pellets. Grids were stained with lead citrate and uranyl acetate and carbon coated.

Electron micrographs show the cells to have most of the typical features observed *in situ*. They retain both a triple layered plasma membrane and nuclear membrane, Golgi apparatus, endoplasmic reticulum, and somewhat swollen mitochondria. The neurons have abundant Nissl substance and 240 Å microtubules are present in the processes. Astrocytes have elongated cell bodies, pale cytoplasm, homogeneous nucleochromatin, and occasional groups of 100 Å neurofilaments. The oligodendroglial cells are light, medium, and dark staining,[29] are rounded with dense cytoplasm, often have clumped nuclear chromatin, and always contain microtubules. Often loops of myelin directly associated with the plasma membrane are found.

[29] S. Mori and C. P. LeBlond, *J. Comp. Neurol.* **139**, 1 (1970).

The neurons and oligodendroglia are over 90% pure as evaluated by both light and electron microscopy. The astrocytes are much less pure because of contamination by processes and occasional broken neurons. The isolated neurons and astrocytes account for 11% of total rat brain DNA.[26] The yield of oligodendroglia is also about 11% of the total DNA of white matter.[24]

The yields and chemical compositions of these three cell types are compared in Table I. On an individual cell basis the astrocytes are heavier than neurons and much heavier than the oligodendrocytes. All of these weights are consistent with the microscopic dimensions of the cells and the fact that the astrocytes retain elaborate processes. Neurons and astrocytes have similar DNA-RNA ratios, but neurons have a higher percentage of RNA, as expected from their abundant Nissl substance. The oligodendroglia have a much higher DNA-RNA ratio than the other two cells,

TABLE I

COMPARISON OF THREE ISOLATED CELL TYPES[a]

	Rat		Bovine Oligodendroglia
	Neurons	Astrocytes	
Yield, 10^6 cells/g fresh tissue	17	3.5	11.4
Dry weight (pg/cell)	178	590	~50
RNA (pg/cell)	24.2	29.1	1.95
DNA (pg/cell)	8.18	11.2	5.14
DNA/RNA ratio	0.34	0.38	2.6
2′,3′-Cyclicnucleotide-3′-phosphohydrolase (μmole of P_i/minute/mg dry weight)	0.42	0.41	3.23
$CHCl_3$–CH_3OH insoluble residue (% of dry weight)	73.9	58.7	69.6
Total lipid (% of dry weight)	24.1	38.9	29.5
Gangliosides (% of dry weight)	0.23	0.60	0.25
Lipids (% total lipid)			
Cholesterol	10.6	14.0	14.1
Total glycolipid	2.1	1.8	9.9
Cerebroside	—	—	7.3
Sulfatide	—	—	1.5
Total Phospholipid	72.3	70.9	62.2
Phosphatidylcholine	39.9	36.3	29.4
Ethanolaminephosphatides	18.2	20.1	14.0
Sphingomyelin	3.2	3.7	7.1
Phosphatidylserine	3.9	5.2	4.7
Phosphatidylinositol	4.9	3.5	4.1

[a] Values taken from Poduslo and Norton.[24]

owing to the low volume of cytoplasm, but the percentage of RNA is actually rather high. The astrocytes have the highest lipid content, and the lipid composition of astrocytes and neurons is very similar, both having little cerebroside or sulfatide. The oligodendroglia, however, have appreciable amounts of these two galactolipids, as would be expected since they are the myelin-generating cells. The ganglioside concentrations are surprising since they are believed to be primarily neuronal constituents. We feel they are probably constituents of all plasma membranes in the nervous system, and are therefore higher in astrocytes because of the very high surface-volume ratio of these isolated cells. However, all of these ganglioside figures are quite low relative to their concentration in gray matter, and none of these data should be taken as an argument against the major concentration of gangliosides being in the dendritic tree and synaptic endings which are lost from the neuron on isolation.

A final word of caution: It is not enough to follow this procedure, as one would a method for myelin or mitochondrial isolation, and be confident that purified fractions will be obtained. Apparently none of these cell isolation procedures is easy to reproduce exactly the first time by inexperienced investigators, or even by experienced investigators in other laboratories. There are apparently critical steps that we are not aware of yet, but in which small deviations can make major differences. Thus, it is essential to monitor the fractions and make minor adjustments in the procedure if necessary. On the other hand, most investigators using the procedures described here have reported excellent success with very few trials.

[55] Isolation, Purification, and Metabolic Characteristics of Rat Liver Hepatocytes

By S. R. WAGLE and W. R. INGEBRETSEN, JR.

Isolation of pure hepatocytes free of other cell types allows a direct approach to study the biochemical properties and its regulation by various nutritional and hormonal factors involved in hepatic metabolism.

There have been many attempts during the past 40 years to produce viable isolated liver cells.[1,2] Early methods depended upon the use of

[1] P. Rous, and J. W. Beard, *J. Exp. Med.* **59,** 577 (1934).
[2] W. C. Schneider and V. R. Potter, *J. Biol. Chem.* **149,** 217 (1943).

mechanical force to dissociate the cells from the liver.[2-5] Later various chelating agents (citrate, tetraphenylboron, and EDTA) in conjunction with mechanical disruption were used.[6-10] These methods are ineffective in producing morphologically intact or metabolically active cells.[11-15]

Shortly after Rodbell[16] demonstrated that adipocytes could be isolated by incubating fat tissue in the presence of collagenase, Howard et al.,[17,18] described a similar method for isolating liver cells by incubating liver slices with collagenase and hyaluronidase (50 mg and 100mg/100 ml, respectively). These enzymatically prepared cells appeared to be morphologically intact and metabolically active.[17,18] Berry and Friend[19] reported that the Howard and Pesch[18] method could be modified by perfusing the liver with collagenase and hyaluronidase in an *in situ* recirculating liver perfusion apparatus prior to incubation. This modification greatly increased cell yields and improved the subcellular integrity of the isolated liver cells. Although this method has been widely accepted, the procedure described here reports some important modifications of the Berry and Friend[19] method which allows for even higher yields of metabolically active hepatocytes that are completely free of other cell types. In contrast to the Howard and Pesch[18] or the Berry and Friend[19] method, the current procedure exposes the liver to low concentrations of collagenase (10–20 mg/100 ml) for a very short period of time (10–15 minutes).

Isolation and Purification of Hepatocytes

Animals. Fed (180–200 g, 6–8 weeks old; and 300–380 g, 16–20 weeks old) or 18–24-hour fasted male Cox rats were used for these studies. All

[3] P. M. G. St. Aubin and N. L. R. Bucher, *Anat. Rec.* **112**, 797 (1952).
[4] J. P. Kaltenbach, *Fed. Proc., Fed. Amer. Soc. Exp. Biol.* **11**, 237 (1952).
[5] J. P. Kaltenbach, *Exp. Cell Res.* **7**, 568 (1954).
[6] N. G. Anderson, *Science* **151**, 627 (1953).
[7] M. V. Branster and R. K. Morton, *Nature (London)* **180**, 1283 (1957).
[8] S. T. Jacob and P. M. Bhargava, *Exp. Cell Res.* **27**, 453 (1962).
[9] C. Rappaport and G. B. Howze, *Proc. Soc. Exp. Biol. Med.* **121**, 1010 (1966).
[10] J. A. Ontko, *Biochim. Biophys. Acta* **137**, 13 (1967).
[11] P. R. Kerkof, S. Smith, H. T. Gagne, D. R. Pitelka, and S. Abraham, *Exp. Cell Res.* **58**, 445 (1969).
[12] M. N. Berry and F. O. Simpson, *J. Cell Biol.* **15**, 9 (1962).
[13] C. Harris and C. A. Leone, *J. Cell Res.* **28**, 405 (1966).
[14] H. Kalant and R. Miyata, *Biochem. J.* **86**, 336 (1963).
[15] G. F. Lata and J. Reinertson, *Nature (London)* **179**, 47 (1957).
[16] M. Rodbell, *J. Biol. Chem.* **239**, 375 (1964).
[17] R. B. Howard, A. K. Christiansen, F. A. Gibbs, and L. A. Pesch, *J. Cell Biol.* **35**, 675 (1967).
[18] R. B. Howard and L. A. Pesch, *J. Biol. Chem.* **243**, 3105 (1968).
[19] M. N. Berry and D. S. Friend, *J. Cell Biol.* **43**, 506 (1969).

rats were maintained on Purina Laboratory Chow and tap water *ad libitum*.

Chemicals. All organic chemicals were of reagent quality. Biochemical reagents were purchased from Sigma Chemical Co., St. Louis, Mo. Radioactive tracers were purchased from INC, Irvine, Calif.

Reagents and Buffers. Modified calcium and glucose-free Hanks buffer[20] was made up in a stock solution (80.0 g NaCl, 4.0 g KCl, 0.6 g $Na_2HPO_4 \cdot 2H_2O$, 0.6 g KH_2PO_4, and 2.0 g $MgSO_4 \cdot 7H_2O$ per liter) and stored at 4°. The stock solution was diluted 1:9 with glass distilled water. Diluted Hanks buffer was gassed for 30 minutes with 95% O_2 and 5% CO_2 at 37° before the addition of $NaHCO_3$ (final concentration 25 mM). Thirty minutes after the addition of $NaHCO_3$ the pH was checked with a Beckman pH meter and adjusted to 7.4 with 1 M $NaHCO_3$. This buffer was used for isolating liver cells. Umbreit Ringer 25 mM bicarbonate buffer was used for all metabolic studies. All buffers were kept equilibrated with 95% O_2 and 5% CO_2 and maintained at 37° throughout the procedure.

Surgery

Portal Vein and Inferior Vena Cava Cannulation. The liver donor is anesthetized with sodium pentobarbital (60 mg/kg, Abbott) and placed on a dissecting board. The peritoneal cavity is opened by making two lateral incisions starting at the midline of the lower abdomen and proceeding anteriorly up both sides of the rat until the diaphragm is reached but not penetrated. The resulting flap of skin and muscle is then lifted, exposing the peritoneal cavity. The portal vein and the inferior vena cava are isolated. Two ligatures (silk 4–0) are loosely placed around the portal vein, one anterior to the entrance of the superior mesenteric vein and the other just posterior to the entrance of the splenic vein. A third ligature is then placed loosely around the inferior vena cava just anterior to the entrance of the right renal vein.

A small incision is made on the ventral surface of the portal vein between the two ligatures and a small glass rod is then rapidly placed in the incision to keep the vein open. The portal vein is rapidly cannulated (within 5–10 seconds) with a 2–3 cm piece of Intramedic tubing (P.E. 205) attached to a 1-liter pediatric microdrip bottle (McGaw) suspended 1 meter above the operating table. The bottle contains 200–400 ml of Ca^{2+}-free modified Hanks buffer maintained at 37° and equilibrated with 95% O_2 and 5% CO_2. The two ligatures around the portal vein are then tied. The inferior vena cava is immediately severed below the right

[20] J. H. Hanks and R. E. Wallace, *Proc. Soc. Exp. Biol. Med.* **71**, 196 (1949).

kidney. This prevents the liver from swelling and allows the perfusate to flow freely through the liver.

The ligamentum teres hepatis, diaphragm, and rib cage are then cut exposing the thorax cavity. A loose ligature is placed around the inferior vena cava just anterior to the diaphragm. An incision is made in the right atrium of the heart and 3–4 cm piece of intramedic tubing (P.E. 205) is used to cannulate the inferior vena cava via the atrium. The ligatures around the inferior vena cava in both the thorax and the abdomen are then tied. From this point until the liver is placed in the perfusion box, the tissue is perfused with oxygenated buffer at a rate of 30–60 ml/minute.

Liver Removal. The diaphragm is completely detached from the rib cage and the inferior vena cava cannula is freed from the surrounding mesentery. The liver is moved anteriorly to expose the caudate lobe of the liver. The mesentery supporting the caudate lobe is carefully removed and the esophagus severed. The liver is lifted by the thoracic inferior vena cava ligature and the portal cannula and dissected out of the rat. The surgery from laparotomy to total hepatectomy should take 4–6 minutes to complete.

Liver Perfusion. The liver is placed in a Miller perfusion apparatus (Fig. 1) and perfused for 15 minutes with 100 ml of modified Hanks buffer containing 1.5 g of bovine serum albumin and 10 mg each of streptomycin sulfate and penicillin G. Following this initial perfusion period, collagenase (10–20 mg, Sigma Type 1, 130 units/mg; Sigma Lot 101C-0160) is added and the perfusion is continued for 10–15 minutes at a flow rate of 25–35 ml/minute.

Cell Isolation. After being perfused with collagenase for 10–15 minutes, the liver is removed, placed in a 250-ml plastic beaker containing 30 ml of modified Hanks buffer, finely minced with scissors, and bubbled gently with 95% O_2 and 5% CO_2 for 1 minute. The resulting tissue suspension is poured into a 15-ml plastic centrifuge tube and centrifuged for 15–30 seconds at low speed (800–1000 rpm; 30–50 g) in a tabletop International clinical centrifuge equipped with swinging buckets. The supernatant is discarded and the cellular sediment is resuspended in modified Hanks buffer by stirring gently with a glass rod. This tissue suspension is allowed to stand until the larger particles settle (35–40 seconds). The cells in the top layer are carefully decanted into a 15-ml plastic conical centrifuge tube and spun at 800–1000 rpm. The supernatant is decanted and the cells are washed twice and brought to a final volume of 35–40 ml (3–4 g of cells) with Umbreit Ringer buffer (37°, equilibrated with 95% O_2 and 5% CO_2). These isolated cells should be used immediately for metabolic studies.

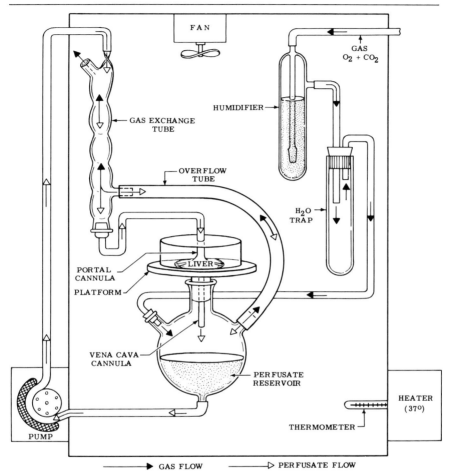

FIG. 1. Perfusion apparatus used in these studies was modified from Miller *et al.*[21] Perfusion apparatus was maintained at 37°, gassed continuously with 95% O_2 and 5% CO_2 at a flow rate of 4 liters/minute. Perfusing pressure was 14–16 cm of H_2O producing a flow rate of 25–35 ml/minute.

Dispersion of Hepatocytes. The cell suspension (35–40 ml; 3–4 g of hepatocytes) is swirled to insure the delivery of a uniform quantity of cells and is then immediately drawn up in a 1-ml, wide mouth, plastic pipette. With this procedure, it is possible to dispense 1 ml aliquots (approximately 70 mg of cells) of the cell suspension into 50 vials within 3–4

[21] L. L. Miller, C. G. Bly, M. L. Watson, and W. F. Bale, *J. Exp. Med.* **94**, 431 (1951).

minutes. Generally cells from at least 2–4 livers can be prepared each day. At least two replicate vials should be incubated in each experiment. Thus, as many as 15–25 variables can be examined at a time with a homogeneous population of parenchymal cells from a single liver.

Studies on the Effects of Enzyme and Calcium Concentration on Cell Yields

Table I summarizes the data on cell yields obtained from fasted and fed 180–200 g rats. Very low cell yields were obtained when a calcium containing buffer was used. Under similar conditions the use of calcium-free buffer resulted in a 10-fold increase in cell yields. The greatest cell yields were obtained when livers from 18–24-hour fasted rats were used in conjunction with calcium-free media. Under these conditions 3.0–4.0 g of liver cells can be isolated from a 200-g rat. This represents a 10-fold increase over that observed by Howard and Pesch[18] and a 2- to 3-fold increase over that obtained with the Berry and Friend method.[19] The use of high concentration of enzymes as suggested by Berry and Friend resulted in a sudden reduction in perfusion rate and in a lower yield of viable cells. Muller *et al.*,[22] using Berry and Friend's method,[19] with *in vitro* liver perfusion found that only 28–50% of the isolated liver cells excluded trypan blue. Howard *et al.*[17,18] found that collagenase in combination with hyaluronidase was more effective than collagenase alone in isolating liver cells. The present study does not confirm these findings. Hyaluronidase was not needed and low concentrations of collagenase were sufficient to isolate large quantities of liver cells. The observation that

TABLE I

EFFECT OF CALCIUM AND VARIOUS CONCENTRATIONS OF COLLAGENASE ON CELL YIELD FROM NORMAL (180–200 G) FED AND FASTED RATS (G OF CELLS)[a]

Animal	Buffer	Amount of collagenase (mg/100 ml)	Cell yield[b] ($N = 10$)
Fed	Ca^{2+} Umbreit Ringer	20	0.48 ± 0.08
	Ca^{2+}-free Hanks	20	2.5 ± 0.11
Starved 24 hours	Ca^{2+}-free Hanks	10	2.14 ± 0.1
		20	3.2 ± 0.14

[a] Liver cells were isolated from fed and fasted rats as described in text.
[b] Values are the mean ± S.E.M.

[22] M. Muller, M. Schreiber, J. Karlenbakc, and G. Schreiber, *Cancer Res.* **32**, 2568 (1972).

90–99% of the isolated liver cells excluded trypan blue indicates that under the conditions used here a high proportion of the cells were viable.

Effects of Collagenase

The effects of 20 mg of collagenase per 100 ml generally followed a predictable time course. Within 5 minutes after the addition of collagenase to the perfusion medium the lateral edge of the liver showed signs of leakage. During the next 5 minutes the leakage rate increased dramatically and the liver flattened and spread over the perfusion platform. The flattening of the liver and subsequent accumulation of fluid around the liver was taken as an indication of complete digestion of the intercellular matrix, and the liver was then removed and minced.

The lot of collagenase was also found to be important in obtaining large quantities of intact liver cells. When enzyme lots were changed in later studies, it was observed that collagenase concentration had to be readjusted to give optimal yields of viable cells. The concentration of the enzyme required for optimal cell yields (10–25 mg/100 ml) under the conditions used was particular to the enzyme lot and was independent of the specific activity (120–180 units/mg). Some lots consistently produced preparations where only 40–60% of the isolated cells were able to exclude trypan blue.

Ca²⁺ Effects

The absence of calcium in the perfusion medium had a dramatic effect on cell yields. Exposure of the perfused liver to calcium-free media influenced the quantity and quality of the liver cells isolated. The highest cell yields were obtained from livers which were perfused with calcium-free media for less than 40–45 minutes. A greater proportion of the cells took up trypan blue or appeared to be morphologically damaged when perfusion periods were extended. Thus, although the use of either 10 or 20 mg of collagenase per 100 ml of perfusion medium resulted in adequate cell yields (Table I) the latter concentration (20 mg) produced a more rapid disruption of the liver within 10–15 minutes.

Effect of Perfusion on Cell Yields

Livers which demonstrated an immediate, rapid, and continuous flow of perfusion medium through the tissue produced larger quantities of liver cells. These livers were uniform in color and texture during perfusion periods. Livers which did not exhibit these characteristics while being perfused did not produce high cell yield. The immediate and continuous flow rate through the liver was, in part, dependent upon the placement of the cannulae during surgery, the rapid removal of the liver from the rat and its subsequent placement in the perfusion apparatus.

The method presented here differs significantly from that originally outlined by Berry and Friend.[19] High concentrations of both collagenase and hyaluronidase were not needed. Collagenase concentrations from 10 to 20 mg/100 ml were sufficient to isolate large quantities of liver cells. Recently, Johnson et al.[23] also reported that 10 mg of collagenase in conjunction with 80 mg of hyaluronidase per 100 ml resulted in cell yields equivalent to that obtained by Berry and Friend.[19] In contrast to these reports hyaluronidase is not needed with the present method.

The prolonged incubation of the cells with enzyme after the liver was minced as recommended by Berry and Friend[19] was not needed. In the present method the liver is exposed to enzyme for only 10–15 minutes in contrast to the 30–45 minutes required with the Berry and Friend method. This minimized the exposure of the cells to various impurities found in commercial collagenase preparations (i.e., trypsin) which have been shown to damage isolated cell preparations and alter hormonal responses.[24] In addition, the cells can be isolated within 40–50 minutes from the start of the surgical procedure as compared to 60–90 minutes needed with the Berry and Friend method.[19]

Microscopic Studies

Early in the isolation procedure at least three types of cells were identified: von Kupffer cells, red blood cells, and damaged and undamaged hepatocytes. Following the completion of the centrifugation procedure, few, if any, red blood cells, von Kupffer cells, or damaged hepatocytes were found by light microscopic examination (Fig. 2). Thus, in contrast to other enzymatic methods,[18,19] homogeneous populations of hepatocytes are differentially separated from other cell types. The hepatocytes shown in Fig. 2 are typical of the cells isolated in the present study. The liver cells were dissociated from each other and demonstrated distinct cell membranes and clearly defined nuclear and cytoplasmic compartments.

Electron microscopic studies (Fig. 3) indicated that the ultrastructure of these cells was normal. The cells demonstrated intact cell membranes, mitochondria, and rough endoplasmic reticulum. In contrast to cells iso-

---

[23] M. E. M. Johnson, N. M. Das, F. R. Butcher, and J. N. Fain, *J. Biol. Chem.* **247**, 3229 (1972).
[24] T. Kono and F. W. Barham, *J. Biol. Chem.* **246**, 6204 (1971)

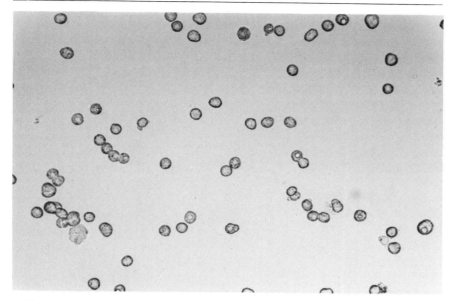

FIG. 2. Light micrograph (×400) of final preparation of liver cells. Few, if any, red blood cells, von Kupffer cells, or damaged hepatocytes are present.

lated by citrate methods[12,25] the mitochondria of the cells isolated in the present study were not swollen and the cell membranes were intact. The electron micrographs shown here are similar to those reported by Berry and Friend[19] but do not show the presence of vacuoles reported by Muller et al.[22] and Capuzzi et al.[26] The isolated cells incubated for 6 hours are still intact and retain much of their intracellular structure.

Metabolic Studies

Studies on Gluconeogenesis. Net glucose production was studied using lactate, alanine, galactose, glycerol, succinate, pyruvate, or fructose as substrates[27-29] by liver cells from fasted and fed rats. In addition, incorporation of ^{14}C bicarbonate into glucose was also measured. The results of these studies are presented in Table II. Fructose is the best

[25] M. T. Zimmerman, T. M. Devlin, and M. P. Pruss, *Nature (London)* **185**, 315 (1960).

[26] D. M. Capuzzi, V. Rothman, and S. Margolis, *Biochem. Biophys. Res. Commun.* **45**, 421 (1971).

[27] S. R. Wagle and J. Ashmore, *J. Biol. Chem.* **236**, 2868 (1961).

[28] S. R. Wagle and J. Ashmore, *J. Biol. Chem.* **238**, 17 (1963).

[29] S. R. Wagle, R. K. Gaskin, A. Jacoby, and J. Ashmore, *Life Sci.* **5**, 655 (1966).

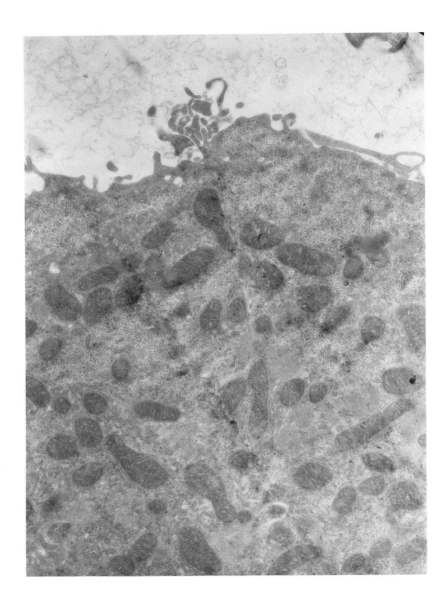

FIG. 3. Electron micrograph (×18,000) of a hepatocyte immediately after isolation. Mitochondria and endoplasmic reticulum are normal. Microvilli are present and the cell membrane is intact. Glycogen granules are also present.

TABLE II

NET GLUCOSE PRODUCTION BY ISOLATED HEPATOCYTES FROM NORMAL FED AND FASTED RATS[a]

Substrate 10 mM	Fasted rats			Fed rats		
	μmoles glucose/g/hr	Net glucose production	μmoles $^{14}CO_2$ into glucose/g	μmole glucose (g/hr)	Net glucose production	μmoles $^{14}CO_2$ into glucose
No substrate	2.3 ± 0.6 (4)	—	0.5 ± 0.07	48.8 ± 5.8	—	0.83 ± 0.09
Lactate	42.5 ± 7.0 (5)	40.2	8.2 ± 1.2	62.3 ± 7.0	14.8 ± 3.5	—
Alanine	18.2 ± 1.2 (5)	15.9	5.1 ± 0.27	56.9 ± 6.0	8.1 ± 1.2	1.5 ± 0.2
Pyruvate	35.2 ± 3.2 (4)	32.9	11.7 ± 1.0	58.1 ± 6.2	9.3 ± 6.2	5.5 ± 0.4
Galactose	16.0 ± 1.6 (6)	13.7	—	—	—	—
Glycerol	25.1 ± 3.7 (6)	22.8	—	—	—	—
Fructose	76.1 ± 6.7 (6)	73.7	—	100.0 ± 11.0	51.3 ± 6.7	—
Succinate	7.0 ± 0.7 (6)	4.7	—	—	—	—

[a] Isolated liver cells were incubated for 1 hour with various substrates and 0.5 μCi of NaH^{14}CO$_3$. Net glucose production was determined by subtracting glucose production in the absence of substrate from that observed in the presence of substrate. Values are means ± S.E.M. of (N) observations, micromoles of glucose per gram per hour.

gluconeogenic substrate used with lactate, pyruvate, glycerol, galactose, alanine, and succinate following in decreasing order.

Isolated liver cells produced net glucose at rates slightly higher than those observed with perfused liver[30] when alanine, pyruvate, or galactose was used as substrate. However, in the presence of lactate, liver cells produced twice as much glucose as did the perfused liver under similar conditions. Liver cells isolated by the present method also exhibited rates of net glucose production in excess of those previously reported for liver cells isolated by other enzymatic methods.[19,23,31] This was most evident when lactate, alanine, pyruvate, or glycerol was used as a substrate (2-fold increase). Rates of glucose production in the presence of fructose were high in all liver cells regardless of the isolation method.

Hems *et al.*[32] have suggested that a stringent test of metabolic integrity of liver cells is the ability to produce glucose from lactate since the operation and regulation of the gluconeogenic system involves the integrated work of different cellular compartments, i.e., mitochondria, cytosol, and microsomes. Gluconeogenesis also involves the controlled interchange of metabolites between various metabolic systems such as the citric acid cycle, respiratory chain phosphorylation, glycolysis, glycogenolysis, protein synthesis, and metabolism as well as fatty acid oxidation. The hepatocytes isolated here show a high degree of cellular and metabolic integrity as judged by these standards.

Glucose production observed with isolated parenchymal cells can be compared to estimated glucose production in the intact fasting rat. It has been previously calculated by isotope dilution[29] and by other methods[33] that *in vivo* glucose production approaches 106 μmoles/g/hour. The studies presented here indicate that with 10 mM lactate or 10 mM fructose, isolated liver cells produce glucose at 50 to 90% of the calculated *in vivo* rate. The agreement between the calculated *in vivo* rate of glucose production and observed glucose production supports the validity of the use of this cell preparation for studying factors which control gluconeogenesis.

Studies on Amino Acid Incorporation into Protein

The incorporation of various [^{14}C]amino acids into protein as a function of time is summarized in Fig. 4. The incorporation of labeled leucine, isoleucine, valine, phenylalanine, or serine was linear for 6 hours in the

[30] E. Struck, J. Ashmore, and D. Weiland, *Advan. Enzyme Regul.* **9**, 219 (1966).
[31] M. N. Berry and E. Kun, *Eur. J. Biochem.* **27**, 395 (1972).
[32] R. Hems, D. Ross, M. L. Berry, and H. B. Krebs, *Biochem. J.* **101**, 284 (1966).
[33] S. H. Exton and C. R. Parks, *J. Biol. Chem.* **242**, 2622 (1967).

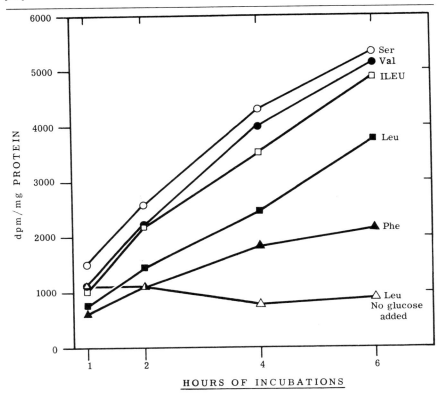

Fig. 4. Amino acid incorporation into protein in isolated liver cells as a function of time. Isolated liver cells from 18–24-hour fasted rats were incubated for various time periods in the presence of 10 mM amino acid (0.5 μCi). Each point is the mean of at least four observations. The standard error of each point is within 10–15% of the mean. Abbreviations: Ser, serine; Val, valine; Ileu, isoleucine; Phe, phenylalanine; and Leu, leucine.

presence of glucose. The observed rate of amino acid incorporation into protein was in the following decreasing order: serine, valine, isoleucine, leucine, and phenylalanine. In the absence of glucose, a decrease in the incorporation of leucine into protein was observed after 4 hours of incubation. This indicated that an energy source was required for maintaining high rates of protein synthesis. It was also observed that more than 50% of the radioactivity incorporated into protein was present in the supernatant fraction. This suggests that newly synthesized protein was being rapidly released into the medium. These results are in agreement with observations obtained using isolated perfused rat liver.[34] The incorpora-

[34] D. W. John and L. L. Miller, *J. Biol. Chem.* **241**, 4817 (1966).

tion of amino acids into protein presumably represents active protein synthesis by isolated hepatocytes and probably was not the result of bacterial growth since both penicillin and streptomycin (750 μg/3 ml) were added to the incubation medium. If antibiotics were not added, large numbers of bacteria were noted at the end of 6 hours of incubation. No bacterial growth was observed when incubation was carried out with antibiotics; hence, all studies were carried out in the presence of antibiotics.

Prolonged viability of these cells was further confirmed by electron microscopic studies and by the fact that cells incubated for 6 hours were still intact and retained much of their intracellular structures.[35] In fact, these cells after 6 hours of incubation appeared to be as good as, if not better than, cells prepared by others before incubation.[12,36,37] In addition, release of glucose-6-phosphatase[38] and malic enzyme[39] in the medium was studied. Less than 2–5% of glucose-6-phosphatase and malic enzyme and 5–10% of lactic dehydrogenase[40] was released in the medium at the end of the 4-hour incubation. It was consistently observed that the isolated cells reaggregated into clumps of 20–100 cells after an hour of incubation. More than 50–70% of the cells excluded trypan blue at the end of 6 hours of incubation. It appears that clumping protected the cells and allowed for continuous anamobic activity. Milder conditions used in the present study for isolating liver cells (low concentrations of enzyme and physiological temperatures) may also account for prolonged viability of these cells.

Hormonal Effects

Studies on the Effects of Glucagon and Epinephrine on Glycogenolysis and of Insulin on Glycogenesis in Isolated Hepatocytes. Effect of *in vitro* addition of glucagon, epinephrine, and insulin on glycogenolysis and glycogen synthesis was studied. The results of these studies are summarized in Table III. Both glucagon and epinephrine at the concentration of 10^{-6} M stimulated glycogenolysis by 80–100%. Addition of 100 μunits of insulin completely abolished epinephrine-stimulated glycogenolysis whereas only 50% inhibition was observed with insulin in glucagon-

[35] S. R. Wagle, W. R. Ingebretsen, Jr., and L. Sampson, *Biochem. Biophys. Res. Commun.* **53**, 937 (1973).

[36] N. G. Anderson, *Science* **151**, 627 (1953).

[37] J. A. Knopp and I. S. Longmuir, *Biochim. Biophys. Acta* **279**, 393 (1972).

[38] D. Monier and S. R. Wagle, *Proc. Soc. Exp. Biol. Med.* **136**, 377 (1971).

[39] R. Y. Hsu and H. A. Lardy, *J. Biol. Chem.* **242**, 520 (1967).

[40] A. Kornberg, *in* "Methods in Enzymology" (S. P. Colowick and N. O. Kaplan, eds.), Vol. 1, p. 441. Academic Press, New York, 1955.

TABLE III
EFFECT OF INSULIN ON GLUCOGEN SYNTHESIS AND EFFECT OF GLUCAGON AND
EPINEPHRINE ON GLYCOGENOLYSIS AND ITS INHIBITION BY INSULIN AND
CONCANAVALIN A IN ISOLATED RAT LIVER HEPATOCYTES

Conditions of incubation	Glycogen synthesis μmoles glucose/g[a]	Hormone addition	Glycogenolysis μmoles glucose release in medium/g/hr[b]
Isolated hepatocytes unincubated	125 ± 15	None	44.8 ± 4.6
Isolated hepatocytes incubated with no substrate	15.2 ± 2.2	Epinephrine (10^{-6})	65.2 ± 7
Isolated hepatocytes incubated with no substrate + 100 μunits insulin	22.0 ± 4	Epinephrine (10^{-6} M) + concanavalin A (100 γ)	42.6 ± 5
		Epinephrine (10^{-6} M) + insulin (100 μunits)	42.2 ± 6
Isolated hepatocytes incubated with 5 mM glucose, 5 mM lactate, 5 mM amino acids mixture	40.5 ± 8	Glucagon (10^{-6} M) Glucagon (10^{-6} M) + insulin (100 μunits)	78.6 ± 8.0 57.0 ± 6.8
Isolated hepatocytes incubated with 5 mM glucose, 5 mM lactate, 5 mM amino acids mixture and 100 μunits insulin	65.8 ± 12	Glucagon (10^{-6} M) + concanavalin A (100 γ)	76.5 ± 9

[a] Approximately 70 mg of isolated hepatocytes were incubated in 3 ml of Ringer bicarbonate medium for 2 hours at 37°. Values are mean ± S.E.M. of six observations.

[b] Approximately 70 mg of isolated hepatocytes were incubated in 3 ml of Ringer bicarbonate medium for 1 hour at 37°. Values are mean ± S.E.M. of six observations.

stimulated glycogenolysis. Concanavalin A completely inhibited epinephrine-stimulated glycogenolysis but had no effect on glucagon-stimulated glycogenolysis. It can also be seen from Table III that addition of insulin stimulated glycogen synthesis and maintained better cellular structure.[35] These results suggest that heptocytes isolated with low concentrations of collagenase retain glucagon, epinephrine, and insulin receptors and that different hormones utilize different receptor sites.

Hepatocytes isolated by the method described here exhibit high rates of glucose production[41] and incorporation of amino acids[41] into protein, are sensitive to hormones at physiological concentrations,[35,42] and reflect the metabolic or disease state[43] of the liver donor. The ability of isolated hepatocytes to retain much of the *in vivo* metabolic characteristics suggests that these cells should prove useful in studying factors involved in the regulation of hepatic metabolism.

[41] W. R. Ingebretsen, Jr. and S. R. Wagle, *Biochem. Biophys. Res. Commun.* **47,** 403 (1972).

[42] S. R. Wagle and W. R. Ingebretsen, Jr., *Biochem. Biophys. Res. Commun.* **52,** 125 (1973).

[43] W. R. Ingebretsen, Jr., M. A. Moxley, D. O. Allen, and S. R. Wagle, *Biochem. Biophys. Res. Commun.* **49,** 601 (1972).

Section IX

In Vivo and Perfusion Techniques

[56] Use of the Perfused Liver for the Study of Lipogenesis

By H. Brunengraber, M. Boutry, Y. Daikuhara,
L. Kopelovich, and J. M. Lowenstein

The last decade has witnessed the renaissance of the perfused liver technique. This has come about in part because it was recognized that tissue slices and preparations of subcellular components often do not constitute suitable models for the study of the regulation of complete metabolic pathways[1]; for example, gluconeogenesis,[2] ketogenesis,[3] and fatty acid synthesis proceed at much slower rates in slices[4] and cell-free extracts[5] than can be observed *in vivo*[6] or in isolated perfused rat liver.[7] On the other hand, *in vivo* studies of liver metabolism are influenced by processes occurring in other organs. The isolated perfused liver is thus an ideal tool for the study of metabolic regulation. The model described below is capable of synthesizing fatty acids at the same rate as the organ *in vivo*, and it responds to the same stimulators and inhibitors.

Preparation of Animals

The rate of fatty acid synthesis varies greatly in relation to food intake, which in turn is related to the nycterohemeral cycle. In order to establish reproducible conditions, it is simplest to impose a feeding schedule on the animals. A simple regimen is to feed the animals once a day for 3 hours. During the meal, the animals are housed in individual cages to avoid contests for access to food. The animals should be weighed every day before the meal; the slowest and fastest weight gainers are discarded. The rate of fatty acid synthesis *in vivo* is maximum between 2 and 5 hours after the start of the meal. It is therefore advisable to start perfusion experiments at the same hour of the day; for example, 3 to 4 hours after the start of the meal.

[1] H. A. Krebs, *in* "Stoffwechsel der isoliert perfundierten Leber" (W. Staib and R. Scholz, eds.), pp. 129–141. Springer-Verlag, Berlin and New York, (1968).

[2] B. D. Ross, R. Hems, and H. A. Krebs, *Biochem. J.* **102**, 942 (1967).

[3] H. A. Krebs, P. G. Wallace, R. Hems, and R. A. Freedland, *Biochem. J.* **112**, 595 (1969).

[4] S. S. Chernick and I. L. Chaikoff, *J. Biol. Chem.* **186**, 535 (1950).

[5] J. A. Watson and J. M. Lowenstein, *J. Biol. Chem.* **245**, 5993 (1970).

[6] J. M. Lowenstein, *J. Biol. Chem.* **246**, 629 (1971).

[7] H. Brunengraber, M. Boutry, and J. M. Lowenstein, *J. Biol. Chem.* **248**, 2656 (1973).

The composition of the diet depends on the investigation, but it should be known and should be reproducible. Commercial pellets often do not meet these requirements. Therefore, it is best to make up the food starting from simple components such as carbohydrate, casein, and oil. One such diet, which induces a high rate of fatty acid synthesis and has been used extensively by us is as follows (in kilograms): glucose, 5.80; casein, 2.20; cellulose, 1.18; salt mixture W,[8] 0.60; Pervinal,[9] 0.02; and liver extract,[10] 0.20 kg; total, 10.

The dry powder is mixed thoroughly before being mixed with 4 liters of 0.5% agar at 40° in a kneading trough. The resulting dough is placed into feeding dishes and is allowed to dry for a week in a cold room or refrigerator.

Choice of Perfusate

It is common practice to perfuse organs with whole blood, diluted blood, or Krebs-Ringer bicarbonate buffer[3] containing albumin and washed erythrocytes. In this type of perfusion, the total red cell mass usually exceeds the weight of the liver, and one is dealing with a model in which one must consider the metabolism of the erythrocytes and of the liver. The high glycolytic capacity of red cells imposes a redox buffer on the liver. Aged red cells which glycolyze at a slower rate than fresh cells are prone to hemolyze. In our experience it is neither desirable nor necessary to use red cells when one wishes to use the perfused liver as a model to study the regulation of fatty acid synthesis. We use Krebs-Ringer bicarbonate buffer containing 4% bovine serum albumin.[11] The albumin is dialyzed before use as a 10% solution against 30 volumes of Krebs-Ringer bicarbonate buffer at 5° with 6 changes of buffer for 48 hours. Dialysis removes various impurities from the albumin such as citrate which is used to prevent clotting of the blood from which the albumin is prepared. We have found that if albumin is omitted from the perfusate the rate of fatty acid synthesis is only about 5% of the rate observed in the presence of albumin.[12]

Perfusion Apparatus

The aim of this section is to describe two types of apparatus which are convenient for either rat or mouse liver perfusion. In both systems

[8] Available from Nutritional Biochemicals Co., Cleveland, Ohio 44128.
[9] Pervinal is produced by Mitchum-Thayer, Inc., Tuckahoe, N.Y. 10707.
[10] Liver extract concentrate was obtained from Nutritional Biochemicals Co.
[11] Fraction V, fatty acid poor, Miles Biochemicals, Kankakee, Ill.
[12] Y. Daikuhara and J. M. Lowenstein, *unpublished observations.*

the perfusate is recirculated at a constant flow rate, and both allow for the use of a nonrecirculating pre-perfusion buffer during surgery. The apparatus is set up in an illuminated cabinet which is kept at 38°.

Rat Liver Apparatus. The rat liver apparatus (Scheme 1) combines various features of the designs described by Hems *et al.*,[13] Scholz,[14] and Staib *et al.*[15] plus modifications of our own, which we have found to possess advantages over other designs. Most of the perfusate is contained in a flat-bottomed glass reservoir of 125 ml capacity,[16] with five top inlets and a bottom outlet. A peristaltic pump[17] sends the perfusate through a Swinnex[18] filter holder containing a glass fiber prefilter and a Millipore filter with a pore size of 1.2 μ (42 and 47 mm in diameter, respectively). The perfusate then passes to a water-jacketed oxygenator[16] which is gassed with 95% O_2–5% CO_2 at a rate of 1 liter/minute. From the oxygenator, the perfusate is sent to the liver by a low pulsing pump.[19] An overflow at the bottom of the oxygenator returns excess perfusate to the reservoir. The liver is held in a block constructed of Lucite by an adjustable clamp in such a manner that it is in part suspended from the inferior vena cava cannula by a residual fragment of the pericaval diaphragm, and in part resting on the perforated platform in the center of the block. The perfusate is passed into a bubble trap which is located in a corner of the block. The bubble trap is connected via a short piece of tubing to the portal vein cannula, which rests in a horizontal slot cut into the block. The perfusate leaves the liver through the inferior vena cava cannula and most of it is drained into the reservoir.

The oxygen concentration in the effluent is measured as follows. A portion of the effluent is diverted immediately after the cannula and is pumped at a constant rate, say 3 ml/minute, past an oxygen electrode and then into the reservoir. The tubing leading from the cannula to the oxygen electrode is made of glass, except for the portion that passes through the peristaltic pump, which is made of thick-walled, narrow internal diameter Tygon tubing ($\frac{5}{32}$ inch o.d., $\frac{1}{32}$ inch i.d.). Oxygen con-

[13] R. Hems, B. D. Ross, M. N. Berry, and H. A. Krebs, *Biochem. J.* **101**, 284 (1966).

[14] R. Scholz, *in* "Stoffwechsel der isoliert perfundierten Leber" (W. Staib and R. Scholz, eds.), p. 25. Springer-Verlag, Berlin and New York, 1968.

[15] W. Staib, R. Staib, J. Hermann, and H. G. Meiers, *in* "Stoffwechsel der isoliert perfundierten Leber" (W. Staib and R. Scholz, eds.), p. 155. Springer-Verlag, Berlin and New York, 1968.

[16] Available from H. Vincent, 14 Lyndworth Close, Headington, Oxford OX 39 ER, England.

[17] Model 1203, Harvard Apparatus Co., Millis, Mass. 02054.

[18] Available from Millipore Corporation, Bedford, Mass. 01730.

[19] Four channel Polystaltic Pump equipped with a 48-rpm motor, Buchler Instruments, Fort Lee, N. J. 07024.

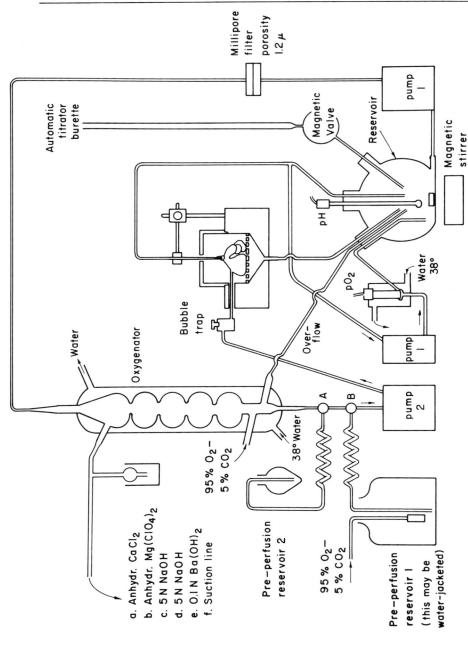

SCHEME 1. Apparatus for rat liver perfusion. Reproduced from ref. 7 with permission of the American Society of Biological Chemists.

sumption can be measured by modifying the apparatus to permit the measurement of either influent or effluent oxygen concentrations. A portion of the influent is diverted immediately after the bubble trap and is pumped at a constant rate, say 3 ml/minute, past the oxygen electrode and then into the reservoir. The electrode (type E5046) is mounted in chamber D616 and is connected to blood analyzer PHM71 (all obtained from Radiometer, Copenhagen). We prefer to use a single oxygen electrode and to use a three-way tap to switch from influent to effluent and back, but the system can readily be adapted for two electrodes.

The bottom drain in the perfusion block serves to drain the perfusate before the inferior vena cava cannula is connected to the outflow tubing. After the circuit has been closed, liquid oozing from the liver passes from the bottom of the drain in the Lucite block to the reservoir, but livers which ooze are rejected.

Two three-way stopcocks are located between the oxygenator outlet and pump 2. They connect the main circuit with two reservoirs containing preperfusion fluids. One reservoir (connected via stopcock B) contains about 500 ml of Krebs-Ringer bicarbonate buffer with or without albumin; it is used for perfusing the liver during surgery. The second reservoir (connected via stopcock A) contains 50 ml of Krebs-Ringer bicarbonate buffer with albumin; it is used when the operation is completed, just before switching the liver to the main perfusate. Both pre-perfusion fluids are equilibrated with 95% O_2–5% CO_2 and pass through small auxiliary preheating coils kept at 38°. (Small coil condensers are suitable for this purpose.) During the pre-perfusion period, stopcock A is used to shut off the oxygenator outlet, and the perfusate running through the oxygenator is returned to the reservoir via the oxygenator overflow.

The contents of the reservoir are mixed with a magnetic stirrer. The reservoir is fitted with a glass electrode connected to a pH meter (Radiometer model 51). The pH meter is linked to a titrator (Radiometer model 11), which operates a magnetic valve (Radiometer, type MNV 1c). The latter operates a burette containing 0.294 N NaOH plus 0.006 N KOH which is connected to the reservoir via polyethylene tubing. The titrator is set to maintain the pH at 7.40. When tritiated water, or a ^{14}C-labeled substrate, or both are present in the perfusate, the effluent gas of the oxygenator is drawn through a drying tube containing anhydrous $CaCl_2$, followed by a drying tube containing magnesium perchlorate. These serve to trap ^3H_2O. The gas is then drawn through two gas washing bottles containing 5 N NaOH, followed by one containing 0.1 N $Ba(OH)_2$. A buildup of gas pressure in the apparatus is avoided by drawing the gas through the train of drying tubes and washing bottles by means of a water pump. A drop in pressure below atmospheric is avoided by means of an oil valve between the oxygenator outlet and the first drying tube. Dilution of the gassing mixture with air is prevented by extending the perfusate overflow line from the oxygenator below the perfusate level in the reservoir.

The apparatus offers the following advantages. The bulk of the perfusate is contained in an easily accessible reservoir. Additions to the reservoir are mixed rapidly and completely. A constant pH is maintained automatically, and the rate of alkali consumption can be recorded as a function of time. The technique permits a constant rate of perfusion to be chosen so that a given effluent oxygen concentration is maintained.

Mouse Liver Perfusion Apparatus. The mouse liver perfusion apparatus (Scheme 2) can be used for the perfusion of livers weighing less than 3–4 g, that is, livers from mice or from very young rats. With such small livers, the perfusion rate must be low (4–16 ml/minute) and the volume of the recirculating perfusate can be kept to 20–40 ml.

The perfusate is sent from the 50-ml reservoir through a filter using channel 1 of the pump.[19] Coming out of the filter, the perfusate meets the gassing mixture coming out of channel 2 of the pump. The bubbled perfusate passes through two water-jacketed coils[20] and then through a debubbler. From the lower outlet of the debubbler, the perfusate is sent to the liver via channel 3 of the pump. From the upper outlet of the debubbler, the gassing mixture plus some excess perfusate passes into the reservoir.

The Lucite liver holder, the pH-stat, and the oxygen concentration monitoring are the same as in the rat liver apparatus. The influent oxygen concentration can be measured by inserting an oxygen electrode in series between channel 3 of the pump and the liver.

The ratio of flow rates through the different channels of the same pump is constant regardless of the pumping speed. This prevents gas bubbles from being sent to the liver by a change in the pumping speed. The tubing running through the pump is made of Tygon of the following diameters. Channel 1: $3/32$ inch i.d., $5/32$ inch o.d.; channel 2: $1/16$ inch i.d., $1/8$ inch o.d.; channel 3: $1/16$ inch i.d., $1/8$ inch o.d.; or $1/32$ inch i.d., $3/32$ inch o.d.; channel 4: $1/32$ inch i.d., $5/32$ inch o.d.

A three-way stopcock (A) is located between the debubbler and channel 3 of the pump. It connects the main circuit with a water-jacketed preperfusion reservoir containing about 250 ml Krebs-Ringer bicarbonate equilibrated with 95% O_2–5% CO_2. It is used for perfusing the liver during surgery.

The pumping system sets up in the oxygenation circuit a column of

---

[20] The bubbler, debubbler, and coils are standard Technicon parts.

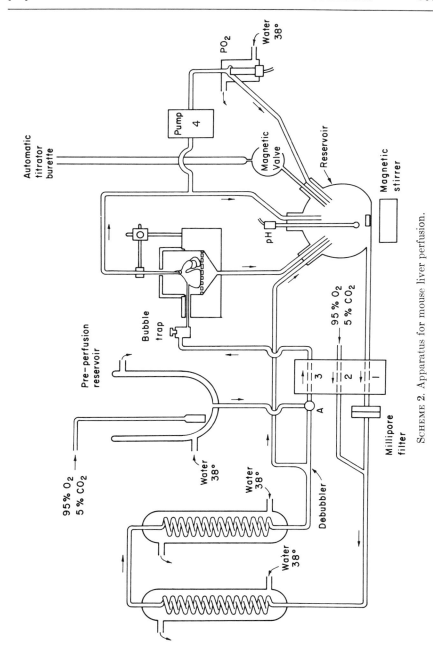

SCHEME 2. Apparatus for mouse liver perfusion.

fluid containing gas bubbles at regular intervals. Foaming of the perfusate returning to the reservoir via the overflow of the debubbler is kept to a minimum by allowing a certain amount of perfusate to pass through this part of the circuit, and by keeping the internal diameter of the overflow tubing to $\frac{1}{16}$ inch or less.

The dead space of the mouse liver perfusion apparatus is about 10 ml. A single passage of the perfusate through the two oxygenating coils insures 95–97% saturation with gas.

Perfusion Technique

The method described below avoids anesthesia and anaerobic conditions. It yields rates of fatty acid synthesis comparable to rates observed *in vivo*.

The rat is killed by a blow on the head (but not on the neck or the back). Immediately thereafter an assistant holds the animal firmly flat on its back, grasping the four limbs to oppose contractions. A large V-shaped incision is made into the abdomen from the pubis toward the flanks with a sharp pair of scissors, cutting through fur and muscle. The portal vein is located, grasped with forceps, an entrance hole is nicked with microscissors, and an 18-gauge plastic cannula is inserted through which oxygenated Krebs-Ringer bicarbonate with or without 4% bovine serum albumin, kept at 38°, is perfused at a rate of 30 ml/minute throughout the entire operation. The Krebs-Ringer solution contains glucose at the same concentration as that used during the experimental period. The time elapsed between the blow on the head and the portal vein cannulation should not exceed 20–30 seconds. The liver should be completely cleared of blood in 30–45 seconds after cannulation. The portal vein and the cannula are held gently but firmly in the correctly aligned positions by the operator until the rat has stopped moving. (The portal vein is held with forceps in one hand, and the cannula is held in the other hand.) As soon as possible the inferior caval vein is cut below the kidneys to prevent the liver from swelling and to facilitate the washing out of blood. Two ligatures are then made and tightened around the portal vein cannula, one between the entry of the duodenal vein and the hilum of the liver, and the second on the mesenteric side of the duodenal vein. The liver is separated from the stomach, the spleen, and the intestine; and the bile duct is cut off at the hilum of the liver. A loose ligature is put around the inferior caval vein (ligature *a*) just above the right renal vein. The thorax is opened, and a loose ligature is put around the inferior caval vein between the heart and the diaphragm (ligature *b*). A 16-gauge plastic cannula is pushed through the right auricle into the atrium and then into the inferior caval vein until it almost reaches the liver. (If desired, an entrance hole for the cannula can be prepared by nicking the right auricle with small scissors.) Ligature *b* is tightened; then a piece of diaphragm is tied with the same thread in a second knot. Ligature *a* is tightened, and the inferior caval vein is cut caudally, close to the ligature. The assistant now raises the rat by the neck. The operator lifts the heart cannula and cuts away the heart, the esophagus, and the posterior part of the diaphragm. (Both ligatures remain with the preparation.) The liver preparation is put into the Lucite perfusion block; the latter is drained by suction. The diaphragm is trimmed except for the tendinous portion, and the esophagus is dissected out. The liver is rinsed with Krebs-Ringer bicarbonate at 38°. A thermocouple[21] is slipped between two lobes, and a Lucite cover is put on the perfusion block.

The perfusate is now changed to 50 ml of Krebs-Ringer bicarbonate containing 4% bovine serum albumin, and the effluent perfusate is discarded. The recirculating perfusion is then started by switching over to the main circuit. This is done by connecting the heart cannula to the line which returns most of the effluent to the reservoir. The bottom drain of the Lucite block is also connected to the reservoir. The closing of the main circuit is taken as zero time; usually 8–10 minutes elapse between starting the operation and zero time. The recirculating perfusate has a volume of 150 ml. A portion of the effluent is pumped at a constant rate (say, 3 ml/minute) past an oxygen electrode and then back into the reservoir. A CO_2 electrode can be mounted in series if desired.

The bile duct is cut purposely to prevent complete loss of the compounds reabsorbed by the gut *in vivo*, in the so-called entero-hepatic cycle. These substances are thus recycled, although they become diluted in the perfusate. If desired, the bile duct can be cannulated just after the portal vein ligatures are tied.

During the perfusion the gross appearance of the liver is watched, and oozing of liquid from the surface of the organ is assessed. In a successful perfusion the liver retains an even and undamaged appearance, and oozing is minimal. The effluent $[O_2]$ is monitored to check the adequacy of the oxygen supply. Samples of perfusate are taken at 5, 10, or 15 minute intervals, and the concentrations of pyruvate, lactate, β-hydroxybutyrate, acetoacetate, and glucose are determined.[22-25] The ratios

[21] Model 42 SC, Yellow Springs Instrument Co., P. O. Box. 279, Yellow Springs, Ohio 45387.

[22] J. Mellanby and D. H. Williamson, *in* "Methods of Enzymatic Analysis" (H. U. Bergmeyer, ed.), p. 454. Academic Press, New York, 1965.

[23] H. J. Hohorst, *in* "Methods of Enzymatic Analysis" (H. U. Bergmeyer, ed.), p. 266. Academic Press, New York, 1965.

[24] D. H. Williamson, J. Mellanby, and H. A. Krebs, *Biochem. J.* 82, 90 (1962).

[25] M. W. Slein, *in* "Methods of Enzymatic Analysis" (H. U. Bergmeyer, ed.), p. 117. Academic Press, New York, 1965.

of [lactate] to [pyruvate] and [β-hydroxybutyrate] to [acetoacetate] are used to monitor the oxidation–reduction status of the extra- and intra-mitochondrial compartments, respectively.[26]

If one wishes to assay tissue metabolites, then at the end of the experiment the liver is lifted out of the perfusion block by the heart cannula with the perfusate still running. The liver is clamped rapidly between aluminum blocks which are precooled in liquid nitrogen.[27] The livers are stored in a liquid nitrogen freezer.

The perfusion technique described above applies to rat liver. It can be adapted to mouse liver by using thinner cannulae (20 or 22 gauge) and slower flow rates of perfusion.

Evaluation of the Model

After the liver has been connected to the perfusion apparatus, it is advisable to let it equilibrate for 1 hour before measuring the rate of fatty acid synthesis. This period allows various metabolic parameters to reach stable levels. The rates of respiration, glycolysis, and ketone body production, and the ratios [lactate]/[pyruvate] and [β-hydroxybutyrate]/[acetoacetate] become stabilized after 45–60 minutes. The cytoplasmic pyridine nucleotide couple is in equilibrium with the cytosolic adenine mononucleotides through the glyceraldehyde-phosphate dehydrogenase reaction,[26] and the [lactate]/[pyruvate] and [β-hydroxybutyrate]/[acetoacetate] ratios reflect the general metabolic status of the liver.

The redox potentials of the two pyridine nucleotide couples can be calculated using the Nernst equation. The couples exhibit a constant difference of potential of 40–60 mV depending on the metabolic state of the liver and also on the values of the midpoint potentials used in the Nernst equation. The midpoint potentials we use[7] are [lactate]/[pyruvate] = —215 mV,[28] and [β-hydroxybutyrate]/[acetoacetate] = —297 mV.[29]

Measurement of Fatty Acid Synthesis

The rate of fatty acid synthesis in the perfused liver can be measured by the incorporation of suitable precursors labeled with radioactive (^{3}H, ^{14}C) or stable (^{2}H, ^{13}C) isotopes. Details are described elsewhere in this volume (Chapter 34).

[26] R. L. Veech, L. Raijman, and H. A. Krebs, *Biochem. J.* **117**, 499 (1970).
[27] A. Wollenberger, O. Ristau, and G. Schoffa, *Pfluegers Arch. Gesamte Physiol. Menschen Tiere* **270**, 399 (1960).
[28] D. H. Williamson, P. Lund, and H. A. Krebs, *Biochem. J.* **103**, 514 (1967).
[29] R. L. Veech, L. V. Eggleston, and H. A. Krebs, *Biochem. J.* **115**, 609 (1969).

The specific radioactivity or the mole fraction of the stable isotope in the labeled precursor does not necessarily remain constant during the perfusion. For example, labeled glucose is diluted by unlabeled glucose which is released by the liver into the perfusate. It follows that to make valid measurements one must determine the specific activity or mole fraction of the labeled precursor several times in the period during which the rate of fatty acid synthesis is measured.

Acknowledgments

This article is based on research supported by grants from the John A. Hartford Foundation, the Fonds National de la Recherche Scientifique (Belgium), and by a NATO travel grant.

[57] Perifused Fat Cells

By D. O. ALLEN, E. E. LARGIS, A. S. KATOCS, JR., and J. ASHMORE

Introduction

In vitro lipolytic experiments with adipose tissue have been conducted by incubating fat pads or isolated fat cells in buffered albumin solution for periods of 5 minutes to 4 hours and measuring the accumulation of free fatty acids and/or glycerol in the incubation medium.[1,2] Rates of lipolysis have been calculated on such experiments. On the basis of these measurements it has been possible to accumulate data concerning hormone-induced changes in the lipolytic process; however, it has been impossible to determine the rate at which these changes take place. It has also been impossible to measure changes in lipolytic rates which followed removal of the hormone. This report describes the perifused isolated fat cell technique by which rates of lipolysis can be measured sequentially at intervals as short as 5 seconds. It is felt that this technique offers a more sensitive method for assessing the effects of hormones on the initiation and/or termination of the lipolytic process.

Apparatus

The apparatus for perifusing isolated fat cells is shown in Fig. 1. The apparatus consisted of the water-jacketed plastic column in which the cells were placed. Both ends of the column were plugged with fibrous plastic filter material (Long-life filter fiber, Sternco Industries) to form

[1] A. Angel, K. S. Desai, and M. L. Halperin, *J. Lipid Res.* **12**, 104 (1971).
[2] J. Moskowitz and J. N. Fain, *J. Clin. Invest* **48**, 1802 (1969).

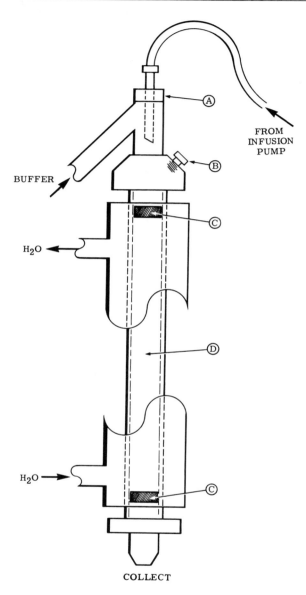

FIG. 1. Apparatus for the perifusion of isolated fat cells: A, injection port for infusing hormones; B, plug to facilitate filling of column with buffer; C, fiber plugs to form ends of cell chamber; and D, cell chamber. The cell chamber was enclosed in a water jacket. Buffer was pumped into the top of the column and collected at the bottom.

a cell chamber with a volume of 6.0 ml (9 × 95 mm) maintained at 37°
by a constant temperature water jacket. Tubing was attached to the top
of the column so that the buffered solution could be pumped through the
cell chamber. An injection port was situated at the top of the column
in such a way that a hypodermic needle could be inserted and through
which drugs or hormones could be infused at a constant rate. Perifusate
was collected from the bottom of the column over known periods of time
(usually 1 minute).

Perifusion of the Isolated Fat Cells

Isolated fat cells were prepared from rat epididymal fat pads by a
modification[3] of the method of Rodbell.[4] Two milliliters of the packed
cells were introduced into the cell chamber. Ringer bicarbonate buffer
(pH 7.4) containing bovine serum albumin (usually 0.5% w/v), which
was maintained at 37° in a constant temperature water bath under an
atmosphere of 95% O_2/5% CO_2, was pumped into the top of the column
at the rate of 10 ml/minute. The flow of the buffer through the cell chamber produced a constant mixing of the fat cell suspension. The perifusate
was collected from the bottom of the column over known intervals of
time and aliquots were taken for determination of the glycerol according
to the fluorometric method of Chernick.[5] A filter fluorometer equipped
with the following filters was used for the assay: primary, C.S. No. 7-37
(Corning Glass Works); secondary, C.S. 5-61, C.S. 4-70, C.S. 3-72 (Corning Glass Works). Glycerol production was used as an index of lipolysis.

Results and Discussion

Fat cells were introduced into the cell chamber and perifused for a
10-minute equilibration period. Over the next 10-minute period several
samples were taken for determination of basal rates of lipolysis (Fig.
2). At 10 minutes an infusion of epinephrine was started. The infusion
rate (0.25 ml/minute) and concentration (4 × 10^{-5} M) of epinephrine
were such that the concentration of epinephrine in the buffer reaching
the cells was 10^{-6} M. Following a 1-minute lag period there was a sharp
increase in the rate of glycerol production. The rate of lipolysis continued
to increase for the next 20 minutes after which it plateaued at a value
10 times that of the basal rate.

To gain some information concerning the rate at which the epinephrine concentration reached equilibrium in the cell chamber and its rela-

[3] J. J. Lech and D. N. Calvert, *J. Lipid Res.* **7**, 561 (1966).

[4] M. Rodbell, *J. Biol. Chem.* **239**, 375 (1964).

[5] S. S. Chernick, Vol. 14, p. 627.

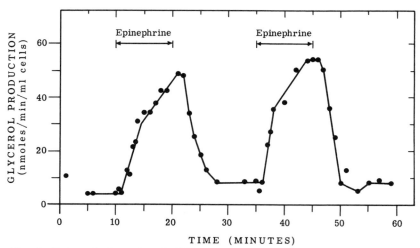

FIG. 2. Rates of glycerol and [³H]epinephrine appearance in perifusate of fat cells. Control period of 10 minutes followed by an infusion of [³H]epinephrine (final concentration 10^{-6} M).

tionship to the stimulation of lipolysis, a tracer amount of [³H]epinephrine was added to the epinephrine infusion solution and the rate of appearance of tritium was followed in the perfusate (Fig. 2). Following the start of the epinephrine infusion, there was a rapid increase in the rate at which tritium appeared in the perfusate. By 5 minutes the rate had plateaued indicating that the concentration of epinephrine within the cell chamber had reached equilibrium. It should be noted that this was a time at which the rate of glycerol production was still continuing to increase. Thus, the development of maximal lipolytic rates was preceded by the development of equilibrium in the epinephrine concentration in the cell chamber.

After 30 minutes of epinephrine stimulation the infusion was stopped. Following a short lag period there was a rapid reduction in the rate of glycerol production with the rates of lipolysis returning to basal levels by 15 minutes after the cessation of the epinephrine infusion. The fall in glycerol production was preceded by a more rapid fall in the rate of appearance of [³H]epinephrine. By 6 minutes the rate of [³H]epinephrine appearance in the perifusate had fallen to essentially zero.

That lipolysis could be repeatedly stimulated in the same fat cell preparation is shown in Fig. 3. Following a 10-minute equilibration period, epinephrine was infused under conditions identical to that described for Fig. 2. At the end of 10 minutes, infusion was stopped and the rate of glycerol production allowed to decay over the next 15 minutes. At that time, an identical infusion of epinephrine was again started and maintained for 10 minutes. The two epinephrine infusions caused rapid, nearly identical, accelerations in the rate of lipolysis. Following cessation of the epinephrine infusion in both cases glycerol production returned to prestimulated rates.

In all of the previous experiments the bovine serum albumin (BSA) concentration in the perifusing medium was maintained at 0.5% (w/v). This would not appear to be an optimal concentration. In the experiment depicted in Fig. 4, fat cells were perfused with 0.5% (w/v) BSA buffer for 10 minutes after which an infusion of epinephrine (final concentration 10^{-6} M) was begun. After lipolytic rates reached a plateau, the BSA con-

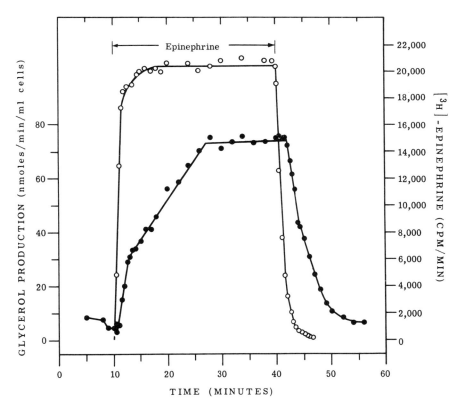

Fig. 3. Glycerol production by perifused isolated fat cells stimulated by two sequential epinephrine infusions. Following a 10-minute control period epinephrine (final concentration 10^{-6} M) was infused for 10 minutes. A second identical infusion was started 15 minutes after the end of the first.

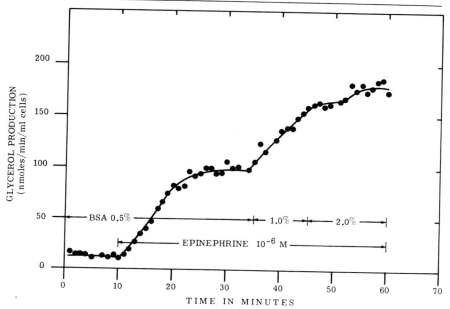

FIG. 4. Effects of BSA concentration on epinephrine-stimulated lipolysis. Fat cells were perifused with 0.5% (w/v) BSA for 10 minutes. Epinephrine infusion (final concentration 10^{-6} M) was then started and maintained for duration of the experiment. At 35 minutes the BSA concentration was increased to 1.0% (w/v) and at 45 minutes BSA concentration was increased to 2.0% (w/v).

centration was increased to 1% (w/v) while continuing the epinephrine. This resulted in an increased rate of lipolysis. Increasing the concentration of BSA to 2% (w/v) produced a small additional increase. One percent (w/v) BSA is now routinely used in all perifusion experiments.

The perifused fat cell preparation provides an exquisitely sensitive system for measuring alterations in rates of lipolysis which occur over a very short period of time. With this technique it is possible to examine the rates at which hormones and drugs initiate the lipolytic response, to follow the time course of the fall in rates of lipolysis following the removal of lipolytic agents, and to study the process by which lipolysis is returned to basal levels following cessation of hormone stimulation.

It would appear that with this simple *in vitro* system *in vivo* regulation may be mimicked without the difficulty of perfusion and vascular flow changes or the added complications of the presence of tissues other than fat cells. The advantage offered is a relatively simple biologic system for elucidation of biochemical events that occur in the initiation and cessation of hormone-induced lipolysis.

Author Index

Numbers in parentheses are reference numbers and indicate that an author's work is referred to, although his name is not cited in the text.

Bardawill, C. J., 26, 34, 38, 47, 54(11), 174, 335
Barden, R. E., 273
Barenholz, Y., 242, 245(1), 246(1, 2), 247, 250, 252(2), 253, 529, 530, 532(5)
Barford, R. A., 321
Barham, F. W., 586
Barker, D. L., 197, 198, 199(1), 201, 224
Barker, H. A., 140
Barnes, E. M., Jr., 102, 103(1, 2), 106(1, 2), 109, 273
Barr, K. L., 288, 289(9)
Bar-Tana, J., 117, 118(3), 121, 122(3, 4)
Barth, C. A., 160, 167
Bartley, D. A., 307, 309(2)
Barton, P. G., 441
Barzilay, I., 439, 440(16), 460, 461, 503(16), 506(16)
Basu, H., 503, 506, 514
Basu, S., 549
Baum, H., 52, 58
Bauman, A. J., 419, 423(44a)
Baumann, W. J., 389, 444, 446(26), 449
Bauminger, S., 206
Beard, J. W., 579
Beck, C., 544
Becker, Y., 227
Beechey, R. B., 389
Beerthius, R. K., 360(13), 361, 376
Behr, J. P., 549
Beijer, K., 386, 387(20), 388(20), 392, 394(20)
Belfrage, P., 184, 187
Bender, M. L., 201
Bennett, E. A., 209, 213(6), 214(6), 216(6), 217(6), 218(6), 220(6)
Benöhr, H. C., 190, 201(6), 206, 208(6)
Benson, A. A., 227
Bentley, R., 340, 395
Berchtold, R., 452, 453, 459, 500
Bergelson, L. D., 472, 474, 475
Bergkvist, H., 393
Bergmeyer, H. U., 298, 302, 303(6)
Bergström, S., 360(21, 22), 361, 363(21), 364(21)
Berry, M. N., 580, 584, 586, 587, 590(19), 592(19), 599
Bersten, A. M., 260, 261(11)
Bhargava, P. M., 580
Bieber, M. A., 377

Biemann, K., 284
Bieth, R., 564
Billimoria, J. D., 456, 488, 491(46), 492(46)
Bird, P. R., 434
Bischel, M. D., 406
Bittman, R., 520
Blakeley, R. L., 197, 223, 224, 225(14)
Blank, M. L., 404, 413
Blau, L., 520
Bloch, K., 48, 49(20), 84, 85, 86, 87, 88, 89, 90, 91, 95(3), 102, 103(3), 106, 107, 109, 254
Blomstrand, C., 562, 563, 564, 574(5)
Bloxham, D. P., 135
Bly, C. G., 583
Bock, R. M., 47, 54(12), 55, 57
Boehm, W., 461
Boguth, W., 198
Bonner, W. M., 102, 103(3), 106, 107, 109
Bonsen, P. P. M., 317, 323(8), 392, 429, 437, 438(13), 439, 454, 455, 461, 465, 466, 469, 477, 520, 525
Borner, K., 198, 199(22), 206(22)
Bortz, W. M., 312
Borum, E. H., 394
Boulter, D., 225
Boursnell, J. C., 220
Boutry, M., 279, 280(9), 282, 597
Bowman, R. E., 564
Boxer, G. E., 279
Bradley, R. M., 312, 550
Bradshaw, R. A., 122, 123(6), 125, 126, 127, 128(13)
Brady, R. O., 48, 74, 96, 312, 542, 544, 545, 546, 547, 548, 549, 550
Branster, M. V., 580
Bray, G. A., 282
Bremer, J., 117
Brenner, R. R., 253, 260(1)
Bressler, R., 45, 58(6), 60
Brimacombe, J. S., 490
Brindley, D., 84, 87(1), 90
Brock, D. J. H., 48, 49(20)
Brock, W. A., 288, 289(8)
Brockerhoff, H., 317, 320, 322, 323, 324, 325(20), 443
Brodie, J. D., 48
Brooks, C. J. W., 380, 393, 395
Brotz, M., 561

Krisch, K., 190, 197, 198, 199, 201(6), 202, 206, 208(3, 6)

Kritchevsky, G., 403, 413, 417, 418(14, 40), 419, 423(14, 44a)

Krivit, W., 422

Kull, A., 60

Kumar, S., 37, 38, 45, 47, 48, 49(10), 53(13), 56, 57(13), 58(10), 73, 273

Kun, E., 590

Kurnberg, A., 592

Kushner, D. J., 469

Kusumi, I., 395

L

Laine, R. A., 423, 425(61a)

Lakshmanan, M. R., 41, 43(10), 44, 274, 313

Lamb, S. I., 535

Lamberts, I., 136

Lands, W. E. M., 273, 317, 321, 360(48), 362, 446, 509(27)

Lane, M. D., 10, 18, 19, 26, 28(3), 29(3), 30, 31, 32, 36, 74, 134, 155, 159(1), 160, 161, 162(1), 163(2), 166, 167, 169(1), 172(1), 173(1)

Lapidot, Y., 439, 440, 443, 460, 461, 503, 506(16)

Larabee, A. R., 45

Lardy, H. A., 592

Larose, J. A. G., 405

Larrabee, A. R., 101

Lata, G. F., 580

LaTorre, J. L., 389

Lauer, W. M., 349, 350, 351(3, 4), 352(3, 4), 353(3)

Lauter, J. L., 530, 532(6)

Law, J. H., 449, 450(33), 459

Lawson, A. M., 340, 342(3), 343(3), 344(3), 367

Layne, E., 163, 174

LeBlond, C. P., 577

Lech, J. J., 182, 609

LeCocq, J., 434, 469(10)

Ledeen, R. W., 550

Lee, J. B., 360(9), 361

Leemans, F. A. J. M., 340, 342(3), 343(3), 344(3)

Lees, M., 227, 243, 244(10), 249, 396, 421(1), 543, 546(3), 549, 550(6)

Legacki, A. B., 298

Lehn, J. M., 549

LeKim, D., 535

Lemesiene, D., 460

Leone, C. A., 580

Lepage, M., 403

Lester, R. L., 264, 408

Levine, L., 206, 288, 289, 290, 293, 296(11, 13, 16), 297

Levy, H. B., 43

Levy, H. R., 11, 13(3), 17

Levy, R. I., 181

Lewin, E., 335

Lewis, K. O., 456, 488, 491(46), 492(46)

Light, R. J., 360(31), 361

Lin, C. H., 38

Lincoln, F. H., 372

Lindberg, O., 560

Lindop, C. R., 389

Lindqvist, B., 383, 390, 391

Liorber, B. G., 521

Lipmann, F., 48, 235, 242, 247, 299

Litman, B. J., 258

Lodin, Z., 562, 563(3)

Loken, S. C., 556

Long, C., 318

Longmuir, I. S., 592

Lorch, E., 560

Lornitzo, F. A., 55, 56(27), 57(27)

Los, M., 363, 364(62), 367(62)

Lowenstein, J. M., 3, 5, 10(4), 11(4), 279, 280(9), 282, 283(10), 285(10), 597, 598

Lowey, S., 48

Lowry, O. H., 34, 35(4), 120, 121, 160, 170, 221, 228, 250, 274, 304, 335

Luchinskaya, M. G., 477

Luddy, F. E., 321

Lukyanov, A. V., 477

Lund, P., 606

Lundquist, F., 302

Lynen, F., 19, 48, 66, 84, 122, 127(2), 128, 129, 134, 135(12), 137, 138(7), 168, 172, 174, 273, 312

Lyutik, A. I., 477

M

McCarthy, E. D., 85, 91, 95(3)

McClelland, M. J., 347

O

P

Subject Index I

In this index acids and their oxygen esters are indexed under the names of their anions; for example, palmitic acid and methyl palmitate will be found under Palmitate, however, palmityl CoA will be found under Palmityl CoA. Chemical synthesis of phospholipids and their analogues is covered in Subject Index II.

D

E

I

Immunology, of prostaglandins, 287–298
Insulin
  effect on fat cells, 560
  effect on hepatocytes, 592–593
  induction of rat liver fatty acid synthase, 44
Iodine vapor, indicator for lipids, 321, 405
Iodoacetamide
  inactivator of acetoacetyl-CoA thiolases, 134, 135
  inactivator of fatty acid synthase, 81
  inactivator of β-hydroxyacyl-CoA dehydrogenase, 128
  inactivator of palmityl thioesterase II, 109
Iodoacetate, inactivator of β-hydroxyacyl-CoA dehydrogenase, 128
O-Isopropylidene derivatives of unsaturated fatty acids, 347

K

3-Ketoacyl-CoA
  absorption at 303 nm (of Mg^{2+} complex), 129
  synthesis, 129
3-Ketoacyl-CoA thiolases, 128–136, see also Acetoacetyl-CoA thiolase
  assay, 129–130
  stability, 132
  tissue contents of individual thiolases, 133
3-Ketodecanoyl-CoA, in assay for 3-ketoacyl-CoA thiolase, 130
Ketogenesis, rate of, by perfused liver, 597, 606
3-Ketohexanoyl-CoA, in assay for 3-ketoacyl-CoA thiolase, 129–130

L

Lactate dehydrogenase, in assay for biotin carboxylase, 26, 28

Laurate, deuterated, mass spectrum of, 342
Lauryl-CoA, substrate for palmityl thioesterase I, 106
Lecithin, see Phosphatidyl choline
Linolenate, in mass spectrometry of triglycerides, 354
Lipase
  lipoprotein, 181
  monoglyceride, 181
  pancreatic, 181–189, 273, 317, 320–322, 324
    assay, 182–185
    in assay of long-chain fatty acyl-CoA, 273
  triglyceride, hormone sensitive, 181–189
    activation by cyclic AMP-dependent protein kinase, 187–189
    assay, 182–185
    inhibitors, 189
    phosphorylation, 188–189
    purification, 185–187
    species differences, 189
Lipid extraction of microsomes, 118–119
Lipids, see also Phospholipids, Triglycerides acetylation, 419–421, 424
  analysis of positional distribution of fatty acids in glycerolipids, 315–325
  column chromatography of, 409–411, 417–421, 423
  Folch partition of, 396–397, 409, 421, 423, 549–551
  indicators for, on thin-layer chromatography, 403–408, 412–413, 415–417
  ion-exchange chromatography of, 419
  lipophilic Sephadex chromatography, 386, 388–391
  thin-layer chromatography, 396–425
Lipofuchsin, in neurons from adult human brains, 575
Lipogenesis, see Fatty acid synthesis
Lipolysis
  in fat pads, 607
  in perifused fat cells, 607–612
    albumin dependence, 611–612
    epinephrine stimulation, 609–612
Lipophilic Sephadex, see Sephadex, lipophilic

U

Urease, jack bean, absorbance at 280 and
278.5 nm, 224–225

V

cis-Vaccenyl-CoA, substrate for palmityl
thioesterase I, 106
Vitamin D, lipophilic Sephadex chromatography, 395
Vitamin K, lipophilic Sephadex chromatography, 395

W

Waxes, lipophilic Sephadex chromatography, 391–392

Y

Yeast, enzymes from
acetyl-CoA:long-chain base acetyltransferase, 242
3-hydroxy-3-methylglutaryl-CoA synthase, 173–177

Subject Index II

This index covers chemical synthesis of phospholipids and their analogues. Other material is covered in Index I.

I

H

L